3D IC Integration and Packaging

About the Author

John H. Lau, Ph.D., has been a senior technical advisor at ASM Pacific Technology in Hong Kong since July 2014. Prior to that, he was a fellow of the Industrial Technology Research Institute in Taiwan for 4½ years; a visiting professor at Hong Kong University of Science and Technology for 1 year; the director of the Microsystems, Modules, and Components Laboratory at the Institute of Microelectronics in Singapore for 2 years; and a senior scientist/MTS at Hewlett-Packard Laboratory/Agilent in California for more than 25 years.

Dr. Lau's professional competences are design, analysis, materials, process, manufacturing, qualification, reliability, testing, and thermal management of electronic, optoelectronic, LED, CIS, and MEMS components and systems, with emphases on solder mechanics and manufacturing, RoHS-compliant products, SMT, fan-in and fan-out flip chip WLP, TSV, and other enabling technologies for 3D IC integration and SiP.

With more than 37 years of R&D and manufacturing experience, Dr. Lau has authored or coauthored more than 425 peer-reviewed technical publications, invented more than 30 issued or pending patents, and given more than 290 lectures/workshops/keynotes worldwide. He has authored or coauthored 18 textbooks on 3D IC integration, TSV for 3D integration, advanced MEMS packaging, reliability of 2D and 3D IC interconnects, flip-chip WLP, area-array packages, high-density PCB, SMT, DCA, lead-free materials, soldering, manufacturing, and solder joint reliability.

Dr. Lau earned a Ph.D. degree in theoretical and applied mechanics from the University of Illinois at Urbana–Champaign, an M.A.Sc. degree in structural engineering from the University of British Columbia, a second M.S. degree in engineering mechanics from the University of Wisconsin at Madison, and a third M.S. degree in management science from Fairleigh Dickinson University. He also has a B.E. degree in civil engineering from National Taiwan University.

Dr. Lau has received many awards from the American Society of Mechanical Engineers (ASME), the Institute of Electrical and Electronics Engineers (IEEE), the Society of Manufacturing Engineers (SME), and other societies, including for the best IEEE/ECTC proceedings paper (1989), outstanding IEEE/EPTC paper (2009), best ASME transactions paper (*Journal of Electronic Packaging,* 2000), and best IEEE transactions paper (*CPMT,* 2010), and the ASME/EEP Outstanding Technical Achievement Award (1998), IEEE/CPMT Manufacturing Award (1994), IEEE/CPMT Outstanding Contribution Award (2000), IEEE/CPMT Outstanding Sustained Technical Contribution Award (2010), SME Total Excellence in Electronics Manufacturing Award (2001), Pan Wen Yuan Distinguished Research Award (2011), IEEE Meritorious Achievement in Continuing Education Award (2000), IEEE Components, Packaging, and Manufacturing Technology Field Award (2013), and ASME Worcester Reed Warner Medal (2015). He is an elected ASME fellow (1999), has been an IEEE fellow since 1994, and has been heavily involved in many of ASME's and IEEE's technical activities.

3D IC Integration and Packaging

John H. Lau, Ph.D.

McGraw Hill Education

New York Chicago San Francisco
Athens London Madrid
Mexico City Milan New Delhi
Singapore Sydney Toronto

Library of Congress Cataloging-in-Publication Data

Lau, John H.

3D IC integration and packaging / John H. Lau.

p. cm.

Includes bibliographical references and index.

ISBN 978-0-07-184806-0 (acid-free paper)

1. Three-dimensional integrated circuits.

2. Microelectronic packaging. I. Title.

TK7874.893.L38 2015

621.3815—dc23 2015013446

3D IC Integration and Packaging

1 2 3 4 5 6 7 8 9 0 QVS/QVS 1 2 1 0 9 8 7 6 5

ISBN 978-0-07-184806-0
MHID 0-07-184806-1

Sponsoring Editor
Michael McCabe

Editing Supervisor
Stephen M. Smith

Production Supervisor
Lynn M. Messina

Project Manager
Asheesh Ratra, MPS Limited

Copy Editor
Rashi Sinha, MPS Limited

Proofreader
A. Nayyer Shamsi, MPS Limited

Indexer
Edwin Durbin

Art Director, Cover
Jeff Weeks

Composition
MPS Limited

Contents

Color versions of many figures are available at www.mhprofessional.com/3DICIP.

Preface

3D IC integration is taking the semiconductor industry by storm. It has been (a) impacting chip suppliers, fabless design houses, foundries, integrated device manufacturers, outsourced semiconductor assembly and test, substrates, electronics manufacturing services, original design manufacturers, original equipment manufacturers, material and equipment suppliers, universities, and research institutes; (b) attracting researchers and engineers from all over the world to go to conferences, lectures, workshops, panels, forums, and meetings to present their findings, exchange information, look for solutions, learn the latest technologies, and plan for their future; and (c) pushing the industry to build standards, infrastructures, and ecosystems for 3D IC integration.

This is a perfect storm! People and companies think that Moore's law is going to take a bow soon and 3D IC integration is the next hot spot. In order to prepare for their future and have a competitive edge, they have been investing heavily in both human and physical resources for 3D IC integration. 3D IC integration is defined as stacking up thin chips/interposers in the third dimension with through-silicon vias (TSVs) and microbumps to achieve high performance and density, low power consumption, wide bandwidth, small form factor, and light weight. Thus TSVs, thin-wafer/chip handling, microbumps, assembly, and thermal management are the most important key enabling technologies for 3D IC integration.

Unfortunately, for most practicing engineers and managers, as well as scientists and researchers, TSVs, thin-wafer strength measurement and handling, microsolder bumping, redistribution layers (RDLs), interposers, chip-to-wafer bonding, wafer-to-wafer bonding, assembly, thermal management, reliability, and 3D IC integration with light-emitting diodes (LEDs), microelectromechanical systems (MEMS), and complementary metal-oxide semiconductor (CMOS) image sensors (CIS) are not well understood. Thus, there is an urgent need, in both industry and research institutes, to create a comprehensive book on the current state of knowledge of these key enabling technologies. This book is written so that readers can quickly learn about the basics of problem-solving methods and understand the tradeoffs inherent in making system-level decisions.

There are 10 major subjects in this book, namely, (1) 3D integration for semiconductor IC packaging (Chap. 1); (2) TSV electrical, thermal, and mechanical modeling and testing (Chap. 2); (3) stress sensors for thin-wafer handling and strength measurement (Chap. 3); (4) package substrate technologies (Chap. 4); (5) microsolder wafer bumping, assembly, and reliability (Chap. 5); (6) 3D Si integration, 2.5D/3D IC integration, and 3D IC integration with passive interposer (Chaps. 6, 7, and 8); (7) thermal management of 2.5D/3D IC integration (Chap. 9); (8) embedded 3D hybrid integration (Chap. 10); (9) 3D IC integration with LEDs, MEMS, and CIS (Chaps. 11, 12, and 13); and (10) 3D IC packaging (Chap. 14).

Chapter 1 briefly discusses 3D IC packaging, 3D IC integration, and 3D Si integration. The supply chains before and for the TSV eras are provided. The status of TSV high-volume manufacturing for CIS and MEMS products is presented.

Chapter 2 presents a high-frequency electrical analytic model and equations for a generic TSV structure. These equations have been verified in the frequency and time domains. Also, the equivalent thermal conductivity equations for a generic TSV are provided. These equations have been verified by 3D simulations of the TSV structure. Finally, Cu pumping and the keep-out-zone of Cu-filled TSVs are discussed.

Chapter 3 details the design, fabrication, and calibration of piezoresistive stress sensors. The application of stress sensors to thin-wafer handling is explored. Also, the application of stress sensors in wafer bumping is shown. Finally, the application of stress sensors in drop tests of embedded ultrathin chips is presented.

Chapter 4 presents the package substrates with build-up layers for flip chip 2.5D/3D IC integration applications. The coreless package substrate is also provided. Finally, the recent advances of package substrates with build-up layers are examined.

Chapter 5 discusses the wafer bumping, assembly, and reliability of 3D IC integration solder bumps at 25-μm, 20-μm, and 15-μm pitches. For each case, the test structure, solder material, under bump metallurgy (UBM), assembly condition, underfill, and reliability assessment are examined.

The next three chapters are specifically for 3D Si integration, 2.5D/3D IC integration, and 3D IC integration with passive interposer. Chapter 6 presents the overview, outlook, and challenges of 3D Si integration. Chapter 7 discusses the potential application of 3D IC integration, such as memory-chip stacking, wide I/O memory or logic-on-logic, wide I/O dynamic random-access memory (DRAM) or hybrid memory cube (HMC), wide I/O 2 and high bandwidth memory (HBM), and wide I/O interface (2.5D IC integration). Also, the fabrication of TSVs and RDLs are detailed. Finally, various thin-wafer handling methods are discussed. Chapter 8 presents three different structures of 3D IC integration with passive interposer. For each structure, the fabrication of the interposer and RDLs and final assembly of chips on both sides of the interposer are provided.

Chapter 9 presents the thermal management of 2.5D/3D IC integration. A new design which consists of an interposer with chips/heat spreader on its top side and chips with or without heat slugs on its bottom side is proposed. Also, a thermal performance comparison between 2.5D and 3D IC integration is provided. Finally, a thermal management system consisting of TSV interposers with embedded microchannels is presented.

Chapter 10 presents embedded 3D hybrid integration. Printed circuit boards using optical waveguides and embedded board-level optical interconnects are examined. Also, an embedded 3D hybrid integration system is proposed. Finally, a semi-embedded TSV interposer with a stress relief gap is presented.

The next three chapters are specifically for 3D IC integration with LEDs, MEMS, and CIS. Chapter 11 presents the status and outlook of Haitz's law and four key segments of LED products. Also, the 2.5D/3D IC and LED integrations are presented. Finally, the thermal management of 3D IC and LED integration is presented. Chapter 12 presents 10 different designs and assembly processes of 3D IC and MEMS integration. Also, a low-temperature bonding of 3D MEMS packaging with solders is provided. Finally, recent developments in advanced 2.5D/3D IC and MEMS integration are examined. Chapter 13 presents the difference between front-illuminated (FI) CIS and back-illuminated (BI) CIS. Two examples (one is chip-to-wafer bonding and the other is wafer-to-wafer bonding) of 3D CIS and IC integration are discussed.

Chapter 14 presents 3D IC packaging, which includes chip stacking by wirebonding, package-on-package, fan-in wafer-level packaging, fan-out embedded wafer-level packaging, and embedded (rigid and flexible) panel-level packaging.

For whom is this book intended? Undoubtedly, it will be of great interest to three groups of specialists: (a) those who are active or intend to become active in research and development of the key enabling technologies of 3D IC integration such as TSVs, interposers, RDLs, thin-wafer handling, microbumps, assembly, and thermal management; (b) those who have encountered practical 3D IC integration problems and wish to understand and learn more methods for solving such problems; and (c) those who have to choose a reliable, creative, high-performance, high-density, low-power-consumption, wide-bandwidth, and cost-effective 3D IC integration technique for their products. This book can also be used as a text for college and graduate students who have the potential to become our future leaders, scientists, and engineers in the electronics and optoelectronics industry.

I hope that this book will serve as a valuable reference source for all those faced with the challenging problems created by the ever-increasing interest in 3D IC integration and 3D IC integration with LEDs, MEMS, and CIS. I also hope that it will aid in stimulating further research and development on key enabling technologies and more sound applications to 3D IC integration products.

The organizations that learn how to design and manufacture TSV, RDL, and microbump interconnects and thermal management in their 3D IC integration and packaging systems have the potential to make major advances in the electronics and optoelectronics industry, and to gain great benefits in performance, functionality, density, power, bandwidth, quality, size, and weight. It is my hope that the information presented in this book may assist in removing roadblocks, avoiding unnecessary false starts, and accelerating design, materials, process, and manufacturing development of key enabling technologies of 3D IC integration and packaging.

John H. Lau, Ph.D.

Acknowledgments

Development and preparation of *3D IC Integration and Packaging* was facilitated by the efforts of a number of dedicated people. I would like to thank them all, with special mention of Asheesh Ratra of MPS Limited and Lynn Messina of McGraw-Hill Education for their unswerving support and advocacy. My special thanks go to Michael McCabe and Steve Chapman, who made my dream of this book come true by approving and effectively sponsoring the project, patiently listening to my excuses for delay, and solving many problems that arose during the book's preparation. It has been a great pleasure and fruitful experience to work with all of them in transferring my messy manuscript into a very attractive printed book.

The material in this book clearly has been derived from many sources, including individuals, companies, and organizations, and I have attempted to acknowledge by citations in the appropriate parts of the book the assistance that I have been given. It would be quite impossible for me to express my thanks to everyone concerned for their cooperation in producing this book, but I would like to extend due gratitude. Also, I would like to thank several professional societies and publishers for permitting me to reproduce some of their illustrations and information in this book, including the American Society of Mechanical Engineers (ASME) conference proceedings (e.g., *International Intersociety Electronic Packaging Conference*) and transactions (e.g., *Journal of Electronic Packaging*), the Institute of Electrical and Electronics Engineers (IEEE) conference proceedings (e.g., *Electronic Components and Technology Conference* and *Electronics Packaging and Technology Conference*) and transactions (e.g., *Components, Packaging, and Manufacturing Technology*), and the International Microelectronics and Packaging Society (IMAPS) conference proceedings (e.g., *International Symposium on Microelectronics*) and transactions (e.g., *International Journal of Microcircuits & Electronic Packaging*).

I would like to thank my former employers, the Industrial Technology Research Institute (ITRI), the Hong Kong University of Science and Technology (HKUST), the Institute of Microelectronics (IME), Agilent, and HP, for providing me with excellent working environments that have nurtured me as a human being, fulfilled my need for job satisfaction, and enhanced my professional reputation. Also, I would like to thank Dr. Don Rice (HP), Dr. Steve Erasmus (Agilent), Professor Dim-Lee Kwong (IME), Professor Ricky Lee (HKUST), and Dr. Ian Yi-Jen Chan (ITRI) for their kindness and friendship while I was at their organizations. Furthermore, I would like to thank Mr. Lee Wai Kwong, CEO, for his trust, respect, and support of my work at ASM Pacific Technology in Hong Kong. Finally, I would like to thank the following colleagues for their stimulating discussions and significant contributions to this book: J. F. Hung, P. Chen, S. Wu, S. Hung, S. Lai, M. Li, S. Sheu, Z. Lin, C. Lin, W. Lo, M. Kao, H. C. Chien, Y. Chao, R. Tain, M. Dai, S. T. Wu, X. Zhang, C. Selvanayagam, P. Tzeng, C. Chen, S. Chen, C. Wu, C. Lee, C. Zhan, J. Chen, Y. Hsu, T. Ku, A. Kumar, Q. Zhang, M. Jong, G. Huang, V. Lee, V. Kripesh, Y. Y. Ong, S. W. Ho, C. H. Khong, A. Yu, W. Hnin, W. Lee, M. Jong, V. Sekhar, D. Pinjala, C. Chan, C. Chao, C. Chiu, C. Huang, C. K. Lee, Y. Huang, H. Fu, J. Huang, Z. Hsiao, S. Huang, C. Fan, Y. Lin, K. Kao, C. Ko, T. Chen, T. Chang, R. Cheng, J. Chang, M. Li, M. Ji, J. Cline, K. Saito, Y. Hsin,

P. Chang, Y. Chang, H. Chang, C. Lin, R. Lee, M. Yuen, J. Wu, C. Lo, H. Fan, H. Chen, L. Li, P. Su, J. Xue, M. Brillhart, G. Y. Tang, S. Tan, N. Khan, T. G. Lim, B. Lee, T. Shioda, H. Kuruveettil, J. Li, K. Suzuki, K. Fujita, K. Yamada, R. Pamidighantam, P. Ramana, L. Lim, C. Teo, H. Yee, C. Tan, O. Chai, Y. Jie, T. Lim, M. S. Zhang, P. Chan, and C. S. Premachandran. Definitely, I would like to thank my eminent colleagues (the enumeration of whom would not be practical here) at ASM, ITRI, HKUST, IME, Agilent, EPS, HP, and Sandia National Lab, and throughout the electronics industry, for their useful help, strong support, and stimulating discussions. Working and socializing with them has been a privilege and an adventure. I learned a lot about life and advanced IC packaging and 3D IC integration technologies from them.

Last, I would like to thank my daughter Judy and my wife Teresa for their love, consideration, and patience by allowing me to work peacefully on this book. Their simple belief that I am making a contribution to the electronics industry was a strong motivation for me. Thinking that Judy has gotten married and has been doing very well at a semiconductor company and that Teresa and I are in good health, I want to thank God for His generous blessings.

CHAPTER 1

3D Integration for Semiconductor IC Packaging

1.1 Introduction

The electronics industry has been the largest industry since 1996 and may well reach 1.6 trillion dollars (i.e., $\$1.6 \times 10^{12}$) by the end of 2015. The most important invention of the electronics industry is, arguably, the transistor (1947), which earned John Bardeen, Walter Brattain, and William Shockley the Nobel Prize in 1956 for Physics. The invention of the IC (integrated circuit) by Jack Kilby in 1958 (which earned him the Nobel Prize in 2000 for Physics) and six months later by Robert Noyce (who did not share the Nobel Prize with Jack Kilby because he passed away in 1990) excited the generations of IC integrations.

The proposal of doubling the number of transistors on an IC chip (for minimum costs and innovations) every 24 months by Gordon Moore in 1965 (also called Moore's law) has been the most powerful driver for the development of the microelectronics industry in the past 50 years. This law emphasizes lithography scaling and integration (in 2D) of all functions on a single chip, perhaps through SoC (system-on-chip). On the other hand, the integration of all these functions can be achieved through 3D integration, which is the focus of this book.

1.2 3D Integration

3D integration consists of 3D IC packaging, 3D IC integration, and 3D Si integration, as shown in Fig. 1.1. They are different and in general the TSV (through-silicon via) separates 3D IC packaging from 3D IC integration and 3D Si integration since the latter two use TSV but 3D IC packaging does not.

TSV is the heart of 3D IC/Si integration. Figure 1.2 shows the U.S. patent (No. 3,044,909) "Semiconductive Wafer and Method of Making the Same" filed on October 23, 1958 and granted on July 17, 1962 by William Shockley. Yes, the same William Shockley who coinvented the greatest invention of all time in the semiconductor industry—the transistor.

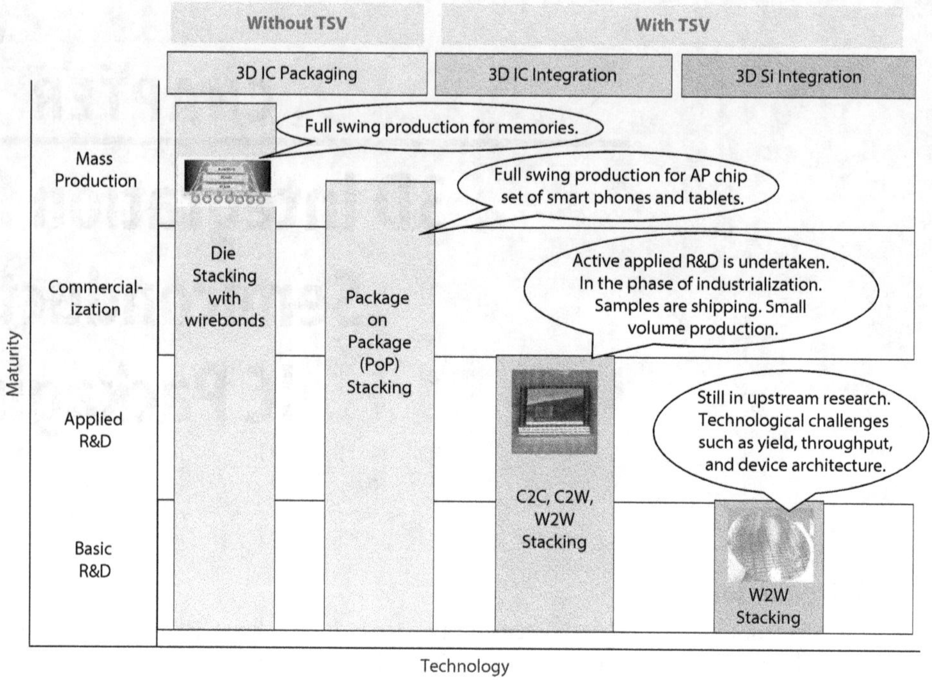

FIGURE 1.1 3D integration technologies vs. maturity.

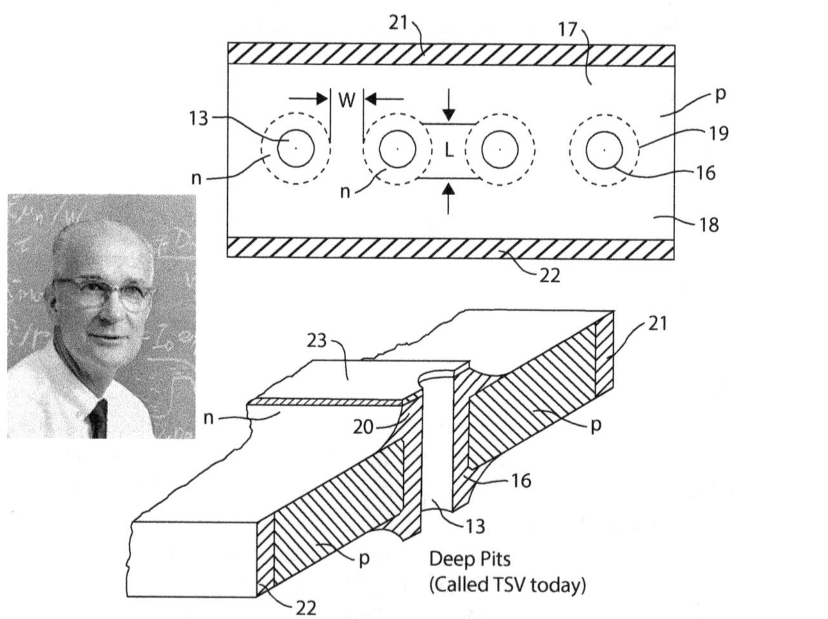

FIGURE 1.2 TSV invented by Shockley (U.S. patent #3,044,909).

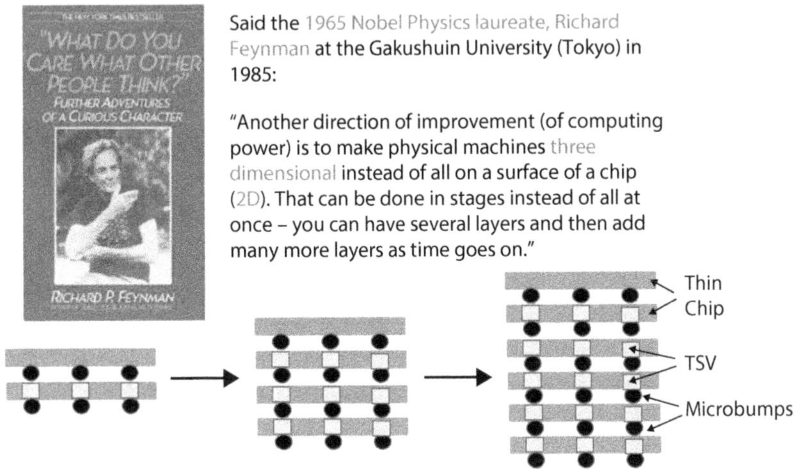

Figure 1.3 Richard Feynman, the most powerful promotor of 3D IC/Si integration.

Shockley invented TSV more than 50 years ago, but it was not meant for 3D IC/Si integration. 3D IC/Si integration [1–96] consists of two or more layers of active electronic components that are integrated vertically through TSVs into a single circuit. It was triggered by the advancement of the SOI (silicon-on-insulator) technology first reported by Gat and his colleagues [97], when people working in the field of semiconductors thought Moore's law would be hitting the wall by the 1990s. Of course, the facts showed otherwise.

The most powerful promotor of 3D IC/Si integration is the 1965 Nobel Physics laureate, Richard Feynman (Fig. 1.3). At the Nishina Memorial Lecture, *Computing Machines in the Future*, at Gakushuin University in Tokyo in 1985, he said, "Another direction of improvement (of computing power) is to make physical machines three dimensional instead of all on a surface of a chip. That can be done in stages instead of all at once—you can have several layers and then add many more layers as time goes on." Even today, many people who are searching for 3D IC/Si integration research funding like to cite his 1985 statement.

In the mid-1980s, the Japan MITI (Ministry of International Trade and Industry) funded and directed the 3D Research Committee of the 3D IC/Si integration project. Their road map is shown in Fig. 1.4 [78]. It was announced then that (1) functional models have been fabricated in stacked double or triple active layers demonstrating the concept of a future 3D structure, (2) the basic technology for stacking active layers will be developed before 1990, and (3) with this technology, various kinds of circuits such as high-packing density memory, high-speed logic, or image processors are expected to be designed and realized in a 3D single chip between 1990 and 2000. History showed that the technology did not materialize within the timeframe stated by the project.

In this chapter, 3D IC packaging, 3D IC integration, and 3D Si integration will be briefly mentioned. The supply chains for manufacturing the 3D IC integration will also be presented. Finally, the TSV manufacturing for CMOS image sensor and MEMS (microelectromechanical systems) will be briefly discussed.

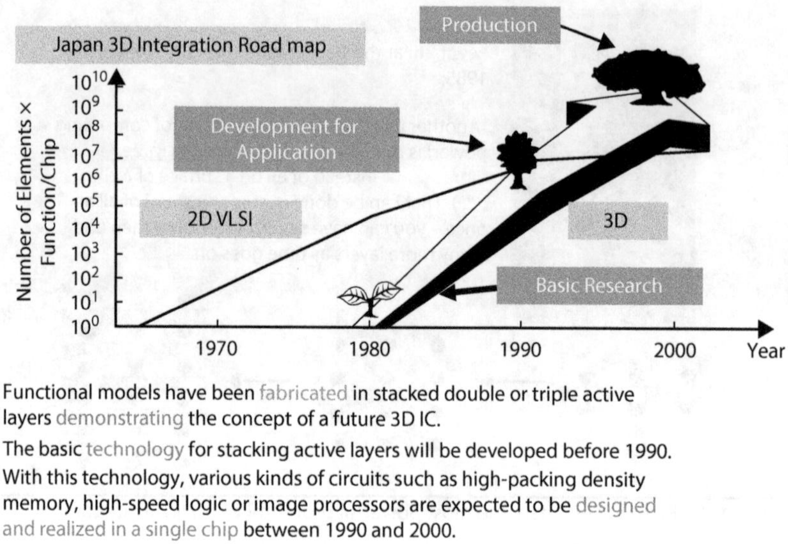

Figure 1.4 Japan 3D integration road map in the mid-1980s.

1.3 3D IC Packaging

3D IC packaging consists of two or more conventional components (packages) stacked in the vertical direction, as shown in Fig. 1.1. The most commons are chip stacking by wirebonding and PoP (package-on-package). One of the examples of chip stacking is stacking up memory chips with wirebonds and die attachments, as shown in Fig. 1.5. Today, 28 memory chips stacking are not uncommon. One of the examples of PoP is shown in Fig. 1.6, where a flip chip with solder bumps attached to a package substrate is supporting another package with cross-die wirebonding in the vertical direction.

3D memory chip stacking and PoP are mature technologies and in high-volume production as shown in the maturity status of 3D integration technology (Fig. 1.1), and thus

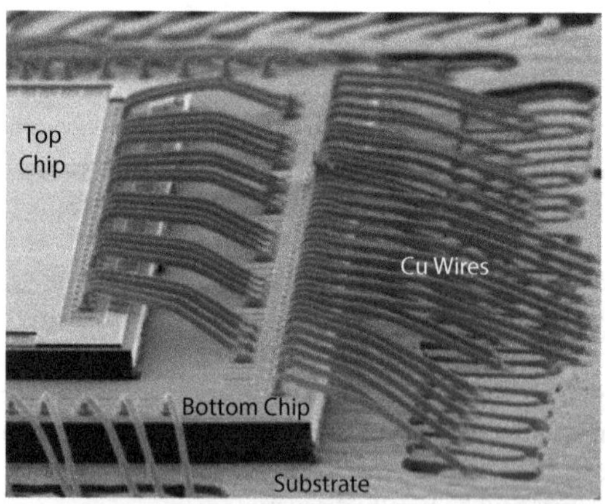

Figure 1.5 Amkor's chip stacking with Cu wirebonding.

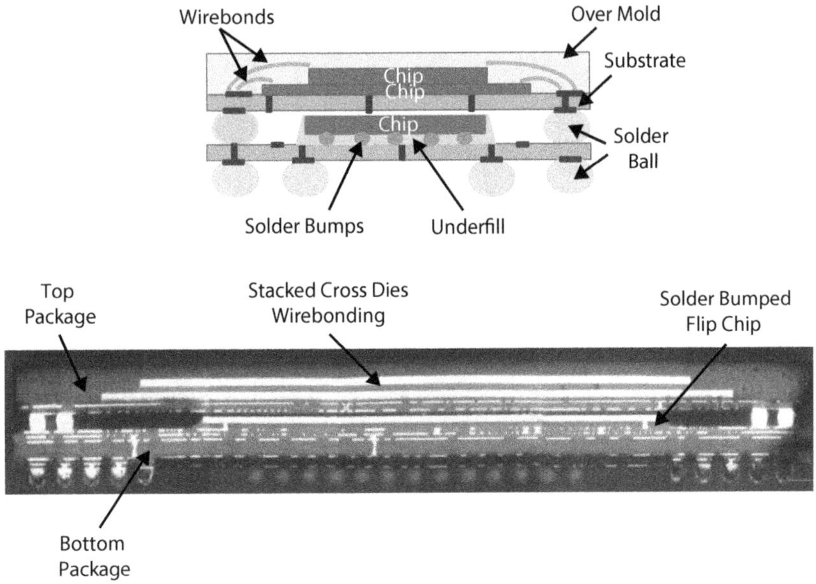

FIGURE 1.6 Package-on-package (PoP) for application processor chip set.

will not be discussed herein. However, 3D IC packaging technologies have been supporting mobile products such as smartphones and tablets and could be the main driver for materials consumption and new materials development for wearable products such as the smartwatch; they will be discussed in the last chapter of this book.

1.4 3D Si Integration

3D Si integration stacking up the wafers/chips in the vertical direction with TSVs but without any bumps (bumpless). Basically, W2W (wafer-to-wafer) is the only way to perform the bonding operation for 3D Si integration and yield is a big issue (e.g., some bad chips are forced to bond on the good chips). In addition, the absence of a gap (or an infinitesimal gap) between wafers and thermal management is a huge problem. Furthermore, the requirements of the bonding conditions, such as the surface cleanness, surface flatness, and class of clean room for 3D Si integration, are very high.

There are at least two different W2W bonding methods for 3D Si integration, namely, Cu-to-Cu bonding and oxide-to-oxide bonding, as shown in Figs. 1.7 and 1.8, respectively. Figure 1.7 shows a cross-section of the interface between the bumpless Cu-to-Cu electrodes (pads) given by the NIMS/AIST/Toshiba/University of Tokyo [79–85]. Figure 1.8 shows a cross-section of MIT's oxide-to-oxide bonding structure of three-layer 3D (ring oscillator) bonded at 275°C [86–92]. It can be seen that (1) the layers are bonded and interconnected with W-plugs, (2) the conventional inter-level connections are in the bottom two layers, and (3) the 3D vias are located in the isolation (field) region between transistors.

Tremendous amount of work still needs to be done before the products can be manufactured using the 3D Si integration technology. Besides thermal management, vias formation, thin-wafer handling, more research and development efforts should also be placed on areas

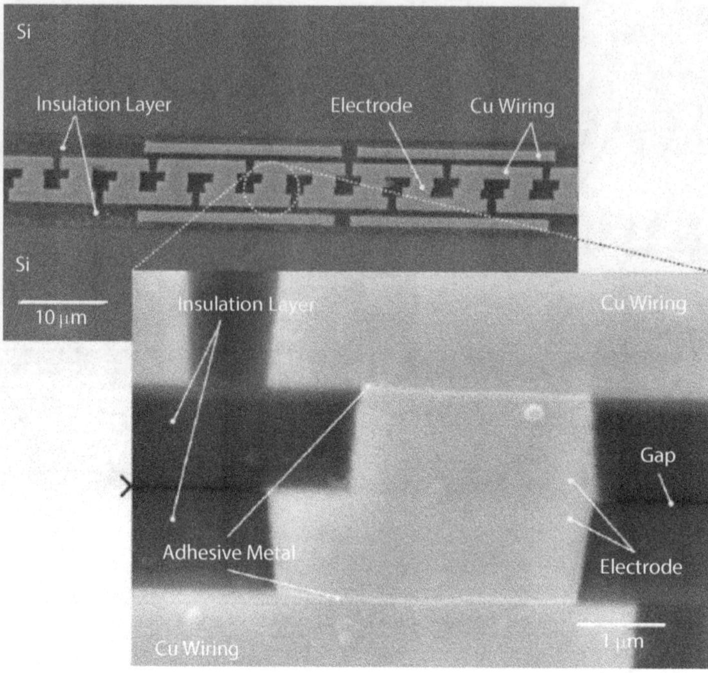

Figure 1.7 3D Si integration. NIMS/AIST/Toshiba/University of Tokyo's Cu-to-Cu bonding.

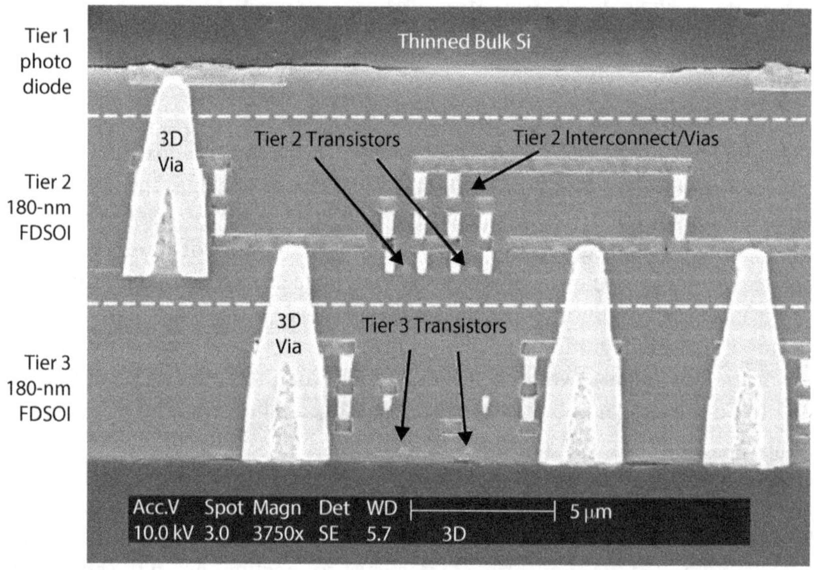

Figure 1.8 3D Si integration. MIT's oxide-to-oxide bonding.

such as cost reduction, design and process parameter optimization, bonding environment, W2W bonding alignment, wafer distortion, wafer bow (warpage), inspection and testing, contact performance, contact integrality, contact reliability, and manufacturing yield issues. In addition, packaging the 3D Si integration module systematically and reliably to the next level of interconnect poses another great challenge. Chapter 6 will present more information of 3D Si integration.

1.5 3D IC Integration

3D IC integration stacks up the thin chips in the vertical direction with TSVs and bumps. Unlike the conventional flip chip bumps (~100 μm), usually the 3D IC integration bumps are very small (<25 μm) and are called microbumps. The ones that are in and going into low-volume production are memory stacking with TSVs, HMC (hybrid memory cube) or wide I/O DRAM (dynamic random access memory), wide I/O DRAM 2, HBM (high bandwidth memory), and 2.5D IC integration (passive interposer).

Samsung mass-produced (August 2014) the industry's first TSV-based 64-GB DDR4 (double data rate type 4) DRAM module (memory stacking with TSVs), which consists of 36 DDR4 DRAM chips, each of which consists of four 4-GB DDR4 DRAM dies. The module performs twice as fast as a module that uses wirebonding packaging, while consuming approximately half the power. The module is for server application.

1.5.1 Hybrid Memory Cube

Figure 1.9 shows the very first sample shipped by Micron/IBM at the end of September 2013. It is an HMC that consists of 4 DRAMs each with 2000+ TSVs sacking on top of a logic controller with TSVs. The TSV DRAM cube is fabricated by Micron and the TSV controller is fabricated by IBM. The microbumps are Cu pillars (20 μm tall) with solder caps. Altera has designed a demonstration board for the HMC and Altera FPGAs.

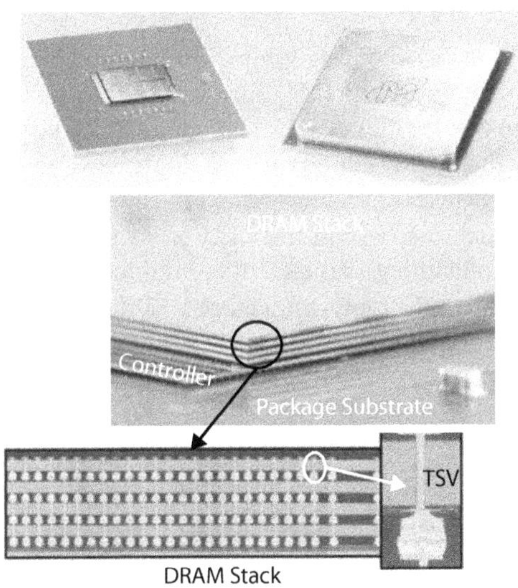

- The hybrid memory cube is a 4-DRAM (each one with 2000+ TSVs) on a logic controller (which size is slightly larger than the DRAMs) with TSVs.
- The hybrid memory cube is on an organic package substrate.
- The TSV-DRAM is ~50 μm thick.
- The TSV-DRAM is with 20-μm (tall) Cu pillar + solder cap.
- The memory cube is assembled one DRAM at a time with thermal compression bonding.
- The heat dissipation is from 10 to 20 W.
- TSV diameter ~5 to 6 μm.

Figure 1.9 Micron's sample on hybrid memory cube (HMC).

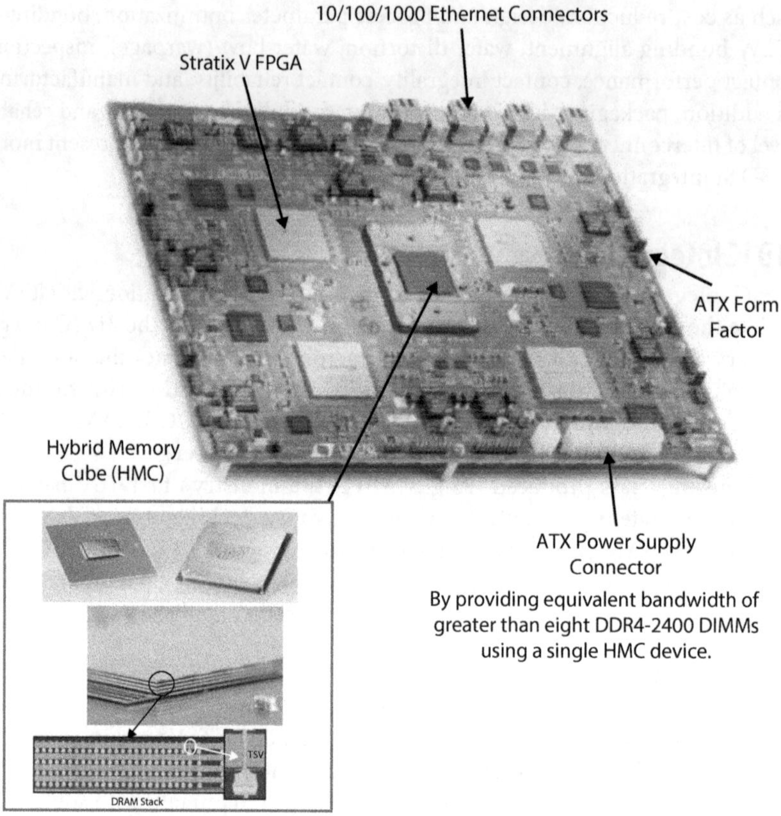

Figure 1.10 Altera FPGAs and HMC.

Figure 1.10 shows the platform using Altera's flagship 28-nm Stratix V FPGA (field programmable gate array) to interoperate with an HMC device [93]. It proves the technology leadership in the memory domain by providing equivalent bandwidth of greater than eight DDR4-2400 DIMMs (dual in-line memory modules) using a single HMC device. Figure 1.11 schematically shows Intel's 72-core "Knights Landing" Xeon Phi coprocessor with the HMCs [94]. Micron reports that having such memory in the CPU package is expected to deliver five times the sustained memory bandwidth versus GDDR5 (graphic double data rate type 5) with one-third the energy per bit in half the footprint. Figure 1.12 shows Fujitsu's microprocessors with HMCs for their supercomputer [95]. The HMCs deliver faster throughout and much better power utilization.

1.5.2 Wide I/O DRAM and Wide I/O 2

JEDEC standard JESD229, *Wide I/O SDR (Wide I/O Single Data Rate)*, was published in December 2011 and JEDEC standard JESD229-2, *Wide I/O 2 (WideI/O2)*, was published in August 2014.

They are meant for a stack of DRAMs with TSVs on a logic controller with TSVs; very similar to the HMC. The microbumps are divided into four quadrants with signal

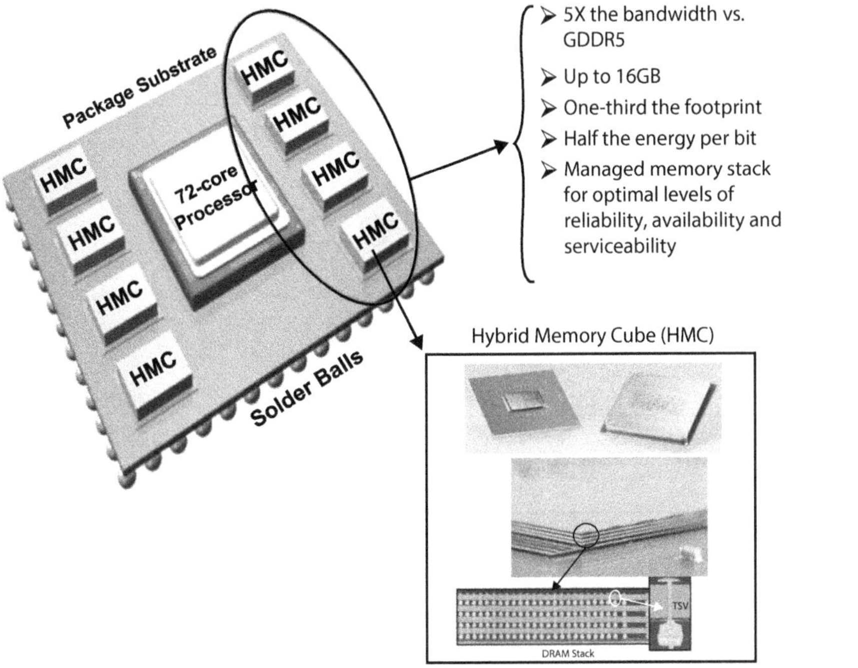

FIGURE 1.11 Intel's "Knight's Landing" with eight HMCs.

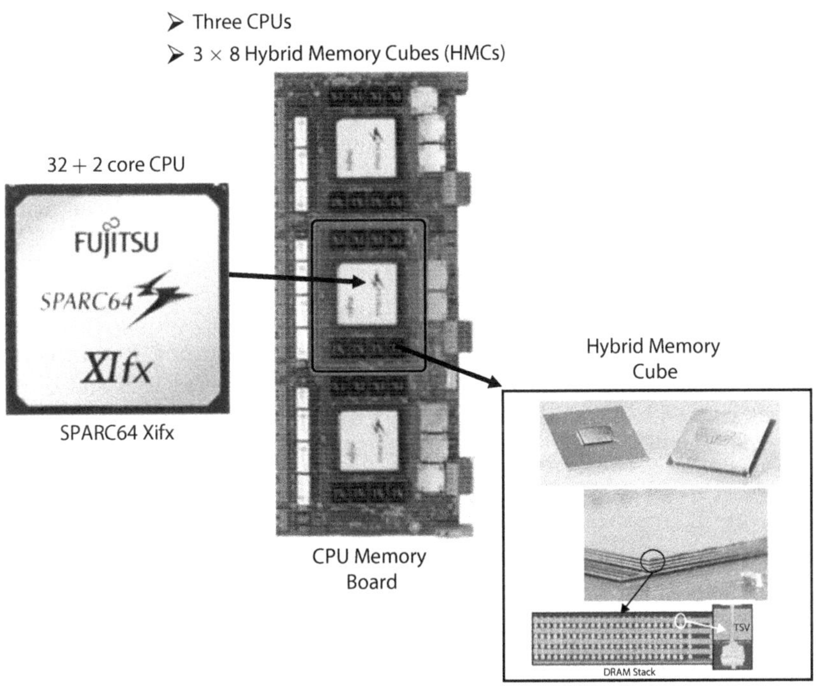

FIGURE 1.12 Fujitsu's CPU with HMCs.

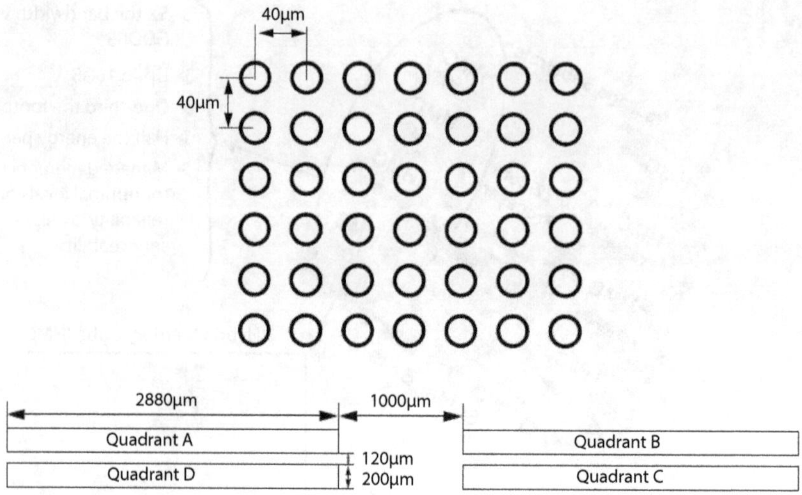

FIGURE 1.13 Wide I/O 2.

assignments mirrored both horizontally and vertically, as shown in Fig. 1.13, where the bump pitch (40 μm) of the area array is also shown. The dimensions of each quadrant are 2880 × 200 μm. There will be a space between quadrants in the x-direction (1000 μm) and in the y-direction (120 μm).

1.5.3 High Bandwidth Memory

Figure 1.14 shows schematically a HBM system by Hynix/AMD, which is based on JEDEC standard JESD235, *HBM DRAM*, published in December 2013. It is meant for graphics applications supporting bandwidth from 128 to 256 GB/s. A TSV/RDL interposer is used to support/connect mainly the lateral communication (HBM interface) between the HBM DRAM memory cube with TSVs and the GPU (graphic processor unit) or CPU (central processor unit) without TSVs. The optional base chip is used for buffering and signal rerouting of the HBM DRAM cube.

1.5.4 Wide I/O Memory (or Logic-on-Logic)

Figure 1.15 shows a wide I/O memory sample made by Samsung. It consists of a wide bandwidth low-power consumption memory chip with 1000+ of I/Os and is right on top of a logic chip with TSVs. The memory chip can also be a logic chip, then it will become a logic-on-logic. The wide I/O memory or logic-on-logic is supporting by an organic package substrate.

1.5.5 Passive Interposer (2.5D IC Integration)

A 2.5D integration is a TSV interposer system that consists of a piece of device-less silicon with TSVs and high-performance, high-density, fine-pitch IC chips without TSV. This piece of device-less silicon (also called a passive interposer) is used to support the chips and has RDLs (mainly) for lateral communication between the chips, as shown schematically in

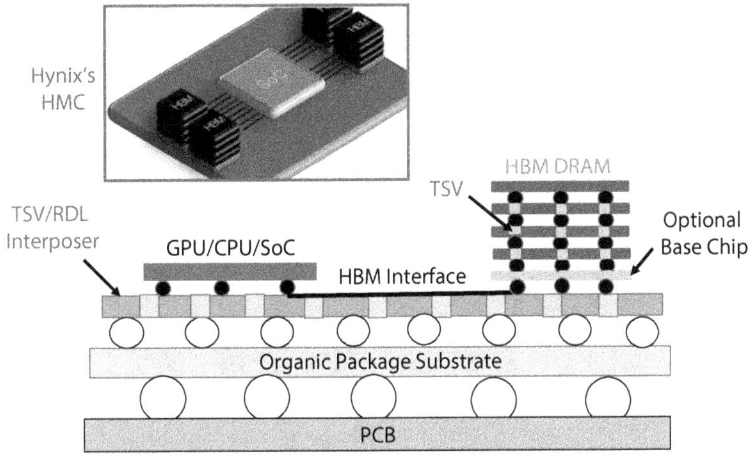

Underfill is needed between the interposer and the organic substrate. Also, underfill is needed between the interposer and the GPU/CPU and the memory cube

Figure 1.14 High Bandwidth Memory (HBM).

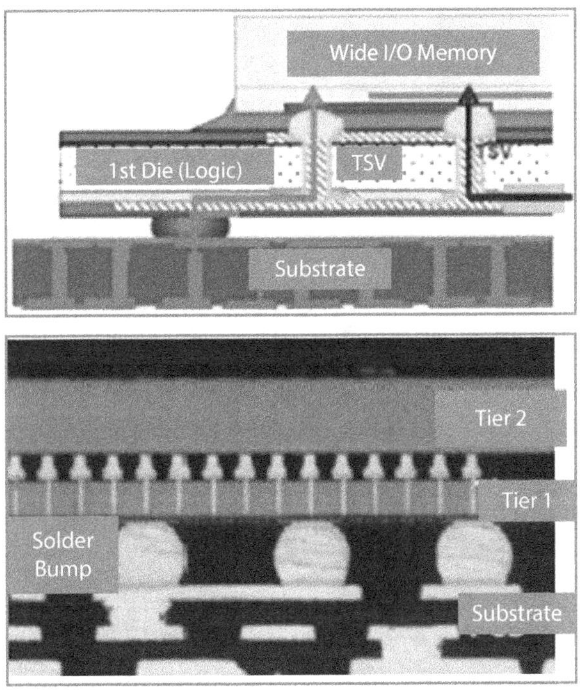

Figure 1.15 Wide I/O memory (or logic-on-logic).

Fig. 1.16. Figure 1.17 shows a sample designed by Xilinx and fabricated by TSMC. It can be seen that even with more than 10 buildup layers on the package substrate, it is still not enough to support the four 28-nm FPGA chips. In addition, a passive TSV interposer with 200,000+ microbumps on 45-μm pitch and four RDLs (three Cu damascene layers and one aluminum layer) at a minimum of 0.4-μm pitch is needed. This type of structure (Figs. 1.16 and 1.17) is called by TSMC as chip on (interposer) wafer on (package) substrate (CoWoS) and has been in small-volume production since early 2013.

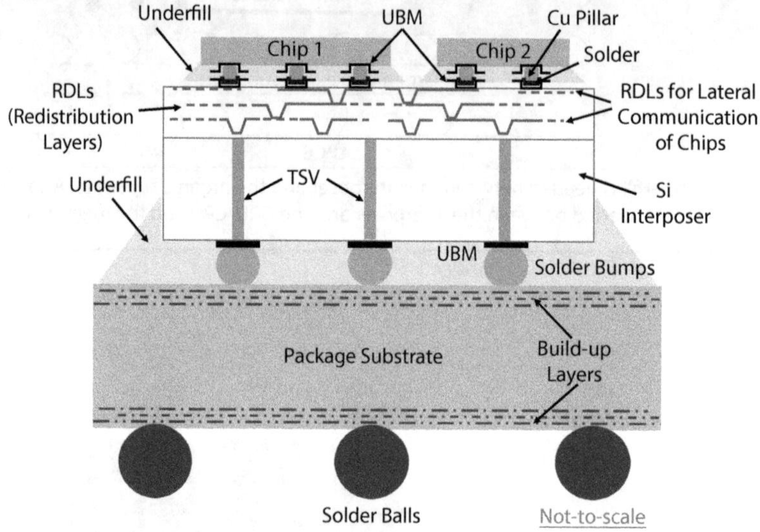

Figure 1.16 TSV/RDL passive interposer supporting chips on package substrate.

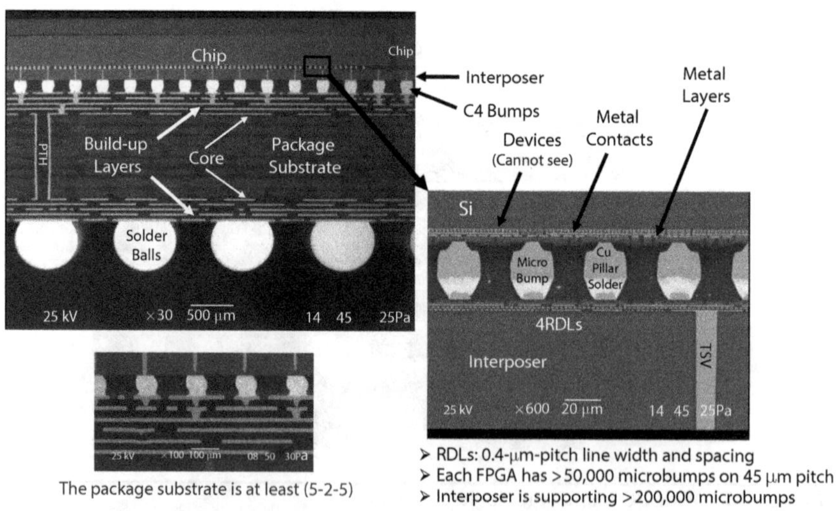

Figure 1.17 Xilinx/TSMC chip on wafer on substrate.

1.6 Supply Chains before the TSV Era

Before the TSV era, the technology supply chains are very well defined and understood. Descriptions of the various entities comprising the supply chain before the TSV era are presented below.

1.6.1 FEOL (Front-End-of-Line)

This is the first portion of IC fabrication where the individual devices such as transistors or resistors are patterned. This process is from a bare wafer to (but not including) the deposition of metal layers. FEOL is usually performed in semiconductor fabrication plants (fabs).

1.6.2 BEOL (Back-End-of-Line)

This is the fabrication in which active devices are interconnected with wiring on the wafer. This process starts from the first layer of metal to bonding pads with passivation. It also includes insulators and metal contacts and is called MOL (middle-of-the-line). The term "MOL" is seldom used and embedded in the BEOL. Again, BEOL is usually performed in the fab.

1.6.3 OSAT (Outsourced Semiconductor Assembly and Test)

This term is also called packaging, assembly and test. The process starts when the passivated wafer is received from the fab and then goes through circuit probing, bumping, thinning, dicing, wiring bonding, flip chip, molding, ball mounting, final testing, and etc.

1.7 Supply Chains for the TSV Era—Who Makes the TSV?

The following steps in the TSV fabrication process impact the various considerations that must be addressed.

1.7.1 TSVs Fabricated by the Via-First Process

The TSVs are fabricated before the FEOL. This can only be done by the fab. However, even in the fab, this seldom happens because the devices (e.g., transistors) are much more important than the TSVs.

1.7.2 TSVs Fabricated by the Via-Middle Process

The TSVs are fabricated right after the FEOL (e.g., transistors) and MOL (e.g., metal contacts), and before the BEOL (e.g., metal layers). In this case, the MOL is no longer embedded in the BEOL because the TSV fabrication process is between them. Owing to logistics and equipment compatibilities, usually the TSV by the via-middle process is done by the fab. For more information, please see Sec. 7.2.3.

1.7.3 TSVs Fabricated by the Via-Last (from the Front Side) Process

The TSVs are fabricated (from the front side of the wafer) after the FEOL, MOL, and BEOL. As of today, there is not a single creditable paper published with this process.

1.7.4 TSVs Fabricated by the Via-Last (from the Back Side) Process

The TSVs are fabricated (from the back side of the wafer) after the FEOL, MOL, and BEOL processes. The CMOS image sensor is an example. Strictly speaking, CMOS image sensors are not examples of 3D IC integration. For CMOS device sample wafers, the only creditable publication is given by LETI et al. [96]. However, due to the technical issues, such as hitting the various embedded alignment targets in the x-, y- and z-directions (to enable the alignment between the metal layers on the top side of the wafer and the positioning of TSVs formed from the back side), TSVs fabricated by the via-last (from the back side) process should be avoided until these issues are resolved. For more information, please see Sec. 7.2.5.

Based on the above discussions, it seems that for active device wafers being used for 3D IC integration applications, TSVs are better fabricated using the via-middle process. Also, the TSVs should be fabricated by the fab, where all the equipment and expertise already exist and the cost to fabricate the TSVs is less than 5% of the cost in fabricating the ($\leq$32 nm) device wafers!

1.7.5 How About the Passive TSV Interposers?

When the industry defined the TSV processes for 3D IC integration, there were no passive interposers yet. Also, since there is no active device in the passive interposers, thus they do not fit into any of the preceding!

1.7.6 Who Wants to Fabricate the TSV for Passive Interposers?

Both the fab and OSAT want to do it! It depends on the layout, design, and fabrication capabilities, especially the line width and spacing of the RDLs. Usually, a few microns of line width and spacing can be done by the OSAT. Otherwise, it should be done by the fab.

1.7.7 Summary and Recommendations

The supply chains for making the TSV have been presented. Some important results and recommendations are summarized as follows:

For device wafers and high volume manufacturing, the TSVs should be fabricated by the foundries and with the via-middle process.

For dummy (deviceless) wafers, TSVs can be done by either the foundries or OSAT. For $\geq$3-μm line width and spacing RDLs and $\geq$5-μm-diameter vias, either the fab or the OSAT can do it. Otherwise, it should be done by the fab.

1.8 Supply Chains for the TSV Era—Who Does the MEOL, Assembly, and Test?

All the TSVs fabricated are blind vias. The blind TSV wafer is followed by solder bumping, temporary bonding, back grinding, TSV revealing, thin wafer handling, debonding, cleaning, etc. that, taken together, are called MEOL (middle-end-of-line). In this section, except for the vertically integrated companies (e.g., TSMC and Samsung), it is better for the MEOL process flow to be performed by the OSAT. The critical steps (including FEOL, MOL, BEOL, TSV, MEOL, assembly and test) and their ownership for making some of the 3D IC integrations are shown in the following sections.

1.8.1 Wide I/O Memory (Face-to-Back) by TSV Via-Middle Fabrication Process

Figure 1.18 shows the critical steps and ownerships for processing the logic wafer. After FEOL (to pattern the devices) and MOL (to make the metal contacts), the TSVs are fabricated by five key steps, namely via formation by deep reactive ion etch, dielectric deposition by plasma enhance chemical vapor deposition, barrier and seed layer by physical vapor deposition, Cu-filling by electroplating, Cu annealing, and CMP (chemical mechanical polishing) to

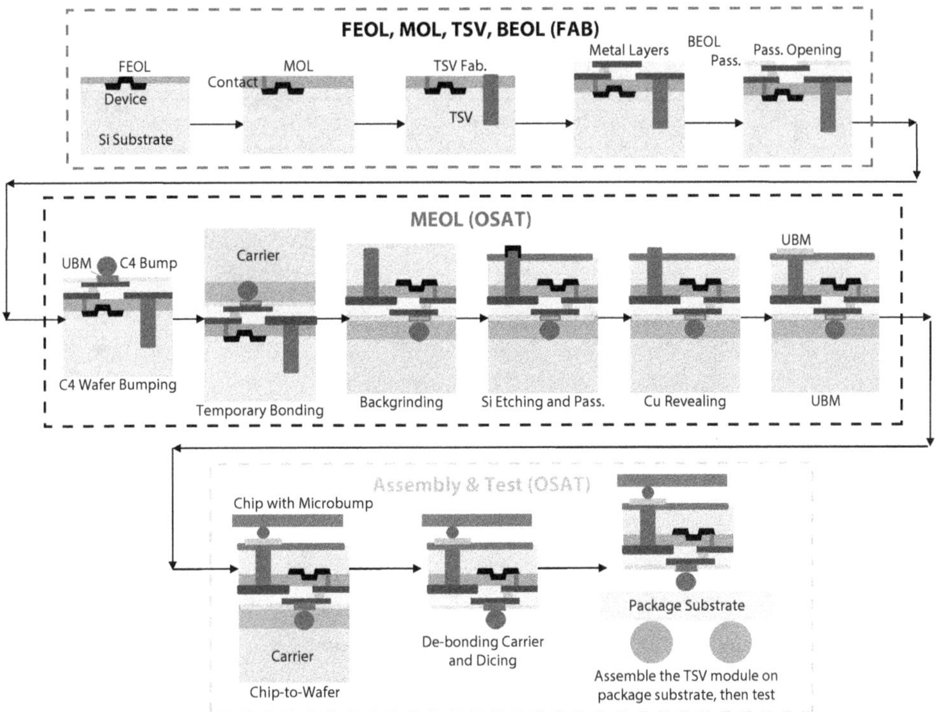

Figure 1.18 Critical steps and ownerships for (face-to-back) wide I/O memory using the TSV via-middle fabrication process.

remove the overburden Cu. These steps are then followed by the buildup of the metal layers, and finally, the passivation/openings (BEOL). All these steps should be done in the fab.

The MEOL starts off by UBM (under bump metallization) and C4 (controlled-collapse chip connection) ordinary wafer bumping with solder to the whole logic wafer. It is followed by temporary bonding of the TSV wafer to a supporting (carrier) wafer with adhesive. Then back grind the TSV wafer to a few microns from the tip of the Cu-filled TSV. It is followed by silicon dry etching to a few microns below the tip of the Cu-filled TSV. After that, a low temperature isolation SiN/SiO_2 deposition is applied to the whole wafer. Then CMP is used to remove the SiN/SiO_2 and the Cu and seed layers of the Cu-filled TSV (Cu revealing). Finally, an UBM is built on the tip of the Cu-filled TSV. All these steps should be done by the OSAT (except for vertical integrated foundries).

Separately, the memory wafer is microbumped with tiny solder bumps or Cu-pillars with solder caps. Then the wafer is diced into individual chips with microbumps/Cu-pillars. These steps also should be done by the OSAT.

Next is C2W (chip-to-wafer) bonding, i.e., the microbumped memory chip is bonded (either by natural reflow or thermal compression) to the TSV wafer with the carrier. After face-to-back C2W bonding, the carrier wafer is debonded from the TSV wafer. It is followed by dicing the TSV wafer into individual TSV modules. This TSV module is solder (natural) reflowed on a package substrate, and then tested. All these C2W bonding, dicing, assembly, and testing steps should be done by the OSAT.

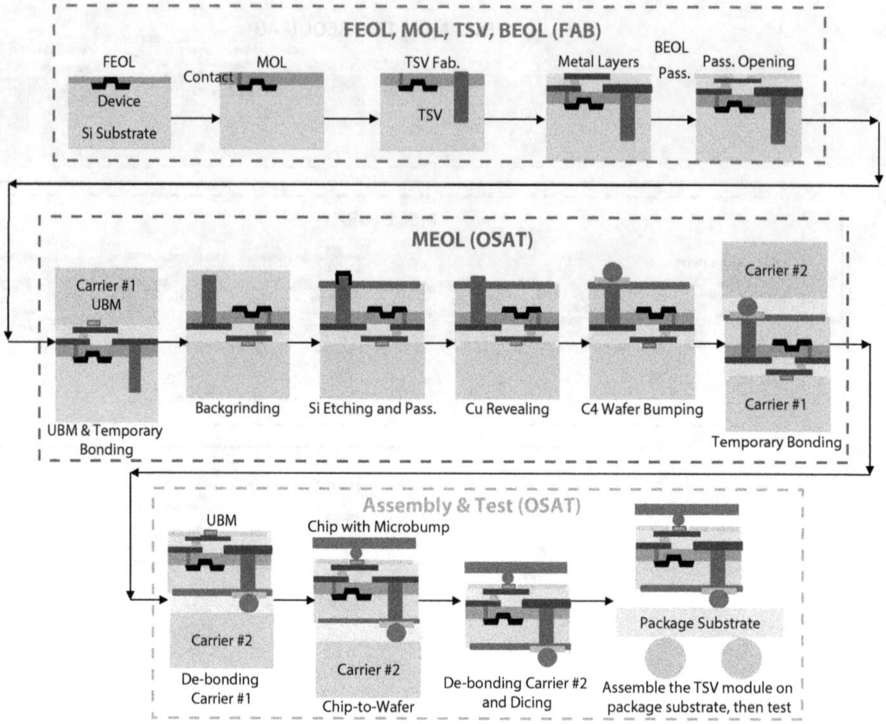

FIGURE 1.19 Critical steps and ownerships for (face-to-face) wide I/O memory using the TSV via-middle fabrication process.

1.8.2 Wide I/O Memory (Face-to-Face) by TSV Via-Middle Fabrication Process

The FEOL, MOL, TSV, and BEOL processes are exactly the same as those done in the TSV via-middle (face-to-back) process. However, the processes that follow it are different. Instead of C4 ordinary wafer bumping with solder after the UBM, the TSV wafer is temporarily bonded to carrier #1. Then, back grind the TSV wafer, reveal the Cu TSV, and fabricate the UBM. Those steps are followed by C4 ordinary wafer bumping with solder and temporary bonding to carrier #2. Then carrier #1 is debonded from the TSV wafer and C2W (face-to-face) bonding is performed. After C2W bonding, the carrier #2 is debonded from the TSV wafer. It is followed by dicing the TSV wafer into individual TSV modules. This TSV module is solder reflowed on a package substrate, then tested. The critical steps and their ownerships are shown in Fig. 1.19.

1.8.3 Wide I/O DRAM by TSV Via-Middle Fabrication Process

After FEOL, MOL, TSV, and BEOL of the logic and DRAM wafers, the SoC/logic wafer will go through the same steps, as shown in Fig. 1.18 for face-to-back, or Fig. 1.19 for face-to-face. For the DRAMs, UBM is done first followed by microwafer bumping of the whole wafer. These processes are then followed by temporary bonding to a carrier wafer, back grinding, Cu revealing, and UBM. These in turn are followed by debonding of the carrier wafer and dicing of the TSV DRAM wafer into individual TSV DRAM chips, as shown in Fig. 1.20.

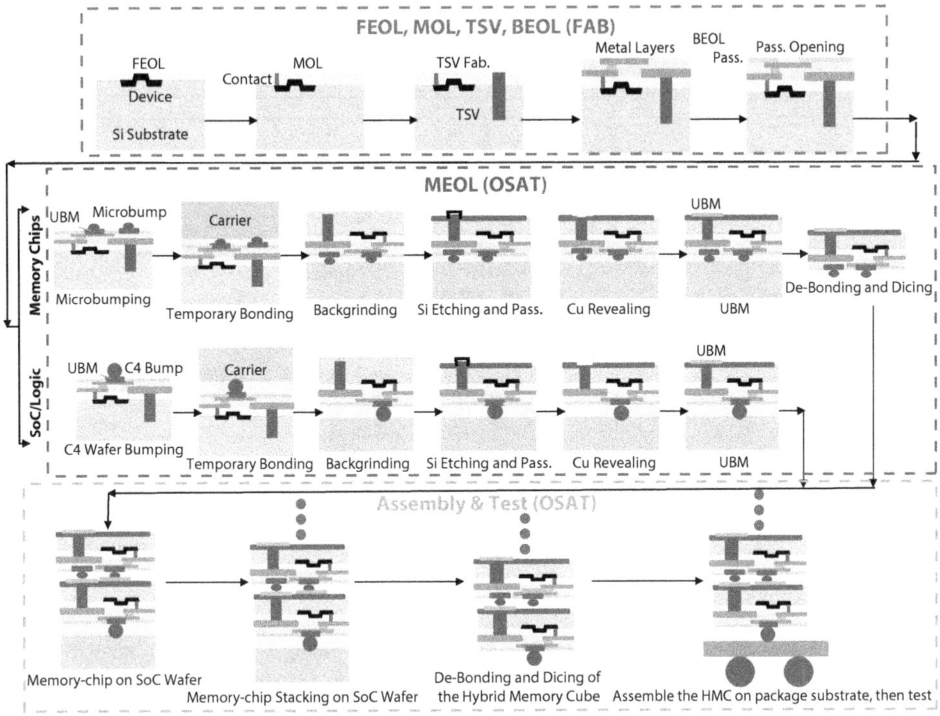

FIGURE 1.20 Critical steps and ownerships for wide I/O DRAM using the TSV via-middle fabrication process.

The next process is C2W (DRAM chip to logic wafer) bonding (e.g., 2-stacking, 4-stacking, 6-stacking or 8-stacking). After C2W bonding, the carrier wafer is debonded from the logic wafer and diced into individual HMCs (DRAM-stacking + logic). These steps are followed by assembly of the HMC with over-mold + underfill on a package substrate, and finally testing.

1.8.4 2.5D IC Integration with TSV/RDL Passive Interposers

Figure 1.21 shows the critical steps and their ownerships. After the deposition of a passivation layer on a piece of dummy silicon (no active devices), the TSV can be fabricated, RDL can be built, and passivation/openings can be made. After the UBM, the TSV wafer is temporarily bonded to carrier #1. It is followed by back grinding, silicon etching, low-temperature passivation, and Cu revealing. Then, UBM, C4 wafer bumping with solder, and temporary bonding to carrier #2 are accomplished.

Separately, the device wafer without TSVs is subjected to either microbumping with tiny solder bumps or Cu-pillars with solder caps. The device wafer is then diced into individual chips with microbumps/Cu-pillars.

The next processes to be accomplished are debonding of carrier #1, performing C2W bonding (the device chip to the TSV wafer). After C2W bonding, carrier #2 is debonded and the TSV wafer is diced into individual TSV modules. Finally, the TSV module can be assembled on a package substrate, and then tested.

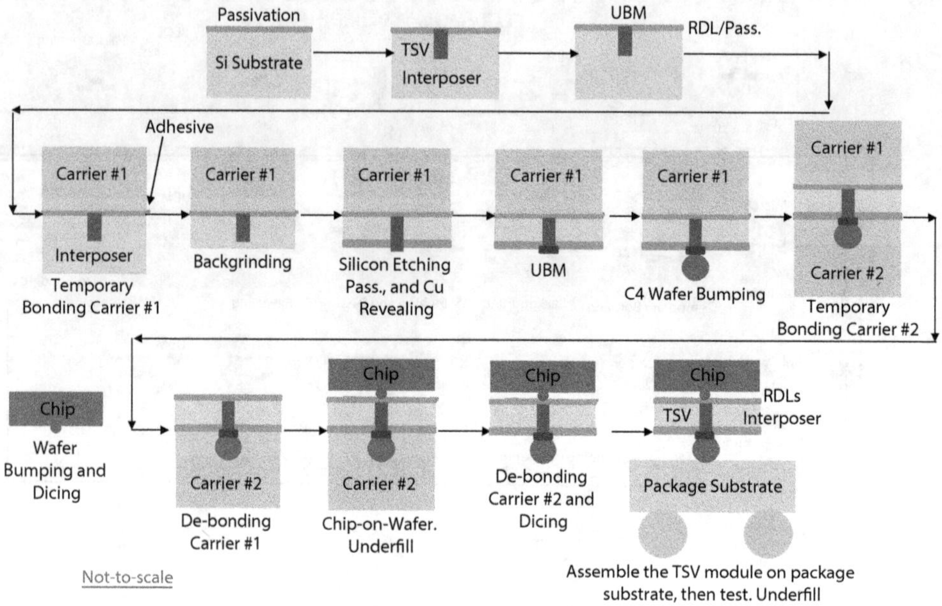

Figure 1.21 Critical steps and ownerships for 2.5D IC integration with a TSV/RDL passive interposer.

It can be seen from Fig. 1.21 that the TSV and RDL can be made by either the fab or the OSAT. It depends on the layout, design, and fabrication capabilities, especially the line width and spacing of the RDLs. Usually, a few microns of line width and spacing can be done by the OSAT. Otherwise, it should be done by the fab.

Except vertically integrated companies, such as TSMC, who want to do the CoWoS process completely in house, most fabless design houses prefer the fab (e.g., UMC and GlobalFoundries) to make the blind TSVs and the RDLs of the passive interposer. Then, the fabs hand the unfinished "TSV interposer" off to an OSAT for MEOL (solder bumping/temporary bonding/thin wafer handling/back grinding/TSV revealing/debonding/cleaning), assembly and test. This is also true for the unfinished TSV device wafers.

1.8.5 Summary and Recommendations

The technology supply chains for 2.5D/3D IC integration manufacturing have been presented. Critical steps such as FEOL, MOL, BEOL, TSV, MEOL, assembly and test, and their ownerships for potential applications and HVM of 2.5D/3D IC integrations such as wide I/O memory (or logic-on-logic), wide I/O DRAM (or HMC), and passive interposer (or 2.5D IC integration) have also been provided and discussed. Some important results and recommendations are summarized as follows:

For both device wafers and passive interposer wafers and high volume manufacturing, the MEOL, assembly and test processes should be done by the OSAT (except vertically integrated companies).

It can be seen from Figs. 1.18 through 1.21 that there are many important steps in the MEOL (solder bumping/temporary bonding/back grinding/TSV Cu revealing/thin wafer handling/debonding/cleaning), assembly and test; therefore, the OSATs should strive to make themselves ready for a robust and high-yielding manufacturing process.

In order to have a smooth hand-off from the fab to the OSAT of the unfinished TSV wafer, more research and development work should be performed on the testing methods for blind TSV wafers for electrical [27, 51], thermal [61], and mechanical performance.

1.9 CMOS Images Sensors with TSVs

1.9.1 Toshiba's Dynastron™

In 2008, Toshiba produced their CMOS image sensor (CIS) with TSVs. Actually, Toshiba did not call it TSV but TCV (through chip via). In order to reduce the size of the product, they replaced the wirebonding COB (chip on board) technology of their CMOS image sensor into solder bumped flip chip technology. In order for the signals from the board to reach the pads of the sensor, they made the TCV, as shown in Fig. 1.22.

1.9.2 STMicroelectronics' VGA CIS Camera Module

In 2009, STMicroelectronics produced their VGA (video graphics array) CIS. They integrated the 2.2-μm pixel CIS into Nokia 2330 camera module, as shown in Fig. 1.23. The CIS

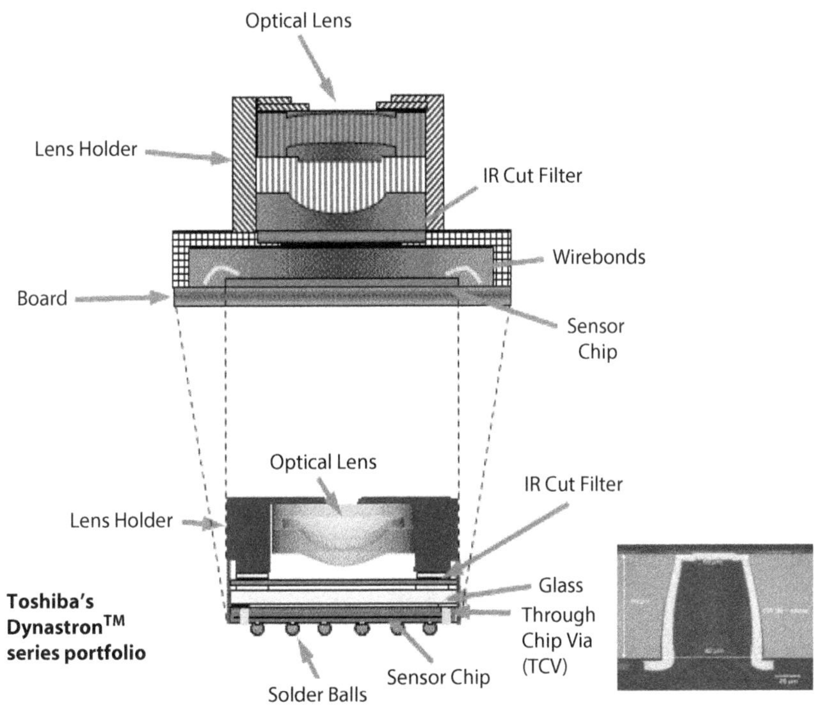

FIGURE 1.22 Toshiba's Dynastron™ with TSV (2008).

chip is manufactured by a CMOS technology with a 0.18-μm process technology. The CIS module is wafer-level packaged using a via-last TSV process.

1.9.3 Samsung's S5K4E5YX BSI CIS

In 2010 Samsung produced their S5K4E5YX 5.1Mp, 1/4.1" optical format 1.4-μm pixel pitch BSI (back side illuminated) CIS. The TSV with a BSI substrate serves to redistribute a back bond pad metallization through the substrate to the front metallization, as shown in Fig. 1.24.

1.9.4 Toshiba's HEW4 BSI TCM5103PL

In 2011 Toshiba produced the HEW4 BSI TCM5103PL 16Mp, 1.4-μm pixel pitch CIS. Its cross-section is shown in Fig. 1.25. It can be seen that (a) the TSV diameter is ~0.5 μm and

Figure 1.23 STMicroelectronics's VGA CIS Camera Module with TSV (2009).

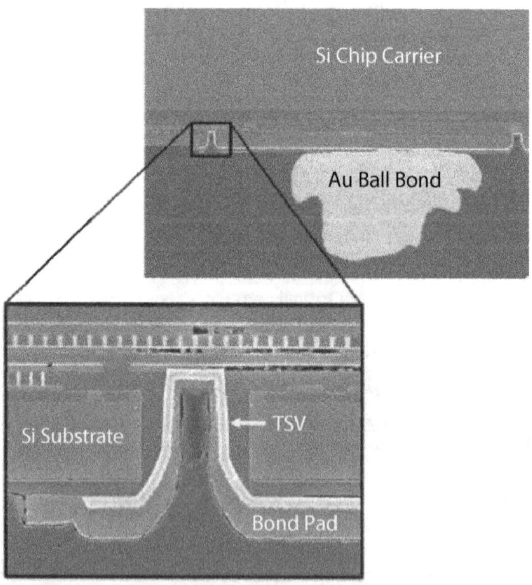

Figure 1.24 Samsung's S5K4E5YX BSI CIS with TSV (2010).

is on ~1.1-μm pitch, (b) the bond pads are isolated by deep trench isolation, (c) the TSVs use polysilicon isolated by annular nitride barrier, and (d) the silicon stain is used, so TSVs stand proud of Si.

1.9.5 Nemotek's CIS

In 2012, Nemotek licensed the technology from Tessera and produced the CIS, as shown in Fig. 1.26. Unlike the CISs made by others, they did not use any of the semiconductor

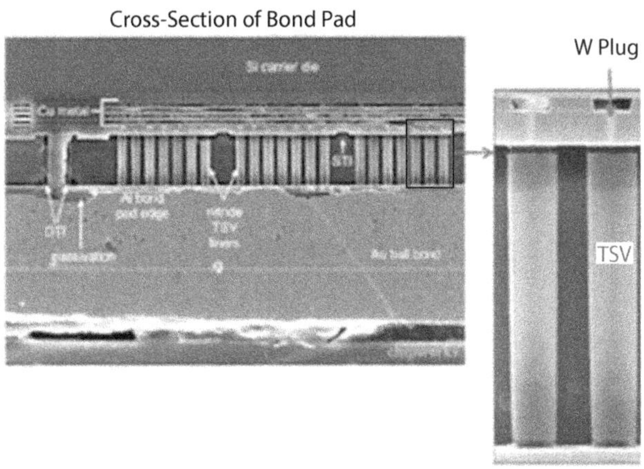

Figure 1.25 Toshiba's HEW4 BSI TCM5103PL with TSV (2011).

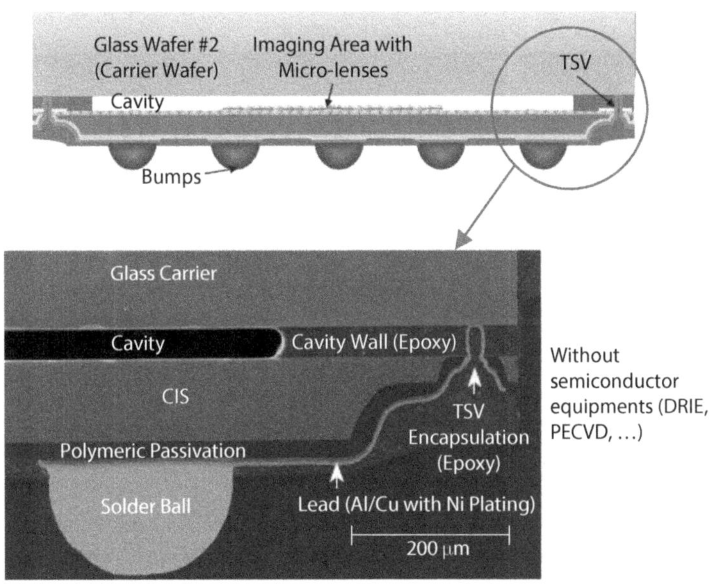

Figure 1.26 Nemotek's CIS with TSV (2012).

Figure 1.27 SONY's ISX014 Stacked Camera Sensor with TSV (2013).

equipments such as the DRIE, PECVD, and CMP to make the TSVs. Instead, they used laser to make the vias and filled it with epoxy.

1.9.6 SONY's ISX014 Stacked Camera Sensor

In 2013, SONY produced the ISX014 stacked camera sensor. It stacks the BSI CIS pixel chip on top of the image processing (logic/analog) chip with TSVs along two opposite edges of the chips, as shown in Fig. 1.27. This is the very first real 3D IC and CIS integration in high-volume production. For more information, please see Sec. 13.3.

1.10 MEMS with TSVs

1.10.1 STMicroelectronics' MEMS Inertial Sensors

In 2011, STMicroelectronics produced the very first MEMS inertial sensors (accelerometers and gyroscopes) with TSVs, as shown in Fig. 1.28. It can be seen that (a) the TSVs are in the Si substrate of the MEMS sensor, and (b) the driver ASIC (application specific IC) is side-by-side with the MEMS sensor and connecting to the TSV by wirebonds.

1.10.2 Discera's MEME Resonator

In 2012, Discera produced the MEMS resonator with TSVs, as shown in Fig. 1.29. It can be seen that (a) the MEMS resonator is right on top of the ASIC, (b) the connections between the ASIC and the TSV are by wirebonds, and (c) the TSVs are in the Si substrate of the MEMS resonator.

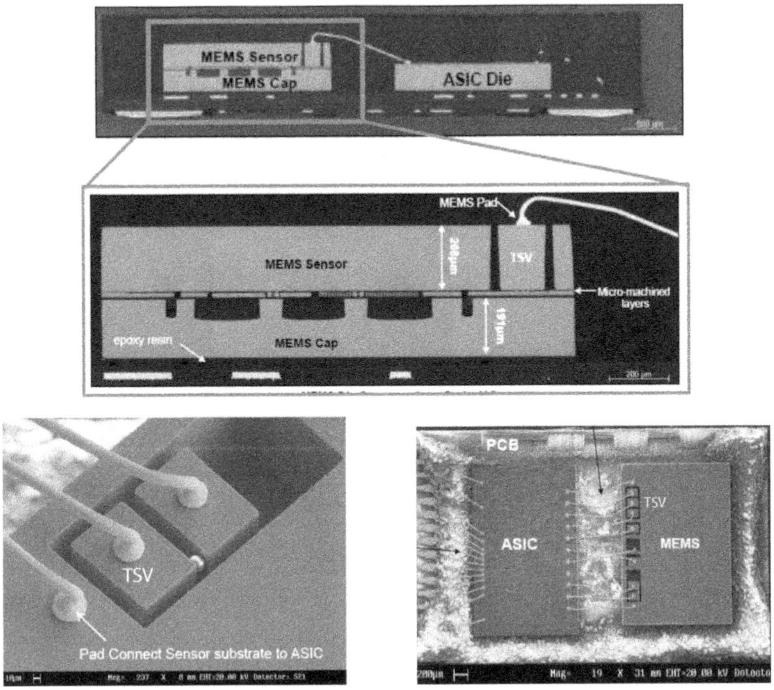

FIGURE 1.28 STMicroelectronics' MEMS Inertial Sensors with TSV (2011).

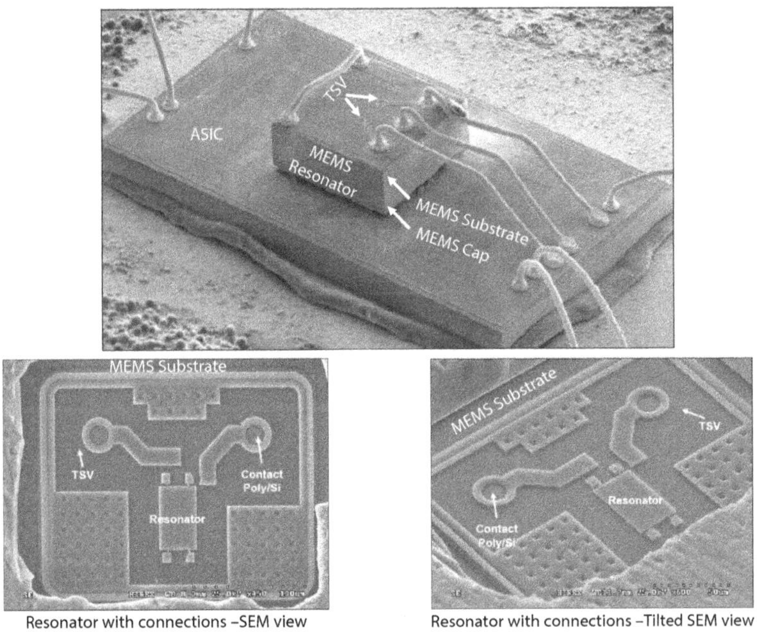

FIGURE 1.29 Discera's MEME Resonator with TSV (2012).

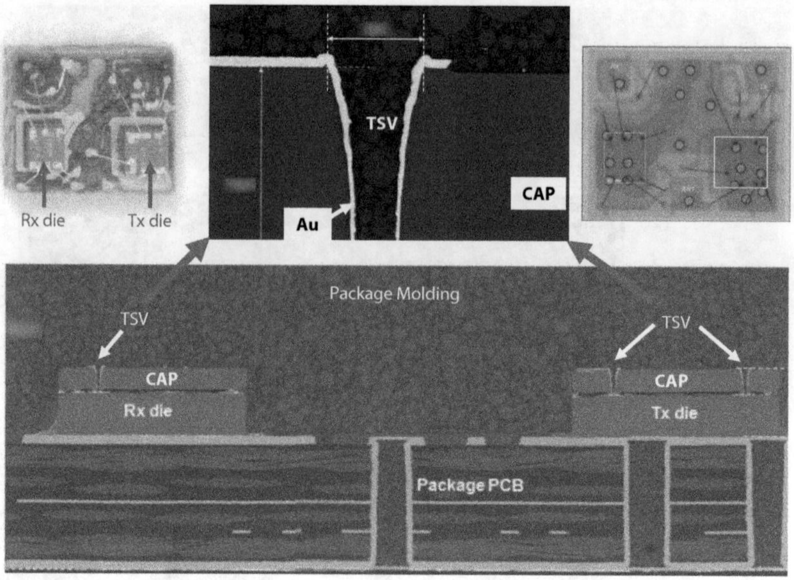

FIGURE 1.30 Avago's FBAR MEMS Filter with TSV (2013).

1.10.3 Avago's FBAR MEMS Filter

In 2013, Avago produced the FBAR (film bulk acoustic resonator) MEMS filter, ACMD-7612: UMTS Band I Duplexer, as shown in Fig. 1.30. It can be seen that (a) TSVs are in the Tx (transceiver) chip and Rx (receiver) chip, (b) unlike the other two MEMS structures shown in Secs. 1.10.1 and 1.10.2, the TSVs are in the cap, and (c) the sidewall of the TSVs is metallized and the TSVs are not filled. For more details of Avago's FRAR MEMS filter, please see Sec. 12.6.4.

1.11 References

[1] Lau, J. H., *Through-Silicon Vias for 3D Integration*, McGraw-Hill Book Company, New York, NY, 2013.

[2] Lau, J. H., *Reliability of ROHS-Compliant 2D and 3D IC Integration*, McGraw-Hill Book Company, New York, NY, 2011.

[3] Lau, J. H., C. K. Lee, C. S. Premachandran, and A. Yu, *Advanced MEMS Packaging*, McGraw-Hill Book Company, New York, NY, 2010.

[4] Lau, J. H., "Overview and Outlook of 3D IC Packaging, 3D IC Integration, and 3D Si Integration," *ASME Transactions, Journal of Electronic Packaging*, Vol. 136, No. 4, December 2014, pp. 1–15.

[5] Lau, J. H., C. Lee, C. Zhan, S. Wu, Y. Chao, M. Dai, R. Tain, et al., "Low-Cost Through-Silicon Hole Interposers for 3D IC," *IEEE Transactions on CPMT*, Vol. 4, No. 9, September 2014, pp. 1407–1419.

[6] Lau, J. H., P. Tzeng, C. Lee, C. Zhan, M. Li, J. Cline, K. Saito, et al., "Redistribution Layers (RDLs) for 2.5D/3D IC Integration," *IMAPS Transactions, Journal of Microelectronic Packaging*, Vol. 11, No. 1, 2014, pp. 16–24.

[7] Lau, J. H., "The Future of Interposer for Semiconductor IC Packaging," *Chip Scale Review*, Vol. 18, No. 1, January–February, 2014, pp. 32–36.

[8] Lau, J. H., "Supply Chains for High-Volume Manufacturing of 3D IC Integration," *Chip Scale Review*, Vol. 17, No. 1, January–February 2013, pp. 33–39.

[9] Khan, N., H. Li, S. Tan, S. Ho, V. Kripesh, D. Pinjala, J. H. Lau, et al., "3-D Packaging With Through-Silicon Via (TSV) for Electrical and Fluidic Interconnections," *IEEE Transactions on CPMT*, Vol. 3, No. 2, February 2013, pp. 221–228.

[10] Lau, J. H., and G. Y. Tang, "Effects of TSVs (Through-Silicon Vias) on Thermal Performances of 3D IC Integration System-in-Package (SiP)," *Journal of Microelectronics Reliability*, Vo. 52, No. 11, November 2012, pp. 2660–2669.

[11] Lau, J. H., "Recent Advances and New Trends in Nanotechnology and 3D Integration for Semiconductor Industry," *The Electrochemical Society, ECS Transactions*, Vol. 44, No. 1, 2012, pp. 805–825.

[12] Chien, H. C., J. H. Lau, Y. Chao, R. Tain, M. Dai, S. T. Wu, W. Lo, et al., "Thermal Performance of 3D IC Integration with Through-Silicon Via (TSV)," *IMAPS Transactions, Journal of Microelectronic Packaging*, Vol. 9, 2012, pp. 97–103.

[13] Chen, J. C., J. H. Lau, P. J. Tzeng, S. Chen, C. Wu, C. Chen, H. Yu, et al., "Effects of Slurry in Cu Chemical Mechanical Polishing (CMP) of TSVs for 3-D IC Integration," *IEEE Transactions on CPMT*, Vol. 2, No. 6, June 2012, pp. 956–963.

[14] Lee, C. K., T. C. Chang, J. H. Lau, Y. Huang, H. Fu, J. Huang, Z. Hsiao, et al., "Wafer Bumping, Assembly, and Reliability of Fine-Pitch Lead-Free Micro Solder Joints for 3-D IC Integration," *IEEE Transactions on CPMT*, Vol. 2, No. 8, August 2012, pp. 1229–1238.

[15] Chai, T. C., X. Zhang, H. Li, V. Sekhar, O. Kalandar, N. Khan, J. H. Lau, et al., "Impact of Packaging Design on Reliability of Large Die Cu/low-κ (BD) Interconnect," *IEEE Transactions on CPMT*, Vol. 2, No. 5, May 2012, pp. 807–816.

[16] Sekhar, V. N., L. Shen, A. Kumar, T. C. Chai, X. Zhang, C. S. Premchandran, V. Kripesh, S. Yoon, et al., "Study on the Effect of Wafer Back Grinding Process on Nanomechanical Behavior of Multilayered Low-k Stack," *IEEE Transactions on CPMT*, Vol. 2, No. 1, January 2012, pp. 3–12.

[17] Zhang, X., R. Rajoo, C. S. Selvanayagam, A. Kumar, V. Rao, N. Khan, V. Kripesh, et al., "Application of Piezoresistive Stress Sensor in Wafer Bumping and Drop Impact Test of Embedded Ultrathin Device," *IEEE Transactions on CPMT*, Vol. 2, No. 16, June 2012, pp. 935–943.

[18] Wu, C., S. Chen, P. Tzeng, J. H. Lau, Y. Hsu, J. Chen, Y. Hsin, et al., "Oxide Liner, Barrier and Seed Layers, and Cu-Plating of Blind Through-Silicon Vias (TSVs) on 300mm Wafers for 3D IC Integration," *IMAPS Transactions, Journal of Microelectronic Packaging*, Vol. 9, No. 1, First Quarter 2012, pp. 31–36.

[19] Lau, J. H., M. S. Zhang, and S. W. R. Lee, "Embedded 3D Hybrid IC Integration System-in-Package (SiP) for Opto-Electronic Interconnects in Organic Substrates," *ASME Transactions, Journal of Electronic Packaging*, Vol. 133, September 2011, pp. 1–7.

[20] Chai, T. C., X. Zhang, J. H. Lau, C. S. Selvanayagam, D. Pinjala, Y. Hoe, Y. Ong, et al., "Development of Large Die Fine-Pitch Cu/low-*k* FCBGA Package with Through-Silicon Via (TSV) Interposer," *IEEE Transactions on CPMT*, Vol. 1, No. 5, May 2011, pp. 660–672.

[21] Lau, J. H., "TSV Interposers: The Most Cost-Effective Integrator for 3D IC Integration," *Chip Scale Review*, Vol. 15, No. 5, September/October, 2011, pp. 23–27.

[22] Sharma, G., V. Rao, A. Kumar, Y. Lim, K. Houe, S. Lim, V. Sekhar, et al., "Design and Development of Multi-Die Laterally Placed and Vertically Stacked Embedded Micro-Wafer-Level Packages," *IEEE Transactions on CPMT*, Vol. 1, No. 5, May 2011, pp. 52–59.

[23] Kumar, A., X. Zhang, Q. Zhang, M. Jong, G. Huang, V. Lee, V. Kripesh, et al., "Residual Stress Analysis in Thin Device Wafer Using Piezoresistive Stress Sensor," *IEEE Transactions on CPMT*, Vol. 1, No. 6, June 2011, pp. 841–851.

[24] Yu, A., J. H. Lau, S. Ho, A. Kumar, W. Hnin, W. Lee, M. Jong, et al., "Fabrication of High Aspect Ratio TSV and Assembly with Fine-Pitch Low-Cost Solder Microbump for Si Interposer Technology with High-Density Interconnects," *IEEE Transactions on CPMT*, Vol. 1, No. 9, September 2011, pp. 1336–1344.

[25] Lau, J. H., "Overview and Outlook of TSV and 3D Integrations," *Journal of Microelectronics International*, V. 28, No. 2, 2011, pp. 8–22.

[26] Lau, J. H., C.-J. Zhan, P.-J. Tzeng, C.-K. Lee, M.-J. Dai, H.-C. Chien, Y.-L. Chao, et al., "Feasibility Study of a 3D IC Integration System-in-Packaging (SiP) from a 300mm Multi-Project Wafer (MPW)," *IMAPS Transactions, Journal of Microelectronic Packaging*, Vol. 8, No. 4, Fourth Quarter 2011, pp. 171–178.

[27] Sheu, S., Z. Lin, J. Hung, J. H. Lau, P. Chen, S. Wu, K. Su, et al., "An Electrical Testing Method for Blind Through-Silicon Vias (TSVs) for 3D IC Integration," *IMAPS Transactions, Journal of Microelectronic Packaging*, Vol. 8, No. 4, Fourth Quarter 2011, pp. 140–145.

[28] Lau, J. H., "Critical Issues of 3D IC Integrations," *IMAPS Transactions, Journal of Microelectronics and Electronic Packaging*, Vol. 7, First Quarter Issue, 2010, pp. 35–43.

[29] Lau, J. H., Y. S. Chan, and R. S. W. Lee, "3D IC Integration with TSV Interposers for High-Performance Applications," *Chip Scale Review*, Vol. 14, No. 5, September/October, 2010, pp. 26–29.

[30] Lau, J. H., "Design and Process of 3D MEMS Packaging," *IMAPS Transactions, Journal of Microelectronics and Electronic Packaging*, Vol. 7, First Quarter Issue, 2010, pp. 10–15.

[31] Lau, J. H., Lee, R., Yuen, M., and Chan, P., "3D LED and IC Wafer Level Packaging," *Journal of Microelectronics International*, Vol. 27, No. 2, 2010, pp. 98–105.

[32] Lau, J. H., "State-of-the-Art and Trends in 3D Integration," *Chip Scale Review*, Vol. 14, No. 2, March/April, 2010, pp. 22–28.

[33] Tang, G. Y., S. Tan, N. Khan, D. Pinjala, J. H. Lau, A. Yu, V. Kripesh, et al., "Integrated Liquid Cooling Systems for 3-D Stacked TSV Modules," *IEEE Transactions on CPMT*, Vol. 33, No. 1, March 2010, pp. 184–195.

[34] Khan, N., V. Rao, S. Lim, H. We, V. Lee, X. Zhang, E. Liao, et al., "Development of 3-D Silicon Module With TSV for System in Packaging," *IEEE Transactions on CPMT*, Vol. 33, No. 1, March 2010, pp. 3–9.

[35] Lim, S., V. Rao, W. Hnin, W. Ching, V. Kripesh, C. Lee, J. H. Lau, et al., "Process Development and Reliability of Microbumps," *IEEE Transactions on CPMT*, Vol. 33, No. 4, December 2010, pp. 747–753.

[36] Yu, D. Q., Y. Li, C. Lee, W. Choi, S. Thew, C. Foo, and J. H. Lau, "Wafer-Level Hermetic Bonding Using Sn/In and Cu/Ti/Au Metallization," *IEEE Transactions on CPMT*, Vol. 32, No. 4, December 2009, pp. 926–934.

[37] Yu, A., N. Khan, G. Archit, D. Pinjala, K. Toh, V. Kripesh, S. Yoon, et al., "Fabrication of Silicon Carriers With TSV Electrical Interconnections and Embedded Thermal Solutions for High Power 3-D Packages," *IEEE Transactions on CPMT*, Vol. 32, No. 3, September 2009, pp. 566–571.

[38] Selvanayagam, C., J. H. Lau, X. Zhang, S. Seah, K. Vaidyanathan, and T. C. Chai, "Nonlinear Thermal Stress/Strain Analyses of Copper Filled TSV (Through-Silicon Via) and Their Flip-Chip Microbumps," *IEEE Transactions on Advanced Packaging*, Vol. 32, No. 4, November 2009, pp. 720–728.

[39] Zhang, X., A. Kumar, Q. X. Zhang, Y. Y. Ong, S. W. Ho, C. H. Khong, V. Kripesh, et al., "Application of Piezoresistive Stress Sensors in Ultra Thin Device Handling and Characterization," *Journal of Sensors & Actuators: A. Physical*, Vol. 156, November 2009, pp. 2–7.

[40] Chen, J., J. H. Lau, T. Hsu, C. Chen, P. Tzeng, P. Chang, C. Chien, et al., "Challenges of Cu CMP of TSVs and RDLs Fabricated from the Backside of a Thin Wafer," *IEEE International 3D Systems Integration Conference*, San Francisco, CA, October 2013, pp. 1–5.

[41] Lau, J. H., H. C. Chien, S. T. Wu, Y. L. Chao, W. C. Lo, and M. J. Kao, "Thin-Wafer Handling with a Heat-Spreader Wafer for 2.5D/3D IC Integration," *Proceedings of the 46th IMAPS International Symposium on Microelectronics*, Orlando, FL, October 2013, pp. 389–396.

[42] Wu, S. T., H. Chien, J. H. Lau, M. Li, J. Cline, and M. Ji, "Thermal and Mechanical Design and Analysis of 3D IC Interposer with Double-Sided Active Chips," *IEEE/ECTC Proceedings*, Las Vegas, NA, May 2013, pp. 1471–1479.

[43] Tzeng, P. J., J. H. Lau, C. Zhan, Y. Hsin, P. Chang, Y. Chang, J. Chen, et al., "Process Integration of 3D Si Interposer with Double-Sided Active Chip Attachments," *IEEE/ECTC Proceedings*, Las Vegas, NA, May 2013, pp. 86–93.

[44] Hung, J. F., J. H. Lau, P. Chen, S. Wu, S. Hung, S. Lai, M. Li, et al., "Electrical Performance of Through-Silicon Vias (TSVs) for High-Frequency 3D IC Integration Applications," *Proceedings of the 45th IMAPS International Symposium on Microelectronics*, September 2012, pp. 1221–1228.

[45] Wu, S. T., J. H. Lau, H. Chien, Y. Chao, R. Tain, L. Li, P. Su, et al., "Thermal Stress and Creep Strain Analyses of a 3D IC Integration SiP with Passive Interposer for Network System Application," *Proceedings of the 45th IMAPS International Symposium on Microelectronics*, September 2012, pp. 1038–1045.

[46] Li, L., P. Su, J. Xue, M. Brillhart, J. H. Lau, P. Tzeng, C. Lee, et al., "Addressing Bandwidth Challenges in Next Generation High Performance Network Systems with 3D IC Integration," *IEEE ECTC Proceedings*, San Diego, CA, May 2012, pp. 1040–1046.

[47] Lau, J. H., S. T. Wu, and H. C. Chien, "Thermal-Mechanical Responses of 3D IC Integration with a Passive TSV Interposer," *IEEE EuroSime Proceedings, Chapter 5: Reliability Modeling*, Lisbon, Portugal, April 2012, pp. 1/8–8/8.

[48] Lau, J. H., S. T. Wu, and H. C. Chien, "Nonlinear Analyses of Semi-Embedded Through-Silicon Via (TSV) Interposer with Stress Relief Gap Under Thermal Operating and Environmental Conditions," *IEEE EuroSime Proceedings, Chapter 11: Thermo-Mechanical Issues in Microelectronics*, Lisbon, Portugal, April 2012, pp. 1/6–6/6.

[49] Wu, S., J. H. Lau, H. Chien, J. Hung, M. Dai, Y. Chao, R. Tain, et al., "Ultra Low-Cost Through-Silicon Holes (TSHs) Interposers for 3D IC Integration SiPs," *IEEE ECTC Proceedings*, San Diego, CA, May 2012, pp. 1618–1624.

[50] Chieh, H. J. H. Lau, Y. Chao, M. Dai, and R. Tain, "Thermal Evaluation and Analyses of 3D IC Integration SiP with TSVs for Network System Applications," *IEEE ECTC Proceedings*, San Diego, CA, May 2012, pp. 1866–1873.

[51] Hung, J. F., J. H. Lau, P. Chen, S. Wu, S. Lai, M. Li, S. Sheu, et al., "Electrical Testing of Blind Through-Silicon Via (TSV) for 3D IC Integration," *IEEE/ECTC Proceedings*, San Diego, CA, May 2012, pp. 564–570.

[52] Zhan, C., P. Tzeng, J. H. Lau, M. Dai, H. Chien, C. Lee, S. Wu, et al., "Assembly Process and Reliability Assessment of TSV/RDL/IPD Interposer with Multi-Chip-Stacking for 3D IC Integration SiP," *IEEE/ECTC Proceedings*, San Diego, CA, May 2012, pp. 548–554.
[53] Tzeng, P., J. H. Lau, M. Dai, S. Wu, H. Chien, Y. Chao, C. Chen, et al., "Design, Fabrication, and Calibration of Stress Sensors Embedded in a TSV Interposer in a 300mm Wafer," *IEEE/ECTC Proceedings*, San Diego, CA, May 2012, pp. 1731–1737.
[54] Huang, S., C. Zhan, Y. Huang, Y. Lin, C. Fan, S. Chung, K. Kao, et al., "Effects of UBM Structure/material on the Reliability Performance of 3D Chip Stacking with 30μm-pitch Solder Micro Bump Interconnections," *IEEE/ECTC Proceedings*, San Diego, CA, May 2012, pp. 1287–1292.
[55] Chien, J., J. H. Lau, Y. Chao, M. Dai, R. Tain, L. Li, P. Su, et al., "Thermal Evaluation and Analyses of 3D IC Integration SiP with TSVs for Network System Applications," *IEEE/ECTC Proceedings*, San Diego, CA, May 2012, pp. 1866–1873.
[56] Sheu, S., Z. H. Lin, C. S. Lin, J. H. Lau, S. H. Lee, K. L. Su, T. K. Ku, et al., "Electrical Characterization of Through-Silicon Vias (TSVs) with an On Chip Bus Driver for 3D IC Integration," *IEEE/ECTC Proceedings*, San Diego, CA, May 2012, pp. 851–856.
[57] Lin, Y., C. Zhan, K. Kao, C. Fan, S. Chung, Y. Huang, S. Huang, et al., "Low Temperature Bonding using Non-Conductive Adhesive for 3D Chip Stacking with 30μm-Pitch Micro Solder Bump Interconnections," *IEEE/ECTC Proceedings*, San Diego, CA, May 2012, pp. 1656–1661.
[58] Lee, C., C. Zhan, J. H. Lau, Y. Huang, H. Fu, J. Huang, Z. Hsiao, et al., "Wafer Bumping, Assembly, and Reliability Assessment of μbumps," *IEEE/ECTC Proceedings*, San Diego, CA, May 2012, pp. 636–640.
[59] Lau, J. H., M. Dai, Y. Chao, W. Li, S. Wu, J. Hung, M. Hsieh, et al., "Feasibility Study of a 3D IC Integration System-in-Packaging (SiP)," *IEEE/ICEP Proceedings*, Nara, Japan, April 13, 2011, pp. 210–216.
[60] Hsin, Y. C., C. Chen, J. H. Lau, P. Tzeng, S. Shen, Y. Hsu, S. Chen, et al., "Effects of Etch Rate on Scallop of Through-Silicon Vias (TSVs) in 200mm and 300mm Wafers," *IEEE ECTC Proceedings*, Orlando, FL, June 2011, pp. 1130–1135.
[61] Chien, J., Y. Chao, J. H. Lau, M. Dai, R. Tain, M. Dai, P. Tzeng, et al., "A Thermal Performance Measurement Method for Blind Through-Silicon Vias (TSVs) in a 300mm Wafer," *IEEE ECTC Proceedings*, Orlando, FL, June 2011, pp. 1204–1210.
[62] Tsai, W., H. H. Chang, C. H. Chien, J. H. Lau, H. C. Fu, C. W. Chiang, T. Y. Kuo, et al., "How to Select Adhesive Materials for Temporary Bonding and De-Bonding of Thin-Wafer Handling in 3D IC Integration?," *IEEE ECTC Proceedings*, Orlando, FL, June 2011, pp. 989–998.
[63] Zhan, C., J. Juang, Y. Lin, Y. Huang, K. Kao, T. Yang, S. Lu, et al., "Development of Fluxless Chip-on-Wafer Bonding Process for 3D chip Stacking with 30μm Pitch Lead-Free Solder Micro Bump Interconnection and Reliability Characterization," *IEEE ECTC Proceedings*, Orlando, FL, June 2011, pp. 14–21.
[64] Huang, S., T. Chang, R. Cheng, J. Chang, C. Fan, C. Zhan, J. H. Lau, et al., "Failure Mechanism of 20μm Pitch Micro Joint Within a Chip Stacking Architecture," *IEEE ECTC Proceedings*, Orlando, FL, June 2011, pp. 886–892.
[65] Lin, Y., C. Zhan, J. Juang, J. H. Lau, T. Chen, R. Lo, M. Kao, et al., "Electromigration in Ni/Sn Intermetallic Micro Bump Joint for 3D IC Chip Stacking," *IEEE ECTC Proceedings*, Orlando, FL, June 2011, pp. 351–357.
[66] Lau, J. H., "The Most Cost-Effective Integrator (TSV Interposer) for 3D IC Integration System-in-Package (SiP)," *ASME Paper no. InterPACK2011-52189*, Portland, OR, July 2011, pp. 1–12.
[67] Lau, J. H., H. C. Chien, and R. Tain, "TSV Interposers with Embedded Microchannels for 3D IC and LED Integration," *ASME Paper no. InterPACK2011-52204*, Portland, OR, July 2011, pp. 1–8.
[68] Lau, J. H., and X. Zhang, "Effects of TSV Interposer on the Reliability of 3D IC Integration SiP," *ASME Paper no. InterPACK2011-52205*, Portland, OR, July 2011, pp. 1–9.
[69] Wu, C., S. Chen, P. Tzeng, J. H. Lau, Y. Hsu, J. Chen, Y. Hsin, et al., "Oxide Liner, Barrier and Seed Layers, and Cu-Plating of Blind Through-Silicon Vias (TSVs) on 300mm Wafers for 3D IC Integration," *Proceedings of IMAPS International Conference*, Long Beach, CA, October 2011, pp. 1–7.
[70] Chang, H. H., J. H. Lau, W. L. Tsai, C. H. Chien, P. J. Tzeng, C. J. Zhan, C. K. Lee, et al., "Thin Wafer Handling of 300mm Wafer for 3D IC Integration," *Proceedings of IMAPS International Conference*, Long Beach, CA, October 2011, pp. 202–207.
[71] Lau, J. H., P.-J. Tzeng, C.-K. Lee, C.-J. Zhan, M.-J. Dai, L. Li, C.-T. Ko, et al., "Wafer Bumping and Characterizations of Fine-Pitch Lead-Free Solder Microbumps on 12" (300mm) wafer for 3D IC Integration," *Proceedings of IMAPS International Conference*, Long Beach, CA, October 2011, pp. 650–656.
[72] Lau, J. H., "TSV Manufacturing Yield and Hidden Costs for 3D IC Integration," *IEEE Proceedings of ECTC*, Las Vegas, NV, June 2010, pp. 1031–1041.
[73] Lau, J. H., "Evolution and Outlook of TSV and 3D IC/Si Integration," *IEEE/EPTC Proceedings*, Singapore, December 2010, pp. 560–570.

[74] Yu, A., J. H. Lau, S. Ho, A. Kumar, Y. Wai, d. Yu, M. Jong, et al., "Study of 15-μm-Pitch Solder Microbumps for 3D IC Integration," *IEEE Proceedings of ECTC*, San Diego, CA, May 2009, pp. 6–10.
[75] Vempati, S. R., S. Nandar, C. Khong, Y. Lim, K. Vaidyanathan, J. H. Lau, B. P. Liew, et al., "Development of 3-D Silicon Die Stacked Package Using Flip Chip Technology with Micro Bump Interconnects," *IEEE Proceedings of ECTC* San Diego, CA, May, 2009, pp. 980–987.
[76] Lau, J. H., and G. Tang, "Thermal Management of 3D IC Integration with TSV (Through-Silicon Via)," *IEEE Proceedings of ECTC*, San Diego, May 2009, pp. 635–640.
[77] Ho, S., S. Yoon, Q. Zhou, K. Pasad, V. Kripesh, and J. H. Lau, "High RF Performance TSV for Silicon Carrier for High Frequency Application," *IEEE Proceedings of Electronic, Components & Technology Conference*, Orlando, FL, May 27–30, 2008, pp. 1946–1952.
[78] Akasaka, Y., "Three-Dimensional IC Trends," *Proceedings of the IEEE*, Vol. 74, No. 12, December 1986, pp. 1703–1714.
[79] Shigetou, A. Itoh, T., Sawada, K., and Suga, T., "Bumpless Interconnect of 6-um Pitch Cu Electrodes at Room Temperature," In *IEEE Proceedings of ECTC*, Lake Buena Vista, FL, May 27–30, 2008, pp. 1405–1409.
[80] Tsukamoto, K., E. Higurashi, and T. Suga, "Evaluation of Surface Microroughness for Surface Activated Bonding," *Proceedings of IEEE CPMT Symposium Japan*, August 2010, pp. 147–150.
[81] Kondou, R., C. Wang, and T. Suga, "Room-Temperature Si-Si and Si-SiN Wafer Bonding," *Proceedings of IEEE CPMT Symposium Japan*, August 2010, pp. 161–164.
[82] Shigetou, A. Itoh, T., Matsuo, M., Hayasaka, N., Okumura, K., and T. Suga, "Bumpless Interconnect Through Ultrafine Cu Electrodes by Mans of Surface-Activated Bonding (SAB) Method," *IEEE Transaction on Advanced Packaging*, Vol. 29, No. 2, May 2006, p. 226.
[83] Wang, C., and T. Suga, "A Novel Moire Fringe Assisted Method for Nanoprecision Alignment in Wafer Bonding," In *IEEE Proceedings of ECTC*, San Diego, CA, May 25–29, 2009, pp. 872–878.
[84] Wang, C., and T. Suga, "Moire Method for Nanoprecision Wafer-to-Wafer Alignment: Theory, Simulation and Application," *IEEE Proceedings of Int. Conference on Electronic Packaging Technology & High Density Packaging*, August 2009, pp. 219–224.
[85] Higurashi, E., D. Chino, T. Suga, and R. Sawada, "Au-Au Surface-Activated Bonding and Its Application to Optical Microsensors with 3-D Structure," *IEEE Journal of Selected Topic in Quantum Electronics*, Vol. 15, No. 5, September/October 2009, pp. 1500–1505.
[86] Burns, J., B. Aull, C. Keast, C. Chen, C. Chen, C. Keast, J. Knecht, et al., "A Wafer-Scale 3-D Circuit Integration Technology," *IEEE Transactions on Electron Devices*, Vol. 53, No. 10, October 2006, pp. 2507–2516.
[87] Chen, C., K. Warner, D. Yost, J. Knecht, V. Suntharalingam, C. Chen, J. Burns, et al., "Sealing Three-Dimensional SOI Integrated-Circuit Technology," *IEEE Proceedings of Int. SOI Conference*, 2007, pp. 87–88.
[88] Chen, C., C. Chen, D. Yost, J. Knecht, P. Wyatt, J. Burns, K. Warner, et al., "Three-Dimensional Integration of Silicon-on-Insulator RF Amplifier," *Electronics Letters*, Vol. 44, No. 12, June 2008, pp. 1–2.
[89] Chen, C., C. Chen, D. Yost, J. Knecht, P. Wyatt, J. Burns, K. Warner, et al., "Wafer-Scale 3D Integration of Silicon-on-Insulator RF Amplifiers," *IEEE Proceedings of Silicon Monolithic IC in RF Systems*, 2009, pp. 1–4.
[90] Chen, C., C. Chen, P. Wyatt, P. Gouker, J. Burns, J. Knecht, D. Yost, et al., "Effects of Through-BOX Vias on SOI MOSFETs," *IEEE Proceedings of VLSI Technology, Systems and Applications*, 2008, pp. 1–2.
[91] Chen, C., C. Chen, J. Burns, D. Yost, K. Warner, J. Knecht, D. Shibles, et al., "Thermal Effects of Three Dimensional Integrated Circuit Stacks," *IEEE Proceedings of Int. SOI Conference*, 2007, pp. 91–92.
[92] Aull, B., J. Burns, C. Chen, B. Felton, H. Hanson, C. Keast, J. Knecht, et al., "Laser Radar Imager Based on 3D Integration of Geiger-Mode Avalanche Photodiodes with Two SOI Timing Circuit Layers," *IEEE Proceedings of Int. Solid-State Circuits Conference*, 2006, pp. 1179–1188.
[93] Altera White Paper, "Addressing Next-Generation Memory Requirements Using Altera FPGAs and HMC Technology," *Altera Corporation*, January 2014.
[94] Myslewski, R., "Intel Teams with Micron on Next-Gen Many-Core Xeon Phi with 3D DRAM Introduces New 'Fundamental Building Block of HPC Systems' with Intel Omni Scale Fabric," *High Performance Computer*, June 2014.
[95] Yoshida, T., "SPARC64 Xifx: Fujitsu's Next Generation Processor for HPC," *Hot Chips: A Symposium on High Performance Chips*, August 11, 2014.
[96] Chaabouni, H., M. Rousseau, P. Ldeus, A. Farcy, R. El Farhane, A. Thuaire, G. haury, et al., "Investigation on TSV Impact on 65nm CMOS Devices and Circuits," *Proceedings of IEEE/IEDM*, December 2010, pp. 35.1.1–35.1.4.
[97] Gat, A., L. Gerzberg, J. F. Gibbons, T. J. Magee, J. Peng, and J. D. Hong "CW Laser of Polyerystalline Silicon; Crystalline Structure and Electrical Properties," *Applied Physics Letter*, Vol. 33, No. 8, October 1978, pp. 775–780.

CHAPTER 2

Through-Silicon Vias Modeling and Testing

2.1 Introduction

TSVs (through-silicon vias) are the heart of 3D IC and Si integration. In this chapter, the electrical, thermal, and mechanical modeling and testing of TSVs are presented and discussed.

2.2 Electrical Modeling of TSVs

In this section, the electrical performance of a generic TSV structure for high-frequency 3D IC integration applications is presented. Emphasis is placed on the proposal of an analytical model and the analytical equations of a TSV with all its key parameters such as TSV diameter, TSV depth, silicon dioxide thickness, pad diameter, and TSV pitch. Also, the model and equations are verified, both in the frequency and time domains, by 3D finite-element EM-simulation up to 30 GHz. Finally, a TSV electrical design guideline is proposed.

2.2.1 Analytic Model and Equations for a Generic TSV Structure

This section proposes a high-frequency model for a generic TSV structure. This analytical model includes a circuit model and a set of analytical equations based on the TSV physical configuration. TSV models and equations have been published by [1–13]. However, they are slightly different from those which will be presented herein.

The generic TSV structure and its structural parameters are shown in Fig. 2.1 [14]. The structure has a signal TSV and a ground TSV that are copper-filled and surrounded by the silicon dioxide (SiO_2) layer in the silicon substrate. Figure 2.1 also shows other important parameters of a TSV—TSV diameter (TSV_D) (d), TSV thickness (h), TSV pitch (p), SiO_2 of TSV side-wall (SiO_2_t) (t), SiO_2 of silicon surface (SiO_2_s) (s), pad diameter (k), and pad thickness (m). The analytical equations are functions of these parameters. Based on the equations provided herein and the circuit model shown in Fig. 2.2, designers can easily design and characterize the electrical performance of TSVs for 3D IC integration applications.

The circuit model for the generic TSV structure shown in Fig. 2.1 is proposed in Fig. 2.2. Each R, L, G, and C element in the circuit model is based on the TSV physical configuration.

The resistance of the TSV (R_{TSV}) is derived from Eq. (2.1), which includes the resistance of the via (R_{Via}) and pad (R_{Pad}). Because of transmit wide bandwidth signal, the skin effect

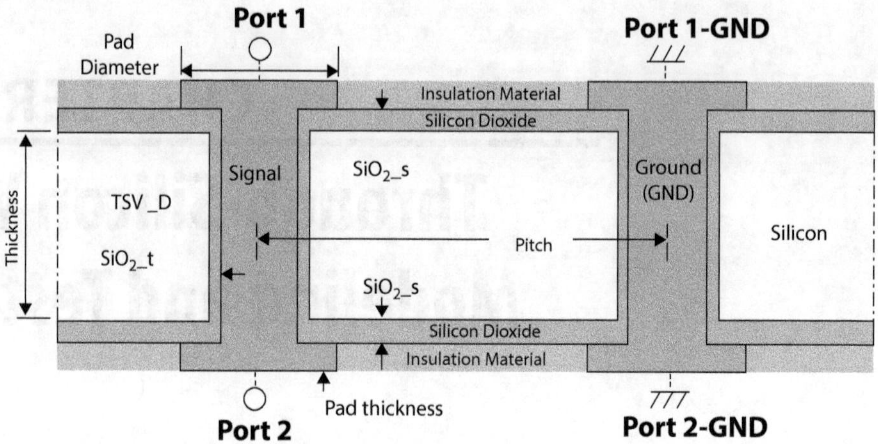

TSV_D	Pad Thickness	Pad Diameter	Si Thickness	Pitch	SiO_2_t	SiO_2_s
d	*m*	*k*	*h*	*p*	*t*	*s*

Figure 2.1 A generic TSV structure and structural parameters.

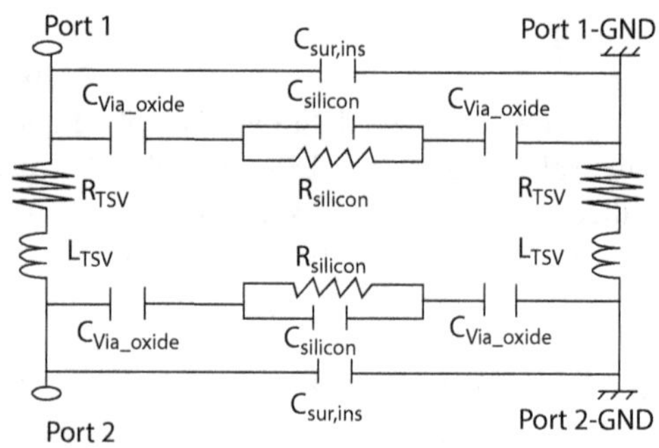

Figure 2.2 Circuit model of the generic TSV structure.

has to be considered for the resistance [15]. R_{Via} is calculated from Eqs. (2.2)–(2.5) [15–17]. There are the DC resistance and AC resistance and ρ_{TSV}, μ, and σ_{TSV} are, respectively, the resistivity, permeability, and conductivity of the via and f is the frequency. R_{Pad} is calculated from the same equation of R_{Via} but h+2s and d need to be replaced by m and k.

$$R_{TSV} = R_{Via} + 2R_{Pad} \tag{2.1}$$

$$R_{Via} = \sqrt{\left(R_{DC,Via}\right)^2 + \left(R_{AC,Via}\right)^2} \tag{2.2}$$

$$R_{DC,Via} = \rho_{TSV} \cdot \frac{h+2s}{\pi\left(d/2\right)^2} \tag{2.3}$$

$$R_{\text{AC,Via}} = \rho_{\text{TSV}} \cdot \frac{h+2s}{\pi \cdot \delta_{\text{skin}} \left(d - \delta_{\text{skin}}\right)} \tag{2.4}$$

$$\delta_{\text{skin}} = \frac{1}{\sqrt{\pi \cdot f \cdot \mu \cdot \sigma_{\text{TSV}}}} \tag{2.5}$$

The inductances of the generic TSV structure, L_{TSV} is derived from Eq. (2.6), which includes the inductance of via and pad, L_{Via} and L_{Pad}. Because of structure similarity, L_{Via} and L_{Pad} are derived from the inductance of two-wire parallel line structure and calculated by Eqs. (2.7) and (2.8) [15, 16], respectively.

$$L_{\text{TSV}} = L_{\text{Via}} + 2L_{\text{Pad}} \tag{2.6}$$

$$L_{\text{Via}} = \frac{\mu_0 \mu_r}{2\pi} \cdot (h+2s) \cdot \ln\left(\frac{p}{d} + \sqrt{\left(\frac{p}{d}\right)^2 - 1}\right) \tag{2.7}$$

$$L_{\text{Pad}} = \frac{\mu_0 \mu_r}{2\pi} \cdot m \cdot \ln\left(\frac{p}{k} + \sqrt{\left(\frac{p}{k}\right)^2 - 1}\right) \tag{2.8}$$

The capacitance caused from the electric field between the signal TSV and ground TSV needs to be considered. Since silicon is a semiconductor and have a silicon conductivity, σ_{silicon}, the electric field when passes through silicon affects the insertion loss of TSV. Thus, the dielectric loss caused by silicon conductivity needs to be considered as well. The capacitance and conductance between TSVs, C_{silicon} and G_{silicon}, formed in the silicon are calculated from Eqs. (2.9) and (2.10) [15, 16], respectively. R_{silicon} is the reciprocal of G_{silicon}. Eqs. (2.9) and (2.10) are derived from the same manner of inductance, two-wire parallel line structure.

$$C_{\text{silicon}} = \frac{1}{2} \cdot \frac{\pi \cdot \varepsilon_0 \cdot \varepsilon_{r,\text{silicon}} \cdot (h+2s)}{\ln\left[\frac{p}{d+2t}\left(1 + \sqrt{1 - \left(\frac{d+2t}{p}\right)^2}\right)\right]} \tag{2.9}$$

$$G_{\text{silicon}} = \frac{1}{2} \cdot \frac{\sigma_{\text{silicon}} \cdot \pi \cdot (h+2s)}{\ln\left[\frac{p}{d+2t}\left(1 + \sqrt{1 - \left(\frac{d+2t}{p}\right)^2}\right)\right]} \tag{2.10}$$

Silicon dioxide surrounds the TSV in order to isolate the silicon and TSV. Therefore, there is capacitance between silicon and TSV, $C_{\text{Via\_oxide}}$, calculated from Eqs. (2.11), (2.12), and (2.13) with the TSV structural parameters. Equation (2.12) is a coaxial-like capacitance formed between via and silicon. Equation (2.13) calculates the capacitance where the pad faces the silicon.

$$C_{\text{Via\_oxide}} = \frac{1}{2}(C_{\text{coaxial}} + C_{\text{Pad}}) \tag{2.11}$$

$$C_{\text{coaxial}} = \pi \cdot \varepsilon_0 \varepsilon_{r,\text{oxide}} \frac{h+2s}{\ln\left(\frac{d/2+t}{d/2}\right)} \tag{2.12}$$

$$C_{\text{Pad}} = \varepsilon_0 \cdot \varepsilon_{r,\text{oxide}} \cdot \frac{\pi \cdot \left[\left(k/2\right)^2 - \left(d/2\right)^2\right]}{s} \tag{2.13}$$

Other parasitic capacitance in the top and bottom side of TSV, $C_{\text{sur,ins}}$, is also derived from Eq. (2.14). Including the capacitance in silicon dioxide at silicon surface, $C_{\text{surface,oxide}}$ and insulation material, $C_{\text{insulation}}$ are calculated by Eqs. (2.15) and (2.16).

$$C_{\text{sur,ins}} = C_{\text{surface,oxide}} + C_{\text{insulation}} \tag{2.14}$$

$$C_{\text{surface,oxide}} = \frac{\pi \cdot \varepsilon_0 \cdot \varepsilon_{r,\text{oxide}} \cdot s}{\ln\left[\frac{p}{d}\left(1+\sqrt{1-\left(\frac{d}{p}\right)^2}\right)\right]} \tag{2.15}$$

$$C_{\text{insulation}} = \frac{\pi \cdot \varepsilon_0 \cdot \varepsilon_{r,\text{insulation}} \cdot m}{\ln\left[\frac{p}{k}\left(1+\sqrt{1-\left(\frac{k}{p}\right)^2}\right)\right]} \tag{2.16}$$

Since the electrical performance of microbumps is similar to the pad of the generic TSV structure, it is the same manner to calculate their parasitic resistance, inductance, and capacitance by using these analytical equations and then adding them into the model.

2.2.2 Verification of the Proposed TSV Model in Frequency Domain

The proposed circuit model and the equations of the generic TSV structure are verified by the 3D finite-element EM-simulation with a wide-range of TSV dimensions and up to frequency = 30 GHz. Based on the structural parameter shown in Fig. 2.1, Table 2.1 shows more than 100 cases of simulations. It can be seen that (a) TSV diameter = 5, 10, 20, 30, 40, and 50 μm; (b) TSV pad diameter = 10, 20, 30, 40, 50, and 60 μm; (c) TSV pad thickness = 3 μm; (d) TSV thickness = 50, 100, and 150 μm; (e) TSV pitch is ranging from 15 to 120 μm for TSV diameter = 5 μm, 25 to 150 μm for TSV diameter = 10 μm, 40 to 150 μm for TSV diameter = 20 μm, 50 to 200 μm for TSV diameter = 30 μm, 60 to 200 μm for TSV diameter = 40 μm, and 70 to 200 μm for TSV diameter = 50 μm; and (f) SiO_2 thickness = 0.2 and 0.5 μm.

Table 2.1 The Analyzed Parameters Matrix for Electrical Responses

Unit: μm	TSV_D = 5 μm	TSV_D = 10 μm	TSV_D = 20 μm	TSV_D = 30 μm	TSV_D = 40 μm	TSV_D = 50 μm
Pad Diameter	10	20	30	40	50	60
Pad Thickness	3	3	3	3	3	3
Thickness	50, 100, 150	50, 100, 150	50, 100, 150	50, 100, 150	50, 100, 150	50, 100, 150
Pitch	15, 30, 60, 90, 120	25, 50, 70, 100, 150	40, 80, 120, 150	50, 100, 150, 200	60, 100, 150, 200	70, 100, 150, 200
SiO_2	0.2, 0.5	0.2, 0.5	0.2, 0.5	0.2, 0.5	0.2, 0.5	0.2, 0.5

For all the cases considered, the 3D finite-element simulation results match very well with the proposed circuit model and equations. Only 18 cases are shown in Figs. 2.3 through 2.8. Figure 2.3 shows the insertion loss of the case: TSV diameter = 5 μm; thickness = 50, 100, and 150 μm; pitch =15 μm; and SiO_2 = 0.2 μm. It can be seen that (a) the insertion loss is a function of frequency, especially at low frequencies; (b) the larger the frequency the larger the insertion loss; and (c) the thicker the TSV the larger the insertion loss. Figure 2.4 shows the insertion loss of the case similar to that shown in Fig. 2.3 except with a very large pitch (120 μm). It can be seen that the effect of TSV pitch is to lower the insertion loss at low frequencies but increase the insertion loss at very high frequencies, especially for thicker TSVs.

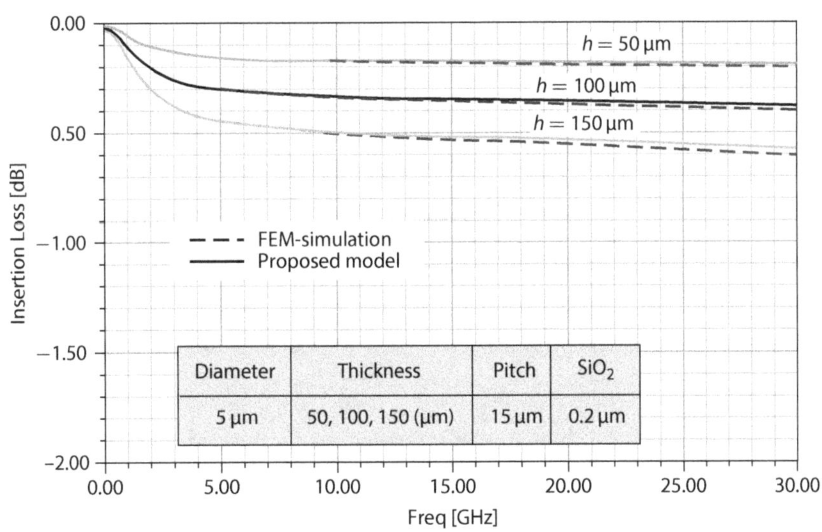

Figure 2.3 Insertion loss for TSV diameter = 5 μm; thickness = 50, 100, 150 μm, pitch = 15 μm, with SiO_2 = 0.2 μm.

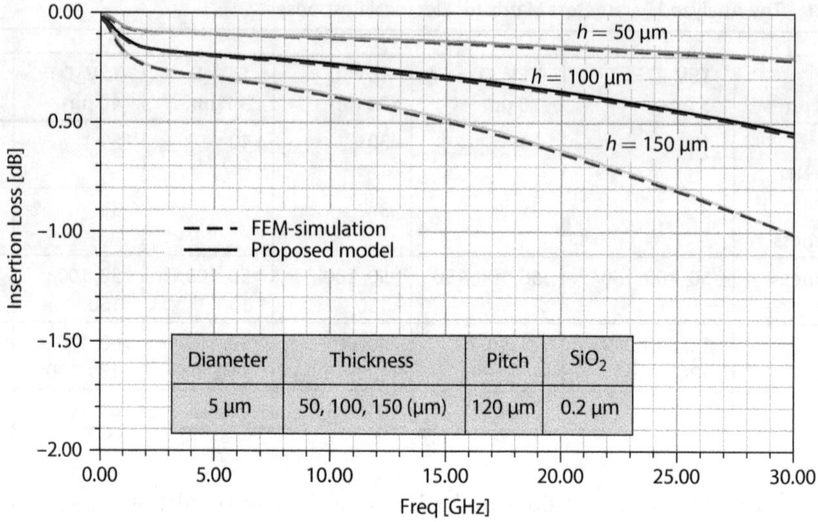

Figure 2.4 Insertion loss for TSV diameter = 5 μm; thickness = 50, 100, 150 μm, pitch = 120 μm, with SiO_2 = 0.2 μm.

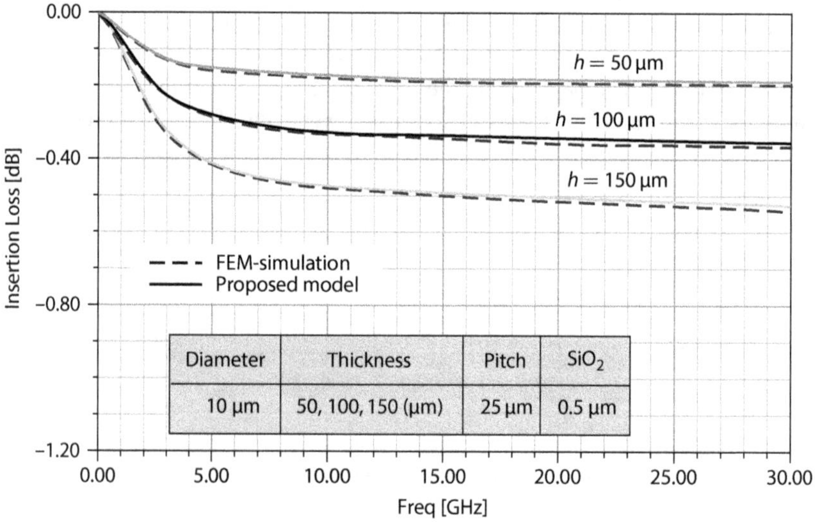

Figure 2.5 Insertion loss for TSV diameter = 10 μm; thickness = 50, 100, 150 μm, pitch = 25 μm, with SiO_2 = 0.5 μm.

Figures 2.5 and 2.6 show similar insertion loss for the cases with TSV diameter = 10 μm and SiO_2 = 0.5 μm. The effect of TSV pitch is to lower the insertion loss at low frequencies, but increase the insertion loss at very high frequencies and thick TSVs.

On the other hand, Figs. 2.7 and 2.8 show the insertion loss of the case with TSV diameter = 20 μm and SiO_2 = 0.5 μm. It can be seen that the effect of TSV pitch is to lower the insertion loss for almost all the frequencies even at very high frequencies and thick TSVs.

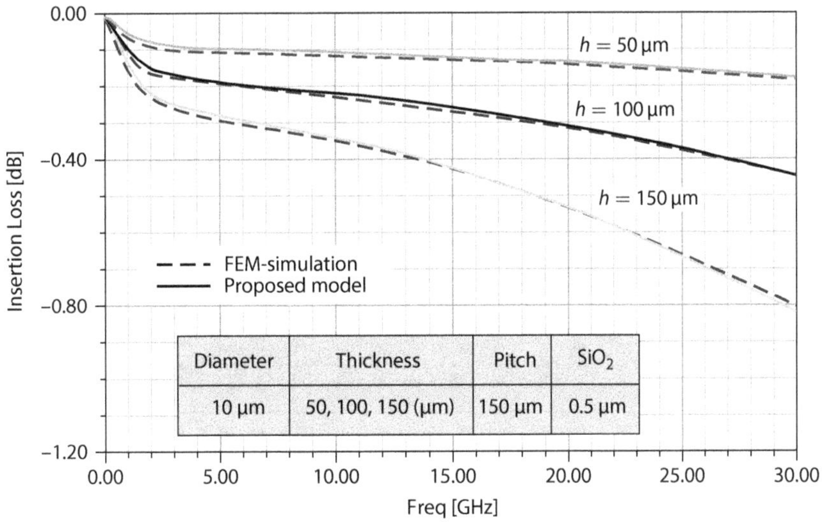

FIGURE 2.6 Insertion loss for TSV diameter = 10 µm; thickness = 50, 100, 150 µm, pitch = 150 µm, with SiO_2 = 0.5 µm.

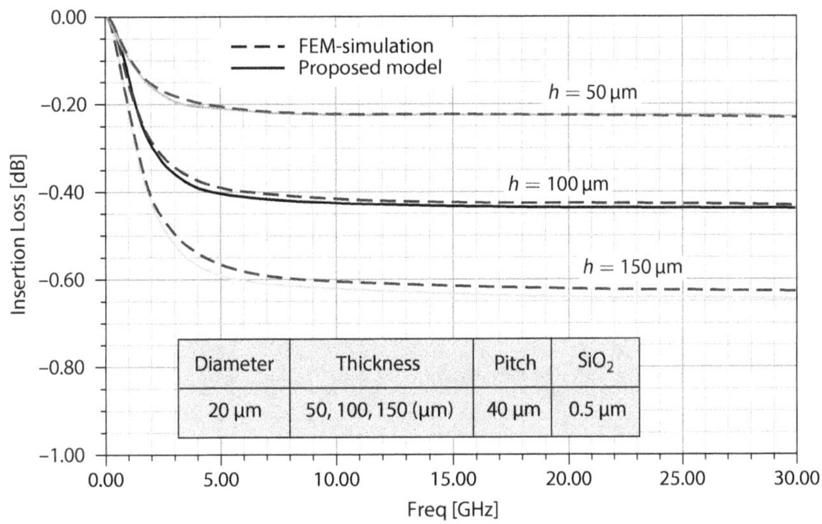

FIGURE 2.7 Insertion loss for TSV diameter = 20 µm; thickness = 50, 100, 150 µm, pitch = 40 µm, with SiO_2 = 0.5 µm.

2.2.3 Verification of the Proposed TSV Model in Time Domain

With the time domain electrical analysis, eye diagram will be simulated by using the insertion loss from the proposed circuit model as well as analytical equations and FEM-simulation. The simulation signal is 11-bits PRBS (Pseudo Random Binary Sequence) with amplitude of 500 mVpp, data rate of 10 Gbps; rising time of 20 ps and loading resistance is 50Ω. Figures 2.9 and 2.10

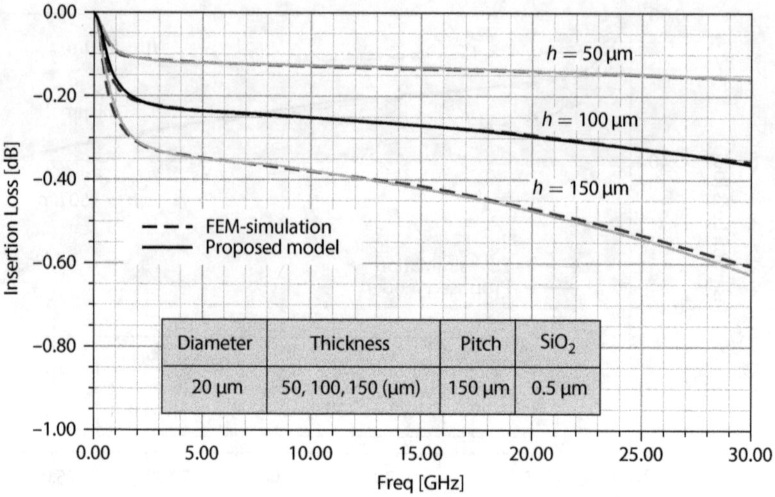

Diameter	Thickness	Pitch	SiO_2
20 μm	50, 100, 150 (μm)	150 μm	0.5 μm

FIGURE 2.8 Insertion loss for TSV diameter = 20 μm; thickness = 50, 100, 150 μm, pitch = 150 μm, with SiO_2 = 0.5 μm.

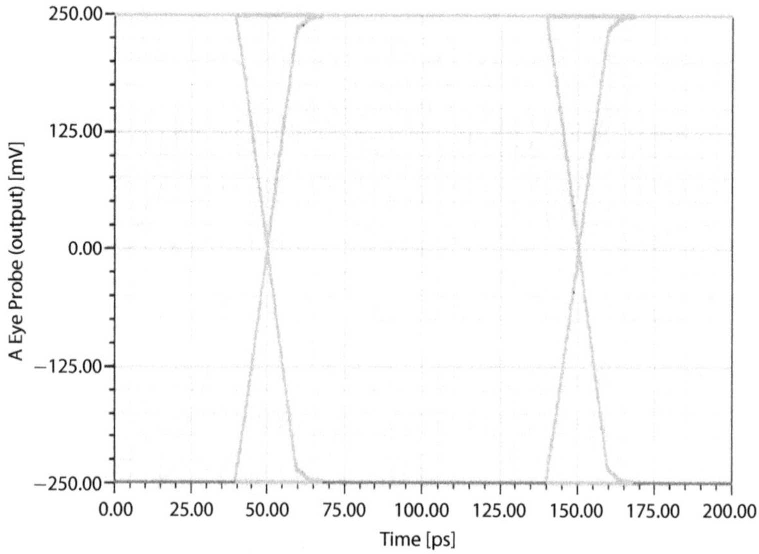

FIGURE 2.9 Eye diagrams from analytical model for TSV diameter = 5 μm; thickness = 50 μm, pitch = 15 μm, and SiO_2 = 0.2 μm.

show, respectively, the analytical model and FEM-simulation for the case with TSV diameter = 5 μm, thickness = 50 μm, pitch = 15 μm, and SiO_2 = 0.2 μm. Figures 2.11 and 2.12 show, respectively, the analytical model and FEM-simulation for the case with TSV diameter = 50 μm, thickness = 150 μm, pitch = 70 μm, and SiO_2 = 0.2 μm. It can be seen that the eye diagrams from the proposed circuit model, equations, and the EM-simulation match very well.

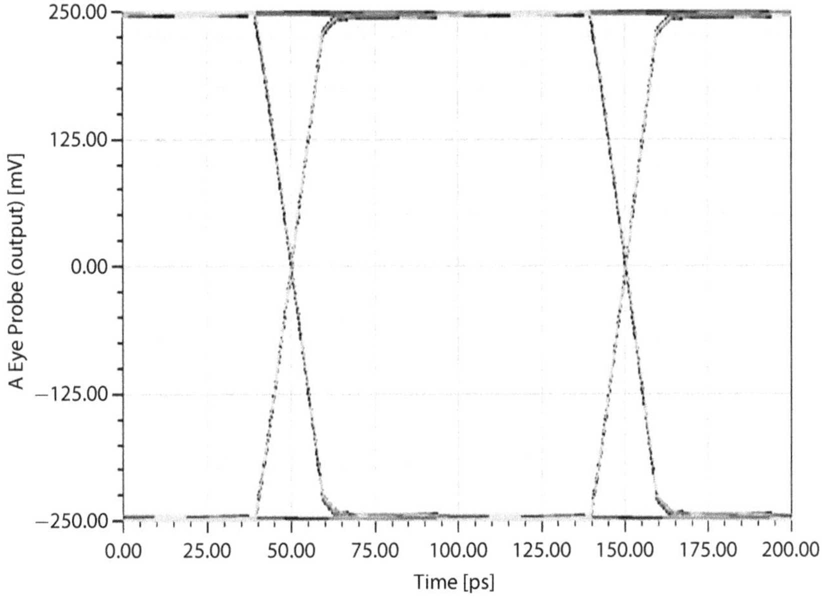

Figure 2.10 Eye diagrams from FEM simulation for TSV diameter = 5 μm; thickness = 50 μm, pitch = 15 μm, and SiO_2 = 0.2 μm.

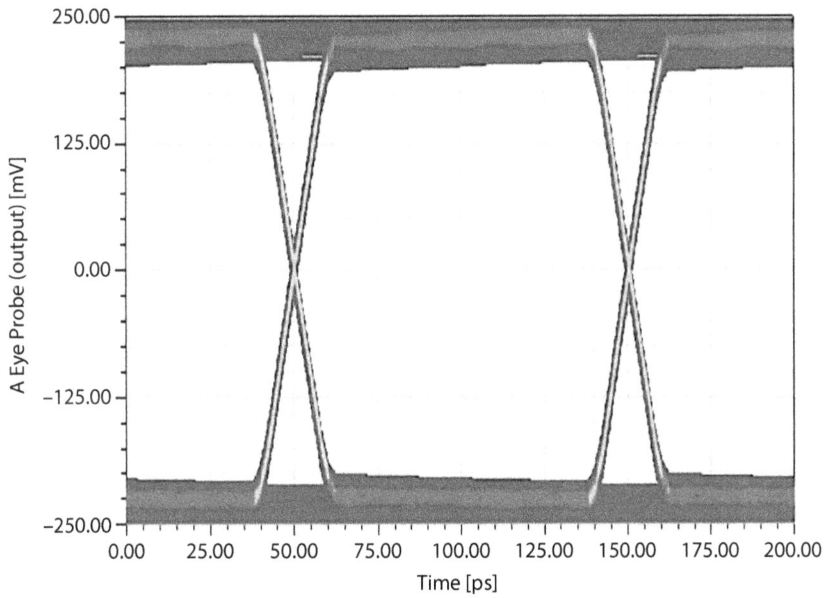

Figure 2.11 Eye diagrams from analytical model for TSV diameter = 50 μm; thickness = 150 μm, pitch = 70 μm, and SiO_2 = 0.2 μm.

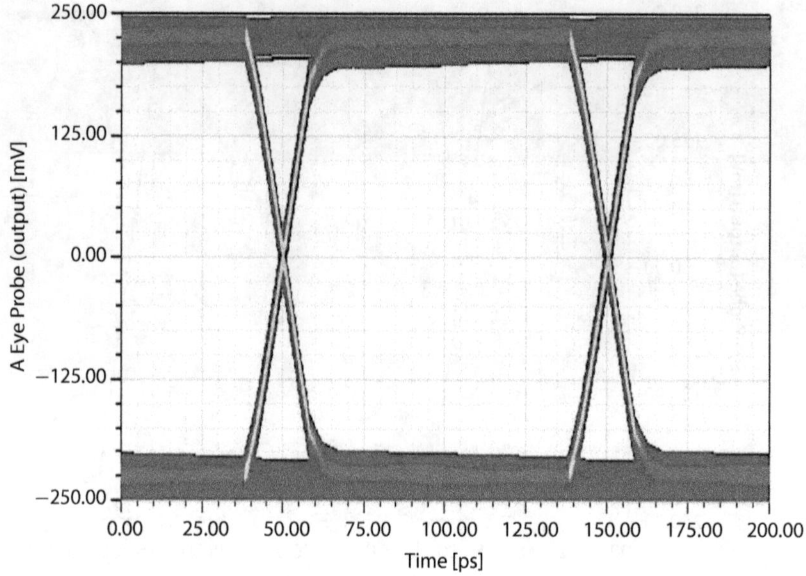

FIGURE 2.12 Eye diagrams from FEM simulation for TSV diameter = 50 μm; thickness = 150 μm, pitch = 70 μm, and SiO_2 = 0.2 μm.

Figure 2.13 shows the eye diagrams (from the analytical model and equations) of 4-chips stacking with the TSV structure: TSV diameter = 40 μm, thickness = 150 μm, pitch = 70 μm, and SiO_2 = 0.2 μm. From the eye diagrams of Figs. 2.9 through 2.13, it can be seen that (a) the TSV with thinner thickness and smaller diameter has a better electrical performance; (b) the TSV with thicker thickness and larger diameter has variation in eye opening; and (c) however, for the multi-TSV chips stacking, the accumulative variation will reduce the eye opening and shrink the design budget because of larger TSV diameter and thickness, as shown in Fig. 2.13.

2.2.4 TSV Electrical Design Guideline

Figure 2.14 shows a simple TSV electrical design guideline. First, input the signal source and structural dimensions of a designed TSV and material property into the proposed analytical circuit and equations to obtain the TSV electrical results. Then, compare the electrical results with a given set of specifications to check whether the results meet the specifications or not. Based on this flow, designers can design and characterize the electrical performance of TSVs in 3D IC integration by using the proposed circuit model and analytical equations.

2.2.5 Summary and Recommendations

The electrical performance of a generic TSV structure for high-frequency 3D IC integration applications has been presented. Some important results are summarized as follows:

- A circuit model and set of equations for the electrical performance of a TSV with all its key elements have been proposed.
- The proposed circuit model and equations have been verified by the 3D finite-element EM-simulations for a wide-range of parameters and frequency as high as 30 GHz.

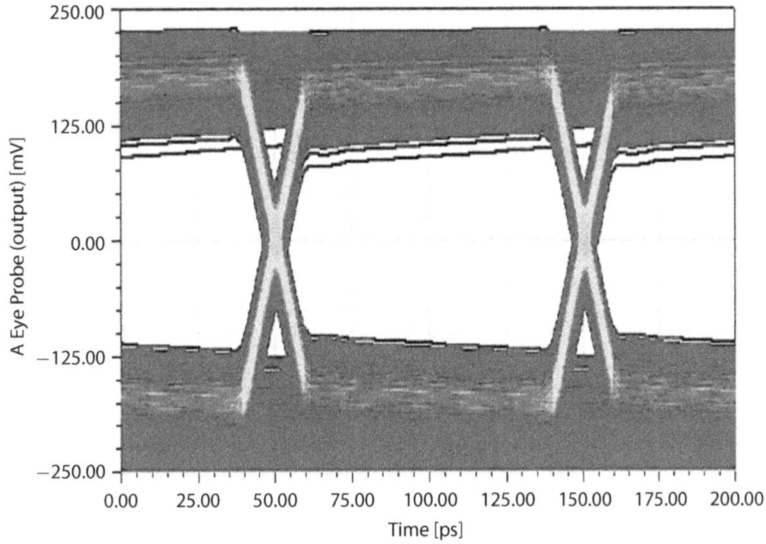

FIGURE 2.13 Eye diagram of 4-chips (TSV diameter = 50 μm, thickness = 150 μm, pitch =70 μm, SiO_2= 0.2 μm) stacking from analytical model.

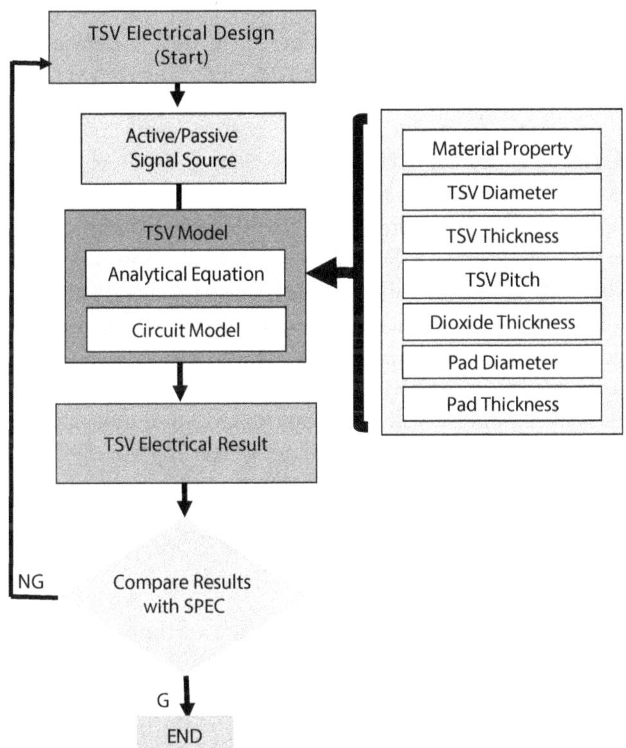

FIGURE 2.14 TSV electrical design guideline.

- The insertion losses from both results of the proposed model and the simulation match very well in the frequency domain.
- The eye diagrams also match very well with the proposed model and the simulation results.
- A simple TSV design guideline has been proposed.

2.3 Thermal Modeling of TSVs

Thermal performance of TSV is presented in this section. Emphasis is placed on the determination of a set of equivalent thermal conductivity equations for Cu-filled TSVs with various TSV diameters, TSV pitches, TSV thicknesses, dielectric thicknesses, and microbump pads. Also, the thermal behavior of a TSV cell is examined. Furthermore, 3D heat transfer simulations are adopted to verify the accuracy of the equivalent equations. Finally, the feasibility of these equivalent equations is demonstrated through a simple 3D IC integration structure.

2.3.1 Cu-Filled TSV Equivalent Thermal Conductivity Extraction

Even with the most advanced software and high-speed hardware, it is very time consuming to model all the TSVs in a 3D IC integration system-in-package (SiP). In [18–20], empirical equations for the equivalent thermal conductivities of TSV chips/interposers with various copper (Cu)-filled TSV diameters, pitches, and aspect ratios have been determined. Unfortunately, the dielectric layer (SiO_2) was not considered. In this section, empirical equations for the equivalent thermal conductivities of Cu-filled TSV chips with various dielectric thicknesses, TSV diameters, TSV pitches, TSV thicknesses, and microbump pads are presented [21–25].

(A) Simulated Model Establishment

Figure 2.15 shows and defines a TSV cell for extracting its equivalent thermal conductivities, and Table 2.2 shows the parameter matrix for simulations. In the figure and table, D_{TSV} is the diameter of the TSV (from 10 to 50 μm); H is the chip thickness (from 10 to 200 μm); P denotes the TSV pitch; and t_{SiO_2} denotes the thickness (from 0.2 to 1.0 mm) of the deposited SiO_2.

The equivalent thermal conductivities to be extracted include k_{xy} and k_z. The subscripts xy and z mean, respectively, the in-plane direction and cross-plane direction of the chip. For Cu-filled TSV in-plane equivalent thermal conductivity (k_{xy}), we apply a positive heat flux (Q''_{in}) to one of its sidewalls (called the hot surface) and a negative heat flux ($-Q''_{out}$) to the opposite sidewalls (called the cold surface) for drawing out the heat, as shown in Fig. 2.16. Based on Fourier's law, we have the equivalent k_{xy}, which can be expressed as

$$Q''_{in} = k_{xy} \cdot \frac{T_{hot} - T_{cold}}{P} \tag{2.17}$$

where T_{hot} and T_{cold} are the average temperature of the hot surface and the cold surface, respectively, and P is the TSV pitch.

For extracting the Cu-filled TSV cross-plane equivalent thermal conductivity (k_z), one buffer block on the top side of the TSV and one at the bottom side are added, as shown in Fig. 2.16. The buffer blocks serve to smooth the heat flow that enters and exits the cell in order to obtain a more reliable and accurate result.

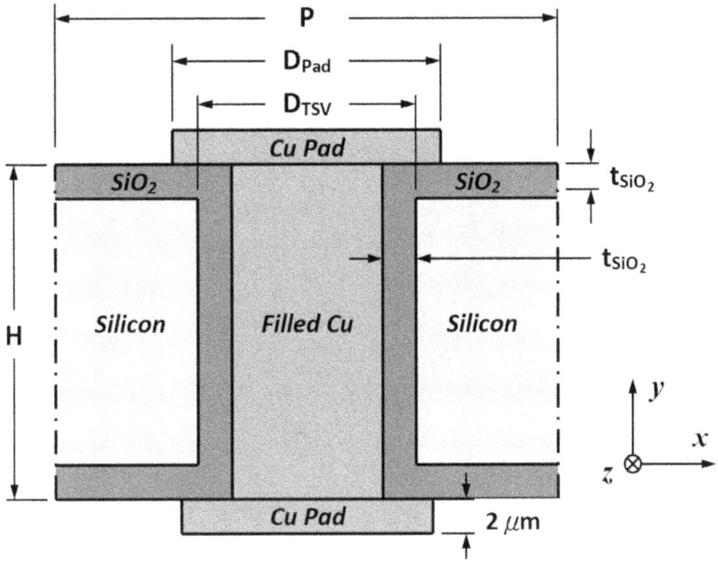

Figure 2.15 Schematic diagram of a TSV cell with important parameters for thermal modeling.

Table 2.2 The Analyzed Parameters Matrix for Thermal Responses

DTsv (μm)	10	20	30	40	50
H (μm)	10, 20, 30, 50, 100, 150	10, 20, 50, 100, 150, 200	10, 20, 50, 100, 150, 250	10, 20, 50, 100, 150, 250	10, 20, 50, 100, 150, 250
P (μm)	25, 35, 40, 50, 70, 100, 130	35, 45, 55, 65, 80, 120, 150	45, 55, 65, 75, 100, 150, 200	55, 65, 75, 85, 100, 150, 200	65, 75, 85, 100, 150, 200
t_{SiO_2} (μm)	0.2, 0.5, 1.0	0.2, 0.5, 1.0	0.2, 0.5, 1.0	0.2, 0.5, 1.0	0.2, 0.5, 1.0

Again, by using Fourier's law, the equivalent thermal conductivity k_z can be expressed as

$$Q''_{\text{in}} = k_z \cdot \frac{T_{\text{hot}} - T_{\text{cold}}}{\Delta H} \tag{2.18}$$

where T_{hot} and T_{cold} are defined as the average temperature of the hot section and the cold section, respectively, and ΔH is the spacing between both sections and is assumed to be 2 μm. The hot and cold sections are mutually parallel and symmetrical to the cell horizontal central line of the Cu-filled TSV.

(B) Boundary Conditions and Material Properties

The Icepak 12.1.6 is adopted as the simulation tool to solve heat conduction problems using the finite volume method. The thermal conductivities of silicon, TSV filled copper, copper

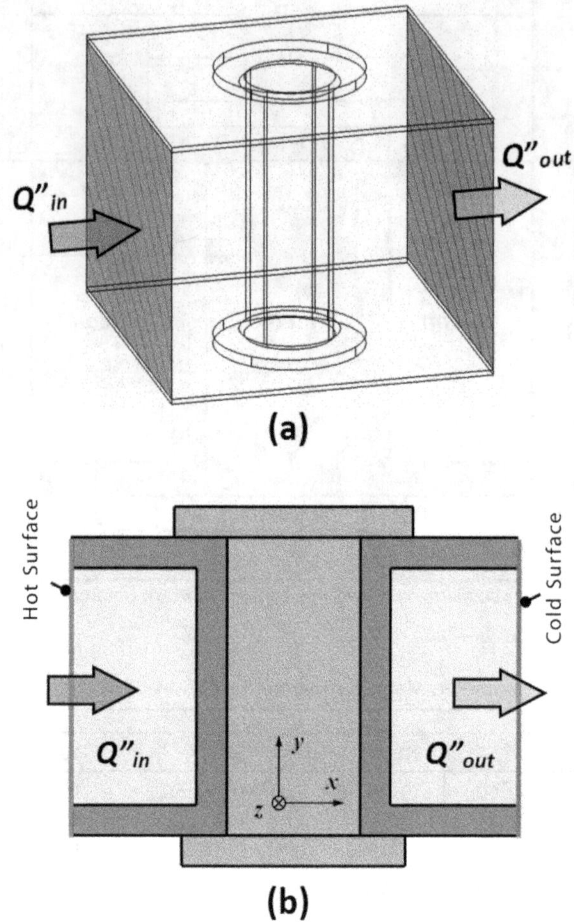

Figure 2.16 TSV cell for k_{xy} extraction. (a) The stereogram of the cell, and (b) the cross-section of the cell.

pad, and SiO_2 are 148, 401, 401, and 1.38 W/m · K, respectively. As for the thermal conductivity of the buffer block, the value is flexible because the purpose of this block is to smoothen the heat flow that enters and exits the TSV cell. Thus, in this study, the thermal conductivity of the buffer blocks is assumed to be 500 W/m · K, and their thickness is 50 μm.

The boundary conditions for the simulations are shown in Figs. 2.16 and 2.17. A constant heat flux (Q''_{in}) is applied to the heating surface and the value is assumed to be 10^8 W/m^2. Also, at the cooling surface, a negative heat flux (Q''_{out}) is applied to draw out the heat. According to the law of energy conservation, the absolute values of both heat fluxes must be equal. As for other boundary conditions, all other surfaces are defined as adiabatic for structure symmetry.

(C) Equivalent Thermal Conductivity Extraction

For demonstration purposes, we consider the following values: $D_{TSV} = 20$ μm, $t_{SiO_2} = 1$ μm, $H = 50$ μm, $P = 65$ μm, and heat flux $Q''_{in} = 10^8$ W/m^2.

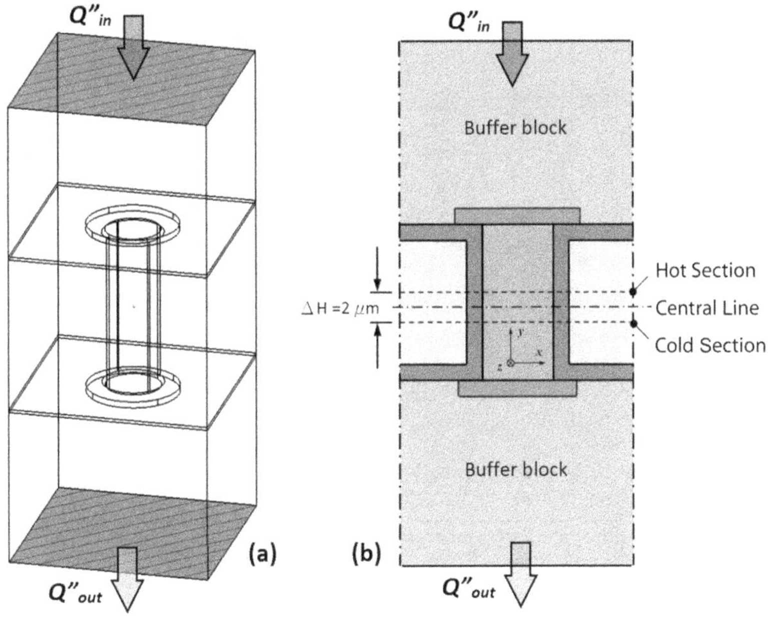

Figure 2.17 TSV cell for k_z extraction. (a) The stereogram of the cell, and (b) The cross-section of the cell.

For equivalent k_{xy} extraction, from the simulation results, we determine the average temperature difference ($T_{\text{hot}} - T_{\text{cold}} = 48.3°\text{C}$) between the hot surface and the cold surface. Since $P = 65$ μm and $Q''_{\text{in}} = 10^8$ W/m², the equivalent k_{xy} can be calculated from Fourier's law and is equal to 134.57 W/m · K.

For equivalent k_z extraction, from the simulation results, we determine the average temperature difference ($T_{\text{hot}} - T_{\text{cold}} = 1.06°\text{C}$) between the hot section and the cold section. Since $Q''_{\text{in}} = 10^8$ W/m² and $H = 2$ μm, the equivalent k_z can be calculated from Fourier's law and is equal to 188.68 W/m · K.

2.3.2 Thermal Behavior of a TSV Cell

(A) The Lateral Heat Transport of the Cell

Figure 2.18 shows the temperature distribution on the horizon central portion of a TSV cell. The length and width of the cell are the same and equal to the TSV pitch. The dielectric layer (SiO_2) and the filled copper is embedded in the cell. It can be observed that, for a thick (>0.2 μm) dielectric layer, the heat flow detours around the TSV, because the dielectric layer has rather poor thermal conductivity ($k = 1.38$ W/m · K for SiO_2) compared with the silicon ($k = 148$ W/m · K). In fact, even though the filled copper of the TSV has a very high thermal conductivity ($k = 401$ W/m · K), the heat flow is still blocked by the dielectric layer, and thus wears down the contribution of copper for thermal performance enhancement in the chip lateral direction.

Based on parameter studies, it is found that the effect of chip thickness (H) on equivalent k_{xy} is insignificant compared to other parameters. In addition, for thicker (>0.2 μm) dielectric (SiO_2) layers, which block the heat from entering the filled Cu, and thus curve

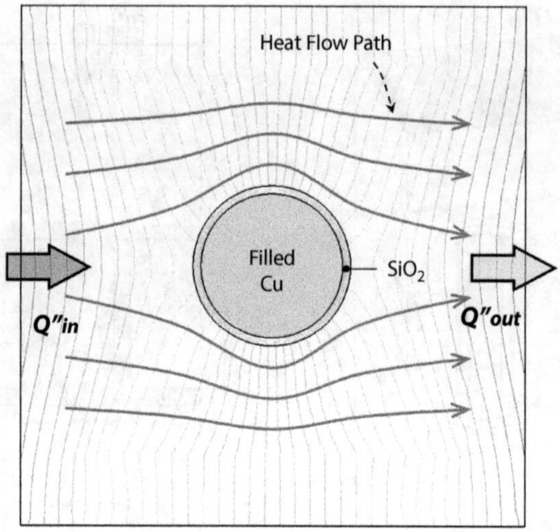

Figure 2.18 Heat flow path and temperature distribution in lateral direction of a TSV cell.

(i.e., make) the heat flow path longer. All the equivalent lateral thermal conductivities (k_{xy}) are smaller than that of silicon (148 W/m · K), especially for larger TSV diameters.

The effect of SiO_2 thickness on equivalent k_{xy} is noticeable, as shown in Fig. 2.19. It can be seen that a thicker SiO_2 layer (>0.2 μm) and a larger TSV diameter (>10 μm) result in a lower equivalent k_{xy}. However, for a thinner SiO_2 layer (≤0.2 μm), regardless of the TSV diameters, the equivalent k_{xy} approaches that the value of silicon (148 W/m · K), as shown in Fig. 2.19.

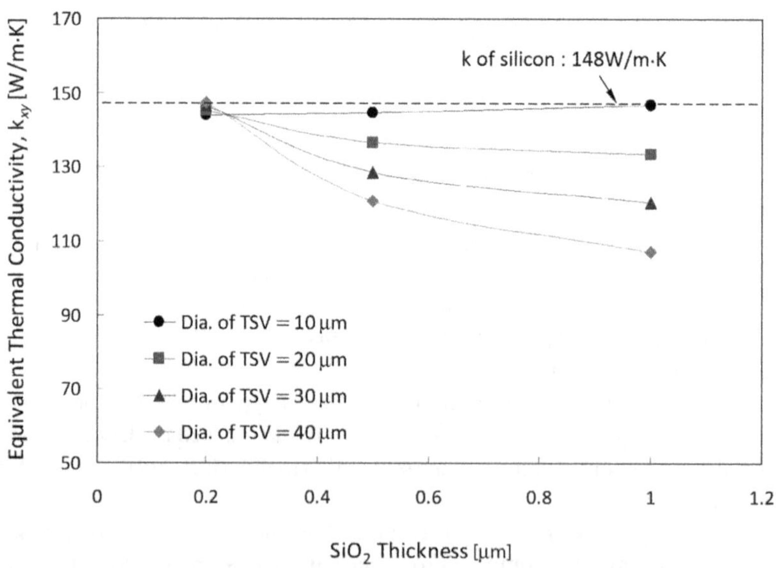

Figure 2.19 Typical correlations between equivalent k_{xy} and SiO_2 thickness: the chip thickness (*H*) is 50 μm and the pitch is 65 μm.

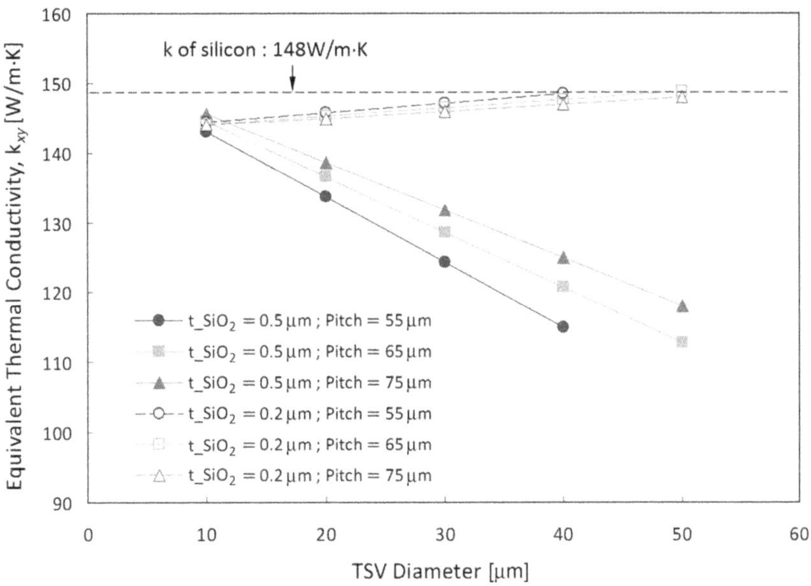

FIGURE 2.20 Typical correlations between equivalent k_{xy} and TSV diameter with different TSV pitch. The chip thickness (H) is 50 μm and the SiO_2 thicknesses (t_{SiO_2}) are 0.2 and 0.5 μm.

The effect of the TSV pitch on the equivalent k_{xy} is noticeable, as shown in Fig. 2.20. In general, a larger TSV density (smaller TSV pitch), a thicker (>0.2 μm) SiO_2 layer, and a larger TSV diameter (>10 μm) result in a lower equivalent k_{xy}. However, when the SiO_2 thickness is ≤0.2 μm, regardless of the TSV diameter and the TSV pitch, the equivalent k_{xy} approaches the value of silicon (148 W/m · K), as shown in Fig. 2.20.

The diameter and pitch of TSV are the other two significant parameters for equivalent k_{xy}. Figure 2.20 shows the correlation between k_{xy} and TSV diameter, and between k_{xy} and TSV pitch. A larger equivalent k_{xy} is caused by a smaller TSV dimension ratio expressed as D_{TSV}/P. In other words, the chip/interposer with higher TSV density will have a worse thermal conductivity in the lateral direction.

(B) The Longitudinal Heat Transport of the Cell

Figure 2.21 shows the temperature distribution in the vertical direction of a Cu-filled TSV cell. Since copper has a higher thermal conductivity than silicon, there is more heat flow inpour and outflow through the cell via the filled copper. Also, the temperature of the filled copper and silicon must equalize at the vertical central portion, due to energy equilibrium. Therefore, some heat flow may cross the dielectric layer from copper to silicon after entering the TSV cell, and the heat flow may cross the dielectric layer again from silicon to the filled copper before exiting the TSV cell.

The dielectric layer (SiO_2) serves as a thermal barrier in the TSV cell. Figure 2.22 shows the effect of SiO_2 thickness and TSV thickness (H) on the equivalent longitudinal thermal conductivity (k_z) for the case where TSV pitch = 50 μm and TSV diameter = 10 μm. It can be seen that (1) for all the SiO_2 thicknesses and TSV thicknesses, the value of k_z is larger than that of silicon (due to the positive effect of Cu and the negative effect of SiO_2); (2) for H = 100 μm, the SiO_2 thickness has no effect on k_z; and (3) for H = 50 μm, the SiO_2

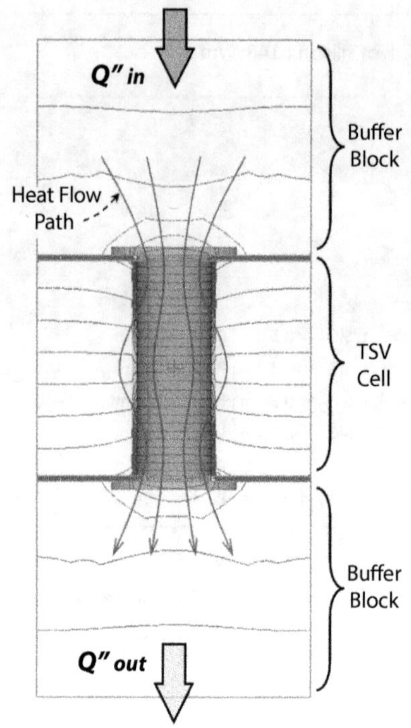

FIGURE 2.21 Temperature distributions in the vertical direction of a TSV cell.

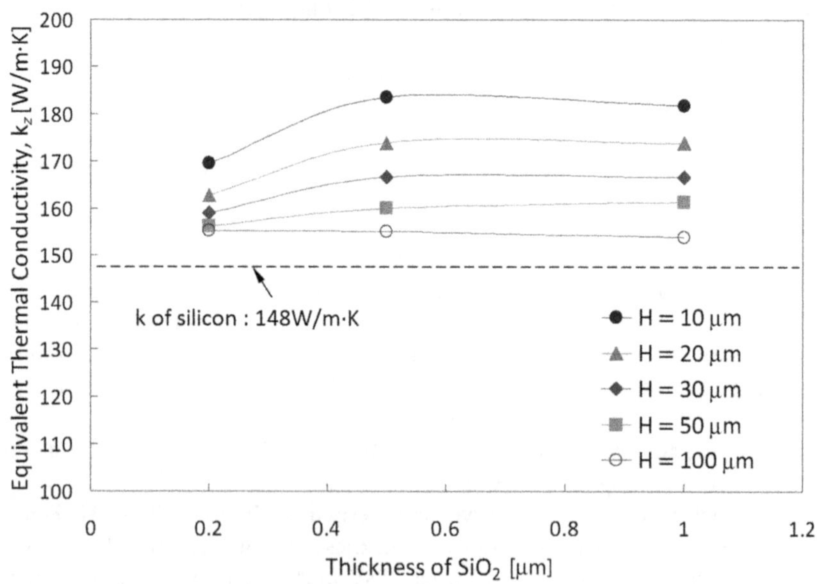

FIGURE 2.22 Typical correlations between equivalent k_z and dielectric layer thickness: the TSV pitch is 50 μm, and the TSV diameter is 10 μm.

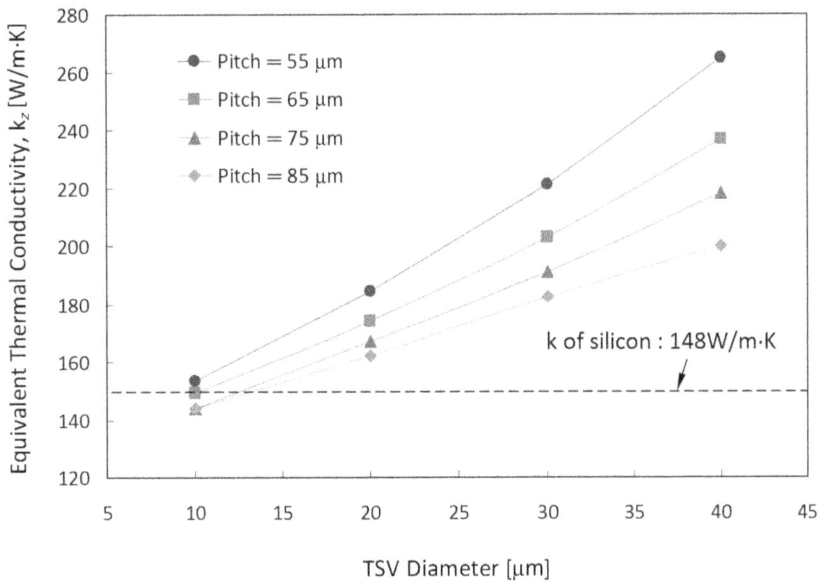

FIGURE 2.23 Correlations between equivalent k_z and the TSV diameter. The chip thickness (*H*) is 50 μm and the SiO_2 thickness (t_{SiO_2}) is 0.2 μm.

thickness (from 0.2 to 0.5 μm) has little effect on k_z; and (4) at SiO_2 = 0.2 μm, the k_z is almost the same for H = 50 μm and H = 100 μm.

Figure 2.23 shows the effect of TSV diameter and TSV pitch on k_z where TSV thickness (H) = 50 μm and SiO_2 thickness = 0.2 μm. It can be seen that (1) the larger the TSV diameter, the larger the k_z; (2) the smaller the TSV pitch the larger the k_z; and (3) again, for all SiO_2 thicknesses, TSV thicknesses, TSV pitches, and TSV diameters, the k_z is larger than that of silicon (due to the positive effect of Cu and the negative effect of SiO_2).

The simple explanation of previous results is that a thicker dielectric layer may obstruct the heat exchange between the filled Cu and Si; thus, more heat flux remains to flow through the Cu, which consequently enhances the weight (influence) of the filled Cu for the cell thermal performance. On the other hand, a TSV with higher aspect ratio (H/D_{TSV}) leads to a lower equivalent k_z, because more heat exchange has occurred between the Cu and Si during the heat flow through the TSV cell, which reduces the influence of the filled Cu for the TSV cell thermal performance.

2.3.3 Cu-Filled TSV Equivalent Thermal Conductivity Equations

(A) Equivalent Thermal Conductivity Equations k_{xy} and k_z

By curve fitting of the simulation results for all the cases considered in Table 2.2, we have the empirical equations of the Cu-filled TSV equivalent thermal conductivity equations.

For equivalent k_{xy}, the empirical equation is

$$k_{xy} = \left(90 \cdot t_{SiO_2}{}^{-0.33} - 148\right) \cdot \left(\frac{D_{TSV}}{P}\right) \cdot H^{0.1} + 160 \cdot t_{SiO_2}{}^{0.07} \tag{2.19}$$

For equivalent k_z, the empirical equation is

$$
\begin{aligned}
0.002 &\le \frac{t_{SiO_2}}{H} \le 0.01 &&\Rightarrow\quad k_z = 128\cdot\exp\left(\frac{D_{TSV}}{P}\right)\\
0.01 &< \frac{t_{SiO_2}}{H} \le 0.02 &&\Rightarrow\quad k_z = 130\cdot\exp\left(1.1\cdot\frac{D_{TSV}}{P}\right)\\
0.02 &< \frac{t_{SiO_2}}{H} \le 0.04 &&\Rightarrow\quad k_z = 258\cdot\left(\frac{D_{TSV}}{P}\right)+113\\
0.04 &< \frac{t_{SiO_2}}{H} \le 0.1 &&\Rightarrow\quad k_z = 298\cdot\left(\frac{D_{TSV}}{P}\right)+121\\
0.1 &< \frac{t_{SiO_2}}{H} \le 0.2 &&\Rightarrow\quad k_z = 136\cdot\ln\left(\frac{D_{TSV}}{P}\right)+383
\end{aligned}
\tag{2.20}
$$

where

$$
\begin{aligned}
&0.2\ \mu\text{m} \le t_{SiO_2} \le 0.5\ \mu\text{m}\\
&10\ \mu\text{m} \le D_{TSV} \le 50\ \mu\text{m}\\
&H \ge 20\ \mu\text{m}\\
&0.1 \le \frac{D_{TSV}}{P} \le 0.77
\end{aligned}
$$

(B) Comparison between the Results Obtained from k_{xy} and k_z and from 3D Heat Transfer Simulations

Figures 2.24 and 2.25 show the comparisons between the 3D simulation data and the equivalent equations. For equivalent k_{xy} and k_z, it can be seen that the largest error of the empirical equations is less than 15%. Most simulation data agree with the equivalent equations, with an error of less than 10%.

(C) How to Use Equivalent Thermal Conductivity Equations (k_{xy} and k_z) for 3D IC Integration?

The equivalent equations can be used for the design and analysis of 3D IC SiPs by establishing an equivalent model. This equivalent model is used to replace the real detailed TSV model for simplifying the thermal simulations. Figure 2.26 shows how a detailed TSV model can be converted to an equivalent model.

The equivalent model regards the TSV group as an equivalent zone (block), which is represented with the calculated equivalent k_{xy} and k_z. The dielectric layers (SiO_2) outside the TSVs are modeled. The solder balls (or bumps) and the traces are converted to other equivalent zones by using the common thermal resistance (series/parallel) calculations. For more information, please read Secs. 8.3.3 and 8.3.4.

2.3.4 Verification of the TSV Equivalent Thermal Conductivity Equations

Figure 2.27 shows a 3D IC integration SiP. There are four memory chips (each with three TSVs and the TSV diameter is D_{TSV} = 10 μm) bonded together (with a 20-μm-diameter microbump) on top of an interposer. Each chip has two heaters (on opposite directions) and each heater generates a heat flux of 1 × 10^6 W/m^2. The chip thickness is 50 μm. The SiO_2

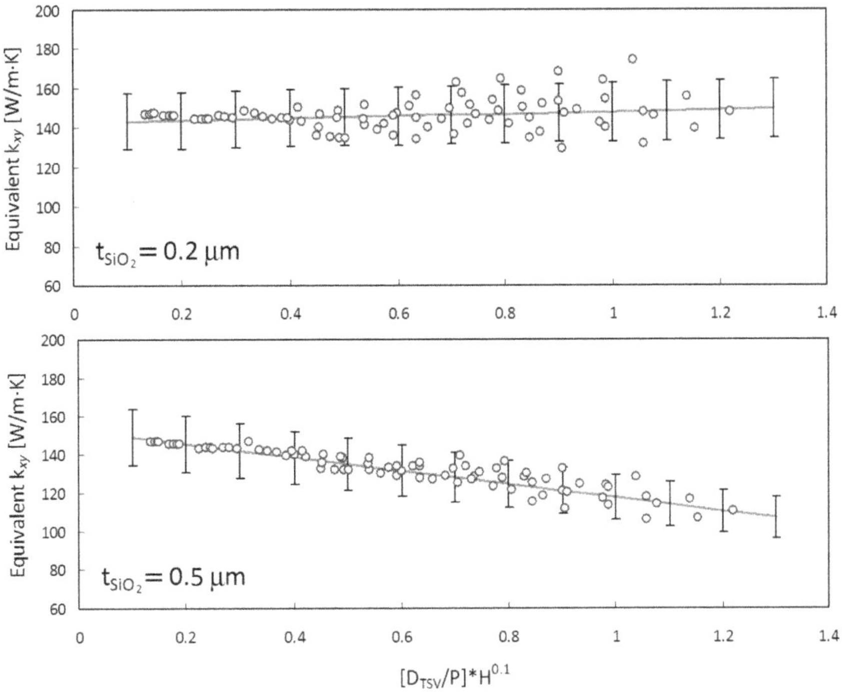

FIGURE 2.24 Comparison between the simulation data (dot) and equivalent k_{xy} equation (curve). The error bar is $\pm 10\%$.

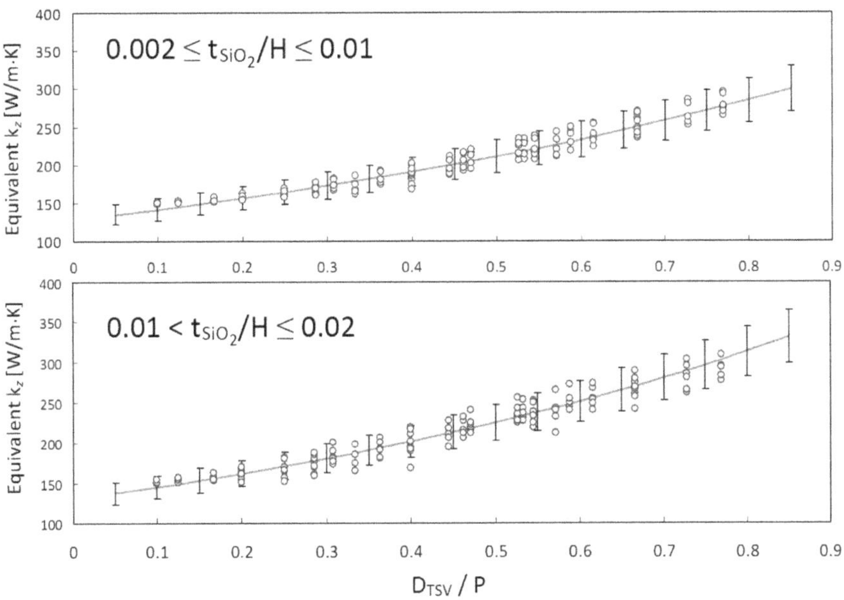

FIGURE 2.25 Comparison between the simulation data (dot) and equivalent k_z equation (curve). The error bar is $\pm 10\%$.

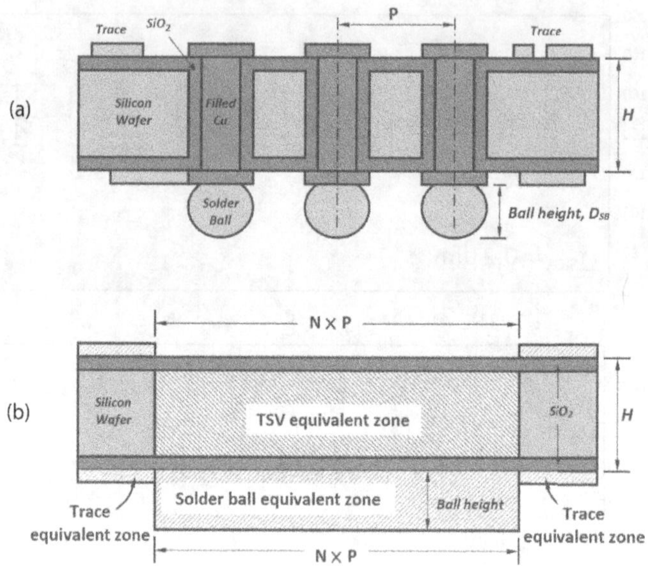

Figure 2.26 The conversion from a detail model to an equivalent model. (a) The detail model, and (b) The equivalent model.

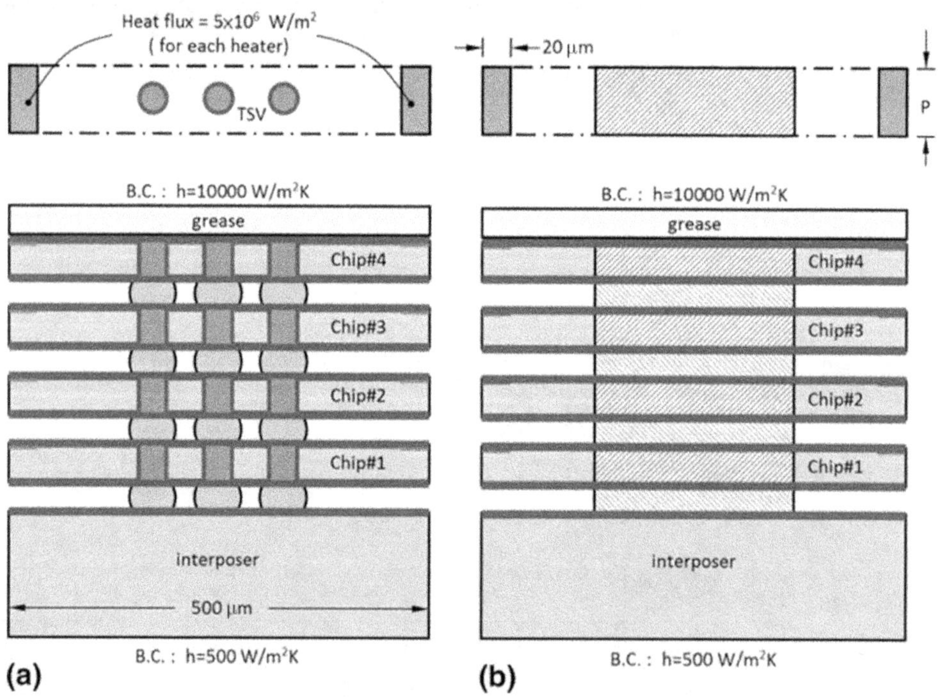

Figure 2.27 Detailed and equivalent models for feasibility verification. (a) The detail model and (b) The equivalent model. There are two heaters on a chip and each heater generates 10^6 W/m^2.

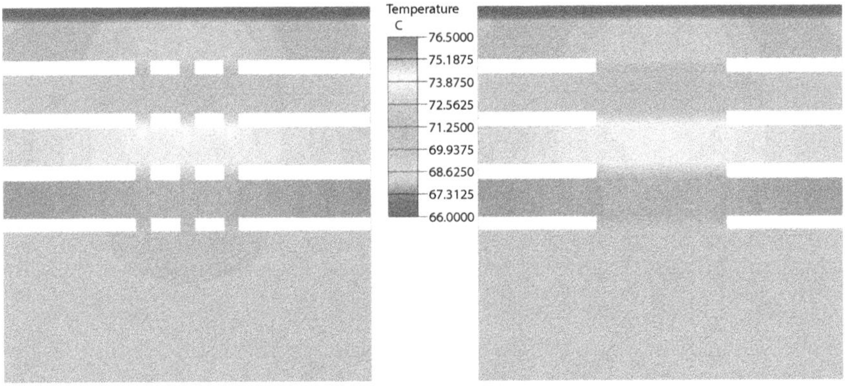

FIGURE 2.28 A typical temperature distribution of the detail model (left) and the equivalent model (right). Here, D_{TSV}, t_{SiO_2}, H and bump diameter are 10, 0.2, 50 and 20 μm, respectively. The boundary conditions and heat generation are shown in Figure 2.27.

thickness = 0.2 μm. Each chip size is 10 × 10 mm and each dissipates 0.8 W of heat. Thus, for the four-chip stacking, the total power generated from the heat is 3.2 W.

The boundary conditions are h = 10,000 W/m^{2}K on the top side of the model and h = 500 W/m^{2}K on the bottom of the interposer. Physically, it means the SiP has an air cooling heat sink on the top side and natural convection cooling on the bottom side. The slice model is introduced: (1) because it is a semi-2D model that can simplify the simulation effort, and (2) for the verification and demonstration of its accuracy and feasibility.

Figure 2.28 shows the temperature distributions of the detailed model (left) and the equivalent model (right). It can be seen that (1) the TSVs are dissipating the accumulated heat from the chips to ambient (left), and (2) even the heat flow within both models is different; however, the temperatures of the heaters of the two models are quite the same.

Figure 2.29 shows the heater temperature comparisons between the detailed model and the equivalent model. It can be seen that (1) for all the TSV pitches, both models predict almost the same heater temperature; (2) for all the TSV pitches, the smallest heater temperature occurs at the top chip #4 and the largest heater temperature occurs at the bottom chip #1 (this is because the main dissipation path of the heat is via the model top side to the ambient); and (3) except for the top chip, the smaller the TSV pitch (a denser TSV arrangement), the lower the heater temperature (this is caused by the higher longitude equivalent thermal conductivity (k_z) of TSV and microbumps).

What would the thermal path and temperature distribution of the memory-chip stacking be with microbumps but without TSVs? The simulation results show that all the chip temperatures are higher than those of the model with TSVs (Cu-filled TSVs provide a better thermal conducting path than Si). The temperature is higher by about 2.5°C for chip #4, 2°C for chip #3, 1.2°C for chip #2, and 0.1°C for chip #1. These results indicate that TSVs do indeed increase the chips' heat dissipation effect, and thermal performance. For this simple example, the thermal performance enhancement is about 6%.

2.3.5 Summary and Recommendations

Empirical equations for the equivalent thermal conductivities of Cu-filled TSV chips/interposers with various dielectric thicknesses, TSV diameters, TSV pitches, TSV

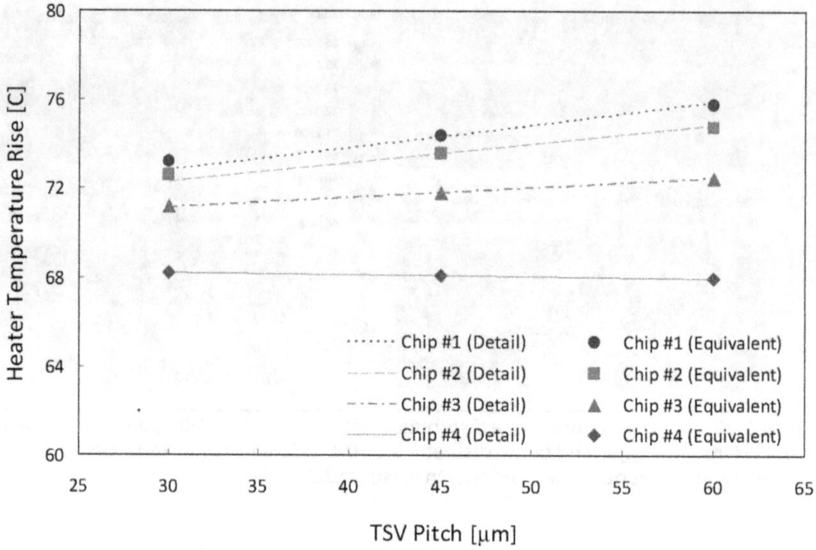

Figure 2.29 Heater temperature comparisons: D_{TSV}, H, t_{SiO_2} and bump diameter are 10, 50, 0.2, 20 μm, respectively. The ambient temperature is 35°C.

thicknesses, and microbump pads have been presented. Also, the thermal behavior of a SiO_2-Cu-filled TSV cell was discussed. Furthermore, the accuracy of the equivalent equations has been verified by 3D heat transfer simulations. Finally, the feasibility of these equivalent equations has been demonstrated through a simple 3D memory-chip stacking structure. Some important results and recommendations are summarized as follows:

- k_{xy} and k_z for SiO_2-Cu-filled TSV memory-chip stacking and interposer have been provided and they are very useful for 3D IC integration design and analysis.
- Based on the 3D simulation results, most errors of k_{xy} and k_z are within 10% and in a few cases the maximum error is 15%.
- For thermal performance of 3D IC integration, the SiO_2 layer serves as a thermal barrier in a TSV, which reduces the value of k_{xy} and k_z. On the other hand, the filled Cu increases the value of k_{xy} and k_z.
- For most cases considered, the value of k_{xy} is smaller than that of silicon, i.e., the effect of the SiO_2 layer is larger than that from the filled Cu (i.e., the SiO_2 layer blocks the heat flow into the filled Cu). However, for the practical cases (real applications) such as $SiO_2 = 0.2$ μm and $H = 50$ μm for memory-chip stacking and $H = 100-200$ μm for interposer, the value of k_{xy} is almost the same as silicon's.
- For all the cases considered, the value of k_z is larger than that of silicon, i.e., the contribution of the filled Cu is larger than the negative effect of the SiO_2 layer.
- The guidelines for using k_{xy} and k_z for 3D IC integration SiP have been provided through examples.

- For the simple 3D IC integration SiP examples considered herein (memory-chip stacking with TSVs and without TSVs), the TSV thermal performance enhancement is about 6%.
- Thermal management is on the *critical path* of 3D IC integration and 3D Si integration. The research institutions and electronic industry should: (a) perform more innovative researches and collect more useful data in this area, and (b) provide some effective thermal management design methods, tools, guidelines, and solutions for the widespread use of 3D Si/IC integrations.

2.4 Mechanical Modeling and Testing of TSVs

Most TSVs are filled with electroplated Cu. The TCE (thermal coefficient of expansion) of Cu is 17.5×10^{-6}/°C and of the surrounding Si is 2.5×10^{-6}/°C. Because of the very large TEM (thermal expansion mismatch) between the Cu and the Si, large stress and strain (deformation) could occur inside them and their surrounding elements while they are subjected to temperature change [26–52]. During the manufacturing of TSVs, this TEM could lead to Cu pumping (protrusion). During environmental operating conditions (e.g., thermal shock or cycling) of products with Cu-filled TSVs, the stresses created by the TEM could affect the hole and electron carrier mobility in devices and even crack the devices, passivations, metal contacts, metal layers, and TSV. In this section, Cu pumping of TSVs and how to avoid it during manufacturing will be presented. The Cu protrusion of TSV under thermal shock cycling will also be provided. The KOZ (keep-out-zone) of Cu-filled TSVs and how to reduce the TEM stress to zero will also be provided.

2.4.1 TEM between the Cu-Filled TSV and Its Surrounding Si

In May 2008, the Cu pumping due to the TEM between the Cu-filled TSV and its surrounding Si was reported in [26, 27]. Nonlinear thermal stress owing to the TEM at the interfaces between the copper, silicon, and dielectric is determined for a wide range of aspect ratios (of the TSV thickness and diameter). Figure 2.30(a) shows a simplified diagram of the TSV for finite-element analysis [26, 27]. The assumptions are (1) since the layer of tantalum (1 kÅ) is much thinner than the dielectric layer (1 μm), its effect is negligible, (2) the redistribution layer is very thin and therefore negligible, (3) silicon and silicon dioxide are not stressed beyond their elastic zone, and (4) copper undergoes elastic deformation, followed by plastic deformation.

One-quarter of the TSV is modeled in an axisymmetric simulation. The boundary conditions used and the mesh at the critical interfaces are shown in Figs. 2.30(b) and 2.30(c), respectively. Note that directions 1 and 2 represent the radial and axial directions, respectively. During temperature ramp-up, the copper expands more than five times as much as silicon and more than 10 times as much as silicon dioxide. The deformation resulting from heating from −40°C to 125°C and cooling from 125°C to −40°C (exaggerated 100 times) is shown in Figs. 2.31(a) and 2.31(b). It can be seen in Fig. 2.31(a) that the silicon dioxide layer is highly strained as it is dragged along by the axial expansion of the copper (the Cu wants to pump up and out of the via) and compressed by the radial expansion of the copper. This deformation, however, is largely not transferred to the bulk of the silicon because silicon is much stiffer than copper and silicon dioxide. Conversely, during temperature ramp-down, a high strain is induced on the material near the interface indicated by points *A* and *B*, which are the potential failure locations.

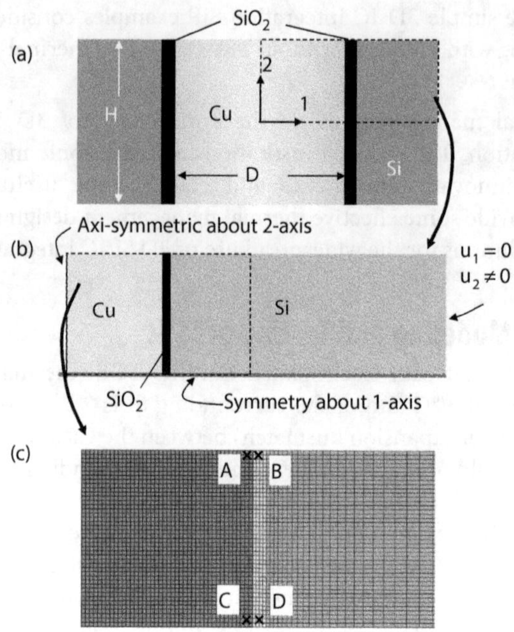

FIGURE 2.30 (a) Schematic of simplified TSV. (b) Quarter model of TSV with applied boundary conditions. (c) Mesh of critical regions.

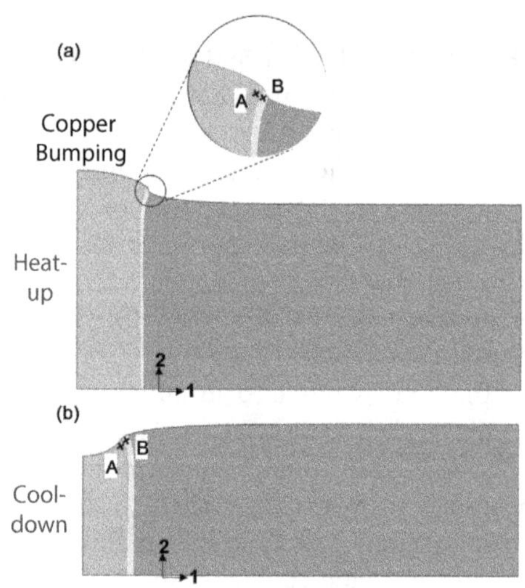

FIGURE 2.31 Deformation of the TSV at critical regions exaggerated 100 times by heat-up (a) and cool-down (b).

Failure may occur at two critical points. First, failure could occur owing to the tearing action during contraction at the interface between copper and silicon dioxide (points *A* and *B* in Fig. 2.30(c)). If there is an overlaying layer such as dielectric layer or Cu pad on top of the TSV, it could be pushed out. Second, cracking of copper or silicon dioxide could occur at the midplane of the TSV (points *C* and *D* in Fig. 2.30(c)).

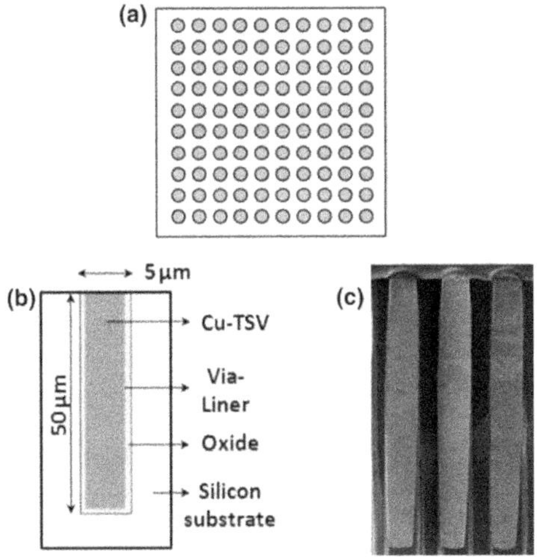

FIGURE 2.32 Schematic diagram of the (a) top view of an array of TSVs (10x10) on 10 μm-pitch and (b) TSV cross section. Cross-sectional SEM image of TSVs.

2.4.2 Experimental Results on Cu Pumping during Manufacturing

IME [36, 37, 38] and IMEC [28, 32, 33, 35, 39, 44] have done lots of works on measuring the Cu pumping. Figure 2.32 shows the test vehicle for IME's experiments, which consists of a piece of 50-μm-thick silicon with 10 × 10 array of 5-μm-diameter TSVs filled with Cu. After CMP of the overburden Cu, the samples are subjected to annealing at temperatures ranging from 250°C to 450°C at nitrogen gas ambient. The duration of each temperature is 30 min.

The SEM (scanning electron microscopy) images of typical Cu pumping are shown in Fig. 2.33. It can be seen that (a) the edge of the TSV pumps out at annealing temperature = 300°C; (b) the center of the TSV pumps out at 350°C; and (c) the edge as well as the center of the TSV pump out at 400°C and 450°C. The Cu pumping measured with AFM (atomic force microscope) versus the annealing temperature is shown in Fig. 2.34. It can be seen

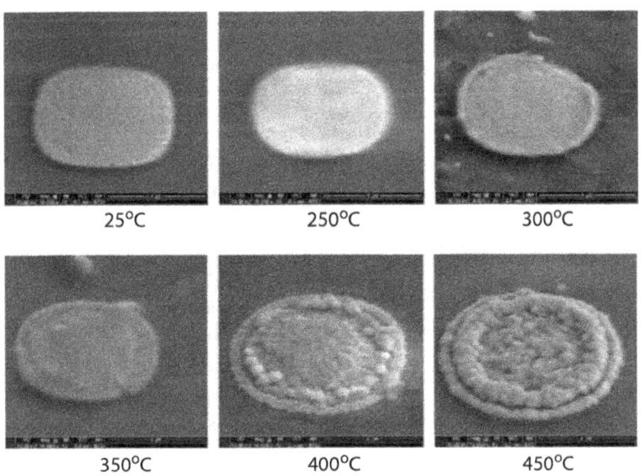

FIGURE 2.33 SEM images of Cu pumping at various annealing conditions and at room temperature as comparison.

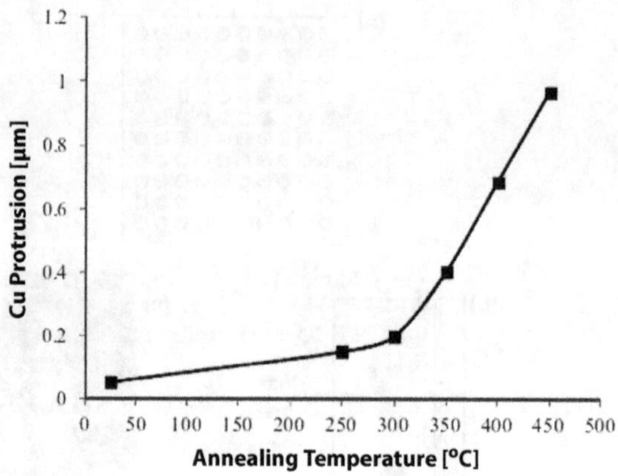

FIGURE 2.34 Cu protrusion (pumping) vs. annealing temperature.

and expected that the larger the annealing temperature the larger the Cu pumping. Thus, a TSV which has not been optimized prior to the BEOL (back-end-of-line) processing, poses a potential yield issue such as damaging (pushing) the overlaying dielectric layer.

Figure 2.35 shows the microstructure of Cu-filled TSVs under different annealing temperatures. It is characterized by a focused ion beam (FIB)–SEM equips with an EBSD (electron backscatter diffraction) system for the grain orientation and grain size evolution of Cu-TSV. It can be seen that the grain sizes are uniformly distributed and exhibit a normal grain growth pattern. The Cu grain size versus temperature is shown in Fig. 2.36. It can be seen that the larger the annealing temperature the larger the Cu grain growth.

Figure 2.37 shows the Cu pumping measurement approach used by IMEC [39]. Unlike IME, IMEC performs the post-plating annealing before the CMP (chemical-mechanical polishing)

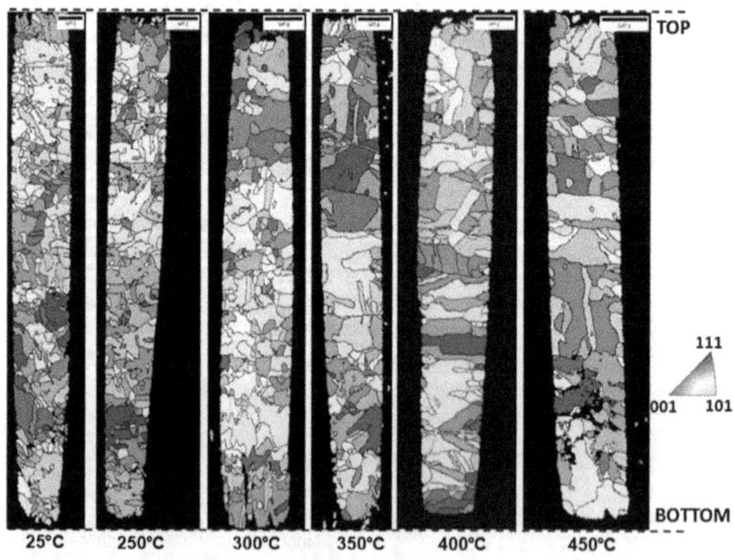

FIGURE 2.35 EBSD FIB/SEM images of TSVs under various annealing conditions and with fresh TSV (25°C).

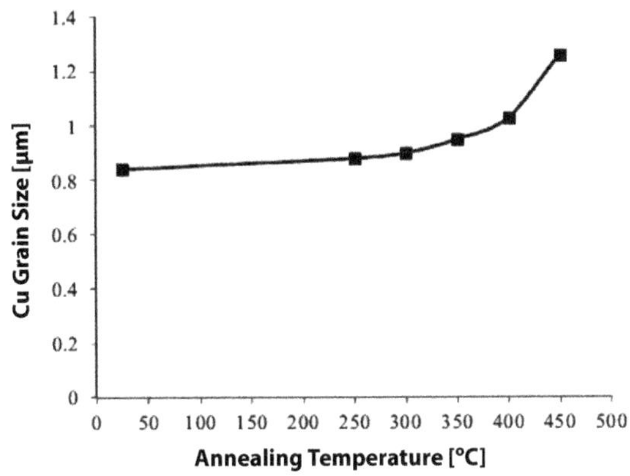

FIGURE 2.36 Grain size vs. annealing temperatures showing the grain growth as temperature increases.

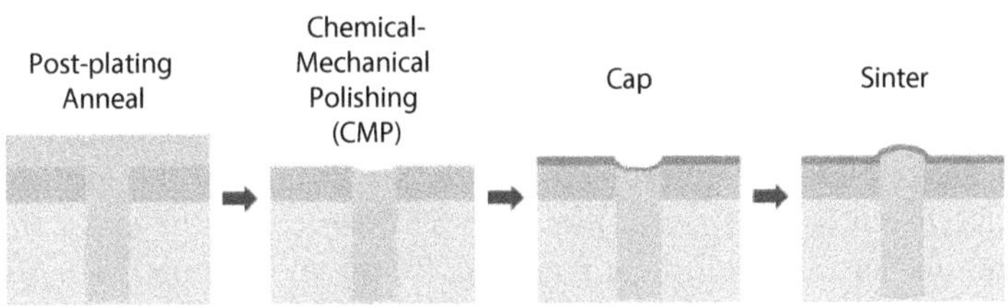

FIGURE 2.37 IMEC's Cu pumping measurement approach.

of the overburden Cu. The SiO_2 cap on top of the Cu-filled TSV aims to mimic the presence of BEOL layer. The sinter (420°C for 20 min) aims to mimic the thermal budget of the final anneal step after BEOL processing at IMEC. Two sets of samples are sued. The diameter and thickness of the first set are 5 µm and 50 µm (5 × 50 µm), and the other set are 10 and 100 µm (10 × 100 µm). The annealing conditions are shown in Table 2.3 and the measurements are performed in a nitrogen atmosphere.

TABLE 2.3 Post-plating Annealing Temperature and Time for Cu Pumping Experiments

Sample	Post-plating Anneal Temperature [°C]	Post-plating Anneal Time [min]
1	—	—
2	180	20
3	300	20
4	350	20
5	420	5
6	420	20
7	420	80
8	500	20

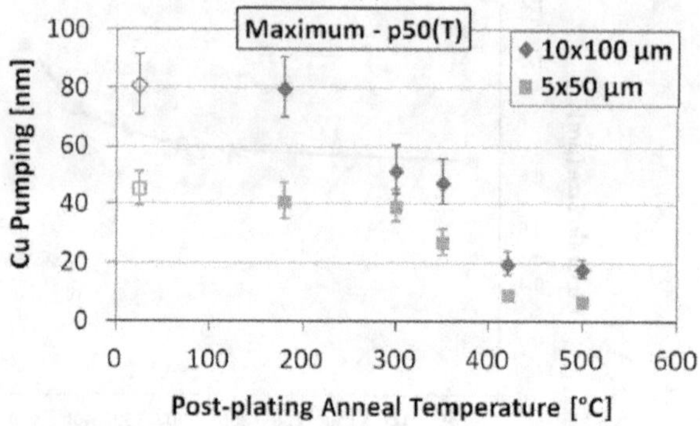

FIGURE 2.38 Maximum residual Cu pumping ($h_1 - h_0$) vs. post-plating anneal temperature at median rank for 5 × 50 μm and 10 × 100 μm TSVs. The open symbols are for the samples without post-plating anneal.

The maximum "residual" Cu pumping is defined as the difference between h_1 and h_0 (Fig. 2.37). h_0 is the maximum Cu pumping before the sinter and h_1 is the maximum Cu pumping after the sinter. In this case, the residual Cu pumping, during the fixed sinter, is a function of the post-plating annealing conditions. The maximum residual Cu pumping versus the post-plating anneal temperature (for 20 min) at the median rank is shown in Fig. 2.38. It can be seen that (a) for both TSV structures the larger the anneal temperature the smaller the residual Cu pumping, (b) for all the anneal temperatures considered the larger the TSV diameter the larger the residual Cu pumping, and (3) the post-plating annealing condition (420°C with 20 min) is optimal for Cu pumping problem.

2.4.3 Cu Pumping under Thermal Shock Cycling

Unlike IME and IMEC, RPI/SEMATECH [40] measure the Cu pumping from the TSV back side, which is exposed after temporary wafer bonding, thinning, TSV revealing, and metallization. Also, the sample is subjected to thermal shock cycling test which is between room temperature and a peak temperature, the ramp-up/cool-down rate is 25°C/s, the dwell time at the peak temperature is 150s, and for a total of 15 cycles. Five peak temperatures (200, 250, 300, 350, and 400°C) are considered, i.e., very high acceleration tests. The test vehicle is shown in Figure 2.39. The TSV diameter and thickness are 5.5 and 50 μm, respectively.

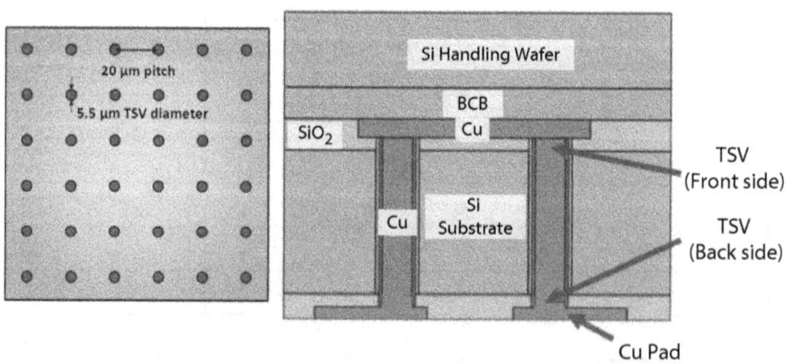

FIGURE 2.39 Schematic cross-sectional view of the TSV structure. TSV diameter = 5.5 μm and the TSV pitch = 20 μm.

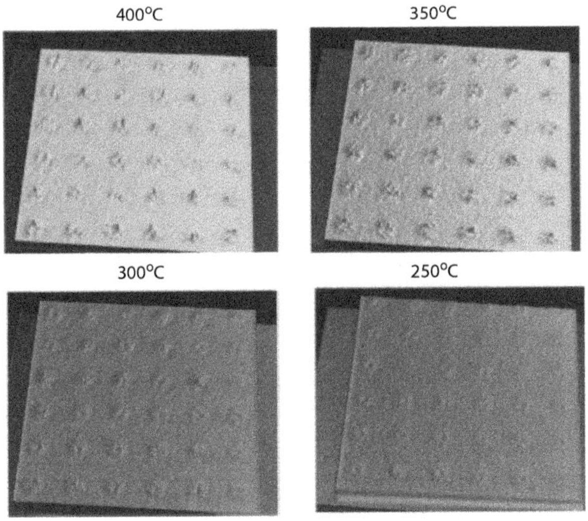

FIGURE 2.40 Optical surface profiles of TSV pumping after the thermal shock cycling tests with peak temperatures ranging from 250°C to 400°C (the Cu pumping is not visible at 200°C).

Figure 2.40 shows the optical surface profiles of the Cu pumping (from the back side) after 15-cycle of thermal shock tests from room temperature to different peak temperatures ranging from 250°C to 400°C (no visible Cu pumping at 200°C). The average Cu pumping after the tests is shown in Fig. 2.41. It can be seen from both figures that the larger the thermal shock peak temperature the larger the Cu pumping from the back side of the TSVs.

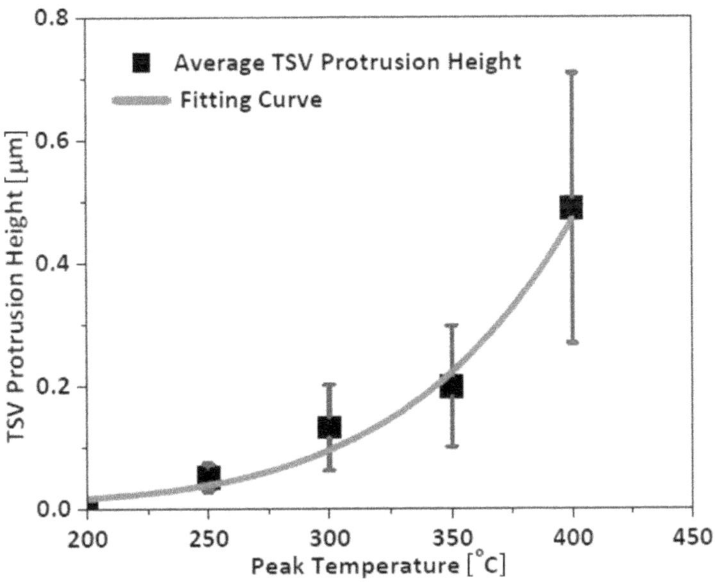

FIGURE 2.41 Average TSV Cu pumping vs. various peak temperatures.

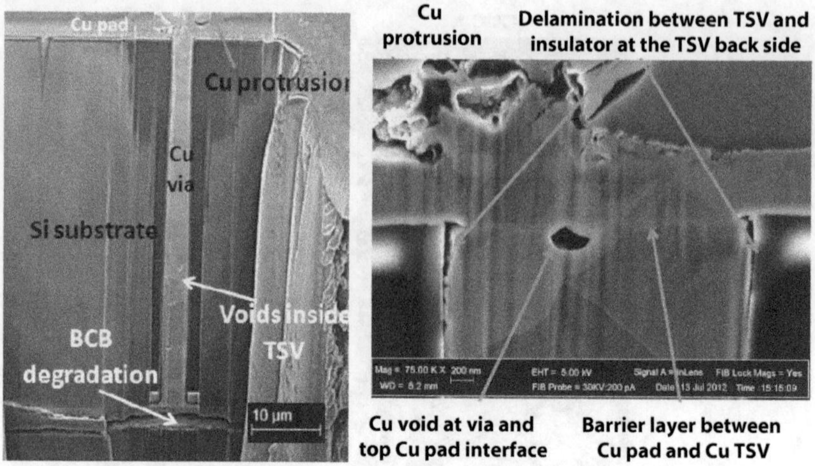

FIGURE 2.42 Cross-sectional SEM images for TSV with a large Cu pumping: (a) voids inside the TSV, (b) void at the interface between the TSV and the back side Cu pad, and (c) interface delamination at the sidewall between the TSV and the insulator at the back side Cu pad.

The cross-sectional SEM images for the Cu pumping from the back side under the most severe thermal shock condition (with a peak temperature = 400°C for 15 cycles) are shown in Figs. 2.42 and 2.43. It can be seen that (a) there are voids inside the TSV, (b) there is void at the interface between the TSV and the back side Cu pad, (c) there are interface delaminations at the sidewall between the TSV and the insulator at the back side Cu pad, (d) large Cu pumping is visible, (e) there is a crack within the Cu TSV, and (f) there are voids at grain boundaries.

The cross-sectional SEM images for Cu pumping from the front side under the most severe thermal shock condition are shown in Fig. 2.44. It can be seen that (a) the Cu pumping

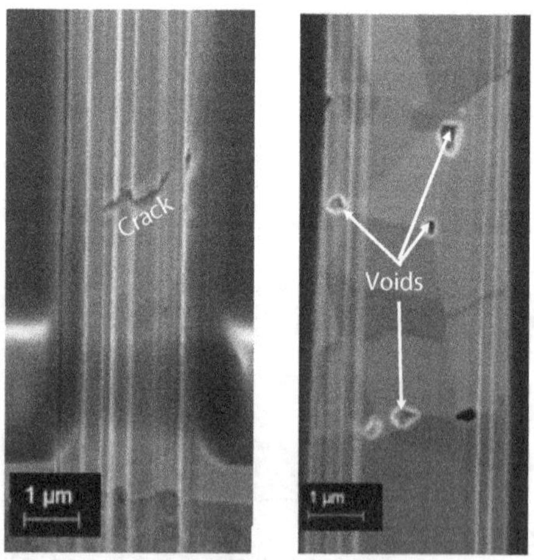

FIGURE 2.43 Cross-sectional SEM images for TSV with a large Cu pumping: (a) a crack within the Cu TSV, and (b) voids at grain boundaries.

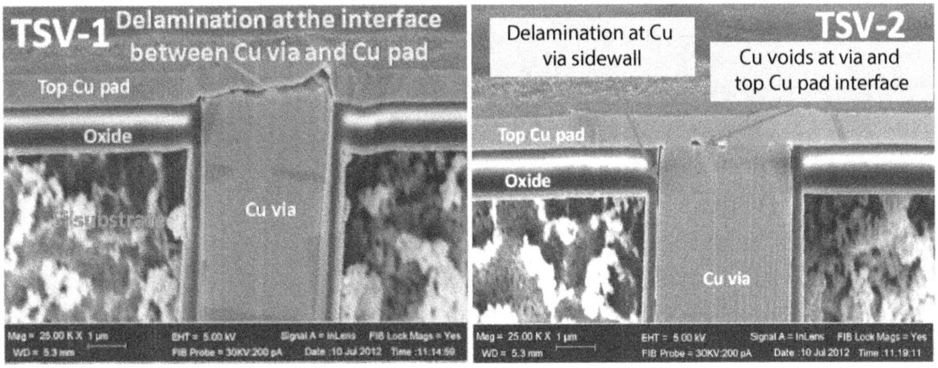

FIGURE 2.44 Cross-sectional SEM images for TSVs with a very small Cu pumping: (a) Cu pumping is not visible, (b) delamination at the interface between the TSV and the frontside Cu pad for TSV-1, (c) delamination at the TSV sidewall near the front side Cu pad, and (d) void at the interface between the TSV and the front side Cu pad.

from the front side is very small and not visible, (b) there is delamination at the interface between the TSV and the front side Cu pad for TSV-1, (c) delamination at the TSV sidewall near the front side Cu pad occurs, and (d) there is Cu void at the interface between the TSV and the front side Cu pad.

2.4.4 Keep-Out-Zone of Cu-Filled TSVs

As mentioned in Sec. 2.4.2, there could be a very large stress due to the TEM between the Cu-filled TSV and the surrounding Si. This stress causes hole and electron mobility variation in CMOS devices, which can result in the degradation of electrical performance. KOZ is the area surrounding each TSV from which all devices must keep out so that they are not affected by the TSV TEM stress.

Figure 2.45 shows the test vehicle (5 μm diameter and 24 μm thick) for IMEC [28] to perform the KOZ measurements. There are two sets of wafers, one is subjected to 250°C for 30 seconds and the other is subjected to 250°C for 30 seconds + 420°C for 20 min. By

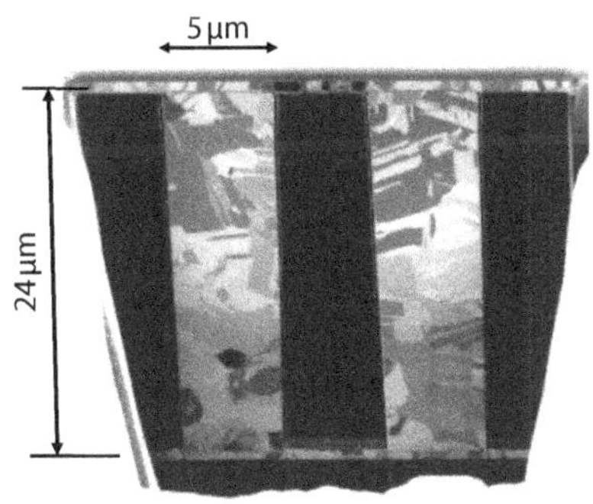

FIGURE 2.45 TSV test vehicle: 5 μm in diameter and 24 μm deep.

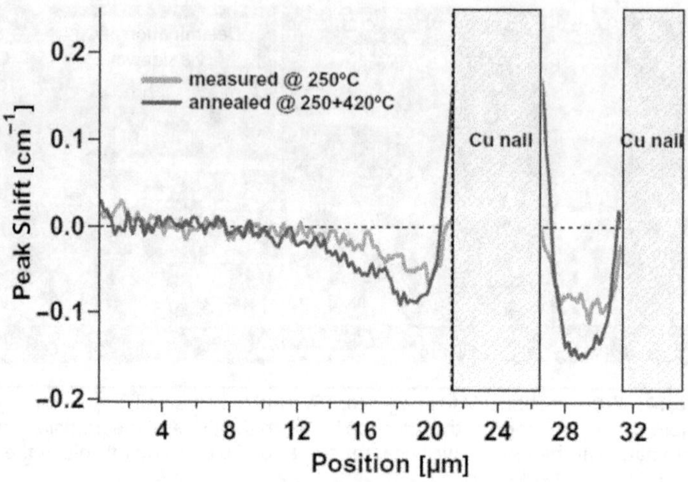

FIGURE 2.46 μ-Raman spectroscopy measurement results for two loading conditions: (1) 250°C for 30 seconds, and (2) 250°C for 30 seconds + 420°C for 20 minutes.

using the μ-Raman method the spectroscopy measurement results are shown in Fig. 2.46. (The stress is related to the Raman peak shift and if there is uniaxial or biaxial stress in the silicon, the Raman frequency shift is then linearly related to the stress.) It can be seen that (1) for both loading conditions, due to stress concentration, the maximum stress occurs at the boundary of the TSV and dies out quickly, and (2) the larger the loading the larger the stress. The KOZ for various Cu-filled TSV diameters, while assuming a carrier mobility change tolerance of 5%, is shown in Fig. 2.47 [28]. It can be seen that the larger the TSV

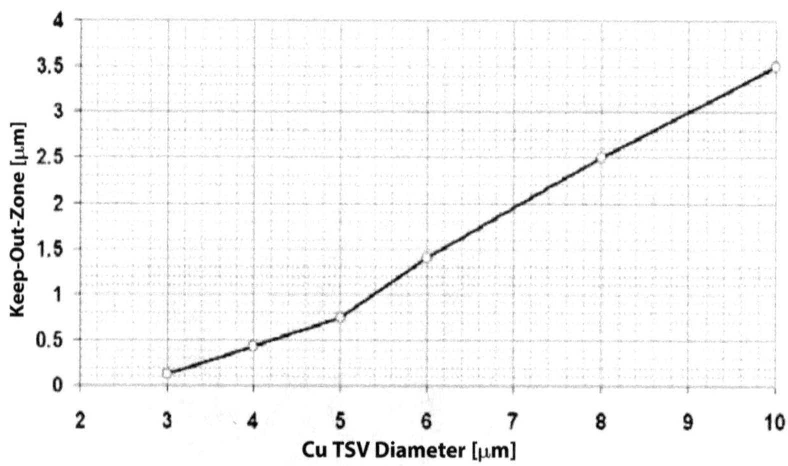

FIGURE 2.47 The KOZ for various Cu-filled TSV diameters, while assuming a carrier mobility change tolerance of 5%.

diameter the larger the KOZ. Thus, in order to save the expensive silicon device chip area, the smaller the TSV diameter the better.

The TSV TEM stress can be reduced to zero if an airgap is fabricated around a TSV [44], as shown in Figs. 2.48 and 2.49. It can be seen from Fig. 2.49 that the circumferential stress for the 5-μm-diameter TSV with an airgap reduces to zero. Cross-sectional FIB/SEM images for a 3-μm-diameter and 50-μm-deep Cu-filled TSV with a 23-μm-deep integrated airgap are shown in Fig. 2.50. It can be seen that the airgap is ~1 μm from the TSV. For more information on the fabrication of the airgap, please read [44].

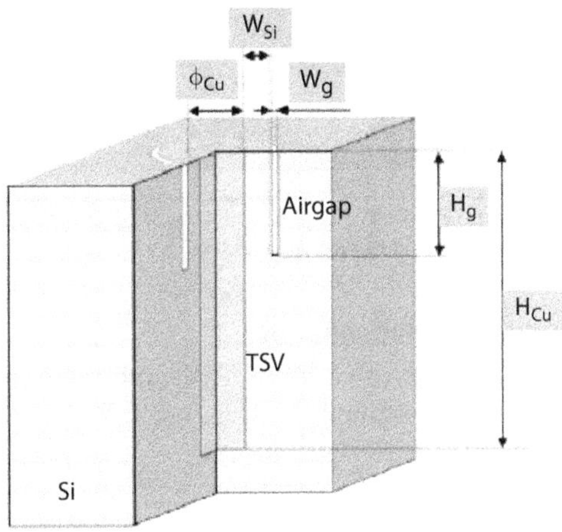

FIGURE 2.48 Schematic of a TSV with a ring-shaped airgap. Φ_{Cu} = TSV diameter; W_s = distance between the TSV and the airgap; W_g = airgap width; H_{Cu} = TSV depth; and H_g = airgap depth.

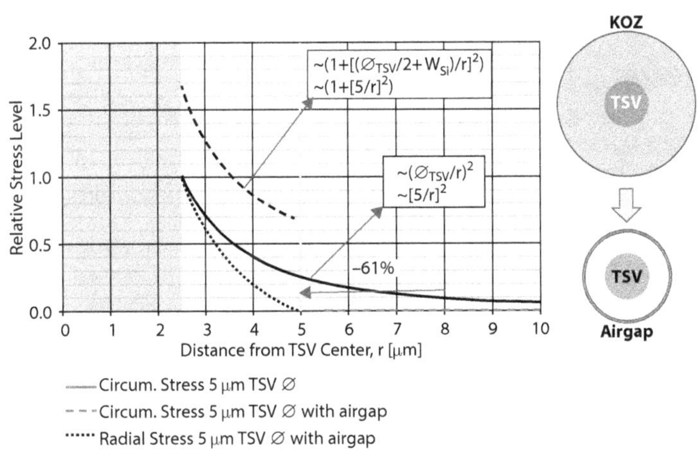

FIGURE 2.49 Circumferential stress reduces when an airgap is added around the 5 μm–diameter TSV.

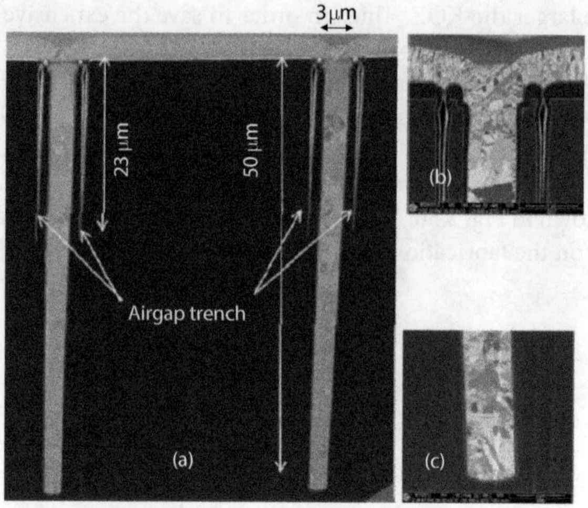

FIGURE 2.50 Cross-sectional FIB/SEM images showing (a) a 3 μm-diameter and 50 μm-deep TSV with a 23 μm-deep airgap, (b) the airgap closure, and (c) the TSV bottom.

2.4.5 Summary and Recommendations

Cu pumping and KOZ issues of Cu-filled TSVs and their solutions have been presented. Some important results and recommendations are summarized as follow:

- The larger the annealing temperature the larger the Cu grain growth.
- The larger the annealing temperature the smaller the residual Cu pumping.
- The post Cu-plating annealing condition (420°C for 20 min) before CMP is optimal.
- The larger the TSV diameter the larger the KOZ, i.e., the larger the chip size. Thus, it is recommended to have the TSV diameter as small as possible.
- An airgap around the TSV can reduce the TEM stress to ~zero.

2.5 References

[1] Ryu, C., J. Lee, H. Lee, K. Lee, T. O, and J. Kim, "High Frequency Electrical Model of Through Wafer Via for 3-D Stacked Chip Packaging," *Proceeding of Electron. Syst. Integration Technol. Conf.*, September, 2006, pp. 215–220.

[2] Savidis, I., and E. G. Friedman, "Electrical Modeling and Characterization of 3-D Vias," *Proc. IEEE Int. Symp. Circuits Syst.*, Seattle, WA, May 18–21, 2008, pp. 784–787.

[3] Bermond, C., L. Cadix, A. Farcy, T. Lacrevaz, P. Leduc, and B. Flechet, "High Frequency Characterization and Modeling of High Density TSV in 3-D Integrated Circuits," *Proc. 13th IEEE Workshop Signal Propagation Interconnects*, Strasbourg, France, May 12–15, 2009, pp. 1–4.

[4] Cadix, L., A. Farcy, C. Bermond, C. Fuchs, P. Leduc, M. Rousseau, M. Assous, et al., "Modelling of Through Silicon Via RF Performance and Impact on Signal Transmission in 3-D Integrated Circuits," *Proc. IEEE 3-D Syst. Integration Conf.*, San Francisco, CA, September 28–30, 2009, pp. 1–7.

[5] Han, K. J., and M. Swaminathan, "Polarization Mode Basis Functions for Modeling Insulator-Coated Through-Silicon Via (TSV) Interconnections," *Proceeding of Signal Propagate. Interconnects*, May 12–15, 2009, pp. 1–4.

[6] Cho, J., K. Yoon, J. Pak, J. Kim, J. Lee, H. Lee, K. Park, et al., "Guard Ring Effect for Through Silicon Via (TSV) Noise Coupling Reduction," *Proceedings of IEEE CPMT Symposium Japan*, August 2010, pp. 151–154.

[7] Kim, J. J. Pak, J. Cho, J. Lee, H. Lee, K. Park, and J. Kim, "Modeling and Analysis of Differential Signal Through Silicon Via (TSV) in 3D IC," *Proceedings of Japan CPMT Symposium*, 2000, pp. 1–4.

[8] Savidis, I., S. M. Alam, A. Jain, S. Pozder, R. E. Jones, and R. Chatterjee, "Electrical Modeling and Characterization of Through-Silicon Vias (TSVs) for 3-D Integrated Circuits," *Journal of Microelectronic*, 2010, pp. 9–16.
[9] Katti, G., M. Stucchi, K. De Meyer, and W. Dehaene, "Electrical Modeling and Characterization of Through Silicon Via for Three-Dimensional ICs," *IEEE Trans. Electron Devices*, Vol. 57, No. 1, Jan 2010, pp. 256–262.
[10] Kim, J., J. S. Pak, J. Cho, E. Song, J. Cho, H. Kim, T. Song, et al., "High-Frequency Scalable Electrical Model and Analysis of a Through Silicon Via (TSV)," *IEEE Transactions on Components, Packaging and Manufacturing Technology*, Vol. 1, No. 2, 2011, pp. 181–195.
[11] Kim, J., W. Lee, Y. Shim, J. Shim, K. Kim, J. S. Pak, and J. Kim, "Chip-Package Hierarchical Power Distribution Network Modeling and Analysis Based on a Segmentation Method," *IEEE Transactions on Advanced Packaging*, Vol. 33, No. 3, 2010, pp. 647–659.
[12] Pak, J., J. Cho, J. Kim, J. Lee, H., Lee, K. Park, and J. Kim, "Slow Wave and Dirlectric Quasi-TEM Modes of Metal-Insulator-Semiconductor (MIS) Structure Through Silicon Via (TSV) in Signal Propagation and Power Delivery in 3D Chip Package," *IEEE Proceedings of ECTC*, May 2011, pp. 667–672.
[13] Hung, J. F., J. H. Lau, P. Chen, S. Wu, S. Lai, M. Li, S. Sheu, et al., "Electrical Testing of Blind Through-Silicon Via (TSV) for 3D IC Integration," *IEEE/ECTC Proceedings*, San Diego, CA, May 2012, pp. 564–570.
[14] Hung, J. F., J. H. Lau, P. Chen, S. Wu, S. Hung, S. Lai, M. Li, et al., "Electrical Performance of Through-Silicon Vias (TSVs) for High-Frequency 3D IC Integration Applications," *Proceedings of the 45th IMAPS International Symposium on Microelectronics*, September 2012, pp. 1221–1228.
[15] Cheng, D. K., *Field and Wave Electromagnetics*, Addison-Wesley, New York, NY, 1989.
[16] Paul, C. R., *Introduction to Electromagnetic Compatibility*, Wiley, Hoboken, NJ, 2006.
[17] Hall, S. H., G. W. Hall, and J. A. McCall, *High-Speed Digital System Design*, Wiley, Hoboken, NJ, 2000.
[18] Lau, J. H., and G. Tang, "Thermal Management of 3D IC Integration with TSV (Through Silicon Via)," *IEEE Proceedings of ECTC*, San Diego, May 2009, pp. 635–640.
[19] Lau, J. H., G. Tang, G. Y. Y. Hoe, X. W. Zhang, C. T. Chong, P. Damaruganath, and K. Vaidyanathan, "Effects of TSV (Through Silicon Via) Interposer/Chip on the Thermal Performances of 3-D IC Packaging," *ASME Paper No. IPACK2009-89380*, July 2009.
[20] Hoe, Y., Y. Tang, D. Pinjala, T. Chai, J. H. Lau, X. Zhang, and V. Kripesh, "Effect of TSV Interposer on the Thermal Performance of FCBGA Package," *IEEE/EPTC Proceedings*, Singapore, December 2009, pp. 778–786.
[21] Chien, H. C., J. H. Lau, Y. Chao, R. Tain, M. Dai, S. T. Wu, W. Lo, et al., "Thermal Performance of 3D IC Integration with Through-Silicon Via (TSV)," *Proceedings of IMAPS International Conference*, Long Beach, CA, October 2011, pp. 25–32.
[22] Chien, H. C., Y. Chao, J. H. Lau, M. Dai, R. Tain, M. Dai, P. Tzeng, et al., "A Thermal Performance Measurement Method for Blind Through Silicon Vias (TSVs) in a 300mm Wafer," *IEEE ECTC Proceedings*, Orlando, FL, June 2011, pp. 1204–1210.
[23] Chien, H. C., J. H. Lau, Y. Chao, M. Dai, R. Tain, L. Li, P. Su, et al., "Thermal Evaluation and Analyses of 3D IC Integration SiP with TSVs for Network System Applications," *IEEE/ECTC Proceedings*, San Diego, CA, May 2012, pp. 1866–1873.
[24] Chien, H. C., J. H. Lau, Y. Chao, R. Tain, M. Dai, S. T. Wu, W. Lo, et al., "Thermal Performance of 3D IC Integration with Through-Silicon Via (TSV)," *IMAPS Transactions, Journal of Microelectronic Packaging*, Vol. 9, 2012, pp. 97–103.
[25] Chien, H. C., J. H. Lau, T. Chao, M. Dai, and R. Tain, "Thermal Management of Moore's Law Chips on Both Sides of an Interposer for 3D IC integration SiP," *IEEE ICEP Proceedings*, Japan, April 2012, pp. 38–44.
[26] Selvanayagam, C., J. H. Lau, X. Zhang, S. Seah, K. Vaidyanathan, and T. Chai, "Nonlinear Thermal Stress/Strain Analysis of Copper Filled TSV (Through Silicon Via) and Their Flip-Chip Microbumps," *IEEE Proceedings of Electronic, Components & Technology Conference*, Orlando, FL, May 2008, pp. 1073–1081.
[27] Selvanayagam, C., J. H. Lau, X. Zhang, S. Seah, K. Vaidyanathan, and T. C. Chai, "Nonlinear Thermal Stress/Strain Analyses of Copper Filled TSV (Through Silicon Via) and Their Flip-Chip Microbumps," *IEEE Transactions on Advanced Packaging*, Vol. 32, No. 4, November 2009, pp. 720–728.
[28] Okoro, C., Y. Yang, B. Vandevelde, B. Swinnen, D. Vandepitte, B. Verlinden, and I. De Wolf, "Extraction of the Appropriate Material Property for Realistic Modeling of Through-Silicon-Vias using μ-Raman Spectroscopy," *Proceeding of IEEE International Interconnect Technology Conference*, June 2008, pp. 16–18.
[29] Liu, Xi, Q. Chen, P. Dixit, R. Chatterjee, R. Tummala, and S. K. Sitaraman, "Failure Mechanisms and Optimum Design for Electroplated Copper Through-Silicon Vias (TSV)," *Proceedings of IEEE/ECTC*, San Diego, CA, May 2009, pp. 624–629.
[30] Lu, K. H., X. Zhang, S.-K. Ryu, J. Im, R. Huang, and P. S. Ho, "Thermo-Mechanical Reliability of 3-D ICs containing Through Silicon Vias," *Proceedings of IEEE/ECTC*, San Diego, CA, May 2009, pp. 630–634.

[31] Chen, Z., X. Song, and S. Liu, "Thermo-Mechanical Characterization of Copper Filled and Polymer Filled TSVs Considering Nonlinear Material Behaviors," *Proceedings of IEEE/ECTC*, San Diego, CA, May 2009, pp. 1374–1380.
[32] Okoro, C., C. Huyghebaert, J. Van Olmen, R. Labie, K. Lambrinou, B. Vandevelde, E. Beyne, et al., "Elimination of the Axial Deformation Problem of Cu-TSV in 3D Integration," *Proceedings of 11th International workshop on stress-induced phenomena in metallization*, April 2010, pp. 214–220.
[33] Okoro, C., R. Labic, K. Vanstreels, A. Franquet, M. Gonzalez, B. Vandevelde, E. Beyne, et al., "Impact of the Electrodeposition Chemistry used for TSV Filling on the Microstructural and Thermo-Mechanical Response of Cu," *Journal of Materials Science*, February 2011, pp. 1–15.
[34] Lu, K. H., S.-K. Ryu, Q. Zhao, X. Zhang, J. Im, R. Huang, and P. S. Ho, "Thermal Stress Induced Delamination of Through Silicon Vias in 3-D Interconnects," *Proceedings of IEEE/ECTC*, Las Vegas, NV, May 2010, pp. 40–45.
[35] De Wolf, I., K. Croes, O. Varela Pedreira, R. Labie, A. Redolfi, M. Van De Peer, K. Vanstreels, et al., "Cu Pumping in TSVs: Effect of pre-CMP Thermal Budget," *Microelectronics Reliability*, Vol. 51, No. 9–11, 2011, pp. 1856–1859.
[36] Che, F., H. Li, X. Zhang, S. Gao, and K. Teo, "Wafer Level Warpage Modeling Methodology and Characterization of TSV Wafers," *IEEE ECTC Proceedings*, Orlando, FL, June 2011, pp. 1196–1203.
[37] Che, F., W. Putra, A. Heryanto, A. Trigg, S. Gao, and C. Gan, "Numerical and Experimental Study on Cu Protrusion of Cu-Filled Through-Silicon Vias (TSV)," Proceedings of *3rd IEEE International 3D System Integration Conference (3DIC)*, 2011, pp. P1-1–13.
[38] Heryanto, A., W. Putra, A. Trigg, S. Gao, W. Kwon, F. Che, X. Ang, et al., "Effect of Copper TSV Annealing on Via Protrusion for TSV Wafer Fabrication," *Journal of Electronic Materials*, Vol. 41, No. 9, 2012, pp. 2533–2542.
[39] De Messemaeker, J., O. V. Pedreira, B. Vandevelde, H. Philipsen, I. De Wolf, E. Beyne, and K. Croes, "Impact of Post-Plating Anneal and Through-Silicon Via Dimensions on Cu Pumping," *Proceedings of IEEE/ECTC*, Las Vegas, NV, May 2013, pp. 586–591.
[40] Zhang, D., K. Hummler, L. Smith, and J. Jian-Qiang Lu, "Backside TSV Protrusions Induced by Thermal Shock and Thermal Cycling," *Proceedings of IEEE/ECTC*, Las Vegas, NV, May 2013, pp. 1407–1413.
[41] Malta, D., C. Gregory, M. Lueck, D. Temple, M. Krause, F. Altmann, M. Petzold, et al., "Characterization of Thermo-Mechanical Stress and Reliability Issues for Cu-Filled TSVs," *IEEE ECTC Proceedings*, Orlando, FL, June 2011, pp. 1815–1821.
[42] Saettler, P., M. Boettcher, and K.-J. Wolter, "Characterization of the Annealing Behavior for Copper-Filled TSVs," *Proceedings of IEEE/ECTC*, San Diego, CA, May 2012, pp. 619–624.
[43] McDonougth, C., B. Backes, W. Wang, R. Caramto, and E. Geer, "Thermal and Spatial Dependence of TSV-Induced Stress in Si," *Proceeding of IEEE International Interconnect Technology Conference*, 2011, pp. 1–3.
[44] Civale, Y., S. Van Huylenbroeck, A. Redolfi, W. Guo, K. B. Gavan, P. Jaenen, A. La Manna, et al., "Via-Middle Through-Silicon Via with Integrated Airgap to Zero TSV-Induced Stress Impact on Device Performance," *Proceedings of IEEE/ECTC*, Las Vegas, NV, May 2013, pp. 1420–1424.
[45] Okoro, C., P. Kabos, J. Obrzut, K. Hummler, and Y. Obeng, " Accelerated Stress Test Assement of Through-Silicon Via Using RF Signals," *IEEE Transactions on Electron Devices*, Vol. 60, No. 6, June 2013, pp. 2015–2021.
[46] Ryu, S., K. Lu, T. Jiang, J. Im, R. Huang, and P. Ho, "Effect of Thermal Stresses on Carrier Mobility and Keeo-Out Zone Around Through-Silicon Via for 3-D Integration," *IEEE Transactions on Device and Materials Reliability*, Vol. 12, No. 2, June 2012, pp. 255–262.
[47] Athikulwongse, K., A. Chakraborty, Jae-seok Yang, D. Pan, Sung Kyu Lim, "Stress-Driven 3D-IC Placement with TSV Keep-Out Zone and Regularity Study," *IEEE International Conference on Computer-Aided Design*, 2010, pp. 669–674.
[48] Moongon, J. J. Mitra, D. Pan, Sung Kyu Lim, "TSV Stress-Aware Full-Chip Mechanical Reliability Analysis and Optimization for 3-D IC," *IEEE Transactions on Computer-Aided Design of Integrated Circuits and Systems*, Vol. 31, No. 8, 2012, pp. 1194–1207.
[49] Mercha, A., G. Van der Plas, V. Moroz, I. De Wolf, P. Asimakopoulos, N. Minas, S. Domae, et al., "Comprehensive Analysis of The Impact of Single and Arrays of Through Silicon Vias Induced Stress on High-K/Metal Gate CMOS Performance," *Proceedings of IEEE/IEDM*, 2010, pp. 2.2.1–2.2.4.
[50] Lau, J. H., *Through-Silicon Vias for 3D Integration*, McGraw-Hill Book Company, New York, NY, 2013.
[51] Lau, J. H., *Reliability of ROHS-Compliant 2D and 3D IC Integration*, McGraw-Hill Book Company, New York, NY, 2011.
[52] Lau, J. H., C. K. Lee, C. S. Premachandran, and A. Yu, *Advanced MEMS Packaging*, McGraw-Hill Book Company, New York, NY, 2010.

CHAPTER 3

Stress Sensors for Thin-Wafer Handling and Strength Measurement

3.1 Introduction

The generation of stress during manufacturing, packaging, and testing of the semiconductor devices is an inevitable phenomenon. Understanding these stress data is important for process selection, devices reliability, and fabrication yield. Piezoresistive stress sensors are known to be a potential device to measure the stresses in the semiconductor device packages [1–37]. Edwards et al. [1] used an n-type piezoresistive stress sensors to qualitatively evaluate the stress states in the plastic packages. For quantitative measurement of stresses, various calibration methods have been developed to determine piezoresistive coefficients [7, 8–12].

The calibrated piezoresistive silicon stress sensors have been widely used to evaluate the stresses at the silicon die surface after several packaging steps, such as die attach, underfill, and encapsulation [6, 7, 13]. The large CTE (coefficient of thermal expansion) mismatch between silicon chip and other packaging materials, such as epoxy material and substrate or board, has been reported to generate large residual stresses ranging from 100 to 200 MPa at the silicon die surface after undergoing different packaging steps [6, 7, 13, 18]. In this chapter, piezoresistive stress sensors will be designed, fabricated, and calibrated for experimental purpose, with which detailed stress can be measured during wafer thinning process. Also, piezoresistive stress sensors will be used to evaluate the stresses in device wafer after wafer level packaging processes, such as UBM (under bump metallurgy) fabrication, dry-film process, and solder bumping. Finally, piezoresistive stress sensors will be used to determine the stresses of an embedded ultrathin chip during drop impact test. The design and fabrication of stress sensors on a wafer will be briefly mentioned first.

3.2 Design and Fabrication of Piezoresistive Stress Sensors

The stress sensors are fabricated in a test vehicle reported in [37–39]. Due to the presence of the RDLs (redistribution layers), one on top and one at the bottom of the TSV (through-silicon via) interposer [37–39], the present stress sensors are embedded in the interposer, i.e., underneath the top RDL of the interposer.

3.2.1 Design of Piezoresistive Stress Sensors

Figure 3.1 shows the layout of the interposer [37–39] in a system-in-package, where the piezoresistive devices as the stress sensor (in red circles) are embedded [37]. Since chips are stacked above the interposer at the center, the measurement pads, which are at the peripheral, are connected with top redistributed layer (Cu metal lines) to the devices. Figure 3.2 shows a detailed layout of a stress sensor with four orientations, which can be used to monitor the stress and warpage by measuring the resistance formed by the implanted-Si-region. The resistor (30 × 100 μm, area in dotted line) is connected on each side with Cu metal lines (TR1, Top RDL) through the body contact (20 × 20 μm, shown in the lighter area as dotted lines).

The location and layout of an n-type stress-sensor design is shown in Figs. 3.1 and 3.2. The sensors R_1, R_2, R_3, and R_4 are fabricated along the [110], [$\bar{1}$00], [$\bar{1}$10], and [010] directions of a p-type (100) silicon wafer. The resistance of the R_1, R_2, R_3 and R_4 sensors is about 233 Ω. The sheet resistance of all the sensors is about 70 Ω. The resistances of sensors are due to their size, which was 30 × 100 μm for each resistance.

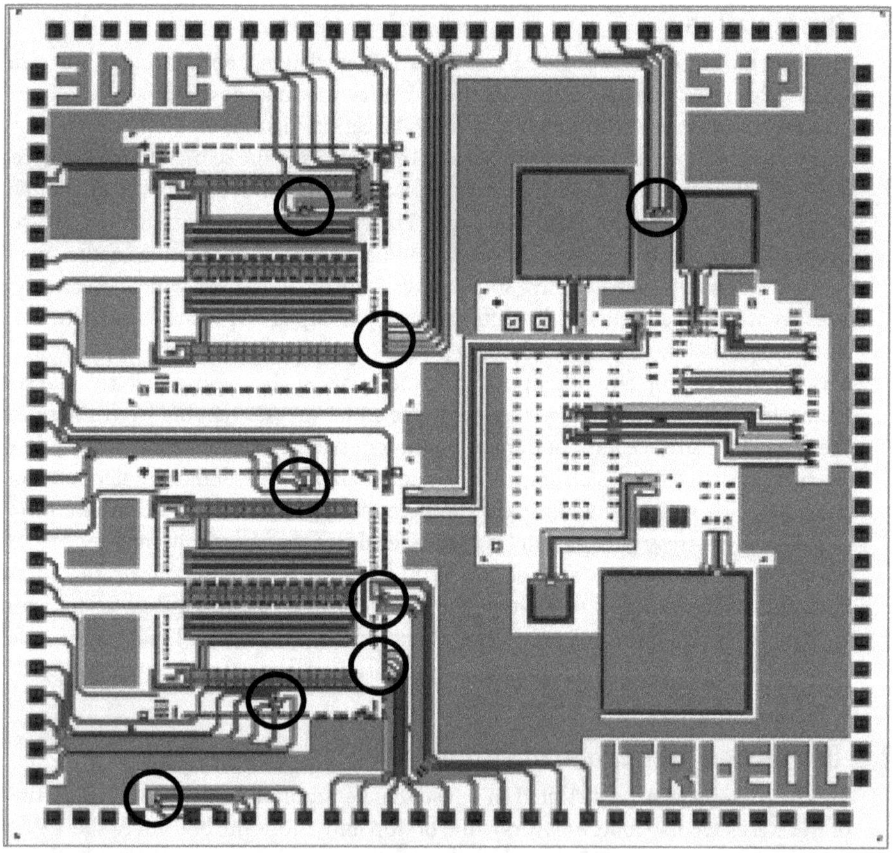

FIGURE 3.1 Location of stress sensors on a TSV interposer.

3.2.2 Fabrication of the Stress Sensors

Figure 3.3 shows the process flow of fabricating a stress sensor. First, the dielectrics over the stress sensor region are defined and etched for ion implantation (P^+, 40 KeV, $1 \times 15\ cm^{-2}$). The postimplantation annealing condition (for dopant activation) is 1100°C for 30 seconds in N_2. After removing the dielectrics for implantation, new interlayer-dielectrics (ILD) with 4000 Å thickness are deposited by PECVD (plasma enhanced chemical vapor deposition). TSV (15 μm in diameter; 100 μm in depth) is formed with TSV liner (5000 Å) by SACVD (subatmospheric chemical vapor deposition) and Cu damascene process with Ta barrier by PVD (physical vapor deposition).

The metal connection for stress sensor is formed with the Cu dual-damascene process: (1) first, the dielectrics for metal trace (TR1) are deposited by PECVD and defined by reactive-ion etch (RIE) (2000 to 2500 Å dielectric remained above TSV Cu surface); (2) second, the dielectrics for body contact (BC) to the implanted-Si are defined by RIE (3000 to 3500 Å dielectric remained above Si surface); (3) third, etching-back is performed to expose TSV Cu surface and implanted-Si area; (4) finally, Cu dual-damascene process is completed by Cu plating and Cu CMP (chemical-mechanical polishing) with Ti barrier.

Figure 3.4 shows the second step of this Cu dual-damascene process with photoresist still on the sample. The subsequent etching-back process as-mentioned earlier should ensure the BC region reach the interface of ILD and Si substrate (to expose Si surface) while TR1 region reach the interface of TR1 dielectric and TSV liner (to expose TSV Cu surface).

Figure 3.5 shows the top view (OM images) after TR1+BC etching-back and Cu dual-damascene. The shallow implanted-Si region is observed between two body contacts of

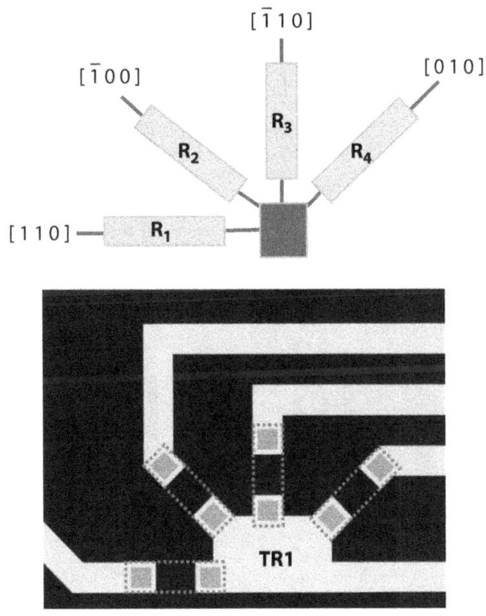

Figure 3.2 Piezoresistive stress-sensor rosette underneath the RDL (TR1) in a TSV interposer.

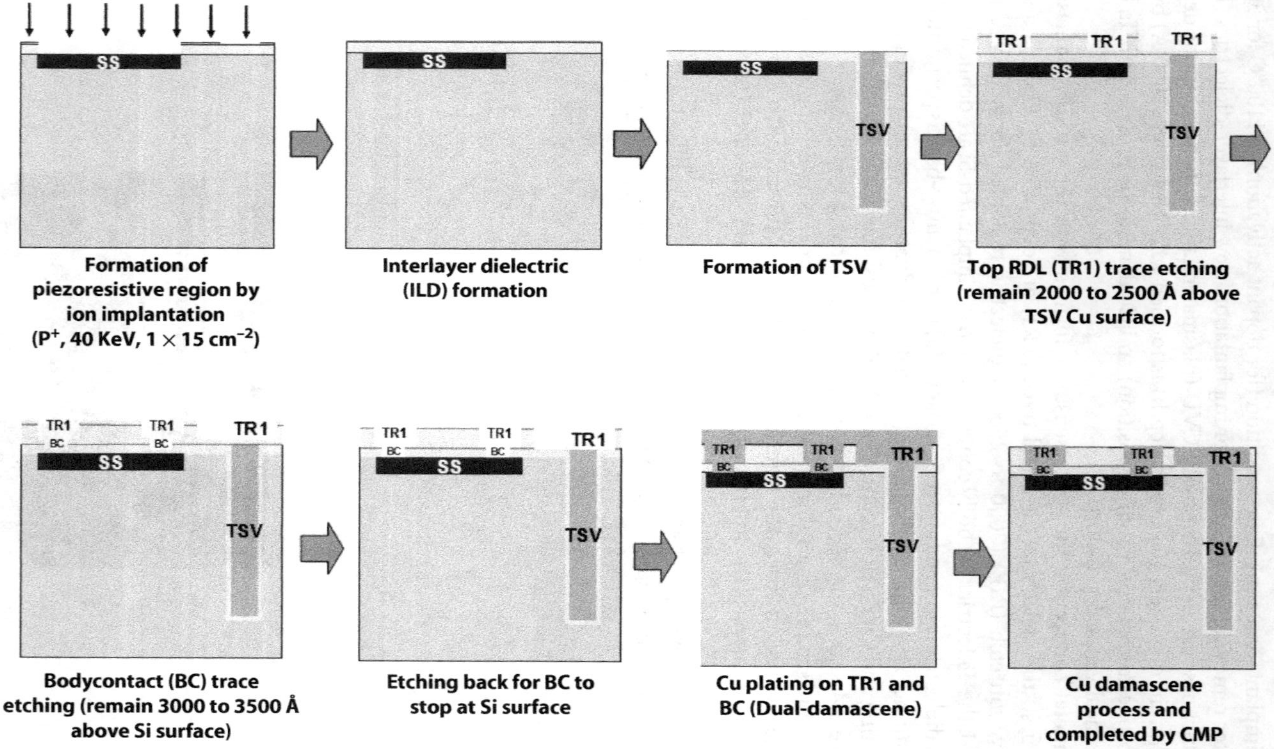

Figure 3.3 Process flow for fabricating an embedded stress sensor in a TSV interposer.

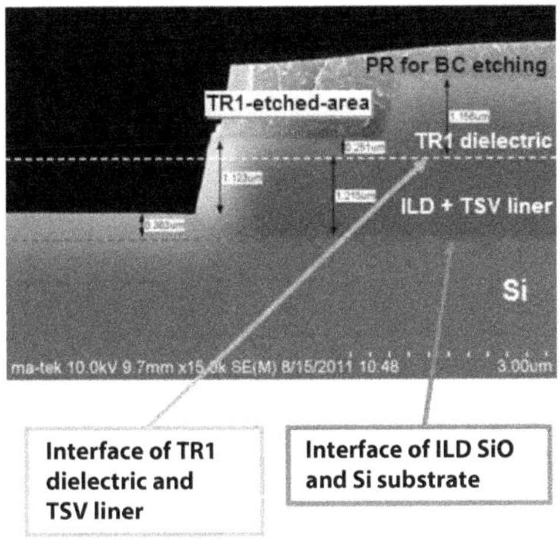

Figure 3.4 SEM image of the cross-section of the second step of the Cu dual-damascene process for the metal connection of a stress sensor.

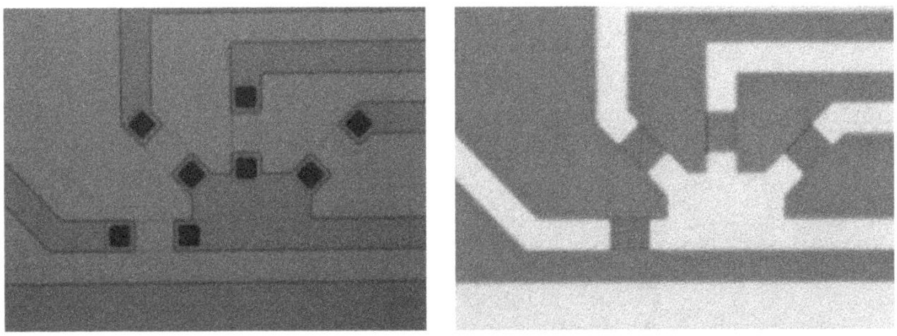

Figure 3.5 Optical-microscopy images after (left) TR1 + BC etching-back, and (right) Cu dual-damascene is completed.

each stress sensor. A high-resolution profiler is used to check the etching profile after TR1 and BC etching (Fig. 3.6). The scan profile can ensure the etching stops at the target (i.e., TR1 etching stops at TSV Cu surface and BC etching stops at Si surface). Figure 3.7 shows the stress sensor fabricated.

3.2.3 Summary and Recommendations

The design and fabrication of the piezoresistive stress sensors have been presented. Some important results and recommendations are summarized as follows:

- A detailed and workable fabrication process of making the embedded stress sensors in a TSV interposer has been provided and discussed.
- OM has been used to observe the shallow implanted-Si region between the two body contacts of each stress sensor.

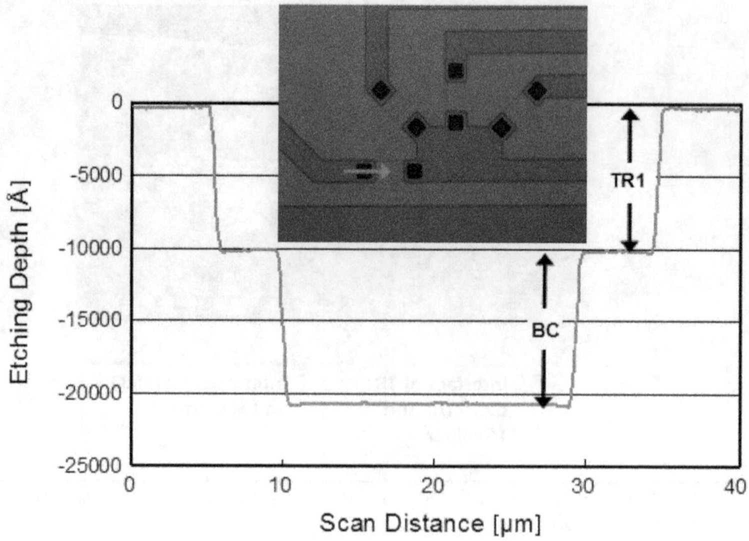

FIGURE 3.6 Profile of TR1 and BC etching by high-resolution profiler.

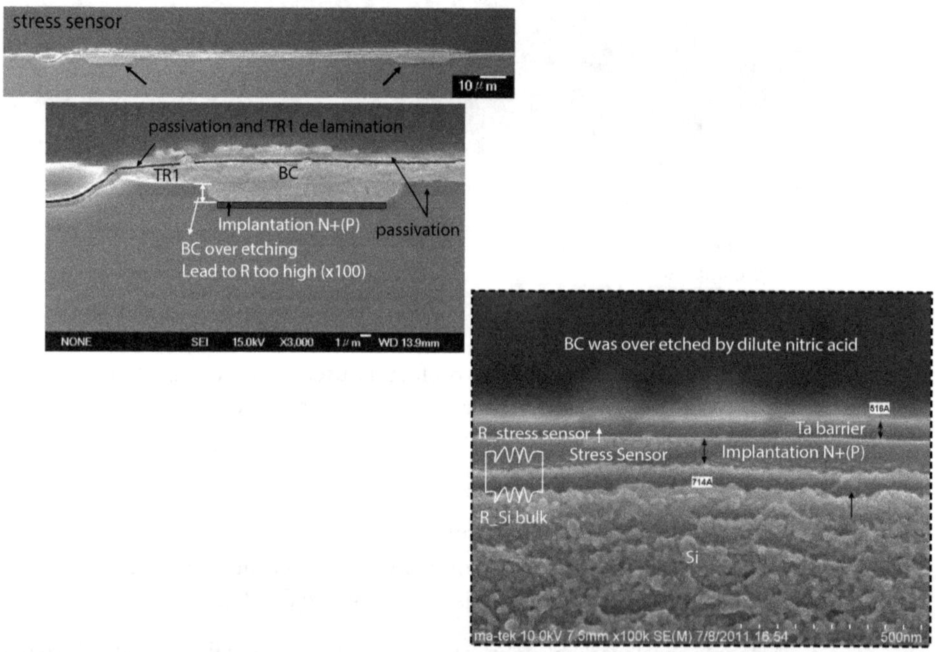

FIGURE 3.7 The fabricated stress sensor.

- HRP has been used to check the (after RDL and body contact etching) profile to make sure the RDL etching stops at TSV Cu surface and the body contact etching stops at the Si surface.
- SEM has been used to identify the fabricated stress sensor.
- Since only in-plane stresses (σ_x and σ_y) are interested herein, thus only the values of R_1 and R_3 are used. If out-of-plane stress (σ_z) is also interested, then the values of R_2 and R_4 should also be used.

3.3 Application of Stress Sensors in Thin-Wafer Handling

In this section, strength measurements of wafer after thinning will be provided [9, 10]. The design, fabrication, and calibration of piezoresistive stress sensors will be briefly mentioned first.

3.3.1 Design, Fabrication, and Calibration of Piezoresistive Stress Sensors

The layout of the stress sensor design for thin-wafer handling is shown in Fig. 3.8. Two stress sensors are placed orthogonally for the measurement of the parallel and perpendicular components of the applied stress. The crystal orientations of these two resistors are [110] and [$\bar{1}$10], respectively. Another pair of the stress sensor is rotated 135° with respect to the axis of the applied stress. The crystal orientations of these two resistors are [010] and [100]. The resistance of R_1 and R_3 sensors is around 0.625 kΩ and the resistance of R_2 and R_3 is around 0.932 kΩ. R_1 has the same size as R_3 with a width of 100 μm and a length of 350 μm while R_2 has the same size as R_4 with a width of 80 μm and a length of 420 μm. These sensors can be fabricated with the processes mentioned in Sec. 3.2.2 and are shown in Fig. 3.9.

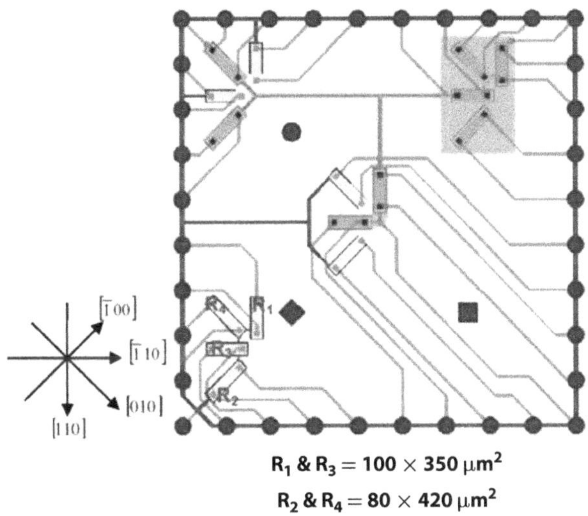

FIGURE 3.8 Design of stress sensors.

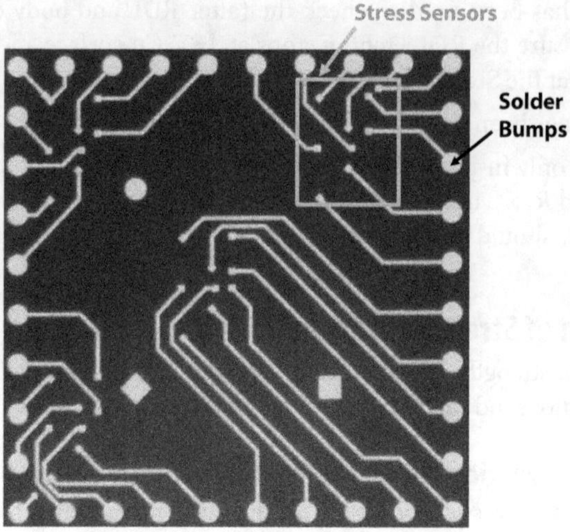

FIGURE 3.9 The fabricated stress sensor.

A sensor calibration process involves the measurement of resistance change of sensor with the application of known stress and then determining the piezoresistive coefficients using the theory of piezoresistivity [5, 6, 8]. According to the theory of piezoresistivity, if a uniform uniaxial stress σ_x is applied in the [110] direction of sensor rosette shown in Fig. 3.8, then the piezoresistive coefficients can be determined by

$$\Pi_{11} + \Pi_{12} = \frac{1}{\sigma_x}\left(\frac{\Delta R_1}{R_{10}} + \frac{\Delta R_3}{R_{30}}\right) \tag{3.1}$$

$$\Pi_{44} = \frac{1}{\sigma_x}\left(\frac{\Delta R_1}{R_{10}} - \frac{\Delta R_3}{R_{30}}\right) \tag{3.2}$$

where ΔR_i and R_{i0}, respectively, are the stress-induced resistance change and the initial resistance of ith sensor. With the known values of piezoresistive coefficients ($\pi_{11} + \pi_{12}$ and π_{44}) and considering the case of in-plane stress, $\sigma_z = 0$, piezoresistive sensors can be used to measure unknown residual in-plane stress components by measuring the change in resistance of sensors and using the following expressions [5, 6, 8]:

$$\sigma_x = \frac{\Pi_{44}\left(\frac{\Delta R_1}{R_{10}} + \frac{\Delta R_3}{R_{30}}\right) + (\Pi_{11} + \Pi_{12})\left(\frac{\Delta R_1}{R_{10}} - \frac{\Delta R_3}{R_{30}}\right)}{2\Pi_{44}(\Pi_{11} + \Pi_{12})} \tag{3.3}$$

$$\sigma_y = \frac{\Pi_{44}\left(\frac{\Delta R_1}{R_{10}} + \frac{\Delta R_3}{R_{30}}\right) - (\Pi_{11} + \Pi_{12})\left(\frac{\Delta R_1}{R_{10}} - \frac{\Delta R_3}{R_{30}}\right)}{2\Pi_{44}(\Pi_{11} + \Pi_{12})} \tag{3.4}$$

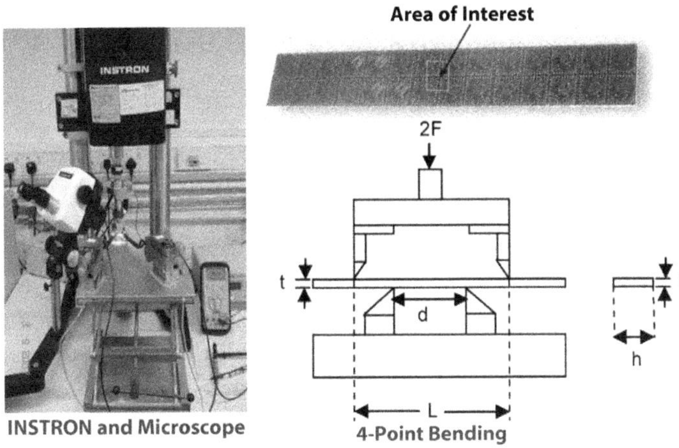

FIGURE 3.10 Test set-up for calibration of stress sensor.

Various methods, such as 4PB (four-point bending) using a strip of sensor chips [5, 11–13], 4PB using a sensor chip [14], wafer level vacuum chuck [12], and hydrostatic pressure [7], have been used to calibrate the piezoresistive stress sensor. In this study, 4PB method using a strip of sensor chips is used for the calibration due to its simple set-up and ease of understanding [7, 11–13]. For the calibration, two sensor wafers are diced into the sensor strips of size 70 × 10 mm, as shown in Fig. 3.10. Each sensor strip contained two rows of 14 sensor test chips, however, only sensor rosette that was located at the center of strip is used in the calibration process to meet the condition of uniform uniaxial stress.

The calibration of stress sensor using the 4PB of sensor strip requires a 4PB fixture and a loading machine to apply uniform uniaxial load on the sensor strip, a microscope to visualize the probing pads, two microprobes to probe the sensor pads, and a multimeter to measure the sensor's resistance. Thus, a simple calibration set-up, as shown in Fig. 3.10, is established to apply a known load and to measure the sensor's resistance. An Instron microtester with a 100 N load cell is used to apply a known load onto the sensor strip. A schematic representation of 4PB fixture is shown in Fig. 3.10. A uniform uniaxial tensile stress "σ," as shown in Fig. 3.11, can be

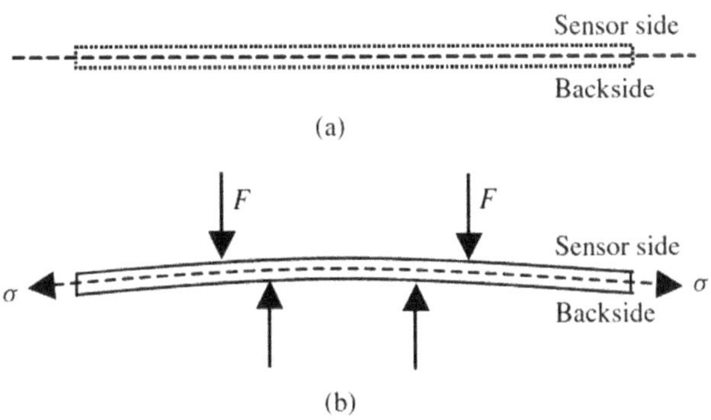

FIGURE 3.11 Schematic of the bending of sensor strip showing (a) zero stress state and (b) tensile stress state at sensor plane.

applied on the sensor plane by using this type of 4PB fixture and the applied stress can be calculated from the applied load "2*F*" considering the standard 4PB theory [8, 11] as

$$\sigma = \frac{3F(L-d)}{t^2 h} \tag{3.5}$$

where L is the loading span, d is the supporting span, h is the width of the sensor strip, and t is the thickness of the sensor strip. This formula works well if the deflection due to the load is not significant and the t and h are small as compared to the d and L [11]. Thus in this study, L, d, h, and t are chosen to be of 50, 20, 10, and 0.73 mm, respectively.

In the calibration process, first the resistance of stress sensor is measured without any applied load. Then, a known load (with an increment of 4 N up to 24 N) is applied on the sensor strip using 4PB fixture and Instron microtester and the corresponding resistance of stress sensor is measured with the help of microprobes and multimeter. Considering the variation in the readings, at least eight sensor strips from two different wafers are taken to calculate the average value. Figure 3.12 shows the resistance of sensors as a function of applied stress. It can be seen that resistance of n-type sensor decreases with an increase in applied stress.

To determine the piezoresistive coefficients from the resistance versus applied stress data, according to Eqs. (3.1) and (3.2), slops of plots between normalized resistance changes, i.e., $(\Delta R_1/R_{10} \pm \Delta R_3/R_{30})$, and applied stress, as shown in Fig. 3.13, are measured. The piezoresistive coefficients, $\pi_{11} + \pi_{12}$ and π_{44}, of n-type sensors have been found to be -1.98×10^{-4} MPa^{-1} and -1.06×10^{-4} MPa^{-1}, respectively. These values of piezoresistive coefficients are in agreement with the previously reported works [2] and [15].

Once the piezoresistive coefficients $\Pi_{11} + \Pi_{12}$ and Π_{44} of the n-type stress sensor are determined, the stresses (σ_x and σ_y) can be calculated from Eqs. (3.3) and (3.4) with the measured change in resistance (ΔR_i) and initial resistance (R_{i0}). In this chapter, the piezoresistive stress sensors will be used to evaluate the stresses in wafers after thinning (backgrinding).

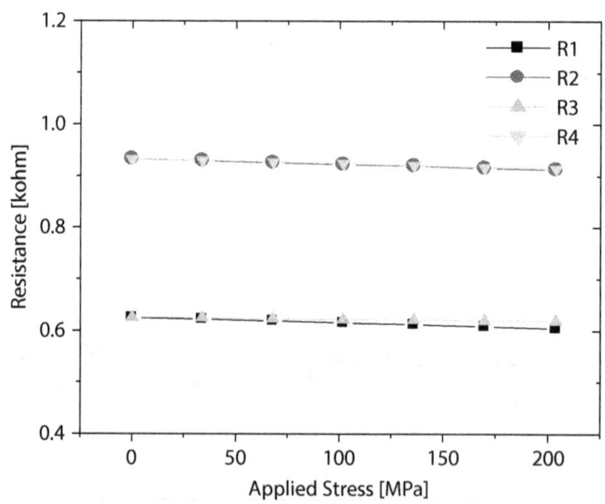

Figure 3.12 Resistance of sensors as a function of applied stress.

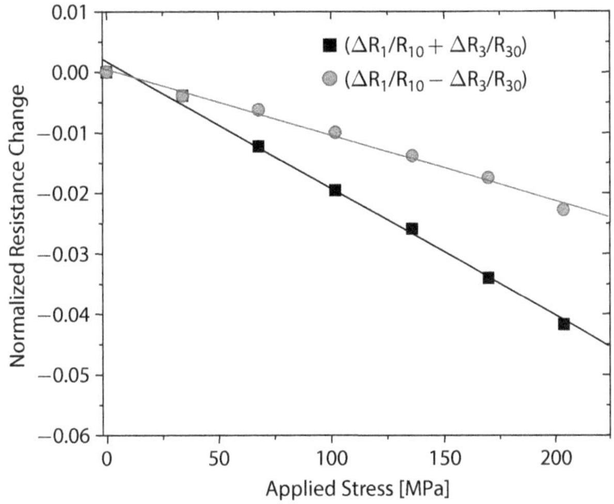

Figure 3.13 Normalized resistance change of sensors as a function of applied stress.

3.3.2 Stress Measurement in Wafer after Thinning

Wafer thinning process induces residual stress or mechanical damages on the back side of the wafer. In this study, stress sensor wafers are first thinned by means of coarse mechanical grinding using particles with a mesh size approximately #600, followed by fine mechanical grinding with a mesh size of approximately #2000. In order to remove the microcracks on the back side of the wafer, an additional process (namely final polishing) is also needed for the thinned stress sensor wafers. In coarse mechanical grinding process, the spindle speed is 3000 rpm, CT (chuck table) speed is 230 rpm, feed rate is 180 μm/min. In fine mechanical grinding process, the spindle speed is 3400 rpm, CT speed is 230 rpm, feed rate is 12 μm/min. In the final polishing process, slurry is 200 cc/min, load is 150 g/cm^2, pad speed is 230 rpm, and CT speed is 210 rpm.

Stress sensor wafers are thinned to 400, 200, and 100 μm. At each wafer, stress measure is performed at nine locations (Fig. 3.14). The residual stresses are determined by Eq. (3.3).

Figure 3.15 shows the stress measurement results after above-mention backgrinding processes. It is found that wafer stress largely increases with reducing the wafer thickness. Almost 7 times increase in stress is observed when wafer thickness decreases from 400 to 100 μm. This is because thinner wafer has a larger warpage than the thick one. Warpage of the 100-μm-thick sensor wafer is 32.33 μm (Fig. 3.16) while warpage of the 200-μm-thick sensor wafer is 24.39 μm.

In this study, wafer warpage is measured using wafer mapping system (FSM) and considering the 33 points measurement, as shown in Fig. 3.16. FSM uses noncontact IR (infrared) interferometric technique to provide a map of wafer thickness and thickness variation. The definition and concept of measuring warpage is given below:

- Min: 61.21 μm (Fig. 3.16)—Minimum distance of wafer median surface from the reference plane.

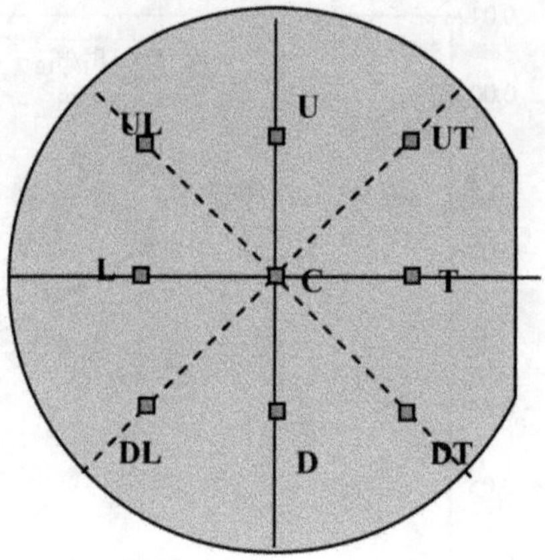

Figure 3.14 Stress measurement locations on a wafer.

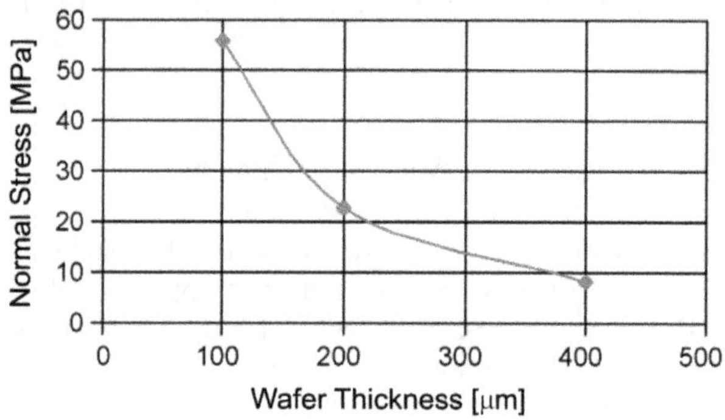

Figure 3.15 Stress (σ_x) measurements after backgrinding.

- Max: 93.54 μm (Fig. 3.16)—Maximum distance of wafer median surface from the reference plane.
- Warpage: The differences between the maximum and minimum distances of the median surface of a free, unclamped wafer from a reference plane = 93.54 − 61.21 = 32.33 μm.

3.3.3 Summary and Recommendations

The piezoresistive stress sensors have been designed, fabricated, and calibrated, with which detailed stress can be measured during wafer thinning process. Some important results and recommendations are summarized as follow [9, 10]:

- The thinner the wafers the more the induced residual stress.
- Almost seven times increase in stress has been observed when wafer thickness decreases from 400 to 100 μm.
- The thinner the wafer the easier to cause wafer damage.

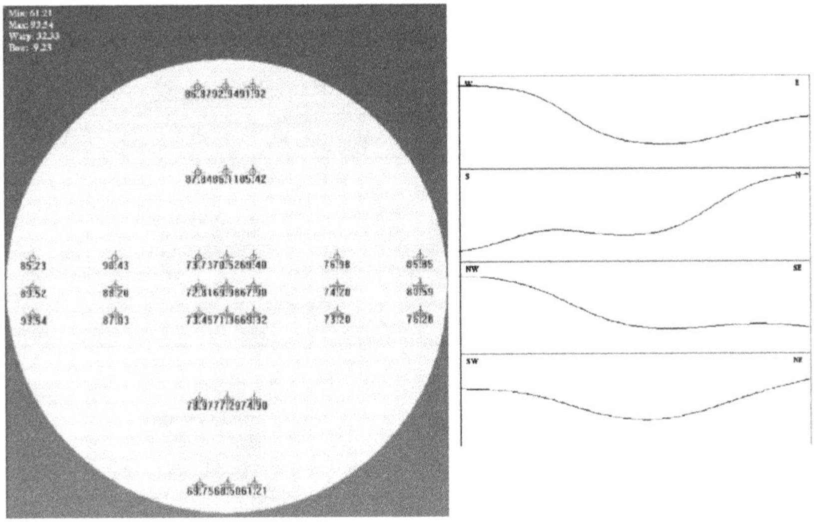

Figure 3.16 Warpage data and profile of 100-μm-thick stress sensor wafer after backgrinding. (Min: 61.21 μm = minimum distance of wafer median surface from the reference plane. Max: 93.54 μm = maximum distance of wafer median surface from the reference plane. Warpage = 93.54 − 61.21 = 32.33 μm.)

3.4 Application of Stress Sensors in Wafer Bumping

The designed, fabricated, and calibrated stress sensors shown in Fig. 3.8 are used to measure the stress in wafer bumping processes such as UBM, dry-film, and solder bumping of wafers [35, 36].

3.4.1 Stresses after UBM Fabrication

UBM fabrication is a conventional process that is done before solder bumping. In this study, stresses after UBM deposition are measured to understand the mechanical influence of UBM fabrication process on device wafer. A thin UBM consisting of Ti (1 kÅ)/Ni (7 kÅ)/ Au (1 kÅ) layers is fabricated on Al or Cu pads of stress sensor devices by sputtering at the temperature ranging from 100°C to 150°C. The process steps of UBM fabrication are shown in Fig. 3.17.

Figure 3.18 shows the optical image of stress sensor wafer after UBM fabrication. Residual stresses in sensor device wafer after UBM fabrication are calculated by measuring the change in resistance of stress sensors due to UBM fabrication process. The resistance measurement is conducted at 17 locations of a device wafer as shown in Fig. 3.19, and then by using Eqs. (3.3) and (3.4), two in-plane stress components (σ_x and σ_y) are determined, as shown in Fig. 3.20. The determined average values of stresses are $\sigma_x = 11.2$ MPa and $\sigma_y = 17.3$ MPa. These stress values indicate that UBM fabrication introduced tensile stress at the surface of device wafer. However, they are too small to be significant.

3.4.2 Stresses after Dry-Film Process

Residual stresses in sensor wafer are evaluated after processing negative-type dry-films of two different thicknesses, 50 and 100 μm. The thickness of sensor wafer was around 0.73 mm.

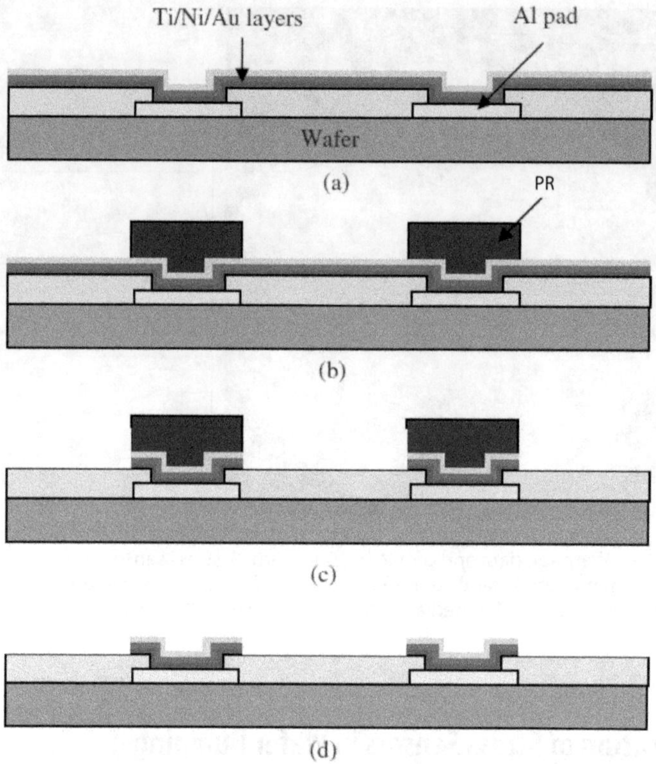

Figure 3.17 Process flow of UBM fabrication. (a) Ti/Ni/Au deposition, (b) Photoresist (PR) coating and patterning, (c) Au/Ni/Ti etching, and (d) PR stripping.

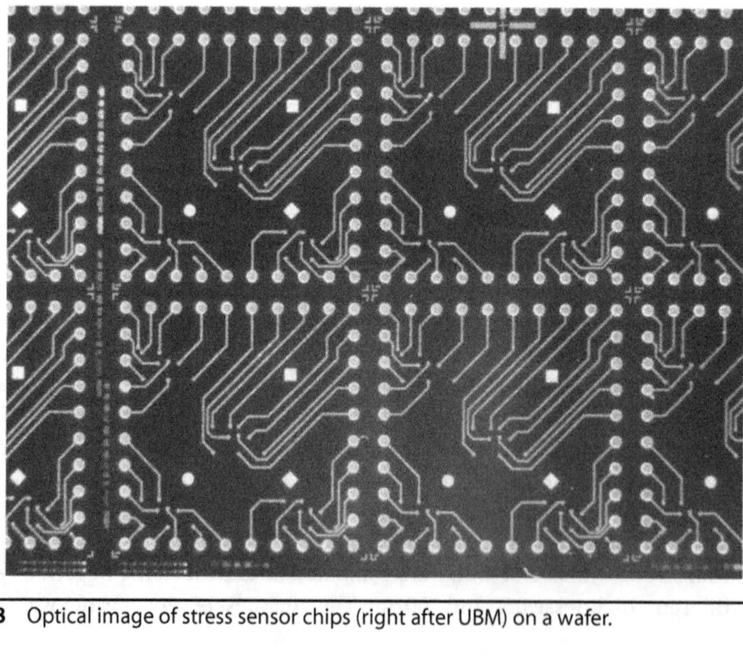

Figure 3.18 Optical image of stress sensor chips (right after UBM) on a wafer.

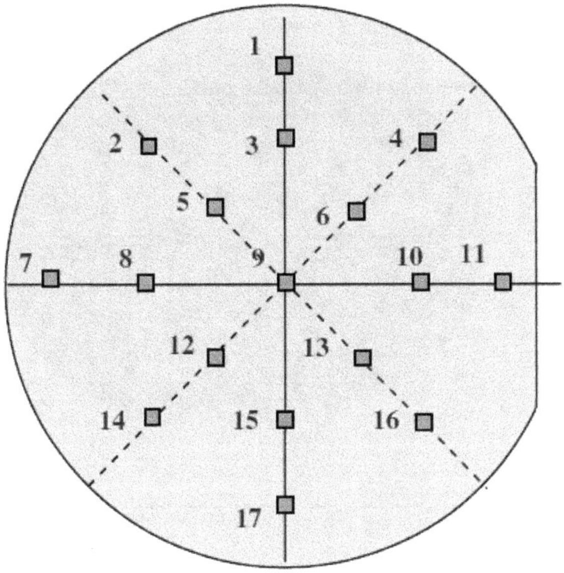

Figure 3.19 Stress measurement locations on a stress sensor wafer.

A complete dry film process comprised dry-film lamination, exposure, and development processes. In dry-film lamination, roller temperature is around 110°C and complete lamination process takes around 1 min. After lamination, 50- and 100-μm dry-films are exposed to UV light of 85 and 200 mJ/cm^2 energies, respectively. After UV exposure, 50- and 100-μm dry-films are developed in a developer for 4 min and 7 to 8 min, respectively.

Figure 3.21 shows the optical images of sensor wafer after 100-μm-thick dry-film processing. Residual stresses in sensor device wafer due to dry-film process are determined by measuring the change in resistance of stress sensors. Residual stresses in sensor device wafer after dry-film process are determined at 17 locations of a device wafer as mentioned

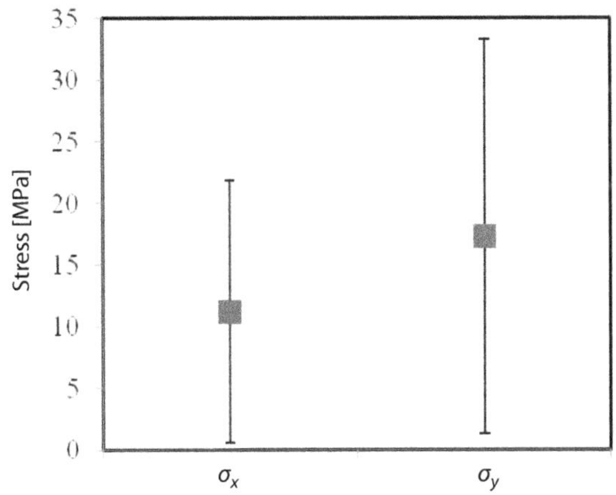

Figure 3.20 In-plane stresses (σ_x and σ_y) measured right after UBM fabrication.

FIGURE 3.21 Optical image of stress sensor device wafer right after 100-μm-thick dry-film process.

in previous section. Figure 3.22 shows the value of residual stresses developed due to dry-film process as a function of dry-film thickness. It can be seen that dry-film process induces compressive stress at the surface of device wafer and the amount of compressive stress increases with the increase in dry-film thickness. The development of the compressive stress is due to the lamination of high CTE dry-film on low CTE Si wafer at high temperatures of 110°C.

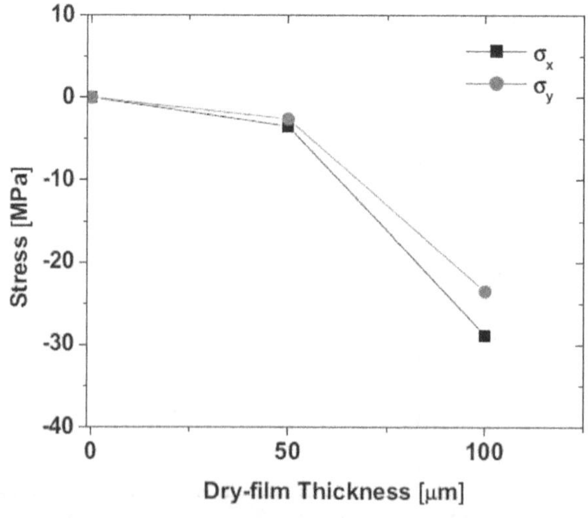

FIGURE 3.22 In-plane stresses (σ_x and σ_y) as a function of dry-film thickness.

3.4.3 Stresses after Solder Bumping Process

After dry-film processing, solder bumping is carried out on stress sensor wafer, and residual stress due to solder bumping process is measured. Figure 3.23 shows the process steps that are involved in the fabrication of solder bumps. A 40-μm-thick layer of dry-film is laminated on sensor wafer at 110°C and then the dry-film is patterned. After pattering, Sn1wt%Ag0.5wt%Cu (SAC 105) solder paste is screen printed and reflowed for 2 min at a peak temperature of 260°C. Finally, the dry-film is stripped to reveal solder bumps. Residual stresses in sensor device wafer after solder bumping are measured at 17 locations of a device wafer as mentioned in previous section. The measured average values of stress are $\sigma_x = 15.4$ MPa and $\sigma_y = 11.7$ MPa, and also shown in Fig. 3.24. The measured values of stress imply that solder bumping generates tensile stress at the surface of device wafer with the maximum value of 15.4 MPa.

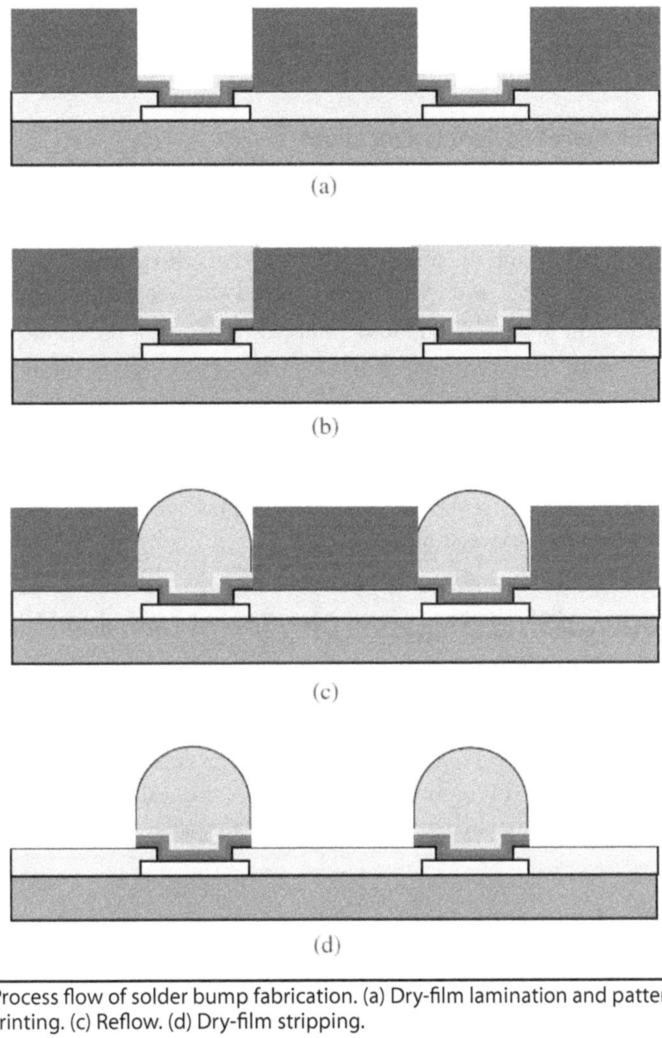

Figure 3.23 Process flow of solder bump fabrication. (a) Dry-film lamination and patterning. (b) Solder paste screen printing. (c) Reflow. (d) Dry-film stripping.

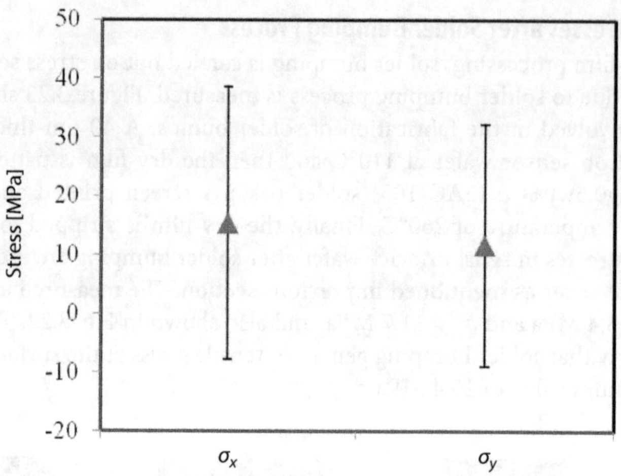

Figure 3.24 In-plane stresses (σ_x and σ_y) measured right after solder bumping.

3.4.4 Summary and Recommendations

The effects of wafer bumping processes on the stress acting on the chip have been presented. Some important results and recommendation are shown as follows:

- About 200 μm diameter of Ti(1 kÅ)/Ni(7 kÅ)/Au(1 kÅ) UBM deposition on piezoresistive sensor wafer only generated a small amount of tensile stress (17.3 MPa).
- Dry-film processing induced compressive stress at the surface of sensor wafer and the value increased from 3.5 to 28.8 MPa with the increased in dry-film thickness from 50 to 100 μm.
- About 40-μm-thick and 200-μm-diameter SAC105 solder deposition on sensor wafer also only generated a small amount of tensile stress (15.4 MPa).
- Since the stress generated from the wafer bumping processes is so small, from now on the effect of wafer bumping on the stress acting on the chip can be neglected.

3.5 Application of Stress Sensors in Drop Test of Embedded Ultrathin Chips

During drop test, there is a differential flexing between the silicon chip and the organic substrate. With a chip of conventional thickness, the differential flexing deforms the solder joint/substrate, eventually resulting in their failure. With the ultrathin silicon chip, however, the increased flexibility of the chip could change the failure mode from joint/substrate failure to chip cracking. This section aims to measure the real-time stress in a thin chip during a drop test to ascertain whether chip cracking is a possible risk when dealing with very thin chips [35, 36].

3.5.1 Test Vehicle and Fabrication

The designed, fabricated, and calibrated stress sensor chip shown in Fig. 3.8 is also used for the process development of chip-last embedded active prototype: square chips of 5 mm in size with one peripheral I/Os (see Figs. 3.8 and 3.9). As interconnections, SAC 105 solder

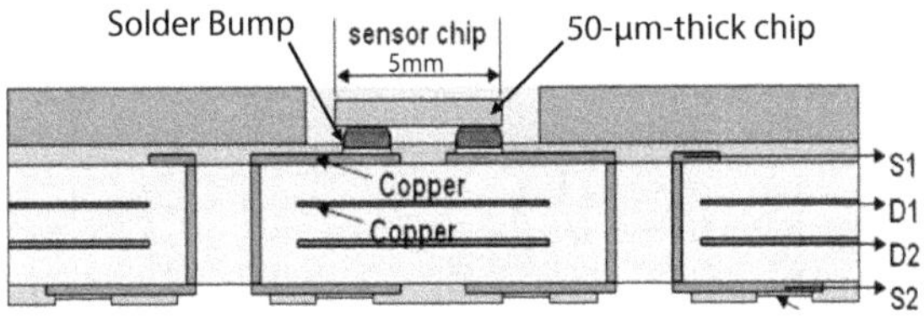

Figure 3.25 Cross-section of an embedded ultrathin (50 μm) stress sensor chip in the JEDEC-JESD22-B111 board.

bumps of 200 μm in diameter and 40 μm in height are used. The chips are fabricated by the above-mentioned wafer bumping processes.

The conventional way of backgrinding includes mounting the wafer to the backgrinding tape, followed by backgrinding. The thinned wafer is then transferred from the backgrinding tape to dicing tape. However, this usual process of backgrinding to 50-μm-thick wafer has resulted in extensive wafer crack on the backgrinded side of the thinned wafer. As a result of extensive wafer crack, DBG (dicing before grinding) is used instead. In DBG, the wafers are diced to 80 μm deep first before backgrinding to 50 μm from the back side of the wafer. After backgrinding to 50 μm, the thinned wafer is transferred to a thermal-release tape. The singulated chip is released from the tape by heating the thermal-release tape to 140°C.

The ultrathin sensor chips are placed onto the pads inside the cavities of a JEDEC-JESD22-B111 [40] board (as shown in Figs. 3.25 and 3.26) using the flip chip machine—FC150 (SUSS MicroTec) with an accuracy of ± 1 μm. No-clean flux is used to reduce flux residue that might be present inside cavities and in the gap between chips and the board. Solder is reflowed in conventional solder reflow oven with the lead-free solder reflow profile. After that, a conventional underfill is filled between the ultrathin chip and the board, and cured. Finally, a glob-top of sorts is applied and cured. Figure 3.26 shows the test board with 50-μm-thick stress senor chips assembled and encapsulated.

3.5.2 Experimental Setup and Procedure

The drop tests performed in this section are carried out according to the JEDEC standard JESD22-B111. The Avexx Shock Tester instrumented with an accelerometer is used. The height and pressure of this Avexx System are varied until the accelerometer attached onto

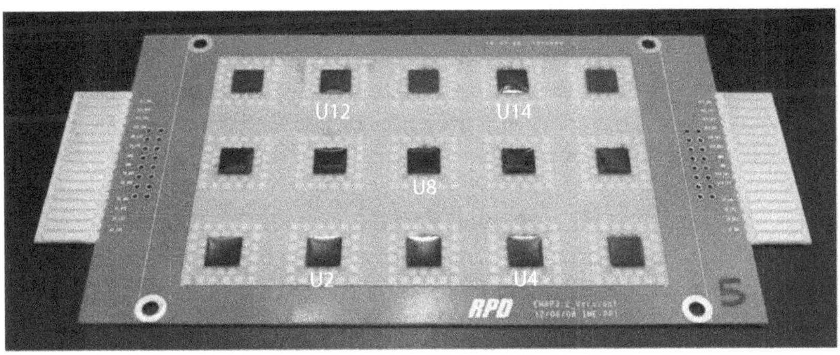

Figure 3.26 The assembled ultrathin (50 μm) stress sensor chip in the JEDEC-JESD22-B111 board.

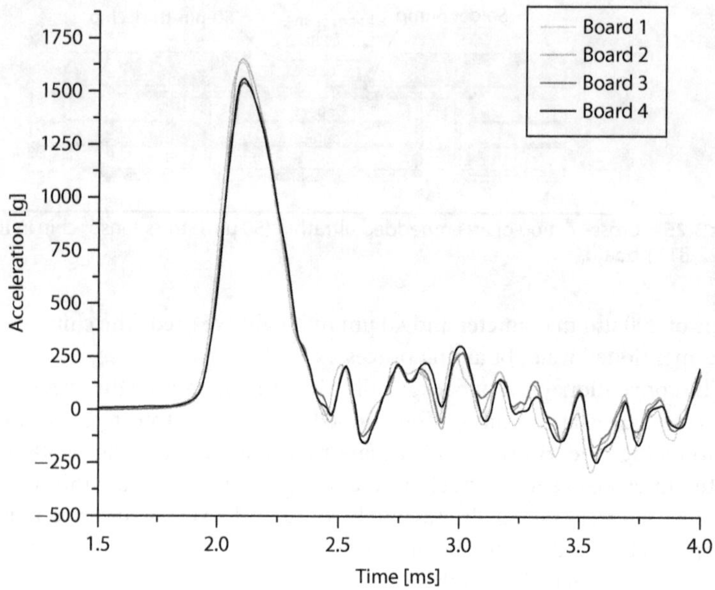

FIGURE 3.27 Acceleration pulse from the 30th drop of four different boards.

the drop table with beeswax detected a half sine shock pulse of amplitude 1500 g and pulse duration (width) 0.5 ms. This acceleration-time plot for the 30th drop of four different tested boards is shown in Fig. 3.27 as an indication of the repeatability of the tests.

The JEDEC test board used in these tests is shown in Fig. 3.26. The electrical connections and instrumentation used for stress measurement are described as follows: two perpendicular n-type sensors (R_1 and R_3) are monitored. These sensors are connected in series with a Keithley source meter (Model 2410). Two high-resolution 16-bit channels of the Yokogawa scope order (Model: DL750) are then connected in-parallel across each resistor. A current of 5 mA is sourced and the resulting potential difference across each resistor is measured. This is converted to resistance using Ohm's law. The change in resistance and initial resistance is then used to determine the stresses based on existing piezoresistive theory as given in Eqs. (3.3) and (3.4). The board is mounted facedown onto the tester. Subsequently, the wires to all the chips at five positions U2, U4, U8, U12, and U14, as indicated in Fig. 3.26, are soldered. The oscilloscope is then switched on and the test is carried out. Two perpendicular sensors (R_1 and R_3) from the chip at position U8 have the stresses in them monitored in real time with the data from every fifth drop being saved. The remaining sensors are probed after every five drops and the resistance values are recorded.

3.5.3 In-Situ Stress Measurement Results

The stress in the chip during the drop is successfully measured for four boards. It is assumed that the stress at the start of the test is zero. Hence the resistance at the start of the test is taken to be the initial resistance of the sensors. The stress in the x and y directions in Board 4 during the fifth drop are shown in Figs. 3.28 and 3.29. The stresses are initially high in magnitude and decrease gradually due to damping. As the flexing of the board is damped

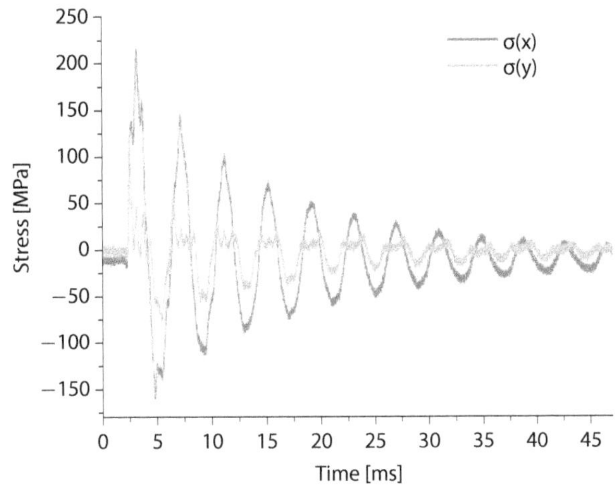

FIGURE 3.28 Stress measured during the fifth drop of Board #4.

out, the stress in the chip reduces as well. The stress in the y-direction, which is smaller in magnitude, seems to be complicated by an additional mode superimposed onto the signal. As this mode is gradually damped out as well, it is suspected that it is related to the flexing frequency.

Comparisons between the stress values obtained from the 5th drop and 30th drop are also shown in Fig. 3.30. In one of the boards tested there is an increase in the maximum stress value from the 5th to the 30th drop, while in rest there is a decrease or slight change. The maximum and minimum stress during the first flexing cycle in the 5th drop and 30th drop is shown in Fig. 3.30 and Table 3.1. The maximum stress experienced by the chip at U8 location is 305 MPa, while the minimum is −220 MPa.

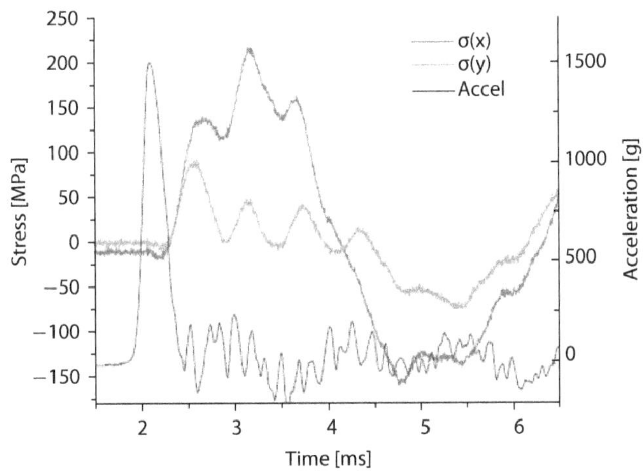

FIGURE 3.29 Stress in the first cycle of flexing of the fifth drop of Board #4.

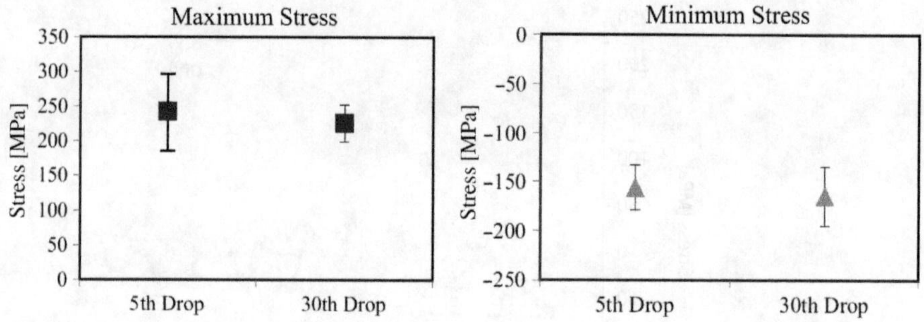

FIGURE 3.30 (a) Maximum stress and (b) minimum stress of the first flexing cycle in four boards after the fifth and 30th drop.

TABLE 3.1 Maximum and Minimum Stresses of the First Flexing Cycle in Four Boards After the 5th and 30th Drop

	5th drop		30th drop		5th drop		30th drop	
Sample Name	**Max (MPa)**	**Deviation**	**Max (MPa)**	**Deviation**	**Min (MPa)**	**Deviation**	**Min (MPa)**	**Deviation**
Board I	199	44	220	6	−199	43	−170	5
Board 2	251	8	191	35	−156	0	−220	55
Board 3	305	62	279	53	−110	46	−106	59
Board 4	217	26	214	12	−158	2	−162	3
Average	243	35	226	27	−156	23	−165	31

3.5.4 Reliability Testing

Reliability tests are carried out on four boards with 20 sensor chips. In other words, each board has five sensor chips at five positions U2, U4, U8, U12, and U14 (as indicated in Fig. 3.26). Each sensor chip has 16 chains. Each chain measured consists of the sensor resistor, traces on the chip, traces on the board, and two solder joints. The failure criterion is the event detector's (i.e., the measured chain) resistance change shall be no more than 20% increase in resistance from the initial resistance (JEDEC standard JESD22-B111). Figure 3.31 shows the typical resistance values at the beginning, after 5 drops and after 30 drops for the board measured. No such failure is observed. In short, the embedded stress sensor samples have passed drop impact test up to 30 drops (JESD22-B111).

3.5.5 Summary and Recommendations

Application of piezoresistive stress sensor in drop impact test of embedded ultrathin chip has been presented. Some important results and recommendations are summarized as follows:

- The real-time characterization of an embedded ultrathin (50 μm) flip chip with piezoresistive stress sensors indicated that the maximum stress experienced by

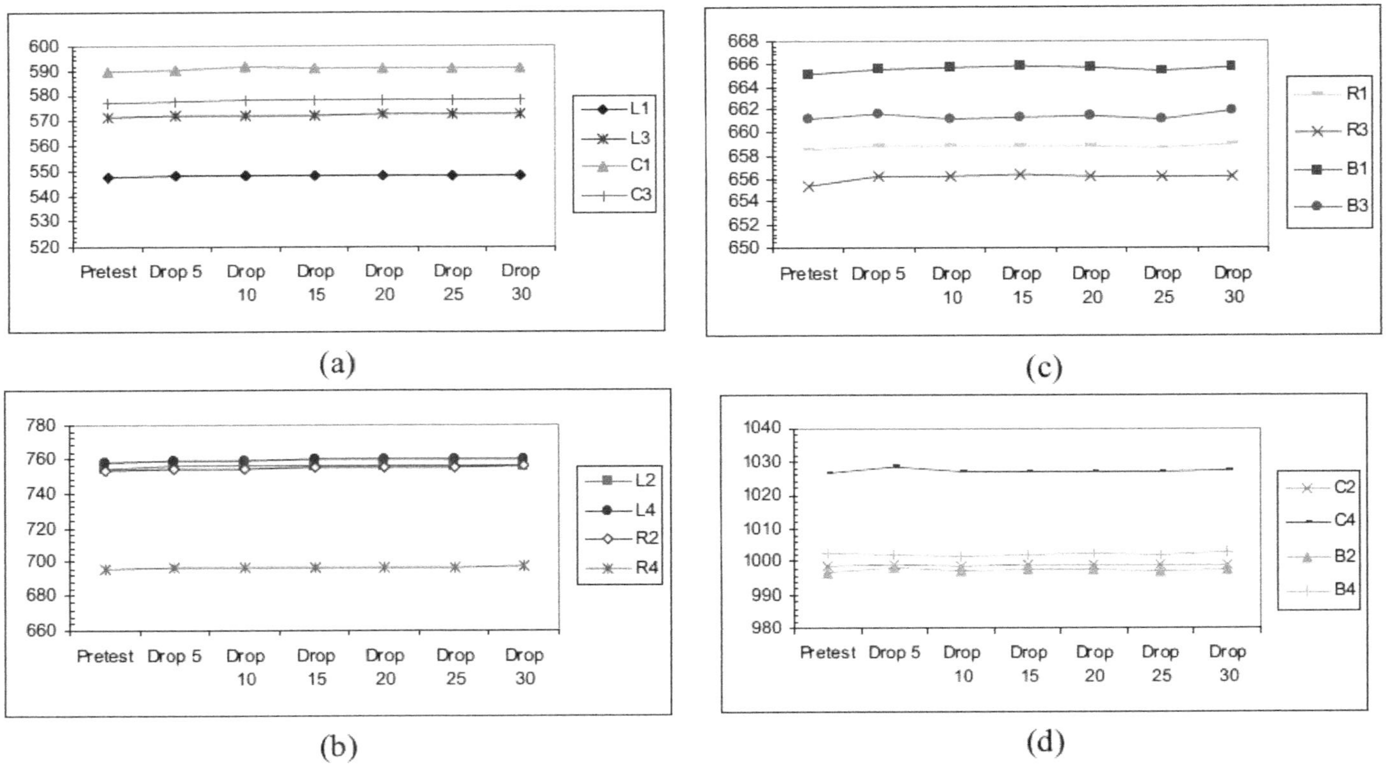

Figure 3.31 Typical resistance values (Ω) at the start, after five drops, and after 30 drops for the board measured.

the chip at U8 location is 305 MPa, while the minimum is −220 MPa during the JESD22-B111 drop test.

- All the embedded stress-sensor samples have passed the drop impact test up to 30 drops (JESD22-B111).
- The real-time characterization of devices with piezoresistive stress sensors during tests such as the thermal cycling test is strongly recommendation.

3.6 References

[1] Edwards, D. R., K. G. Heinen, S. K. Groothuis, and J. E. Martinez, "Shear Stress Evaluation of Plastic Packages," *IEEE Transactions on Components, Hybrid, and Manufacturing Technology*, Vol. 10, No. 4, December 1987, pp. 618–627.

[2] Miura, H., A. Nishimura, S. Kawai, and K. Nishi, "Development and Application of Stress-Sensing Test Chip for IC Plastic Packages," *Transactions of the Japan Society of Mechanical Engineers (Series A)*, Vol. 53, 1987, pp. 1826–1832.

[3] Miura, H., A. Nishimura, S. Kawai, and K. Nishi, "Residual Stress in Resin-Molded IC Chips," *Transactions of the Japan Society of Mechanical Engineers (Series A)*, Vol. 55, 1989, pp. 1763–1770.

[4] Miura, H., M. Kitano, A. Nishimura, and S. Kawai, "Thermal Stress Measurement in Silicon Chips Encapsulated in IC Plastic Packages under Thermal Cycling," *Journal of Electronic Packaging*, Vol. 115, No. 1, 1993, pp. 9–15.

[5] Bittle, D. A., J. C. Suhling, R. E. Beaty, R. C. Jaeger, and R. W. Johnson, "Piezoresistive Stress Sensors for Structural Analysis of Electronic Packages," *Journal of Electronic Packaging*, Vol. 113, No. 3, September 1991, pp. 203–215.

[6] Sweet, J. N., "Die Stress Measurement Using Piezoresistive Stress Sensor," in *Thermal Stress and Strain in Microelectronics Packaging*, J. H. Lau, Ed., Van Nostrand, New York, NY, 1993, pp. 221–268.

[7] Rahim, M. K., J. C. Suhling, D. S. Copeland, M. S. Islam, R. C. Jaeger, P. Lall, and R. W. Johnson, "Die Stress Characterization in Flip Chip on Laminate Assemblies," *IEEE Transactions on Components and Packaging Technologies*, Vol. 28, No. 3, September 2005, pp. 415–429.

[8] Lo, T. C. P., and P. C. H. Chan, "Design and Calibration of a 3-D Microstrain Gauge for In Situ on Chip Stress Measurements," in *Proceedings of the International Conference Semiconductors Electronics*, Penang, Malaysia, November 1996, pp. 252–255.

[9] Kumar, A., X. Zhang, Q. Zhang, M. Jong, G. Huang, V. Lee, V. Kripesh, et al., "Residual Stress Analysis in Thin Device Wafer Using Piezoresistive Stress Sensor," *IEEE Transactions on CPMT*, Vol. 1, No. 6, June 2011, pp. 841–851.

[10] Zhang, X., A. Kumar, Q. X. Zhang, Y. Y. Ong, S. W. Ho, C. H. Khong, V. Kripesh, et al., "Application of Piezoresistive Stress Sensors in Ultrathin Device Handling and Characterization," *Sensors and Actuators A: Physics*, Vol. 156, November 2009, pp. 2–7.

[11] Beaty, R. E., R. C. Jaeger, J. C. Suhling, R. W. Johnson, and R. D. Butler, "Evaluation of Piezoresistive Coefficient Variation in Silicon Stress Sensor Using a Four-Point Bending Test Fixture," *IEEE Transactions on Components, Hybrid, and Manufacturing Technology*, Vol. 15, No. 5, October 1992, pp. 904–914.

[12] Suhling, J. C., R. A. Cordes, Y. L. Kang, and R. C. Jaeger, "Wafer-Level Calibration of Stress Sensing Test Chips," in *Proceedings of the IEEE Electronic Components and Technology Conference*, June 1994, pp. 1058–1070.

[13] Suhling, J. C., and R. C. Jaeger, "Silicon Piezoresistive Stress Sensors and their Applications in Electronic Packaging," *IEEE Sensors Journal*, Vol. 1, No. 1, June 2001, pp. 14–30.

[14] Slattery, O., D. O. Mahoney, E. Sheehan, and F. Waldron, "Source of Variation in Piezoresistive Stress Sensor Measurements," *IEEE Transactions on Components, Packaging, and Manufacturing Technology*, Vol. 27, No. 1, March 2004, pp. 81–86.

[15] Zou, Y., J. C. Suhling, R. W. Johnson, R. C. Jaeger, and A. K. M. Mian, "In Situ Stress State Measurements during Chip-on-Board Assembly," *IEEE Transactions on Electronics Packaging Manufacturing*, Vol. 22, No. 1, January 1999, pp. 38–52.

[16] Lwo, B.-J., K.-F. Tseng, C.-H. Kao, T.-S. Chen, and J.-S. Su, "LFBGA Packaging Stress Measurement with Piezoresistive Sensors," in *Proceedings on the 4th International Symposium Electronic Materials Packaging*, December 2002, pp. 482–487.

[17] Rahim, M. K., J. Roberts, and J. C. Suhling, "Continuous In-Situ Die Stress Measurements during Thermal Cycling Accelerated Life Testing," in *Proceedings on the 57th Electronic Components Technology Conference*, Reno, NV, May 2007, pp. 1478–1489.

[18] Pustan, D., E. Rastiagaev, and J. Wilde, "In Situ Analysis Of The Stress Development during Fabrication Process of Micro-Assemblies," in *Proceedings on the 59th Electronic Components Technology Conference*, San Diego, CA, May 2009, pp. 117–124.

[19] Jaeger, R. C., J. C. Suhling, and R. Ramani, "Errors Associated with the Design, Calibration of Piezoresistive Stress Sensors in (100) Silicon," *IEEE Transactions on Components, Packaging, and Manufacturing Technology - Part B: Advanced Packaging*, Vol. 17(1), 1994, pp. 97–107.

[20] Palmer, D., D. Benson, D. Peterson, and J. Sweet, "IC Chip Stress During Plastic Package Molding", *Proceedings of IEEE/ECTC*, May 1998, pp. 1326–1331.

[21] Husstedt, H., U. Ausserlechner, and M. Kaltenbacher, "In-Situ Measurement of Curvature and Mechanical Stress of Packaged Silicon," in *Proceedings of the IEEE Sensors Conference*, Waikoloa, HI, November 2010, pp. 2563–2568.

[22] Husstedt, H., U. Ausserlechner, and M. Kaltenbacher, "In-Situ Analysis of Deformation and Mechanical Stress of Packaged Silicon Dies with an Array of Hall Plates," *IEEE Sensors J.*, Vol. 11, No. 11, November 2011, pp. 2993–3000.

[23] Lemke, B., R. Baskaran, S. Ganapathysubramanian, and O. Paul, "Stress Distribution under Electroless Nickel Bumps Extracted using Arrays of 7×7 Piezo-FETs," in *Proceedings of IEEE Sensors Conference*, Waikoloa, HI, November 2010, pp. 2573–2576.

[24] Lemke, B., R. Baskaran, S. Ganapathysubramanian, and O. Paul, "Experimental Determination of Stress Distributions under Electroless Nickel Bumps and Correlation to Numerical Models," *IEEE Sensors Journal*, Vol. 11, No. 11, November 2011, pp. 2711–2717.

[25] Niehoff, K., T. Schreier-Alt, F. Schindler-Saefkow, F. Ansorge, and H. Kittal, "Thermo-Mechanical Stress Analysis," in *Proceedings of the European Microelectronics and Packaging Conference and Exhibition*, Rimini, Italy, June 2009, pp. 1–5.

[26] Barlian, A., R. Narain, J. T. Li, C. E. Quance, A. C. Ho, V. Mukundan, and B. L. Pruitt, "Piezoresistive MEMS Underwater Shear Stress Sensors," in *Proc. Micro Electro Mech. Syst.*, Istanbul, Turkey, January 2006, pp. 626–629.

[27] Park, S.-J., J. C. Doll, N. Harjee, and B. L. Pruitt, "Optimization with Process Limits and Application Requirements for Force Sensors," in *Proceedings of IEEE Sensors Conference*, Waikoloa, HI, November 2010, pp. 1946–1949.

[28] Park, S.-J., J. C. Doll, and B. L. Pruitt, "Piezoresistive Cantilever Performance - Part I: Analytical Model for Sensitivity," *Journal of Microelectromechanical Systems*, Vol. 19, No. 1, February 2010, pp. 137–147.

[29] Park, S.-J., J. C. Doll, and B. L. Pruitt, "Piezoresistive Cantilever Performance - Part II: Optimization," *Journal of Microelectromechanical Systems*, Vol. 19, No. 1, February 2010, pp. 149–161.

[30] Gieschke, P., B. Sbierski, and O. Paul, "CMOS-Based Piezo-FET Stress Sensors in Wheatstone Bridge Configuration," in *Proceedings of IEEE Sensors Conference*, Limerick, Ireland, November 2011, pp. 93–96.

[31] Rowe, C. H., A. Donoso-Barrera, C. Renner, and S. Arscott, "Giant Room-Temperature Piezoresistive in a Metal-Silicon Hybrid Structure," *Physical Review Letters*, Vol. 100, No. 14, 2008, pp. 145501-1–145501-4.

[32] Gee, S. A., V. R. Akylas, and W. F. Bogert, "The Design and Calibration of a Semiconductor Strain Gauge Array," *Proceedings of IEEE International Conference Microelectronics Test Structures*, Vol. 1. February 1988, pp. 185–191.

[33] Zhong, Z. W., X. Zhang, B. H. Sim, E. H. Wong, P. S. Teo, and M. K. Iyer, "Calibration of a Piezoresistive Stress Sensor in [100] Silicon Test Chips," *Proceedings of the 4th Electronics Packaging and Technology Conference*, December 2002, pp. 323–326.

[34] Kumar, A., X. Zhang, Q. Zhang, M. Jong, G. Huang, V. Lee, V. Kripesh, C. Lee, J. H. Lau, D. Kwong, V. Sundaram, R. R. Tummula, and G. Meyer-Berg, "Residual Stress Analysis in Thin Device Wafer Using Piezoresistive Stress Sensor", *IEEE Transactions on CPMT*, Vol. 1, No. 6, June 2011, pp. 841–851.

[35] Zhang, X., R. Rajoo, C. S. Selvanayagam, A. Kumar, V. S. Rao, N. Khan, V. Kripesh, et al., "Application of Piezoresistive Stress Sensor in Wafer Bumping and Drop Impact Test of Embedded Ultra-Thin Device", *IEEE/ECTC Proceedings*, Orlando, FL, May 2011, pp. 1276–1282.

[36] Zhang, X., R. Rajoo, C. S. Selvanayagam, A. Kumar, V. Rao, N. Khan, V. Kripesh, et al., "Application of Piezoresistive Stress Sensor in Wafer Bumping and Drop Impact Test of Embedded Ultrathin Device", *IEEE Transactions on CPMT*, Vol. 2, No. 16, June 2012, pp. 935–943.

[37] Tzeng, P., J. H. Lau, M. Dai, S. Wu, H. Chien, Y. Chao, C. Chen, et al., "Design, Fabrication, and Calibration of Stress Sensors Embedded in a TSV Interposer in a 300mm Wafer", *IEEE/ECTC Proceedings*, San Diego, CA, May 2012, pp. 1731–1737.
[38] Lau, J. H., C.-J. Zhan, P.-J. Tzeng, C.-K. Lee, M.-J. Dai, H.-C. Chien, Y.-L. Chao, et al., "Feasibility Study of a 3D IC Integration System-in-Packaging (SiP) from a 300mm Multi-Project Wafer (MPW)", *IMAPS Transactions, Journal of Microelectronic Packaging*, Vol. 8, No. 4, Fourth Quarter 2011, pp. 171–178.
[39] Zhan, C., P. Tzeng, J. H. Lau, M. Dai, H. Chien, C. Lee, S. Wu, et al., "Assembly Process and Reliability Assessment of TSV/RDL/IPD Interposer with Multi-Chip-Stacking for 3D IC Integration SiP", *IEEE/ECTC Proceedings*, San Diego, CA, May 2012, pp. 548–554.
[40] JESD22-B111, *Board Level Drop Test Method of Components for Handheld Electronic Products*, JEDEC Standard, July 2003.

CHAPTER 4

Package Substrate Technologies

4.1 Introduction

For 3D IC integration, either it is a memory cube on logic, memory-on-SoC (system-on-chip), logic-on-logic, or interposer supporting chips, the package substrate is a must! Why? Because if the 3D IC integration module is mounted directly on a PCB (printed circuit board), due to the thermal expansion mismatch between the Si chips or Si interposer and the FR-4 epoxy PCB, the solder joints are subjected to very large stresses and strains and an underfill is usually needed for the solder joint reliability. Even with the reworkable underfills, however, the quality of the reworked PCB is degraded. Thus, the underfill is usually applied between the chips (or interposer) and the package substrate and not between the package substrate and the PCB. In this case, when the 3D IC integration module is bad, then reflow the solder joints between the package substrate and PCB and remove the whole 3D IC module and substrate. Also, the package substrate is helping to fan-out the circuitries from the chips or interposer, otherwise, very high-density and expensive PCBs are needed.

In this chapter, because of its much lower cost than the ceramic substrate, only the organic package substrate with build-up layers for flip chip 3D IC integration are considered. Also, because of the demand of low-profile for mobile and wearable products, the coreless package substrates are presented. Finally, the most recent advances in organic package substrates with build-up layers are provided.

4.2 Package Substrate with Build-up Layers for Flip Chip 3D IC Integration

4.2.1 Surface Laminate Circuit Technology

More than 20 years ago, IBM in Japan at Yasu invented the SLC (surface laminar circuit or surface laminate circuit) technology, Fig. 4.1 [1–3], which formed the basis of today's very popular low-cost organic package substrates with build-up layers vertically connected through microvias [4] to support solder bumped flip chips. There are two parts of the SLC technology: one is the core substrate and the other is the SLC for the signal wiring. The core substrate is made by the ordinary glass epoxy panel. However, the SLC layers are sequentially built up with the dielectric layers made of photo sensitive epoxy and the conductor plane of copper plating (semi-additive technique and the process flow is schematically shown in Fig. 4.2).

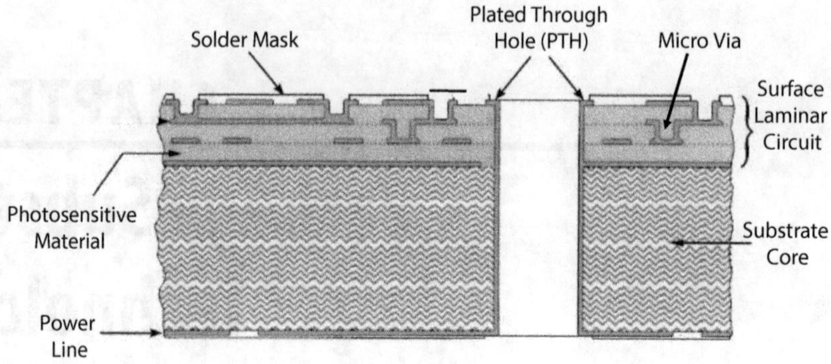

FIGURE 4.1 SLC technology.

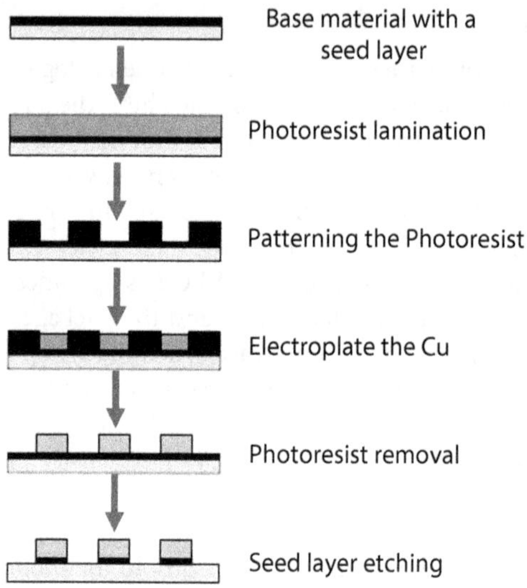

FIGURE 4.2 Conventional semi-additive process.

Figure 4.3 shows a build-up package substrate supporting solder bumped flip chips. It can be seen that the substrate consists of the core which is usually two layers (or four layers as shown in Fig. 4.4) with PTHs (plated through holes). The build-up layers can be on the top side as well as on the bottom side of the core of the substrate and are connected through microvias. The diameter of the microvias is no more than 150 μm. The build-up package substrate shown in Fig. 4.3 is called 4-2-4. The first 4 means there are four build-up layers on the top side of the core of the substrate. The last 4 means there are four build-up layers on the bottom side of the core of the substrate. The middle 2 means there are two layers of the core. On the other hand, the package substrate shown in Fig. 4.4 is called 3-4-3. This means there are three build-up layers on top, three build-up layers on the bottom, and four layers for the core.

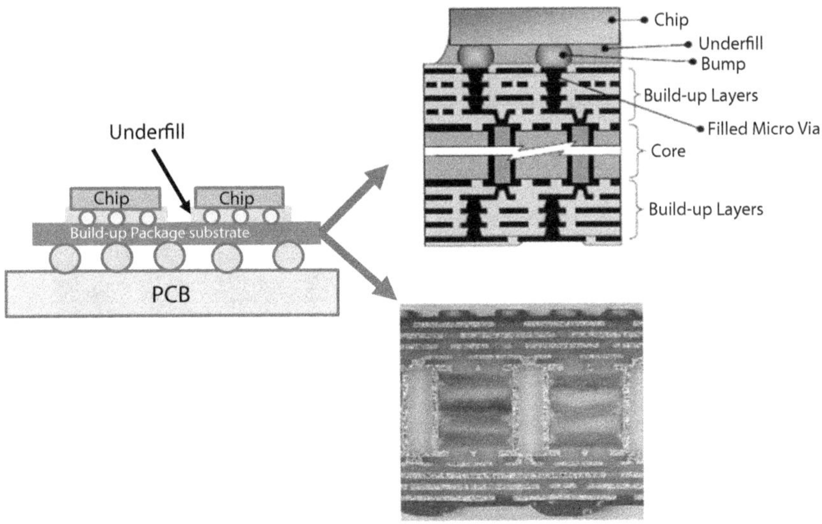

FIGURE 4.3 Package substrate with build-up layers (4-2-4) supporting solder bumped flip chips.

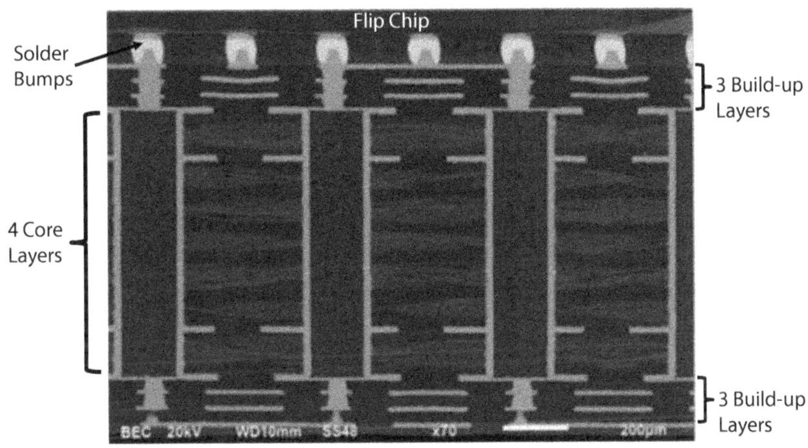

FIGURE 4.4 Tessera's package substrate with build-up layers (3-4-3) supporting solder bumped flip chip.

4.2.2 The Trend in Package Substrate with Build-up Layers

In general, a package substrate with eight build-up layer (4-2-4) and 15-μm-line width and spacing is more than adequate to support most of the chips. However, in the past few years, because of the very high-density, high I/Os, and fine-pitch requirements such as the sliced FPGA (field programmable gate array) and networking systems, much higher build-up layers and finer line-width and spacing are in demanded. For example, Fig. 4.5 shows [5] the 6-2-6 package substrate for Altera/TSMC's FPGAs. A road map for organic package substrate with build-up layers is shown in Table 4.1.

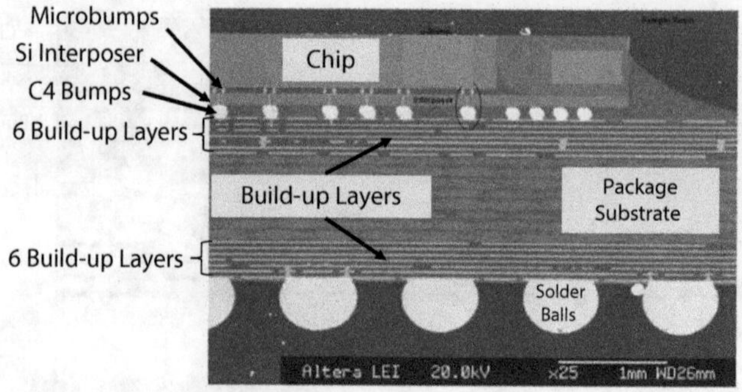

FIGURE 4.5 Altera's package substrate with build-up layers (6-2-6) supporting the Si-interposer and solder bumped flip chips.

TABLE 4.1 Road map of Package Substrate with Build-up Layers

	2014		2015	2016
	Production	**Sample**		
Build-up Structure	6/n/6	8/n/8	8/n/8	8/n/8
Minimum L/S (μm)	8/8	6/6	5/5	3/3
Minimum Pitch (μm), Area Array	150	125	100	100

4.2.3 Summary and Recommendations

Some important results and recommendations are summarized as follows:

- For 3D IC integration, package substrate is a must!
- In general, a package substrate with eight build-up layer (4-2-4) and 15-μm-line width and spacing is more than adequate to support most of the chips.
- A road map of package substrate with build-up layers has been provided.

4.3 Coreless Package Substrates

4.3.1 Advantages and Disadvantages of Coreless Package Substrates

Figure 4.6 shows the comparison between the conventional organic package substrate with build-up layers and the organic coreless package substrate. It can be seen that the biggest difference is that there is not a core of the coreless package substrate and all the layers of the coreless package substrate are the build-up layers [6–24].

The advantages and disadvantages of the coreless package substrate are shown in Table 4.2. It can be seen that (a) because of eliminating the core, the cost of coreless substrate is lower; (b) by eliminating the core, higher wiring ability can be achieved; (c) better electrical performance because of good high-speed transmission characteristic; and (d) definitely smaller form factor. On the other hand, the disadvantages are (a) because of eliminating the core, the warpage of the coreless substrate is larger; (b) easier to have laminate chipping; (c) poor solder joint yield because of less substrate rigidity; and (d) new manufacturing infrastructure is necessary. In 2010, Sony manufactured the first coreless package substrate for the cell processor of their PlayStation 3, as shown in Fig. 4.7 [6].

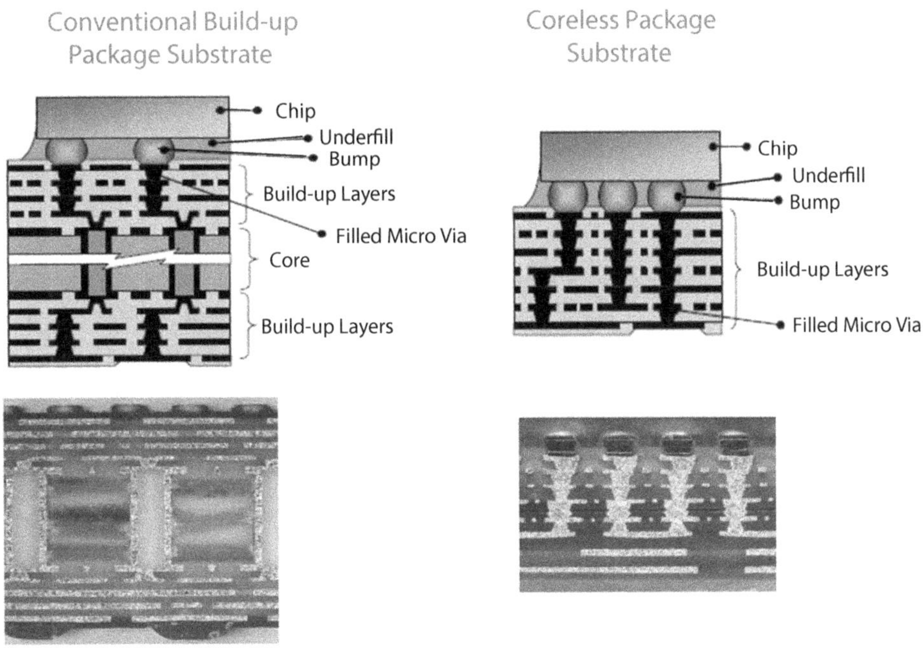

FIGURE 4.6 Comparison between the substrate with build-up layers and coreless substrate.

TABLE 4.2 Advantages and Disadvantages of Coreless Substrate

Advantages	Disadvantages
Lower cost by eliminating the core	Larger warpage (because of low rigidity)
Better electrical performance (good high-speed transmission characteristic)	New manufacturing infrastructure is necessary
Higher wiring ability (by eliminating the core)	Easier to have laminate chipping
Smaller form factor	Poor solder joint yield

4.3.2 Substitution of Si Interposer by Coreless Substrates

Shinko also started manufacturing the DLL3 coreless substrate, as shown in Fig. 4.8, in 2011 with their DLL (direct laser and lamination) process [7]. The special features of their coreless substrate are (a) CO_2 gas laser is used for blind micro via forming, (b) semi-additive process is applied for build-up layer, (c) enhanced electrical performance and high-design flexibility due to no core layer, (d) high-wiring density and high performance IC package, (e) available thinner structure, (f) applied green materials, (g) total package support including substrate design and manufacturing, (h) in devices such as multiprocessor units (MPUs) and application specific integrated circuits (ASICs) for where high-density routing is required, and (i) in devices where a thinner structure is required, as shown in Fig. 4.9.

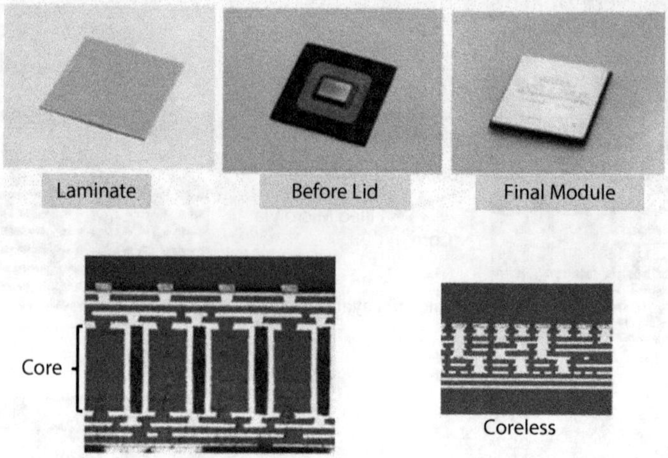

Figure 4.7 Sony's coreless package substrate for the cell processor of their PlayStation 3.

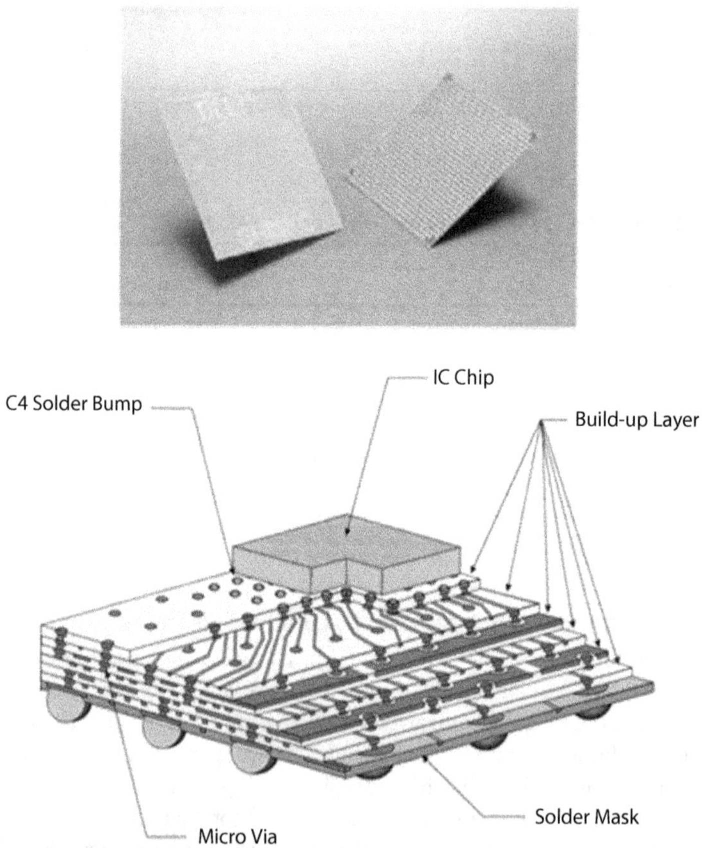

Figure 4.8 Shinko's DLL3 coreless substrate.

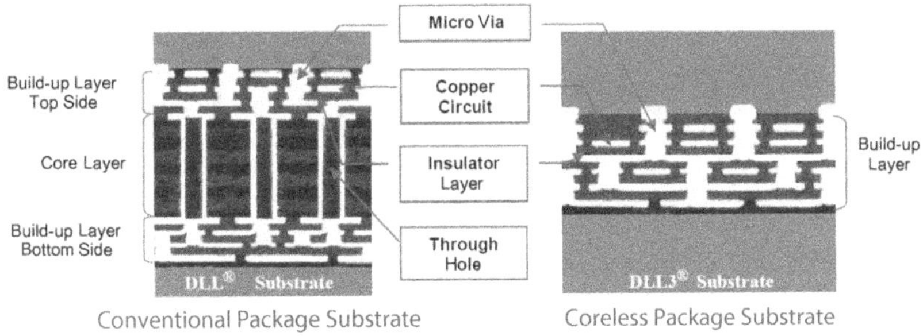

FIGURE 4.9 A comparison between the build-up substrate and coreless substrate.

In 2012, Shinko proposed to use the coreless package substrate to replace the Si interposer (which is supposed to be used to support the high bandwidth memory and 2.5D IC integration) as shown in Fig. 4.10. For sure, the cost in making the coreless substrate is much lower than that in making the TSV/RDL (through-silicon via/redistribution layers) interposer (which requires semiconductor equipment).

4.3.3 Warpage Problem and Solution of Coreless Substrates

Even though coreless substrates have many advantages, they are not popular because of the warpage control issue. One of the key factors affecting the warpage is the CTE (coefficient of

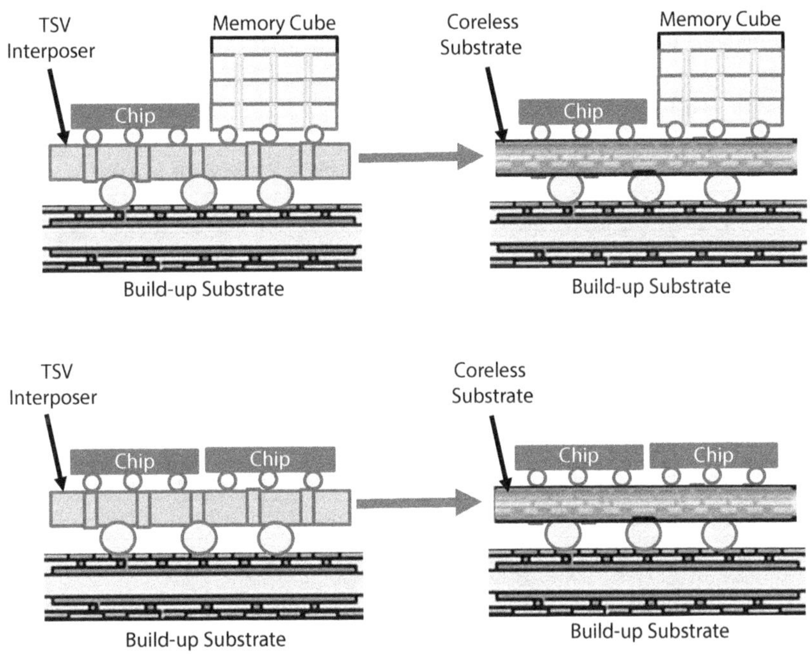

FIGURE 4.10 The Si interposer is replaced by the coreless substrate.

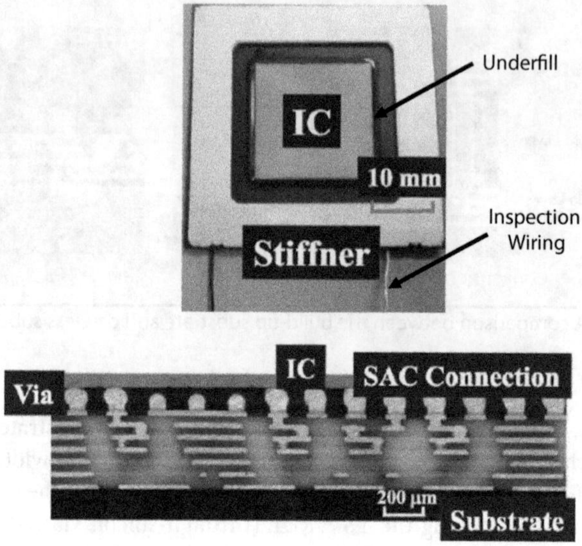

FIGURE 4.11 Fujitsu's test vehicle for minimum warpage of coreless substrate.

thermal expansion) mismatch of substrate materials. Thus, a proper control of this factor will help reduce the warpage issue of coreless substrates.

The CTE of Cu is 17.5×10^{-6}/°C and of resin is 35×10^{-6}/°C. By mixing with a prepreg with a CTE of 15×10^{-6}/°C will lead to less CTE mismatch. Figure 4.11 shows the test vehicle provided by Fujitsu with various combinations of resin and prepreg coreless substrates, as shown in Table 4.3 and Fig. 4.12. It can be seen that (a) for case 1, it is all prepreg; (b) for case 2,

TABLE 4.3 Fujitsu's Test Matrix of Coreless Substrates

Layer	LEGI All-PP	LEG2 Outer-PP	LEG3 Inner-PP	LEG4 None-PP
L1	Cu	Cu	Cu	Cu
V2	prepreg	prepreg	resin	resin
L2	Cu	Cu	Cu	Cu
V3	prepreg	prepreg	prepreg	resin
L3	Cu	Cu	Cu	Cu
V4	prepreg	resin	prepreg	resin
L4	Cu	Cu	Cu	Cu
V5	prepreg	resin	prepreg	resin
L5	Cu	Cu	Cu	Cu
V6	prepreg	prepreg	prepreg	resin
L6	Cu	Cu	Cu	Cu
V7	prepreg	prepreg	resin	resin
L7	Cu	Cu	Cu	Cu

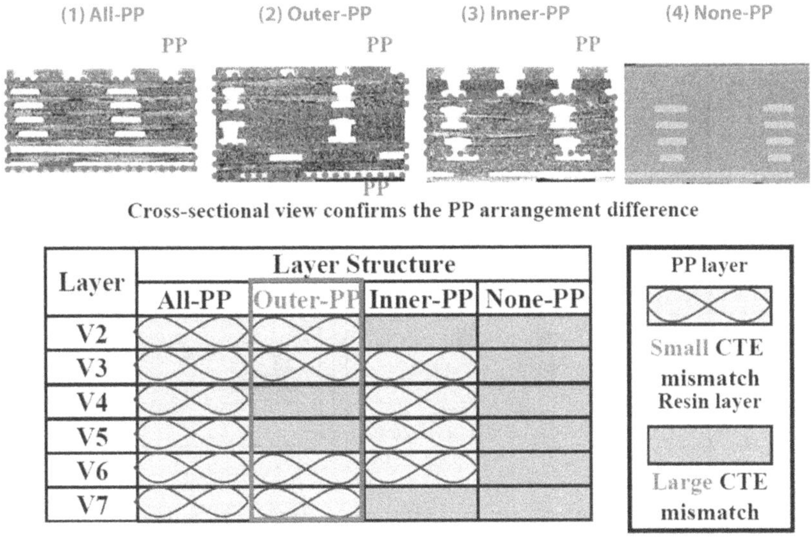

FIGURE 4.12 Fujitsu's combination and relocation of materials for testing the minimum warpage of coreless substrate.

the resin layers are at the middle and the prepreg layers are at the outers of the substrate; (c) for case 3, the prepreg layers are at the middle and the resin layers are at the outers of the substrate; and (d) for case 4, it is all resin. The measurement results [8, 9] are shown in Fig. 4.13. Because of the balance and position of the resin and prepreg, case 2 (the resin layers are at the middle and the prepreg layers are at the outers of the substrate) yields the best results (the least warpage).

Another factor affecting the warpage is the package assembly. Thus, a proper package assembly warpage correction control will help improve the warpage problem of coreless substrate. Figure 4.14 shows a package assembly process with a Cu stiffener for warpage

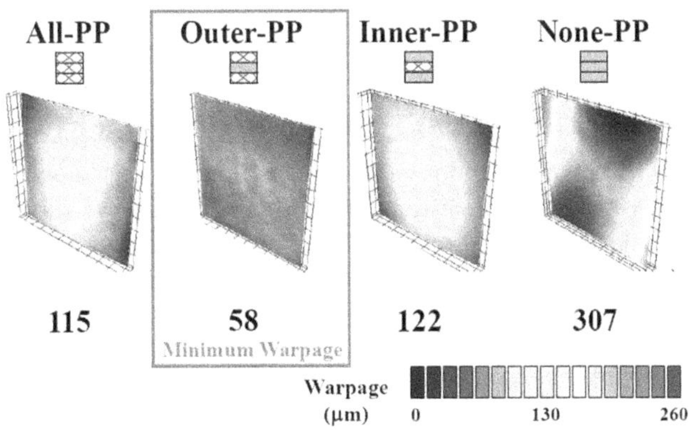

FIGURE 4.13 Warpage measurement results of various combination and relocation of materials of coreless substrate.

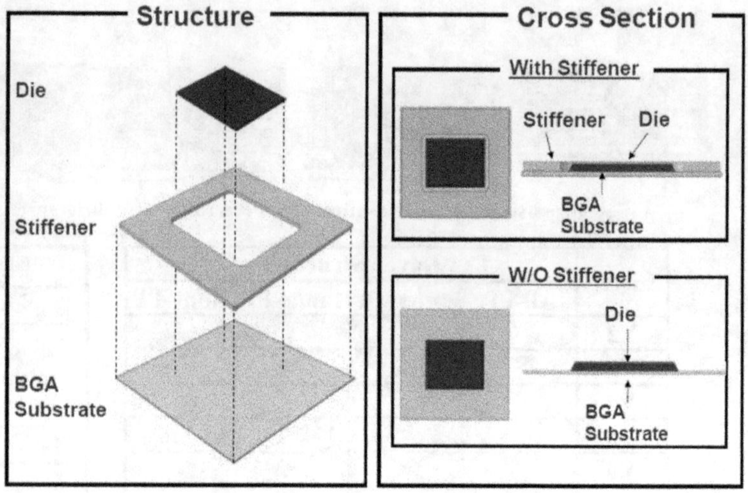

Figure 4.14 Warpage control during package assembly with a Cu stiffener.

control. One controlling parameter on stiffener assembly is the applied pressure, which must be large enough to prevent the deformation (warpage) due to the thermal expansion mismatch between the Cu stiffener and the coreless substrate.

4.3.4 Summary and Recommendations

Some important results and recommendations are summarized as follows:

- The advantages and disadvantages of coreless substrates have been clearly stated.
- For cost reason, the Si interposer can be replaced by the coreless substrate for very high-density, high I/Os, and find-pitch applications.
- Warpage control of coreless substrate by selecting and combining materials which lead to the minimum thermal expansion mismatch has been provided.
- Warpage control of coreless substrate by an optimal package assembly process with a Cu stiffener has been presented.

4.4 Recent Advance of Package Substrate with Build-up Layer

In the past few years, tremendous efforts have been devoted to enhance/advance the capabilities of the conventional low-cost build-up organic package substrates by increasing the number of build-up layers, shrinking the dimensions of the metal line width and spacing, and reducing the pad size and pitch. The most exciting and promising one is Shinko's thin-film RDLs on top of a build-up package substrate.

4.4.1 Thin-Film Layers on Top of Build-up Layer of Package Substrate

In October 2013, Shinko proposed to replace the coreless substrate by thin-film layers on top of the build-up layer of a package substrate. This is a further cost reduction from a TSV/RDL interposer to an organic coreless substrate. It can be seen from Fig. 4.15 that the solder

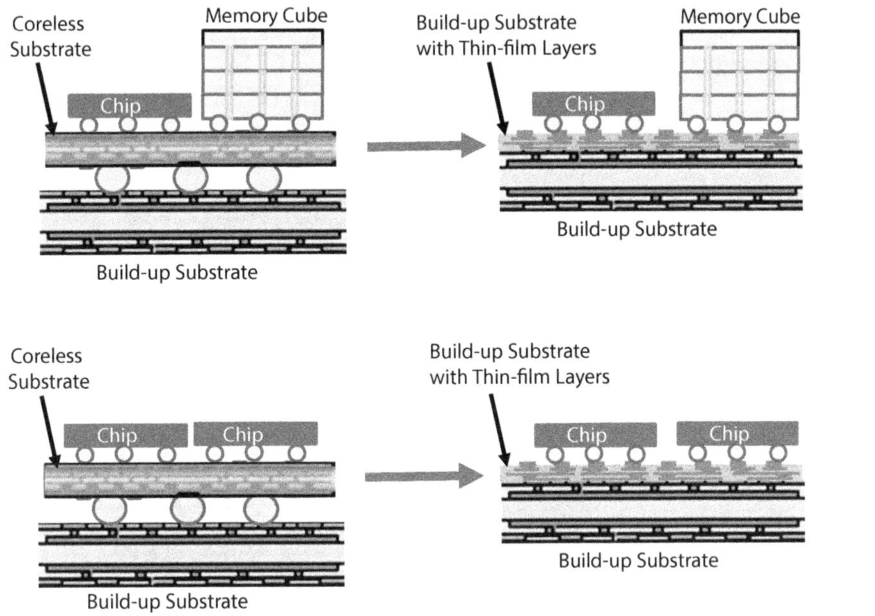

FIGURE 4.15 The coreless substrate is replaced by the thin-film layers on top of the build-up layer.

bumps between the coreless substrate and the build-up substrate are gone (a huge cost saving)! Also, the package substrate profile and weight are lower and lighter, respectively!

Figure 4.16 shows Shinko's i-THOP (integrated thin-film high density organic package) substrate [10] for high performance applications. It is a 4+(2-2-3) test vehicle (Fig. 4.16, top- and middle-left), which means there is a two-layer metal core, three build-up metal layers at the bottom (PCB) side, two build-up metal layers on the top (chip) side, and the first number "4" represents there are four thin-film Cu wiring layers (RDLs) on the surface of the top build-up layer (Fig. 4.16, top- and middle-left). The thickness, line width and spacing of the Cu RDLs can be as small as 2 μm (Fig. 4.16, bottom-left). The thin-film Cu RDLs are vertically connected through a 10 μm via, as shown in Fig. 4.16, middle-right. The surface Cu pad-pitch is 40 μm (Fig. 4.16, top-right) and the Cu pad diameter is 25 μm with a height of 10 to 12 μm (Fig. 4.16, bottom-right). Figure 4.17 shows the key thickness of the structure.

The assembly process of Shinko's substrate is followed and shown in Fig. 4.18. First, the conventional build-up layers are laminated on both sides of the 0.8-mm-thick core (with 100 μm PTH) of the package substrate and the Cu metal build-up layers are formed by the normal semi-additive techniques and vertically connected through a ~50 μm build-up microvia (Fig. 4.16, middle-right). After the back side is coated with a (25-μm-thick) solder mask layer for the ball grid array (BGA) solder balls, the top side surface is finished by CMP (chemical-mechanical polishing) to flatten the laser-drilled and Cu filled vias and to smooth the surface in preparation for applying the fine-wiring insulating resin layers.

A thin-film process is used to deposit the insulating resin layers. Then normal processes are applied to open small-diameter vias and a Ti/Cu seed layer is sputtered on the resin layer. A photoresist is then spun on and exposed by a stepper to make the 2-μm wiring trace (RDL)

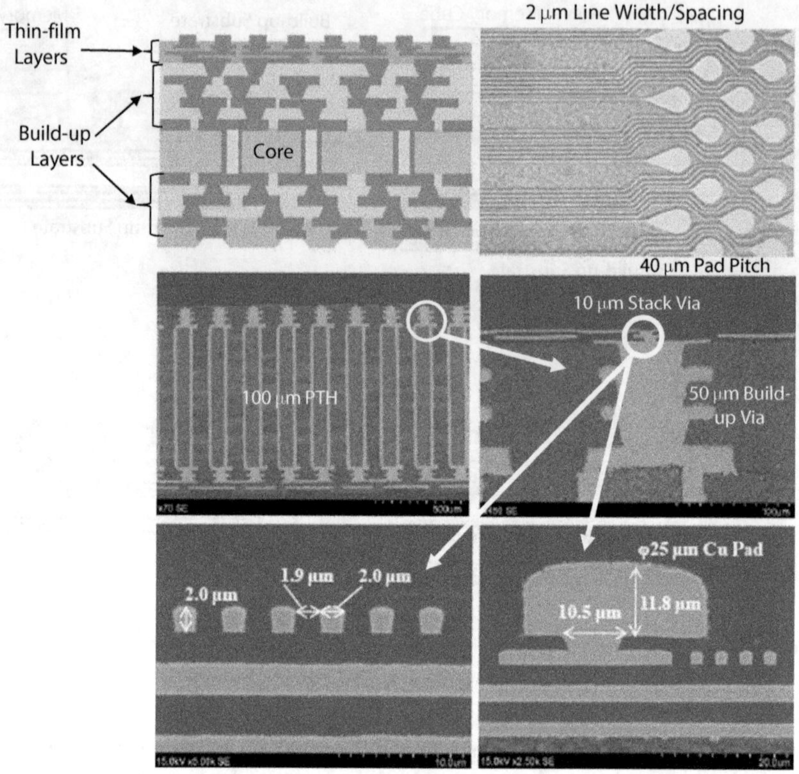

Figure 4.16 Shinko's thin-film RDLs on build-up package substrate test vehicle (TV). (Top-left) schematics of the 4+(2-2-3) TV. (Top-R) 2 μm line width and spacing RDLs and 40 μm pad pitch. (Middle-left) Cross section of the TV: 100 μm PTH at 0.8 mm thick. (Middle-right) Build-up layers are vertically connected by microvias (~50 μm) and thin-film RDLs are vertically connected by 10 μm stack via. (Bottom-left) 2 μm line width, spacing, and thickness RDLs. (Bottom-right) 25 μm diameter pad on 40 μm pitch and the pad thickness is 10 to 12 μm.

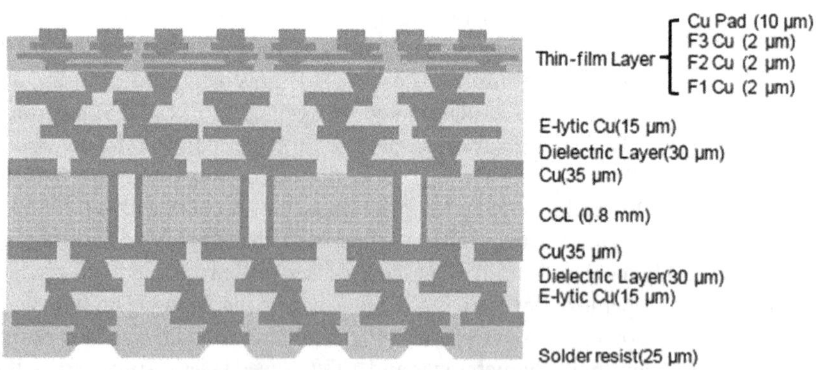

Figure 4.17 Shinko's cross-sectional structure.

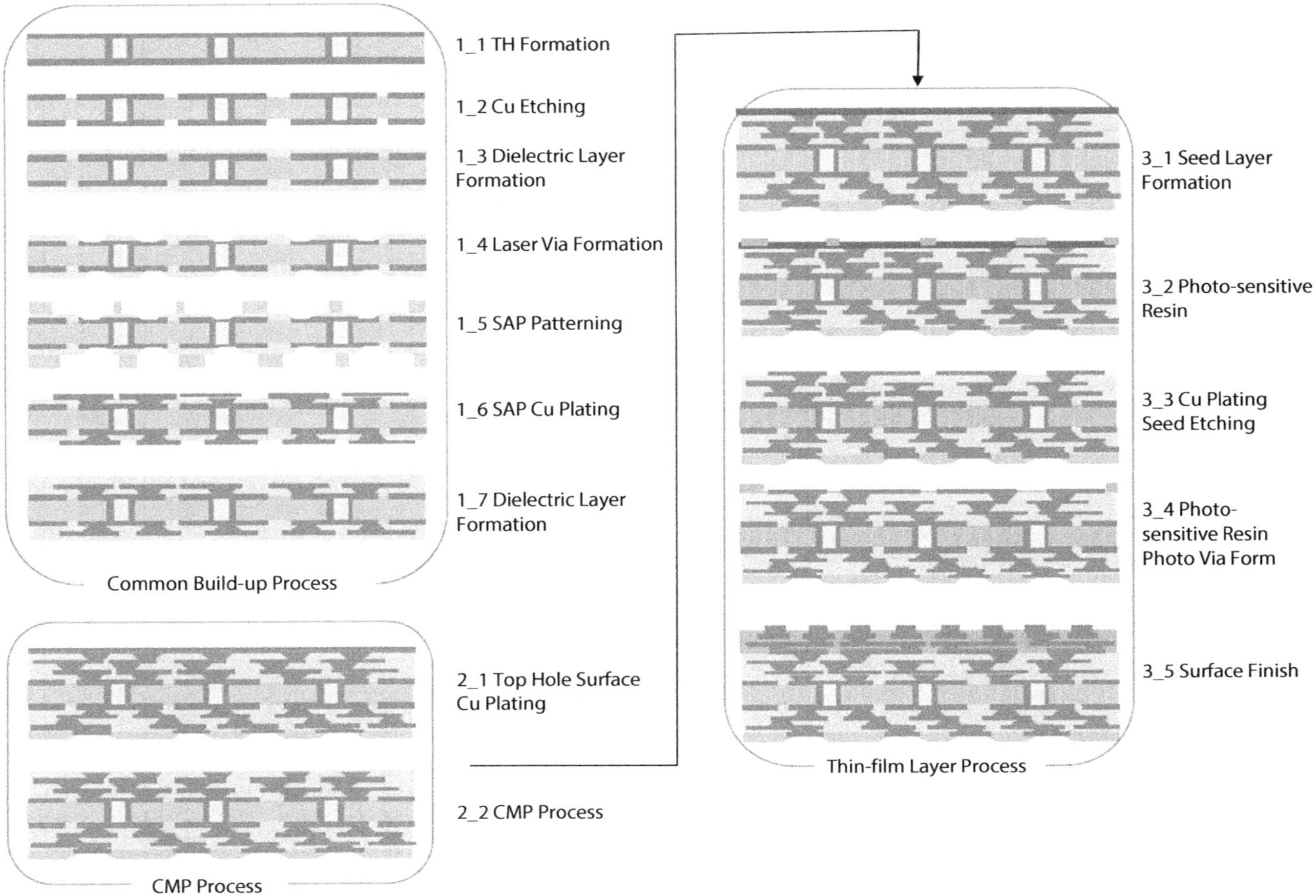

Figure 4.18 Manufacturing process flow.

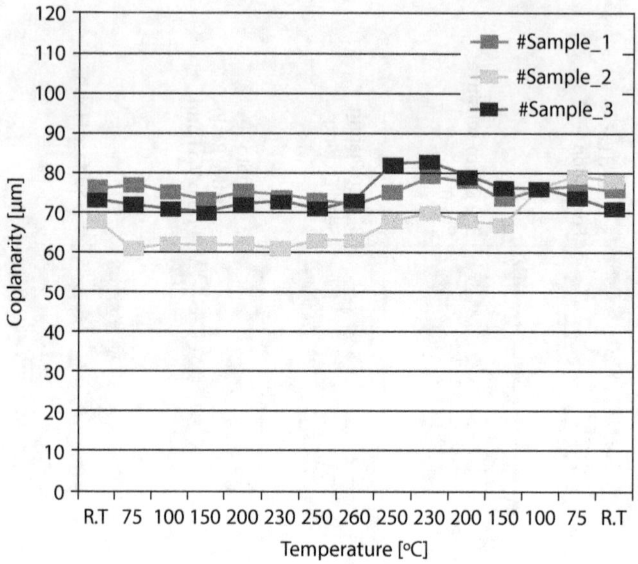

FIGURE 4.19 Warpage measured by the Shadow Moire.

pattern. The wiring thickness is formed by electrolytic Cu plating. Finally, the Cu pads on the top layer are treated by organic solderability preservative (OSP).

4.4.2 Warpage and Qualification Results

Figure 4.19 shows the warpage measurement results of the 4+(2-2-3) i-THOP substrate (40 × 40 mm) under a temperature variation from RT (room temperature) to 260°C and then back to the RT. It can be seen that (a) for all the temperatures the deformed shape (warpage) of the package substrate is convex; (b) when the substrate is heated to the lead-free temperature (260°C) the warpage increases to only 10 μm; and (c) overall, the warpage is stable.

Qualification tests (Table 4.4) have also been conducted on the i-THOP substrate structure. These tests are (a) precondition: MSL (moisture sensitivity level) 3A and reflow (to 260°C) three times; (b) thermal cycling (−55°C ↔ 125°C) for 1000 cycles; and (c) HAST (highly accelerated stress test): 130°C/85%RH/3.5V for 150 h. i-THOP substrate passed the tests and there was no via delamination observed.

TABLE 4.4 Qualification Test Conditions for the Thin-film Layers on Top of the Build-up Layer Package Substrate

Test	Conditions
Precondition	MSL3A[*1)] and reflow (260°C × 3 times)
TS	−55°C to −125°C 1000 Cycles
HAST	130°C/85%RH 150 hours. 3.5 V

[*1)]MSL3A: HTS (125°C/24 hours) + THS (60°C/60%RH/40 hours)

4.4.3 Summary and Recommendations

Some important results and recommendations are summarized as follows [10, 25, 26]:

- As a results of the data discussed above, substrate/packaging houses should follow Shinko's lead and commercialize the thin-film RDLs on top of the package substrate with CMP (to perform the planarization) and stepper (to form the 2 μm or less RDL pattern) technology.
- It should be noted that package substrate with build-up layers is a must. However, Si interposer is an addition (to increase cost) and slows down the electrical performance. Try not to use the Si interposer unless the build-up package substrates are not adequate to support the very high I/O, high-density, and fine-pitch chips.

4.5 References

[1] Tsukada, Y., S. Tsuchida, and Y. Mashimoto, "Surface Laminar Circuit Packaging," *Proceedings of IEEE/ECTC*, May 1992, pp. 22–27.

[2] Tsukada, Y., and S. Tsuchida, "Surface Laminar Circuit, A Low Cost High Density Printed Circuit Board," *Proceedings of Surface Mount International Conference*, Vol. 1, August 1992, pp. 537–542.

[3] Tsukada, Y., "Solder Bumped Flip Chip Attach on SLC Board and Multichip Module," in *Chip On Board Technologies for Multichip Modules*, edited by J. H. Lau, Van Nostrand Reinhold, New York, NY, 1994, pp. 410–443.

[4] Lau, J. H., and S. W. R Lee, *Microvias for Low Cost, High Density Interconnects*, McGraw-Hill Book Company, New York, NY, 2001.

[5] Xie, J., H. Shi, Y. Li, Z. Li, A. Rahman, K. Chandrasekar, D. Ratakonda, et al., "Enabling the 2.5D Integration," *Proceedings of IMAPS International Symposium on Microelectronics*, September 2012, San Diego, CA, pp. 254–267.

[6] Nishitani, Y., "Coreless Packaging Technology for High-performance Application," *IEEE/ECT/CMPT Seminar on Advanced Coreless Package Substrate and Material Technologies*, May 2012, pp. 1–23.

[7] Kimura, M., "Shinko Officially Announces Volume Production of Coreless Substrate," *Nikkei Electronics*, June 21, 2011, pp. 1–2.

[8] Kurashina, M., D. Mizutani, M. Koide, and N. Itoh, "Precision Improvement Study of Thermal Warpage Prediction Technology for LSI Packages," *Proceedings of IEEE/ECTC*, May 2009, pp. 529–534.

[9] Kurashina, M., D. Mizutani, M. Koide, M. Watanabe, K. Fukuzono, and H. Suzuki, "Low Warpage Coreless Substrate for Large-size LSI Packages," *Proceedings of IEEE/ECTC*, May 2012, pp. 1378–1383.

[10] Shimizu, N., W. Kaneda, H. Arisaka, N. Koizumi, S. Sunohara, A. Rokugawa, and T. Koyama, "Development of Organic Multi Chip Package for High Performance Application," *IMAPS Proceedings of International Symposium on Microelectronics*, October 2013, pp. 414–419.

[11] Manusharow, M., S. Muthukumar, E. Zheng, A. Sadiq, and C. Lee, "Coreless Substrate Technology Investigation for Ultra-Thin CPU BGA Packaging," *Proceedings of IEEE/ECTC*, May 2012, pp. 892–896.

[12] Koide, M., K. Fukuzono, H. Yoshimura, T. Sato, K. Abe, H. Fujisaki, "High-Performance Flip-Chip BGA Technology Based on Thin-Core and Coreless Package Substrate," *Proceedings of IEEE/ECTC*, May 2006, pp. 1869–1873.

[13] Kim, J., S. Lee, J. Lee, S. Jung, and C. Ryu, "Warpage Issues and Assembly Challenges Using Coreless Package Substrate," *Proceedings of IPC APEX EXPO*, Las Vegas, NV, 2012, pp. 1–13.

[14] Savic, J., P. Aria, J. Priest, N. Dugbartey, R. Pomerleau, B. Shanker, M. Nagar, et al., "Electrical Performance Assessment of Advanced Substrate Technologies for High Speed Networking Applications," *Proceedings of IEEE/ECTC*, May 2009, pp. 1193–1199.

[15] Kurashina, M., D. Mizutani, M. Koide, M. Watanabe, K. Fukuzono, N. Itoh, and H. Suzuki, "Low Warpage Coreless Substrate for IC Packages," *Transactions of the Japan Institute of Electronics Packaging*, Vol. 5, No. 1, 2012, pp. 55–62.

[16] Fujimoto, D., K. Yamada, N. Ogawa, H. Murai, H. Fukai, Y. Kaneko, and M. Kato, "New Fine Line Fabrication Technology on glass-cloth prepreg without insulation films for package substrate," *Proceedings of IEEE/ECTC*, May 2011, pp. 387–391.

[17] Sakuma, K., E. Blackshear, K. Tunga, C. Lian, S. Li, M. Interrante, O. Mantilla, et al., "Flip Chip Assembly Method Employing Differential Heating/Cooling for Large Dies with Coreless Substrates," *Proceedings of IEEE/ECTC*, Las Vegas, NV, May 2013, pp. 667–673.
[18] Wang, J., Y. Ding, L. Liao, P. Yang, Y. Lai, and A. Tseng, "Coreless Substrate for High Performance Flip Chip Packaging," *Proceedings of IEEE 11th International Conference on Electronic Packaging Technology & High Density Packaging*, August 2010, pp. 819–823.
[19] Kim, G., S. Lee, J. Yu, G. Jung, J. Kim, N. Karim, H. Yoo, et al., "Advanced Coreless Flip-chip BGA Package with High Dielectric Constant Thin Film Embedded Decoupling Capacitor," *Proceedings of IEEE/ECTC*, May 2011, pp. 595–600.
[20] Chang, D., Y. P. Wang, and C. S. Hsiao, "High Performance Coreless Flip-Chip BGA Packaging Technology," *Proceedings of IEEE/ECTC*, May 2007, pp. 1765–1768.
[21] Sung, R., K. Chiang, Y. Wang, and C. Hsiao, "Comparative Analysis of Electrical Performance on Coreless and Standard Flip-Chip Substrate," *Proceedings of IEEE/ECTC*, May 2007, pp. 1921–1924.
[22] Lin, E., D. Chang, D. Jiang, Y. Wang, and C. Hsiao, "Advantage and challenge of coreless Flip-Chip BGA," *Proceedings of IEEE/IMPACT*, October 2007, pp. 346–349.
[23] Sun, Y., X. He, Z. Yu, L. Wan, "Development of Ultra-thin Low Warpage Coreless Substrate," *Proceedings of IEEE/ECTC*, May 2013, pp. 1846–1849.
[24] Nickerson, R., R. Olmedo, R. Mortensen, C. Chee, S. Goyal, A. Low, and C. Gealer, "Application of Coreless Substrate to Package on Package Architectures," Proceedings on IEEE/ECTC, May 2012, pp. 1368–1371.
[25] Lau, J. H., "The Future of Interposer for Semiconductor IC Packaging," *Chip Scale Review*, Vol. 18, No. 1, January–February, 2014, pp. 32–36.
[26] Lau, J. H., "The Role and Future of 2.5D IC Integration," *IPC APEX EXPO Proceedings*, March 2014, pp. 1–9.

CHAPTER 5

Microbumps: Fabrication, Assembly, and Reliability

5.1 Introduction

As mentioned earlier, 3D IC integration is defined as stacking up whatever Moore's law IC thin chips in the third dimension with TSVs and solder microbumps [1–75] to achieve high performance, low power consumption, wide memory bandwidth, and small form factor. Thus solder microbumps are one of the important enabling technologies for 3D IC integration. The ordinary solder bumps ($\sim$100 μm) for flip chip applications [76, 77] are too big for 3D IC integration applications, which require much smaller bumps ($\leq$25 μm) and are called *microbumps*.

Most microbumps consist of a Cu pillar with/without a solder cap. Electroplating of the Cu pillar and the Sn or SnAg solder cap is a relatively simple process and has been widely used to form low-cost flip chip interconnections. In this chapter, fabrication of Sn (or SnAg) capped Cu pillar (with/without Ni barrier) microbumps and ENIG (electroless nickel/immersion gold) UBM (under bump metallurgy) pads will be presented. The TCB (thermal compression bonding) assembly of the microbumps will also be provided. Finally, some reliability data of the microsolder joints will be discussed. The 25-μm-pitch microbumps will be presented first, followed by the 20 μm pitch, and finally the 15 μm pitch.

5.2 Fabrication, Assembly, and Reliability of 25-μm-Pitch Microbumps

5.2.1 Test Vehicle

Figure 5.1 schematically shows a Si chip stacked on a Si carrier [1–4]. The solder microbumps, consisting of a Cu pillar and a Sn solder cap, are fabricated on the Si chip, and the ENIG UBM pads are fabricated on the Si carrier. The pitch for both the solder microbumps and the ENIG pads is 25 μm. TSVs (through-silicon vias) with the same pitch (25 μm) are fabricated on the Si carrier. After the bonding of the Si chip and the Si carrier, the whole stacked chips can then be attached onto a BT (bismaleimide triazine) substrate with ordinary solder bumps. The main focus of this study is to demonstrate the fabrication and TCB processes of the solder microbumps and the ENIG pads.

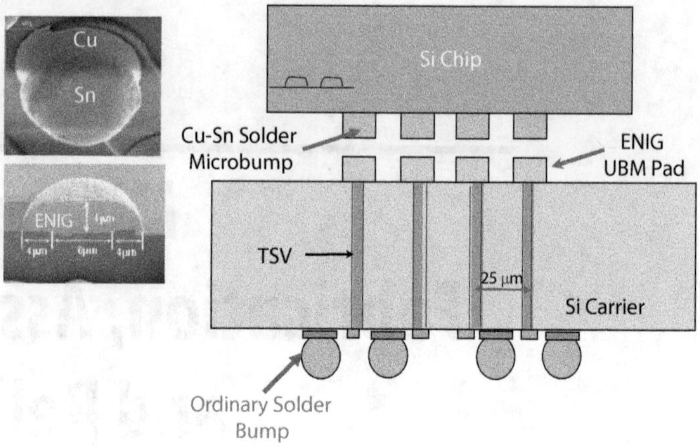

FIGURE 5.1 Schematic drawing of the 3D stacking of the Si chip on Si carrier with TSVs.

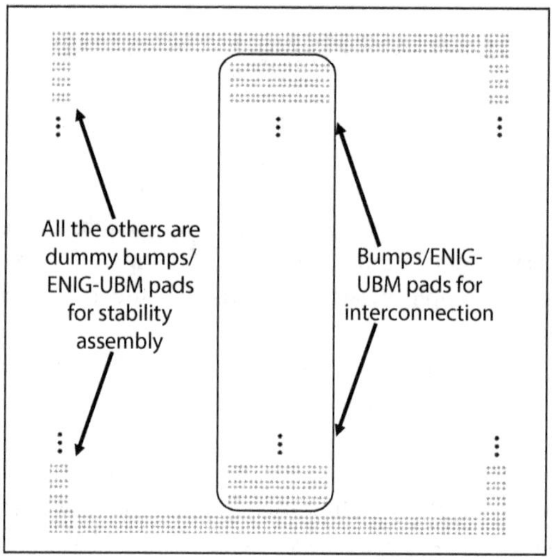

FIGURE 5.2 Distribution of solder microbumps and ENIG pads on the Si chip and Si carrier (not to scale).

Figure 5.2 shows the distribution of the microbumps/ENIG pads on the Si chip/Si carrier. The center columns of the microbumps/ENIG pads are used for signal interconnection, whereas the microbumps/ENIG pads at the periphery are used as dummy bumps/pads for mechanical balance during the bonding process as well as for enhancing the microsolder joints reliability. In total, there are more than 4000 microbumps/ENIG pads on the Si chip/Si carrier.

5.2.2 Structure of the Microbumps

Figure 5.3 shows the structure of the microbumps on the Si chip. It consists of a Cu pillar and a Sn solder cap with a total thickness of 10 μm. In order to achieve a reliable joining, the Sn

a. Bump Height
b. Bump Width
c. Ti Adhesion
d. Cu Seed Layer
e. Cu
f. Sn
g. Passivation:SiO_2

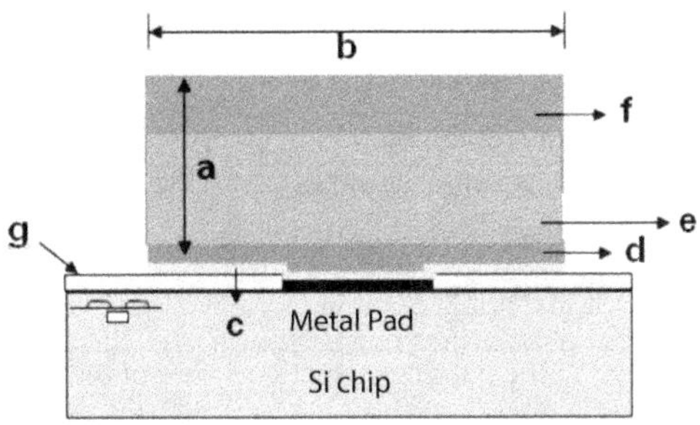

Figure 5.3 Schematic cross-sectional view of the Cu pillar with Sn cap on the top Si chip (not to scale).

cap needs to be thick enough so that after reflow, there is still pure Sn left for the following bonding process. It can be seen in the later results that after one-time reflow, the thickness of the IMCs (intermetallic compounds) is around 1.5 μm (as shown in Fig. 5.10 in Sec. 5.2.4). Therefore, the Sn cap has to be thicker than 1.5 μm.

However, during the TCB process, the thick Sn cap results in the bridging of the adjacent microjoints, especially when misalignment exists. Figure 5.4 shows a fracture surface on the ENIG pad after die shear. The plated Sn cap's thickness is 8 μm. It can be clearly seen that the adjacent microjoints are bridged to each other. To avoid the bridging defect, in this study, the thickness of the Sn cap layer is designed to be between 3 and 4 μm.

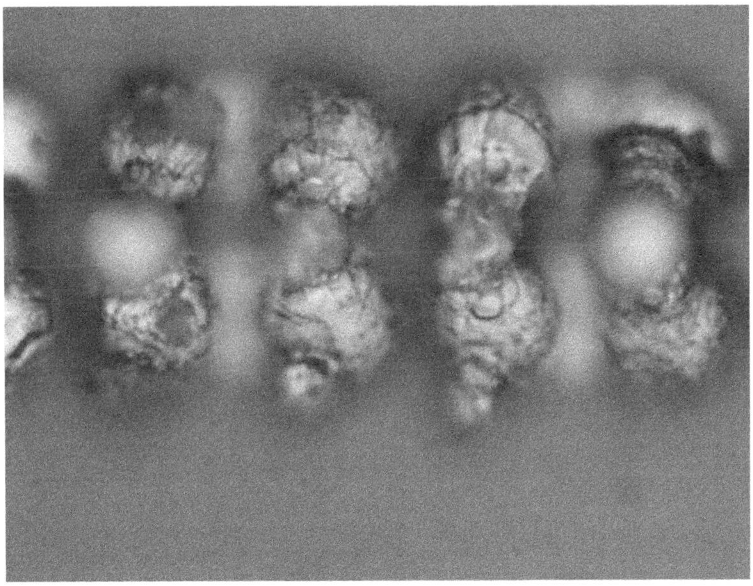

Figure 5.4 Fracture surface on the ENIG pad side after die shear test showing solder bridging between the adjacent microjoints. The thickness of the plated Sn solder layer is 6 to 8 μm.

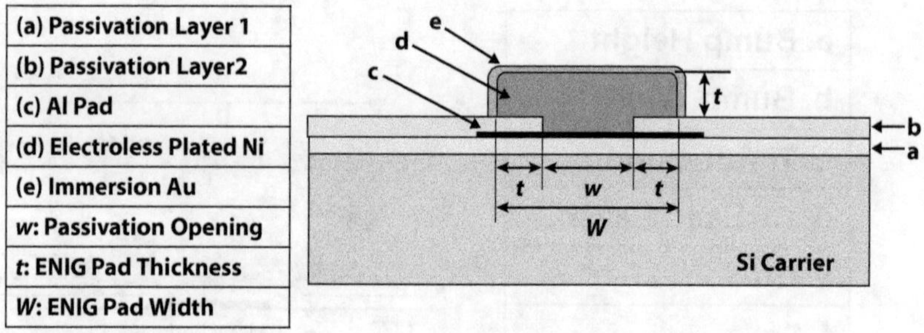

Figure 5.5 Schematic cross-sectional view of the ENIG pad on the Si carrier (not to scale).

5.2.3 Structure of the ENIG Pads

Figure 5.5 shows a schematic cross-sectional view of the ENIG pad. For very fine-pitch metal bonding pads, commonly used metal films, such as TiCuNiAu or AlNiVCu, are difficult to pattern with traditional photolithography and wet etching process. This is because the undercut caused by the wet etching process will damage the small metal pads and reduce the shear strength of the bumps. ENIG is a commonly used metallization layer for lead-free solder bonding. It has the advantage of not requiring any physical vapor deposition or photolithography process steps. Therefore, it is a simple and low-cost process for high-density and high-pin-count IC packages.

In order to determine the optimized thickness of the ENIG pad, thermomechanical simulation is conducted using ANSYS software to investigate the effects of the thickness of the ENIG pad on the stress in the underneath Al metal pad [34].

Figure 5.6 shows a global finite-element model for the assembly (half of the assembly is modeled). The stacking of the Si chip and the Si carrier is mounted onto an eight-layer BT substrate using eutectic solder bump (63Pb37Sn). All of the material properties are assumed to be constant except for those of the Cu filled in the TSVs. The copper is modeled as elastic–plastic material. The temperature change is assumed to be the same throughout the model and perfect adhesion is assumed on all material interfaces. The simulation is performed for different ENIG pad thicknesses in order to find out the stress effect on the Al pad under a temperature loading condition (−55°C→125°C).

Figure 5.7 presents the relationship between the maximum shear stress in the Al metal pad and the thickness of the ENIG pads. The maximum shear stress (τ_{xy}) decreases with the increase of the ENIG pads' thickness. This is because when the bump height increases, the

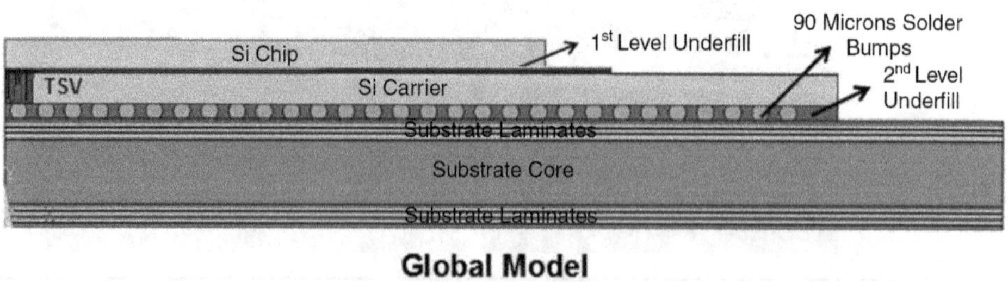

Figure 5.6 Finite-element model of the Si chip and the Si carrier mounted onto an organic substrate.

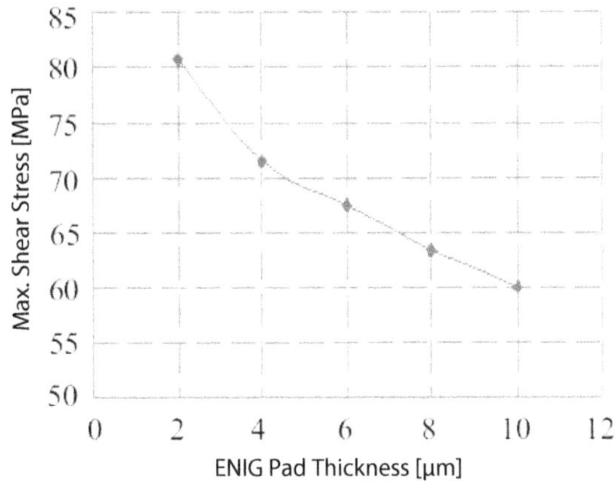

Figure 5.7 Maximum shear stress in the Al metal pad for different ENIG thicknesses.

bending module of the chip decreases, which results in smaller maximum shear stress in the Al pad. As the ENIG pads' thickness increases from 2 to 4 μm, τ_{xy} decreases 5 MPa/μm, whereas τ_{xy} decreases only 1.7 MPa/μm when the ENIG pads' thickness increases from 4 to 10 μm. Therefore, the ENIG pads should be thicker than 4 μm at least.

The thickness of the ENIG pad is also limited by the fabrication process. Considering the isotropic plating characteristics of the ENIG plating, the thicker the ENIG pads, the wider it becomes. Assuming that the opening of the Al pad is w as shown in Fig. 5.5, the relationship between the thickness of the ENIG pads t and the width of the ENIG pads W can be described as $W = w + 2t$ (as shown in Fig. 5.5). Therefore, if the thickness of the ENIG pads is selected as 4 μm, the width of the ENIG pads will be 16 μm (the opening of the Al pad w is 8 μm), and it will be 20 μm if the ENIG pads' thickness increases to 6 μm. As the pitch size is only 25 μm, thicker ENIG pads easily result in solder bridging of adjacent pads during the bonding process. After considering the above thermomechanical simulation results and the ENIG plating process, the targeted ENIG pad thickness is 4 μm.

5.2.4 Fabrication of the 25-μm-Pitch Microbumps

Figure 5.8 shows the process flow for fabrication of Cu pillars and Sn caps on the top die. First, a passivation layer of SiO_2 and a metallization layer of 1-μm-Al film are deposited on the wafer, Fig. 5.8(a). Then a layer of 2-μm-thick photoresist is spin coated and patterned, Fig. 5.8(b). The Al film is etched to form the metal pad and the photoresist is removed, Fig. 5.8(c). Another passivation layer of SiO_2 with 3000 Å thickness is deposited and patterned, Figs. 5.8(d)–5.8(f). Then Ti(500 Å)/Cu(1000 Å) of adhesion/seed layers are sputtered, Fig. 5.8(g). To form a plating mold, a 10-μm-thick layer of photoresist is coated and patterned, Fig. 5.8(h). After that, the Cu pillar and the Sn cap are electroplated in sequence on the Si chip, Fig. 5.8(i). The plating current density is 0.2 A/cm^2 for both the Cu and Sn plating. The plating is at room temperature, and the plating time is 5 min for Cu plating and 3 min for Sn plating. After plating, the photoresist is stripped and the Ti/Cu adhesion/seed layers are etched back by a wet etching process.

Figure 5.9 shows the SEM photos of the fabricated Cu pillar with Sn cap. After plating, a reflow is performed to reshape the microbumps. The reflow peak temperature is 265°C.

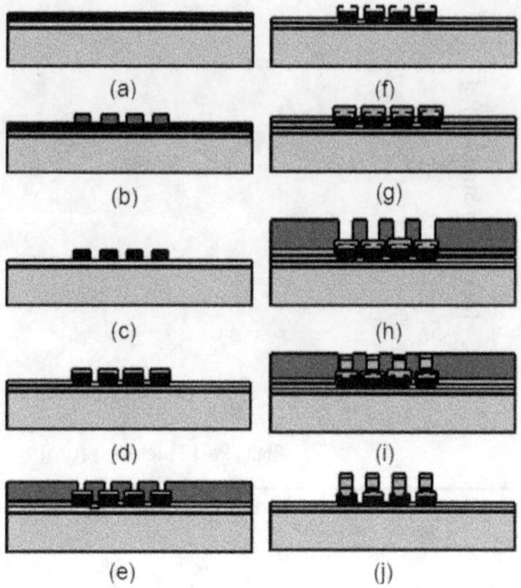

Figure 5.8 Process flow for CuSn solder microbump fabrication on the Si chip. (a) SiO_2 and Al deposition. (b) Photoresist coating and patterning. (c) Al etching to form the metal pad and photoresist stripping. (d) SiO_2 deposition. (e) Photoresist coating and patterning. (f) SiO_2 etch and photoresist stripping. (g) Ti/Cu of adhesion layer and seed layer deposition. (h) 10-μm-thick photoresist coating and patterning as plating mold. (i) Cu/Sn plating sequentially. (j) Photoresist stripping and the Ti/Cu adhesion/seed layers etch back.

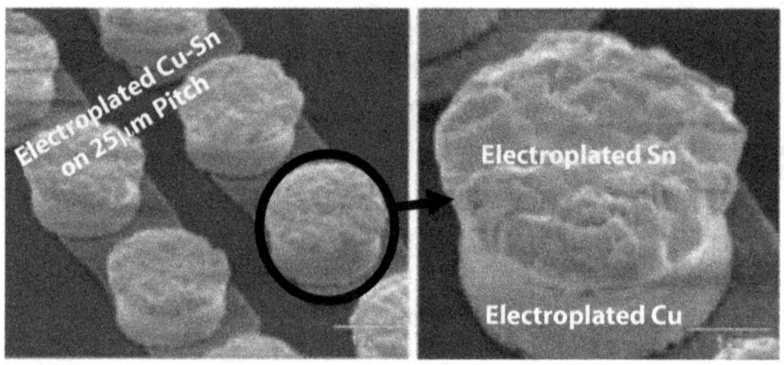

Figure 5.9 SEM images of the Cu pillar with Sn solder cap right after electroplating on the Si chip.

Figure 5.10 presents the SEM photos of the microbumps after reflow and Fig. 5.11 shows the FIB (focus ion beam) image of the microbumps. Energy disperse X-ray results indicate that the formed IMCs include Cu_6Sn_5 and Cu_3Sn after one-time reflow. During reflow, a water cleanable flux is used, and after reflow, the wafer is cleaned in deionized water to remove the flux residues. Figure 5.11 also indicates that after reflow, the average thickness of the IMCs is around 1.5 μm. However, the remaining Sn is still sufficient for bonding the Si chip and the Si carrier. From the reliability results (discussed in Sec. 5.2.8), it can be seen that most of the solder microjoints pass the reliability tests.

5.2.5 Fabrication of ENIG Bonding Pads on Si Carrier

Figure 5.12 shows the process flow for fabrication of ENIG UBM pads. First, a layer of dielectric and Al thin film with 1 μm thick are deposited on top of the wafer, Fig. 5.12(a). Then a thin layer of photoresist with 2 μm thick is spin coated and patterned, Fig. 5.12(b). After that,

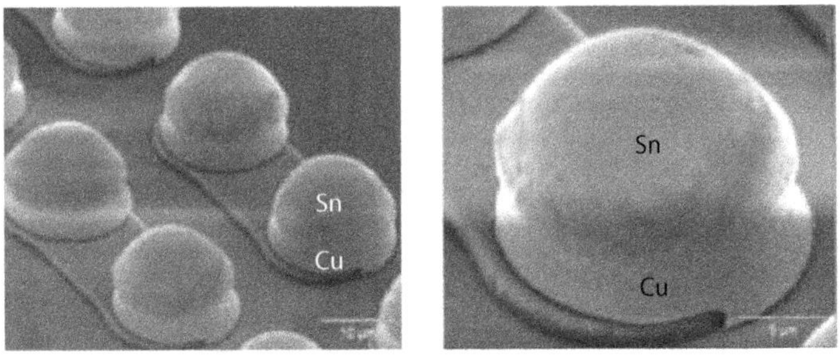

Figure 5.10 SEM images of the Cu pillar with Sn solder cap right after reflowing on the Si chip.

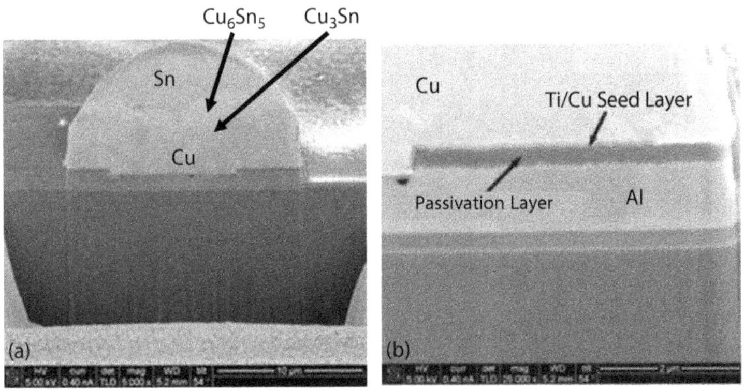

Figure 5.11 (a) FIB/SEM images of the Cu pillar with Sn solder cap on the Si chip. (b) Zoom-in view of the IMC, seed layer, and passivation.

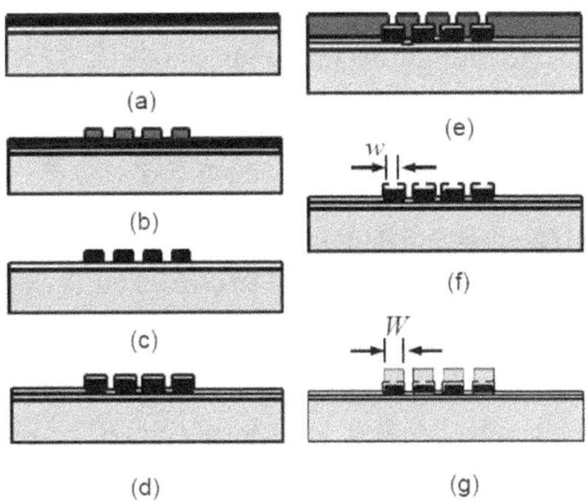

Figure 5.12 Process flow for ENIG plating on the Si carrier.

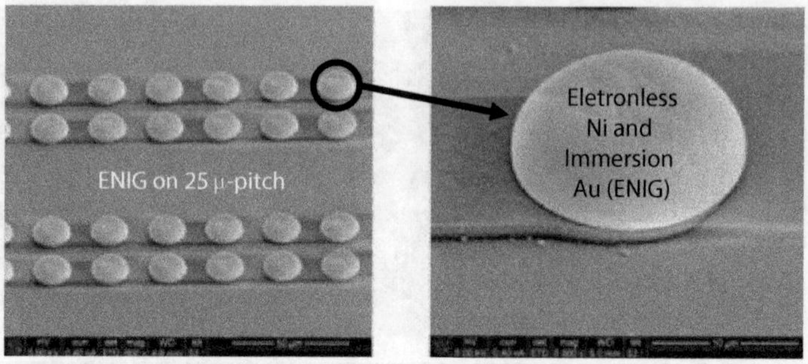

Figure 5.13 SEM images of the ENIG pads of Ni-Au on the Si carrier.

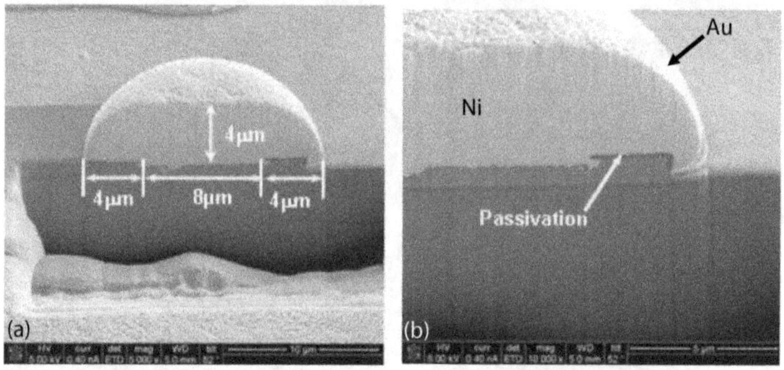

Figure 5.14 (a) FIB/SEM images of the ENIG pads on the Si carrier. (b) Zoom-in view of the Ni, Au, and passivation.

the Al layer is etched and the photoresist is stripped, Fig. 5.12(c). After that, another layer of (3000 Å) SiO_2 dielectric layer is deposited, Fig. 5.12(d). Then another layer of photoresist is spin coated and patterned, Fig. 5.12(e). The dielectric layer is etched to expose the Al pad, Fig. 5.12(f). The opening width is 8 μm. Finally, an electroless Ni-P plating with 7% of P content and immersion Au plating are preceded sequentially. The thickness of the immersion Au is 0.1 μm and the purpose of this layer is to protect the Ni layer from oxidation and improve solder wetting during the bonding process. The temperature for electroless Ni plating is 90°C ± 5°C, Fig. 5.12(g). Figure 5.13 shows the SEM images of the ENIG pads and Fig. 5.14 shows the FIB images of the ENIG pads. The thickness of the ENIG pad is 4 μm, whereas the diameter of the ENIG pad is 16 μm.

5.2.6 Thermal Compression Bonding Assembly

After the fabrication of the Cu pillars with Sn caps on the Si chip and the ENIG UBM bonding pads on the Si carrier, TCB of the Si chip and the Si carrier is conducted on the FC150 flip chip bonder. The bonding process is a fluxless process, because after joining, the gap between the Si chip and the Si carrier is only around 15 μm. If a flux is used, it is difficult to clean the flux residue after bonding, and a lot of voids will exist inside the bond line during the following underfilling process and may cause long-term reliability problems.

Different bonding conditions (bonding temperature, time, and pressure) are evaluated to achieve the best joining result. Quick assessments by checking the shear strength and

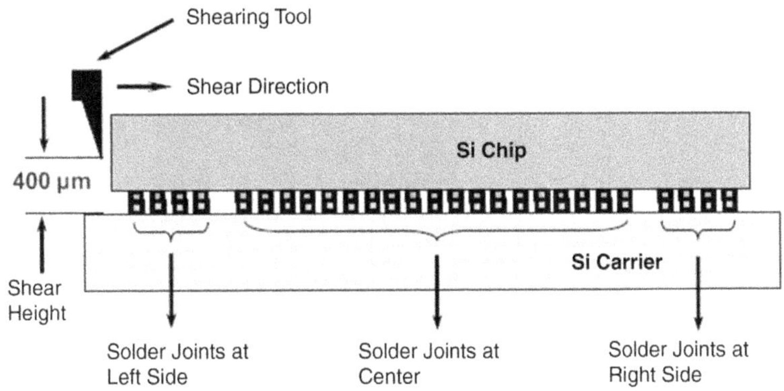

FIGURE 5.15 Schematic drawing of the shear test after assembly.

the fractural surface are conducted to select proper bonding conditions. To evaluate the shear strength of the microjoints, die shear test of the bonded samples is performed on a commercially available shear tester (DAGE-SERIES-4000-T, Dage Precision Industries Ltd., Aylesbury, UK). The force applied to the samples provides shear in the direction parallel to the surface of the test vehicle by the shear probe. The shear speed for this evaluation is 100 μm/s and the height of the shear probe is 400 μm from the Si-carrier surface (the thickness of the Si chip is 725 μm), as shown in Fig. 5.15.

The effects of bonding temperature and bonding time on the shear strength are evaluated. Figures 5.16(a) and 5.16(b) show that the die shear strength changes with various bonding temperature and bonding time. The results show that the die shear strength increases with bonding temperature and bonding time. The reason is that the IMC thickness increases with higher bonding temperature and longer bonding time, which results in higher die shear strength. Figure 5.17 shows the fracture surfaces of the Si chip and the Si carrier that joined under the conditions of 2.45-mN/bump bonding pressure, 250°C bonding temperature, and 20 s bonding time. All fractural surfaces are along the joining interface between the Sn cap

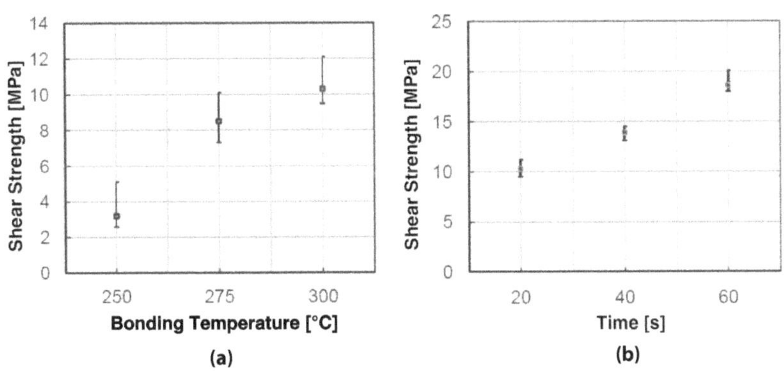

FIGURE 5.16 Effects of bonding temperature and bonding time on the die shear strength with 2.45-mN/bump bonding pressure. (a) Effect of bonding temperature (20 s bonding time). (b) Effect of bonding time (300°C bonding temperature).

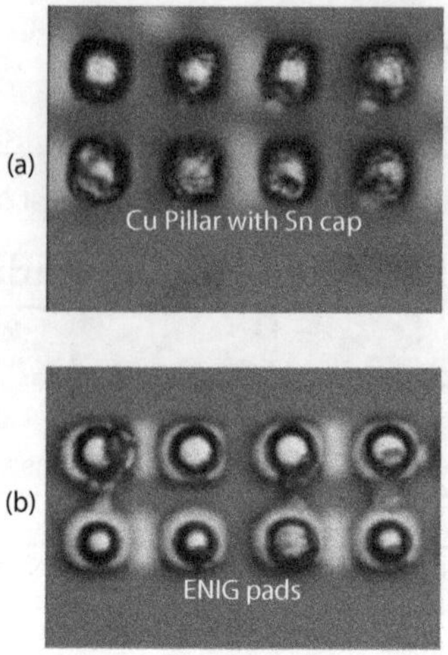

FIGURE 5.17 Fracture surface after die shear tests with bonding condition of 2.45-mN/bump bonding pressure, 250°C bonding temperature, and 20 s bonding time. (a) Cu pillar with Sn cap. (b) ENIG pads.

and the ENIG pad, which indicates a weak joining. The average shear strength is 3.1 MPa, as shown in Fig. 5.16(a).

Figure 5.18 shows the cross-section of the microjoints that joined under the conditions of 2.45-mN/bump bonding pressure, 300°C bonding temperature, and 60 s bonding time. Figure 5.18(a) shows a row of 19 joined solder joints and Fig. 5.18(b) shows the cross-section

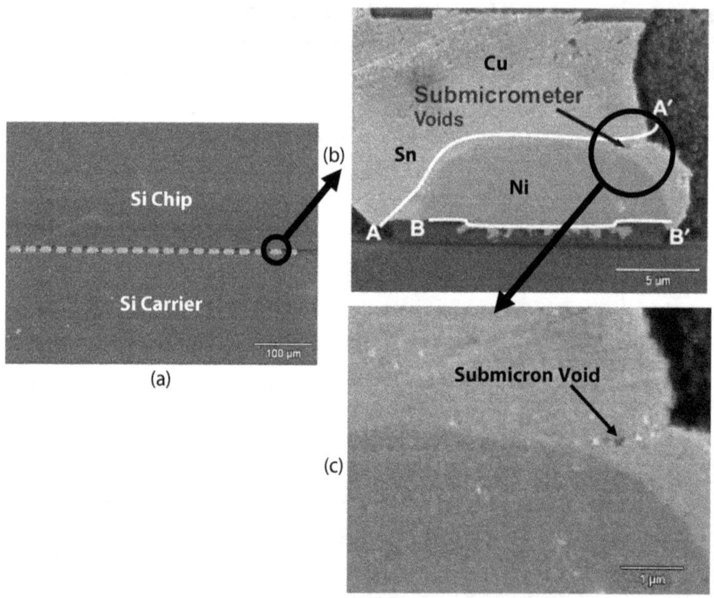

FIGURE 5.18 Cross-section of the microjoints with bonding condition of 2.45-mN/bump bonding pressure, 300°C bonding temperature, and 60 s bonding time. (a) One row of microjoints. (b) Overview of a single microjoint. (c) Zoomed in view of the submicrometer void at the joining interface.

of the second solder joint. It can be seen that successful joining is achieved with few submicrometer voids, observed in some portions of the bonding interface.

A misalignment of around 1 μm is observed between the Si chip and the Si carrier after bonding, as shown in Fig. 5.18(b). In order to avoid misalignment that results from the litho patterning offset on either the Si chip or the Si carrier, alignment during the assembly process is done by aligning the microbumps to the respective ENIG pads directly. Therefore, the misalignment shown in Fig. 5.18(b) has mainly resulted from the bonding process and the limitation (±1 μm post-bond accuracy) of the FC150 flip chip bonder. This misalignment can be reduced by conducting the calibration of the FC150 flip chip bonder before bonding. Figure 5.19 shows the X-ray photos after bonding without/with equipment calibration. The photos indicate that after the calibration of the FC150 flip chip bonder, the misalignment can be reduced; from the photo, it can also be seen systematically that the misalignment/alignment is constant with respect to different bump locations. In order to compensate for the misalignment, the size of the solder microbumps could be designed smaller than that of the ENIG pads.

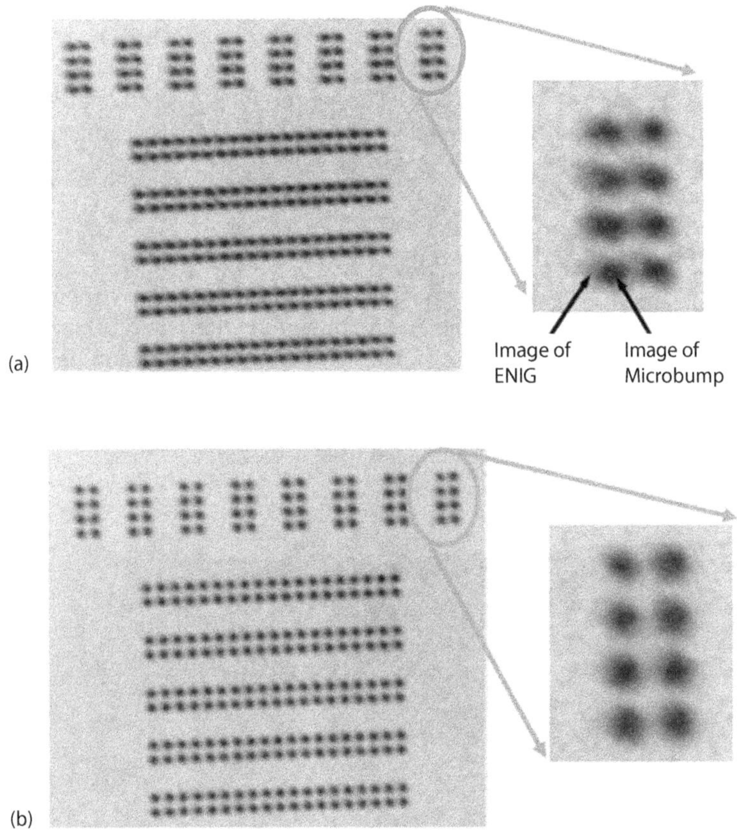

Figure 5.19 X-ray images showing misalignment/alignment with respect to bump location. (Images taken from the Si chip side.) (a) Bonding without the calibration of the FC150. (b) Bonding with the calibration of the FC150.

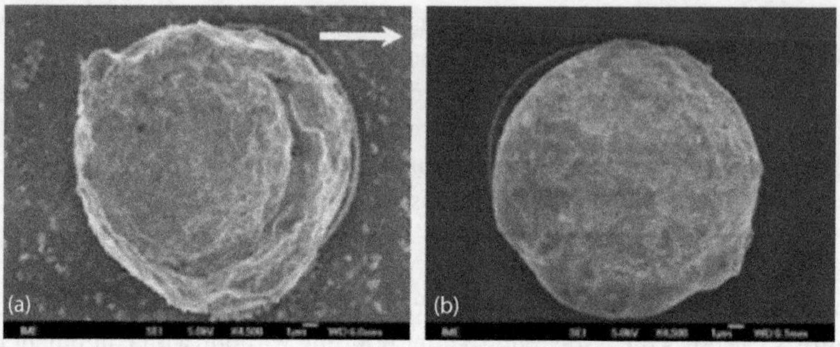

Figure 5.20 Fracture surfaces of the microsolder joint at the left side (Fig. 5.15) after shear (a) on the Si chip and (b) on the Si carrier.

Figures 5.20 to 5.22 show the fracture surfaces after shear tests (the arrow indicates shear direction). Sufficient Sn wetting with ENIG can be observed. The fractural surface is inside the Sn cap and near the interface between the Sn cap and the ENIG pad [along the line AA' shown in Fig. 5.18(b)] for the microjoints at left and center as shown in Fig. 5.15, whereas the microjoints at right are separated mostly along the bottom of the ENIG pads [along the line BB' shown in Fig. 15.18(b)]. During the shear test, the microjoints are subjected to a combination of shear and bend loading, which effectively results in the twisting of the microjoints. The loading is different for the microjoints at different locations, which results in different failure modes.

5.2.7 Evaluation of the Underfill

To protect the microjoint, capillary underfill material is filled in the bond line. Considering that the bond line is only around 10 μm, underfills with fine filler size have to be used. In this study, three underfills are evaluated. The average filler size and the maximum filler size are less than 1 μm and 5 μm, respectively. Figure 5.23 shows the CSAM (C-mode scanning acoustic microscopy) images of the stacking chips with those underfills. It can be seen from the images that voidless underfilling has been achieved.

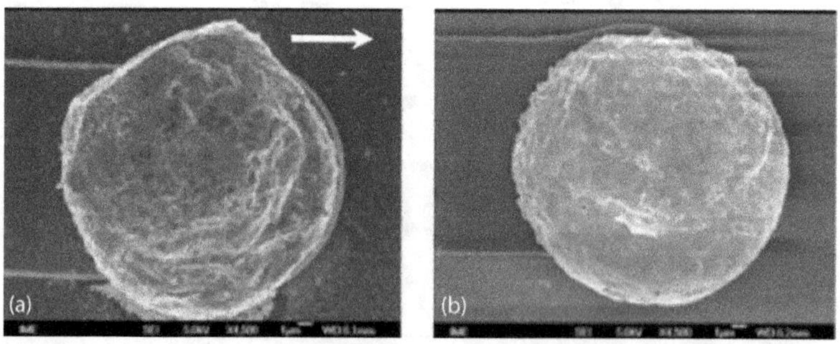

Figure 5.21 Fracture surfaces of the microsolder joint at the center (Fig. 5.15) after shear (a) on the Si chip and (b) on the Si carrier.

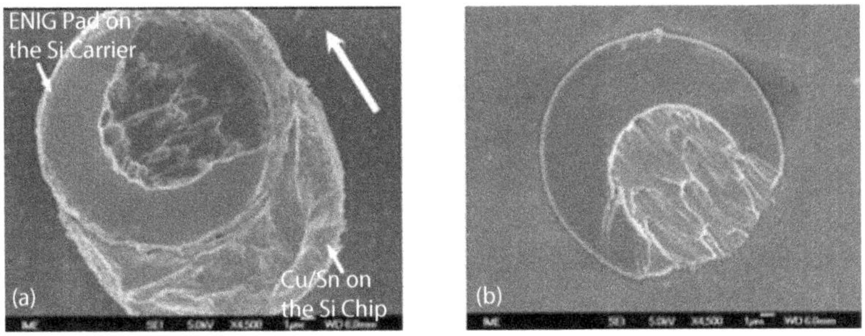

FIGURE 5.22 Fracture surfaces of the microsolder joint at the right side (Fig. 5.15) after shear (a) on the Si chip and (b) on the Si carrier.

FIGURE 5.23 CSAM images after underfilling with different underfills showing no major voids in the gap.

5.2.8 Reliability Assessment

To assess the reliability of the microjoints, bonded samples under joining conditions of 2.45-mN/bump bonding pressure, 300°C bonding temperature, and 60 s bonding time are first tested with JEDEC moisture sensitivity level-3 (30°C/60%RH, 192 h) with 3×reflow (260°C peak temperature). After this preconditioning, additional TCT (temperature cycle test) (−45°C ⇆ +125°C, 1000 cycles) and unbiased HAST (highly accelerated temperature and humidity stress test) (130°C/85%RH, 96 h) are performed. For all read-out points, a pass criterion of less than 10% resistance rise versus time zero is applied for the resistance of a row of microjoints.

Table 5.1 lists the test results. For both TCT and HAST, 1 out of 11 samples failed. Figure 5.24 shows the cross-sectional images of a failed microjoint after HAST. The breakage

TABLE 5.1 Reliability Assessment Test Results

Reliability test	Condition	Sample size	Failed samples
MST-level 3	30°C, 60% RH, 192 h, 3 × 260°C reflow	22	0
HAST	130°C, 85% RH, 96 h	11	1
TCT	−40/+125°C, 5-min dwell/bath, ramp 15°C/min, 1000 cycles	11	1

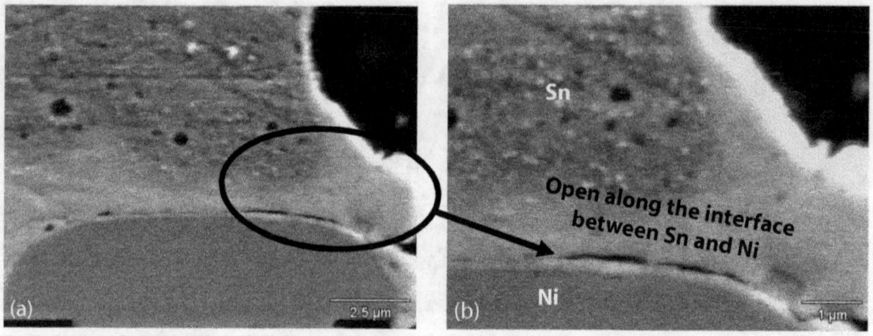

Figure 5.24 Cross-sectional microstructural analysis of the failed sample showing opening or voiding along the Sn/ENIG interface. (a) Overview. (b) Zoom-in view.

is clearly observed at the joining interface between the Sn cap and the ENIG pad. As mentioned earlier in Sec. 5.2.6, there are a few submicrometer voids existed along the initial bonding interface, which lead to breakage after accelerated reliability test.

5.2.9 Summary and Recommendations

The development of fine-pitch and high-density microbumps for 3D stacking has been reported in this section. Some important results and recommendations are summarized as follows:

- Microbumps consisting of a Cu pillar and Sn solder cap with 25 μm pitch have been developed successfully. The ultrafine-pitch and high-density microbump array can be used for 3D chip stacking.
- When the ENIG pads' thickness increased from 2 to 4 μm, the maximum shear stress (τ_{xy}) decreased 5 MPa/μm, whereas τ_{xy} decreased only 1.7 MPa/μm when the ENIG pads' thickness increased from 4 to 10 μm. The plating of thick ENIG resulted in wider ENIG pads, which increased the risk of solder bridging between adjacent microjoints.
- At bonding conditions of 2.45-mN/bump bonding pressure, 300°C bonding temperature, and 60 s bonding time, good joining with average die shear strength of 18.2 MPa has been achieved. After shear test, the fractural surface has been found to be within the Sn solder cap and near the interface between the Sn cap and the ENIG pads for the microjoints at the left side and at the center, and for the microjoints at the right side the fractural surface has been found to be separated mostly along the bottom of the ENIG pads.
- As the gap between the Si chip and the Si carrier is only around 10 μm, fine filler size underfill has been used in this study, and good filling without any apparent voids has been achieved.
- Reliability test results showed that 90% samples can pass TCT and HAST. In the failed sample, there is a breakage along the bonding interface between the Sn caps and the ENIG pads.

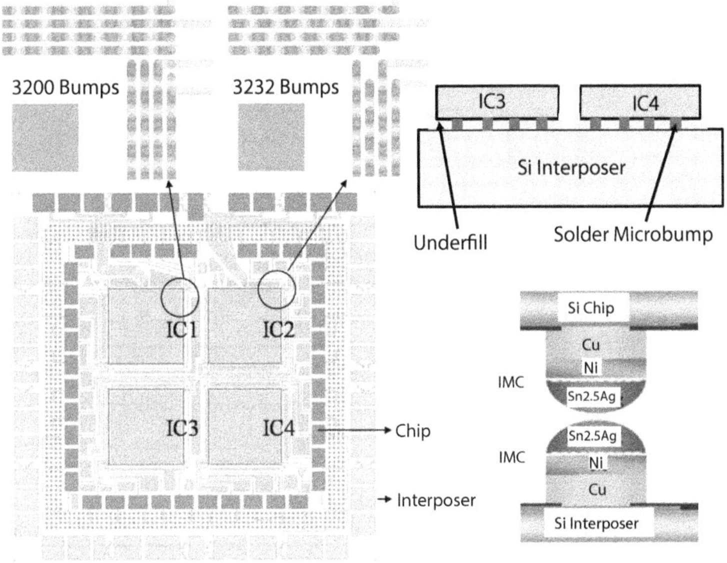

Figure 5.25 Schematic drawing of the test vehicle. The chips and interposer have the same Cu pillar, Ni barrier, and SnAg solder cap.

5.3 Fabrication, Assembly, and Reliability of 20-μm-Pitch Microbumps

5.3.1 Test Vehicle

Figure 5.25 shows the schematic of the test vehicle [42]. There are four test chips (IC1, IC2, IC3, and IC4) that are interconnected to the Si interposer by microbumps with array (3200 I/Os) (IC1 and IC3) and stagger (3232 I/Os) (IC2 and IC4) arrangements. Daisy chains and test pads on the interposer are designed-in for probing the electrical open/short of the circuits after assembly, precondition, and TCT. The chips and interposer specifications are given in Table 5.2.

The formation of the electroplated microbumps is the same on the chips and the interposer with a structure of Cu (5 μm) Ni (3 μm) Sn2.5Ag (5 μm) solder alloy, as shown in

Table 5.2 Specifications of the Test Vehicle

	Chip	Interposer
Dimension (mm)	4.6 × 4.6	20 × 20
Thickness (μm)	100	300
Passivation opening (μm)	6	6
Bump diameter (μm)	12	12
Bump height (μm)	13	13
Number of bumps	≧3200	≧12,000

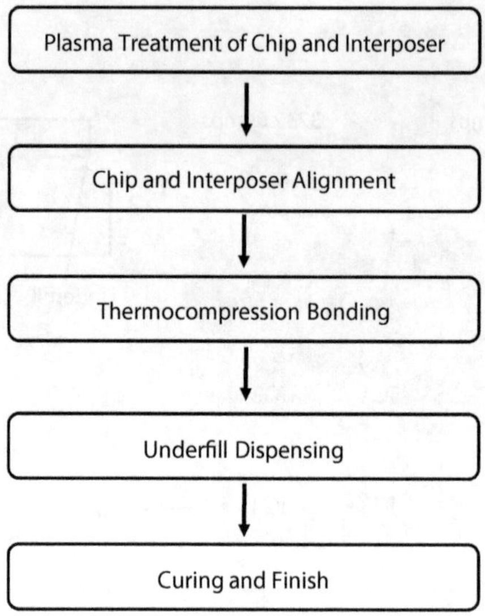

Figure 5.26 Assembly process flow.

Fig. 5.25. In order to reduce the undercut of Cu post, the thickness of Cu seed layer is 2000 Å, which is removed after bump plating and photoresist (PR) stripping by dry etching.

5.3.2 Assembly of Test Vehicle

The process flow of assembly is shown in Fig. 5.26. The microbumps are pretreated by plasma to remove the oxides and the contamination on the surface, and then four chips are interconnected to the interposer by Toray's FC-3000 WS thermocompression bonder at 280°C for 15 s. Subsequently, two different underfill materials are used to seal the microgaps between the chips and the interposer, and then cured at 150°C for 30 min (which is called Underfill A hereafter) and at 165°C for 120 min (which is called Underfill B hereafter), respectively. The characteristics of those two capillary underfill materials for microgap filling are listed in Table 5.3. Finally, a SAM (scanning acoustic microscope) with a 75 MHz probe is used to determine the voids in the underfill.

5.3.3 Formation of Microjoints by Thermocompression Bonding

The conventional bumping condition (5000-Å-thick seed layer) is not able to produce the 20-μm-pitch microbumps because of the undercut of Cu posts, as shown in Fig. 5.27(a). When the thicknesses of the sputtered Cu seed layer is reduced from 5000 to 2000 Å and a dry etching is used to remove the seed layer, the undercut of Cu posts is refined within 10%, as shown in Fig. 5.27(b). Comparing the microjoints shown in Figs. 5.27(c) and 5.27(d), the excess undercut of Cu posts is easier to make the microbumps deform, which may cause the fracture of Cu post after TCT, as shown in Fig. 5.28. However, this defect is no longer found in the reformed microbumps when the undercut issue is improved by decreasing the thickness of Cu seed layer to 2000 Å and using dry etching to substitute for chemical etching.

The conventional reflow with flux is seldom used for the assembly of microbumps because the gap size between chips and interposer is just around 15 μm, and the flux residue

Figure 5.27 SEM images of (a) microbumps with undercut Cu post; (b) refined microbumps; (c) joints of bumps with undercut Cu post; (d) joints of refined microbumps.

Figure 5.28 Fracture of the interface between Cu post and Ni layer in the microsolder joint with the undercut of Cu post.

is difficult to be removed by water or organic solvent, which may cause the formation of voids within the underfill and degrade the microjoints' reliability. Thus TCB is the most popular method to stack chips with microbumps.

5.3.4 Microgap Filling

The microgap filling by underfill is not easy. In the case of the ultra-fine pitch interconnections, the formation of voids can be significantly influenced by the flowability of underfill, which is affected by the factors such as viscosity, filler size, and distribution of solder bumps. Furthermore, the local CTE (coefficient of thermal expansion) mismatch among silicon, Cu in TSV, solder alloy, and underfill results in a high local stress field, which may initiate a fracture in the solder interconnects if there are defects existed. Fillet cracking is another main defect that may result in the delamination between underfill and passivation layer on

Figure 5.29 SAM images of the microgaps sealed by underfill A (a) and underfill B (b) before reliability test.

Table 5.3 Characteristic of Underfill Materials

Item	Underfill A	Underfill B
Tg (°C) by TMA	135	117
Viscosity at 25°C	9 Pa.s	12.3 Pa.s
CTE (ppm/°C)	$\alpha_1 = 42, \alpha_2 = 125$	$\alpha_1 = 38, \alpha_2 = 106$
Filler content (%)	50	60
Filler size (μm)	0.3	0.5
Modulus (GPa)	6.5	7.1

chip, while bulk cracking can initiate solder joint cracking and solder bridging. Therefore, the selection of underfill material is important to enhance the lifetime of the microjoints of a 3D IC integration structure.

Figures 5.29(a) and 5.29(b) show the SAM images of the microgaps respectively sealed by underfill A and underfill B (Table 5.3) before TCT. It can be seen that there are no voids and other defects within underfill A. However, there are some voids within underfill B.

5.3.5 Reliability Test

As those shown in Table 5.4, the test items and conditions of reliability assessment are indicated. All samples are experienced preconditioning test (JESD22-A113D, LV3) to screen the samples of early failure, and there are 72 qualified ones chosen to study their reliability

Table 5.4 Reliability Assessment Test Condition

Item	Test condition
Preconditioning	Baking (125°C, 24 h) Soaking (30°C /60%RH, 192 h) Reflow (260°C, 3 times)
Thermal cycling testing	−55°C ~ 125°C, 3000 cycles, Dwell time = 5 min, Ramp rate =15°C/min

Figure 5.30 SAM images of the microgaps sealed by underfill A (a) and underfill B (b) after reliability test.

characterization. Temperature cycling test (JESD22-A104B) is performed under the condition of $-55°C \leftrightarrows 125°C$ with a dwell time of 5 min. The resistance is measured regularly and the test is stopped when the number of cycle reaches 3000.

The failure criterion in those tests is considered an increase of 20% or more from the initial resistance value. Then, the samples are evaluated by SAM for the detection of voids in underfill within the microgap. Finally, the failed samples are mounted in epoxy, grinded by SiC papers, and polished using Al_2O_3 paste. The cross-sections of the samples are observed by a SEM to learn the morphology of the microjoints, the thickness of IMC and the interfacial morphologies among chip, underfill and interposer. The chemical composition of the intermetallic phase is identified by an EDS (energy dispersive spectrometer).

Figures 5.30(a) and 5.30(b) show the SAM images of the microgaps respectively sealed by underfill A and underfill B after TCT. It can be seen that there are no voids and other defects within underfill A after TCT. However, the voids within underfill B don't expand after TCT. According to the material properties of those two underfill materials listed in Table 5.3, the void issue of underfill B is considered to be caused by its higher filler content and larger filler size, the higher viscosity deteriorates the flowability of underfill B during dispensing and makes the voids form within it finally.

For the comparison of the growth kinetics of the IMCs at both chip-side and interposer-side, other assemblies are aged at 100, 125, and 150°C for 100, 250, 500, 750, and 1000 h, respectively. Then cross-section and observe by the same method mentioned previously. Afterward, the IMC thickness is measured and averaged, the growth kinetic is acquired then by the Arrhenius equation.

5.3.6 Reliability Test Results and Discussion

TCT is chosen to evaluate the reliability of the assemblies which are survived after the preconditioning test, and the relationship between the percent failure and the cycles-to-failure is given in Fig. 5.31. In both cases, no delamination defect is found at the interface between the underfill and the passivation, which means that those two underfills well protect the test vehicle to reduce the strains and stresses. Therefore, the failure of the microjoints shall be dominated by defects and stress/strain caused by CTE mismatch.

Figure 5.31 shows the thermal cycling results of the microjoints with underfill A and underfill B. The failure criterion is defined as the resistance increases by 20% of the original

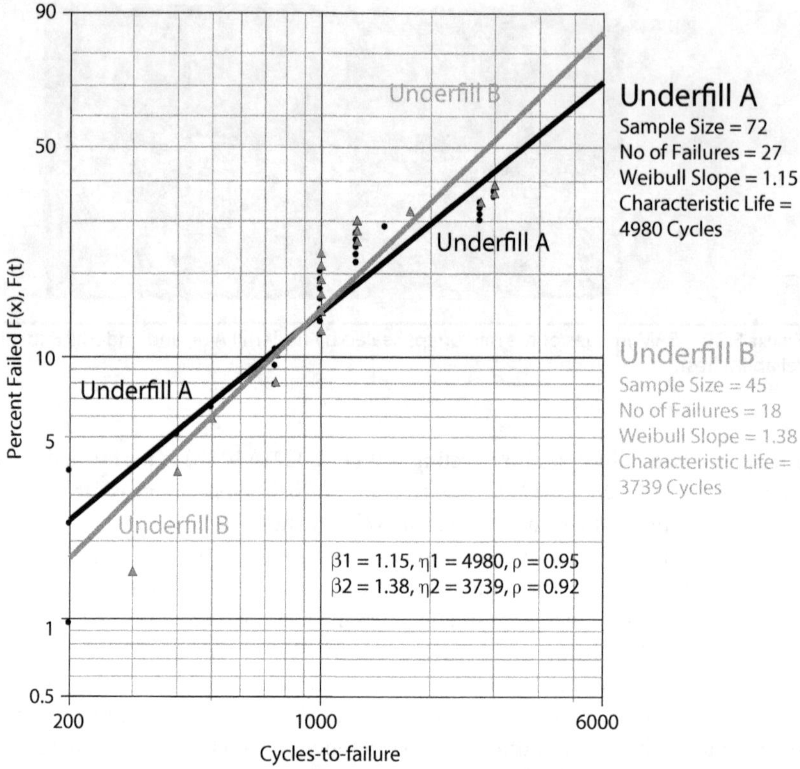

FIGURE 5.31 Life distributions of microjoints with underfill A and underfill B (−55°C ⇆ 125°C and failure criterion = infinitive resistance change).

resistance. It can be seen that the sample size and number of failures of the microjoints with underfill A, respectively, are 72 and 27, and with underfill B are 45 and 18. For both cases, the sample failures are much less than 50%. The TCT stopped at 3000 cycles.

The Weibull slop (h) and characteristic life (b) of underfills A and B are, respectively, 1.15 and 4980 cycles, and 1.38 and 3739 cycles. The mean-time-to-failure (M, or mean life) can be determined by $[h\Gamma(1+1/b)]$, for underfills A and B and are given by respectively, $M_A = 4740$ cycles and $M_B = 3416$ cycles. The percent failed corresponding to the mean life of underfills A and B can be determined by $[1 - \exp(-(x/h)^b)]$ and is given by 60.9% and 58.7% (Table 5.5).

TABLE 5.5 Summary of Reliability Assessment Test Results for Underfills A and B

Item	Sample size	No. of failure	Weibull slope	Characteristic life, cycles (63.2% failed)	MTTF (mean life), cycles	% failed at MTTF
Underfill A	72	27	1.15	4980	4740	60.9
Underfill B	45	18	1.38	3739	3416	58.7

One of the difficult tasks in life testing is to draw conclusions about a population from a small sample size. It is even more difficult to compare the populations of two products based on the knowledge of their limited test data. If one product is found to be superior to another, how confident can it be that the same is true of their populations? Herein, a simple approach is used to determine whether one product's mean life is better than the others, without inquiring what the actual differences are [74]

$$P = \frac{1}{1 + \frac{\log 1/q}{\log 1/(1-q)}}$$

where

$$q = 1 - \frac{1}{\left[1 + \left(\frac{t + 4.05}{6.12}\right)^5\right]^{40/7}}$$

$$t = \frac{\sqrt{1 + \sqrt{T}}\,(\rho - 1)}{\rho\Omega_2 + \Omega_1}$$

$$T = (r_A - 1)(r_B - 1)$$

$$\Omega_1 = \sqrt{\frac{\Gamma(1 + 2/b_B)}{\Gamma^2(1 + 1/b_B)} - 1}$$

$$\Omega_2 = \sqrt{\frac{\Gamma(1 + 2/b_A)}{\Gamma^2(1 + 1/b_A)} - 1}$$

$$\rho = \frac{M_A}{M_B}$$

In these equations, P is the confidence level, M_A=4740 cycles and M_B=3416 cycles are the sample mean lives with underfills A and B, b_A=1.15 and b_B=1.38 are the sample Weibull slopes, r_A=27 and r_B =18 are the number of failures (again, they are much less than 50% of their sample sizes), and T is called the total degrees of freedom. After simple calculation we have, P = 51%. This means that in 51 out of 100 cases, the mean life of the microjoints with underfill A is better than that with underfill B. The other 49 cases no one knows. But this confidence level is too low for us to say the microjoint mean life with underfill A is better than that with underfill B. At best, we can only say that their mean lives are about the same!

The microbumps with both underfills have early failure which is attributed to a worse bonding quality, some microbumps do not interconnect as bonded, which might be stemmed from that the oxidation layer on the surface of the microbumps is too thick to be removed entirely by plasma.

Figure 5.32 Cross-sectional SEM image of the survive microjoint sealed by underfill B after reliability test.

Figure 5.33 Cross-sectional SEM image of the failed microjoint sealed by underfill B after reliability test.

Figure 5.32 shows the cross-sectional image of the survival microjoint sealed by underfill B after TCT for 3000 cycles, and no fracture is found in it. On the other hand, no matter in the cases of underfill A or underfill B, the fracture formed at the interface between the intermetallic phase and the Sn2.5Ag solder alloy is always found at the top chip, as shown in Fig. 5.33, which means that this area is the weak point of the microjoints under TCT. This could be due to the thickness difference between the chips (100 μm) and the interposer (300 μm). Perhaps some defects are formed here as aged, and a proposed failure mechanism is going to be given in the next paragraph.

5.3.7 Failure Mechanism of the Microjoints

Figure 5.34 exhibits the microstructure of the as-bonded microjoint. The intermetallic phase formed at the interface is identified as Ni_3Sn_4. Normally, Ni is used as a diffusion barrier to inhibit the reaction of Sn and Cu because it is an interstitial solute and is classified as a fast diffuser in Sn by small activation energy and very low solubility of 10 ppm [78]. During the bonding process, the bottom interposer is kept at 100°C, and then the bonding head

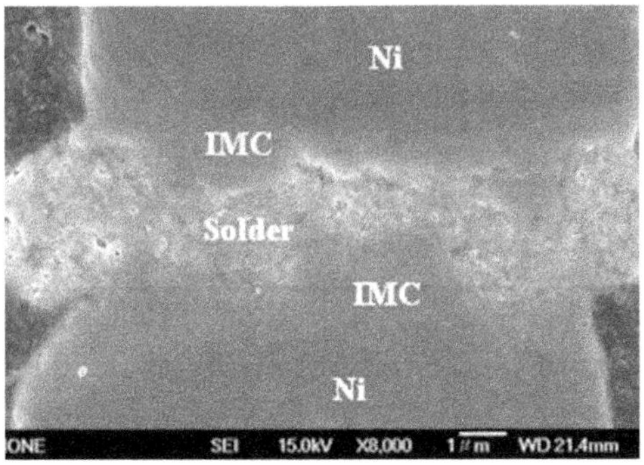

FIGURE 5.34 Cross-sectional SEM image of the as-bonded microjoint.

heats the top chip up to 280°C to melt the Sn2.5Ag solder alloy to form the microjoint. The bonding temperature drives abundant Ni atoms to dissolve into the melted solder, and soon saturates the Sn matrix. Because the density of Ni is higher than that of Sn, the driving force so-called gravity segregation [79, 80] makes Ni atoms accumulate at the interface close to the bottom interposer and result in the formation of a thicker Ni_3Sn_4.

The failure mechanism of the microjoints is presented in Fig. 5.35. Initially, the Ni atoms at chip-side dissolve into the solder alloy as bonding and results in a thicker Ni_3Sn_4 at the interface of interposer-side. Therefore, the reaction of $3Ni + 4Sn \rightarrow Ni_3Sn_4$ at this interface

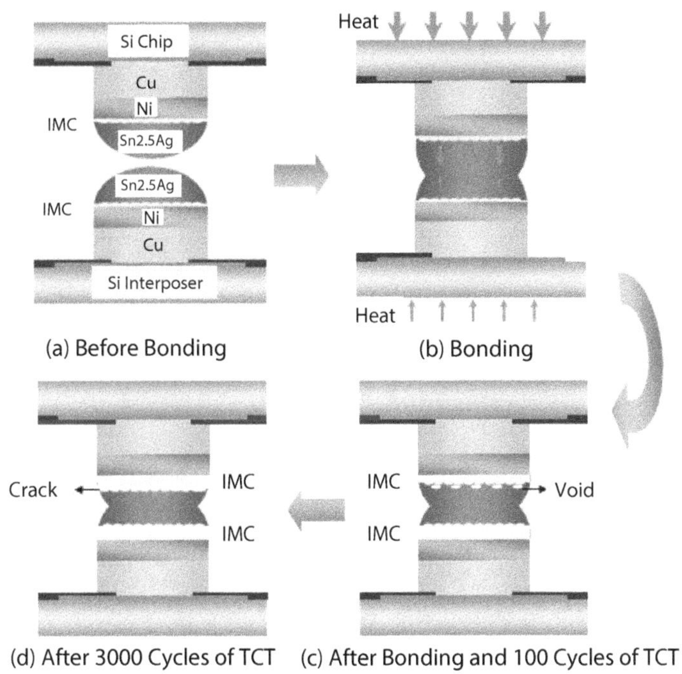

FIGURE 5.35 Schematic drawings of the failure mechanism of microjoints.

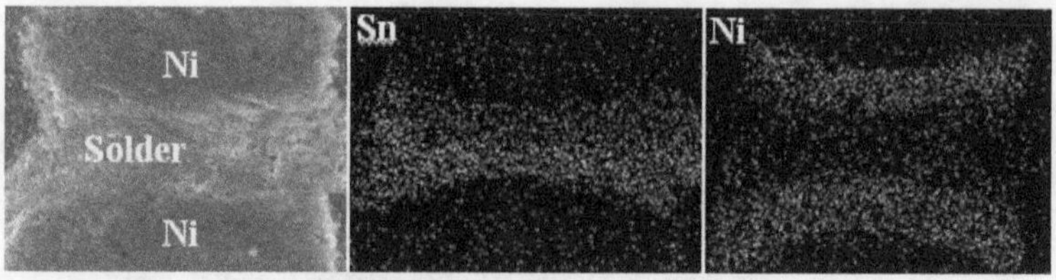

FIGURE 5.36 Elemental mapping analyses of Sn and Ni in a microjoint after reliability test.

is inhibited as heated in TCT because the flux of Sn atoms through the Ni_3Sn_4 layer decrease [81], and the interfacial reaction at the chip-side would be severer. Then a Sn depletion zone, as shown in Fig. 5.36, is formed at the interface between the Ni layer and the Sn2.5Ag solder alloy of the top chip due to the consumption of Sn in the reactions of $3Ni + 4Sn \rightarrow Ni_3Sn_4$ and $Ag + 3Sn \rightarrow Ag_3Sn$. Eventually, the crack propagates along the interface and then the microjoint fails when the stress induced by the Z-axial expansion of underfill enforces on the interface.

The thickness and growth of intermetallic is a very important factor to affect the device reliability. Figure 5.37 is the relationship of the thicknesses of the Ni_3Sn_4 at the chip-side and the interposer-side versus aging time under different temperatures. It can be seen that the growth of Ni_3Sn_4 followed a linear law but the growth rate of the Ni_3Sn_4 at the top chip is significantly higher than that of the bottom interposer, it means the growth of the Ni_3Sn_4 in the microjoints is assumed to be interface reaction-controlled, not diffusion-controlled. Hence the thickness (X) of the Ni_3Sn_4 subject to the aging time can be described by, $X = Dt$, where D is diffusion coefficient, t is the aging time. The diffusion coefficient is given by

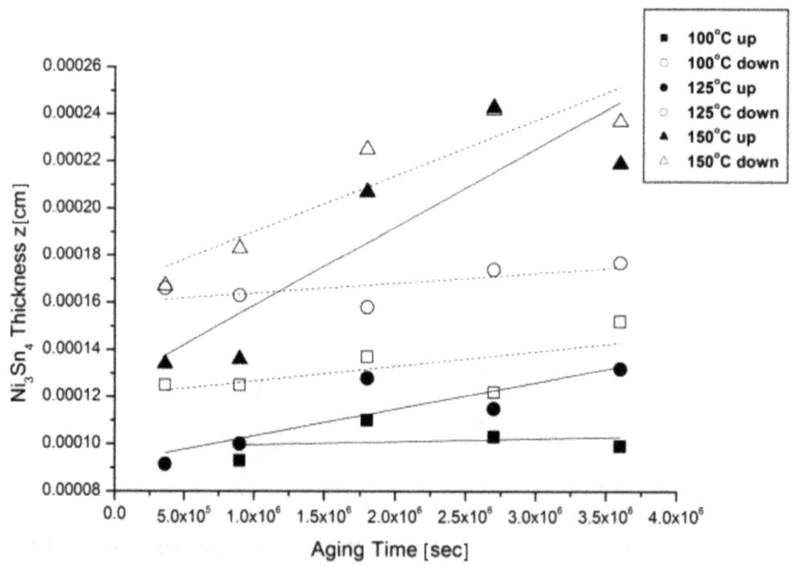

FIGURE 5.37 Relationship of the thickness of Ni_3Sn_4 vs. aging time under different aging conditions.

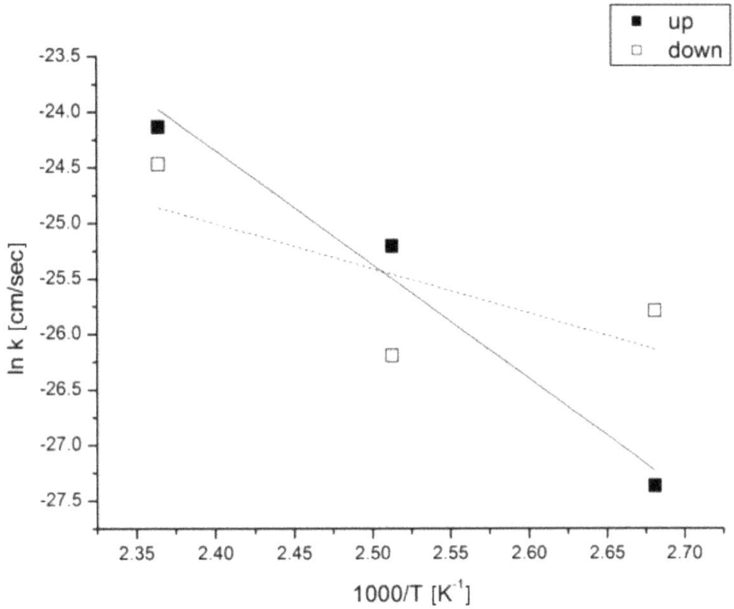

FIGURE 5.38 Arrhenius plot of the growth of Ni_3Sn_4 at various interfaces.

an Arrhenius expression, $D = D_0 \exp(-Q/RT)$, where D_0 is the diffusion constant, Q is the activation energy, R is universal gas constant, and T is the absolute temperature. The activation energy is calculated from the slope of the Arrhenius plot in Fig. 5.38. The apparent activation energies calculated for the growth of the Ni_3Sn_4 at the top chip and the bottom interposer are 84.8 and 33.2 kJ/mol, respectively.

Generally speaking, the growth of Ni_3Sn_4 layer is controlled by the diffusion of Ni in Sn [82]. However, those two values expound that the growth of Ni_3Sn_4 in this system is dominated by the reaction of Sn and Ni at the interface between Ni layer and Ni_3Sn_4 because Ni atoms have saturated the Sn matrix after bonding. Further evidences have been given in Figs. 5.33 and 5.34, in which the thickness of Ni layer after bonding is around 2.4 μm but decreases to 1.8 μm after TCT. The growth of Ni_3Sn_4 at the top chip is more rapid, which results in the formation of Sn depletion zone close to the interface, and induces the fracture during TCT.

5.3.8 Summary and Recommendations

The failure mechanism of microbumps for 3D IC integration by thermocompression bonding has been investigated in this study. Some important results and recommendations are summarized as follows [42]:

- With both underfills, the characteristic life (at 63.2% failures) of the microjoints has been found to be larger than 3700 thermal cycles (−55°C to 125°C) even the bonding time is as short as 15 s.
- With limit amount of thermal cycling data, it has been found that the mean life of microjoints with underfill A is better than that with underfill B with only 51% confidence. Basically, they behave the same.

- The failure of microjoints has been found to be attributed to the formation of Sn depletion zone close to the interface between the bulk-solder and Ni-layer of the top chip.
- The growth of the Ni_3Sn_4 has been found to be controlled by the reaction of Sn and Ni at the interface between Ni-layer and Ni_3Sn_4. This growth has been found to be more rapid at the top chip, which speeds up the formation of Sn depletion zone close to the interface, and induces the fracture during TCT.

5.4 Fabrication, Assembly, and Reliability of 15-μm-Pitch Microbumps

5.4.1 Microbumps and UBM Pads of the Test Vehicle

Figure 5.39 shows a schematic drawing of the test vehicle [5]. It can be seen that it is almost the same as the one (Fig. 5.1) in Sec. 5.2 except the pitch of the chips is reduced from 25 to 15 μm. Also, the thickness of the ENIG is reduced from 4 to 2 μm, and the barrier/adhesion layer has been changed from Ti to Ta and is dry etched with plasma. Finally, bonding of chips with SnCu + Sn solder cap is presented.

The thickness of the ENIG pad is limited by the fabrication process. Assuming the plating of ENIG is isotropic [10], and the opening of Al pad (w as shown in Fig. 5.12) is 4 μm, then the relationship between the thickness of the ENIG pads and the width of the ENIG pads (W as shown in Fig. 5.12) can be described as Fig. 5.40. From Fig. 5.40, when the thickness of the ENIG pads is selected as 2 μm, the width of the ENIG pads will be 8 μm. As the pitch size is only 15 μm, thicker ENIG pads may easily result in solder bridging of adjacent pads during assembly. Thus, 2-μm-thick ENIG is selected, as shown in Fig. 5.41.

In fabrication of CuSn solder microbumps, most frequently used barrier/adhesion layer is Ti and after plating, this adhesion layer is etched away by wet etching method. However, it is found that if Ti is used as the adhesion layer and later etched away by wet etching method as in the case of the 25 μm pitch (Sec. 5.2), then after wet etching, lots of solder microbumps are lost at the same time, as shown in Fig. 5.42(a). The reason is that when using wet etching process to etch the Ti adhesion layer, undercut happens and since the diameter of the bump is only 8 μm, most of the solder microbumps peel off. In order to resolve this problem, Ta is used here as the adhesion layer and later is etched away with dry etch process. As a result, it can avoid peeling off of the solder microbumps, as shown in Fig. 5.42(b). Figure 5.43 shows SEM images of the fabricated CuSn microbumps and Al pad. After plating, a reflow is performed to reshape the microbumps and uniform bump heights are obtained. The reflow temperature is 265°C.

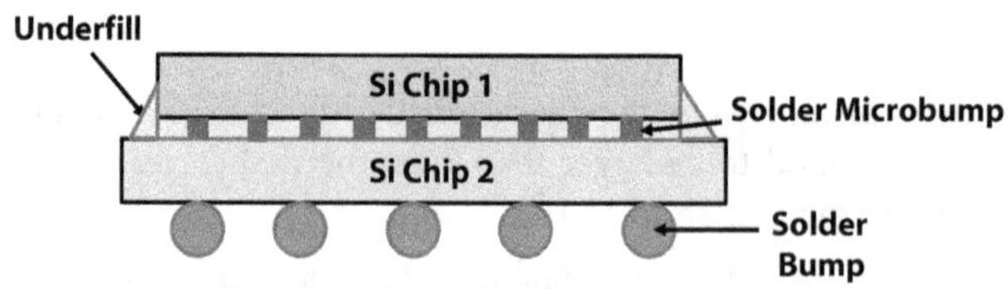

FIGURE 5.39 Schematic drawing of 3D stacking of Memory + CPU package architectures, Fig. 5.1 (not to scale).

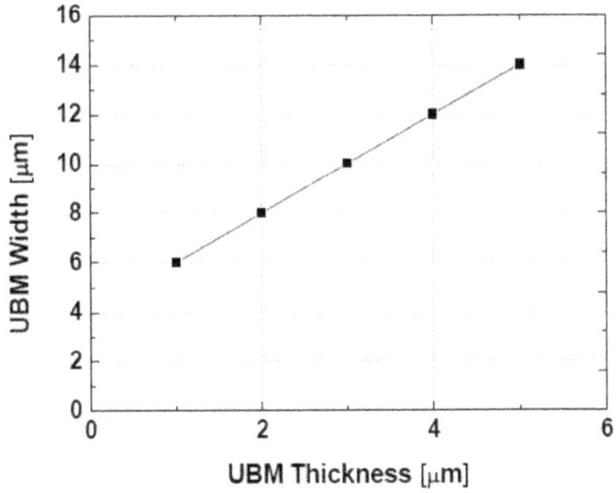

Figure 5.40 Relationship between the ENIG UBM thickness and the ENIG UBM width (assuming ENIG plating is isotropic and the opening of the Al pad is 4 μm).

Figure 5.41 SEM images of ENIG UBM pads.

5.4.2 Assembly

After fabrication of the CuSn solder microbumps and the ENIG UBM pads, the Si chips are joined together using FC150 flip chip bonder. The assembly process is a non-flux process because after joining, the gap between the Si chip and the Si carrier is only about 10 ~ 15 μm. Therefore, if flux is used, it is very difficult to clean it and then after underfill process, a lot of voids are formed inside the gaps. As mentioned, on the Si chip 2, either ENIG or

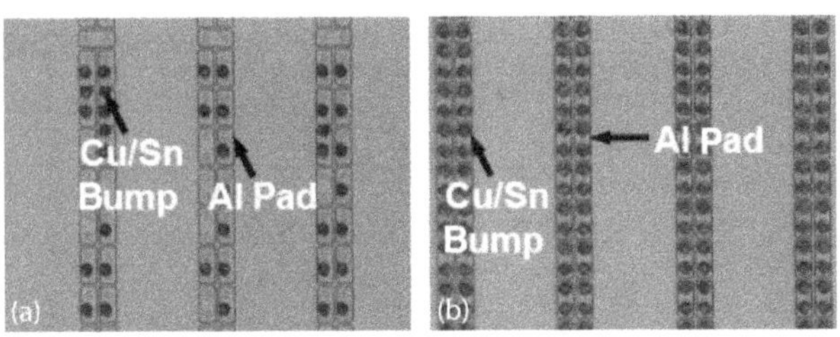

Figure 5.42 (a) Fabrication results of CuSn solder microbumps with Ti used as seed layer and etched away with wet etching; lots of solder microbump missed after fabrication. (b) Fabrication results of CuSn solder microbumps with Ta used as seed layer and etched with dry etch process.

FIGURE 5.43 SEM images of CuSn solder microbumps.

CuSn solder microbumps are fabricated. Therefore, assemblies are done by either joining CuSn solder microbump to ENIG pad or joining CuSn solder microbump to CuSn solder microbump.

5.4.3 Assembly with CuSn Solder Microbump and ENIG Pad

The first assembly is done with CuSn solder microbump and ENIG UBM pad. The joining conditions are pressure = 20 MPa; bottom substrate temperature = 300°C; up arm temperature = 350°C; and time = 60 s. Figure 5.44 shows the cross-section of microjoints with CuSn solder microbump and ENIG pad jointed together. It can be seen that all the Sn is consumed into IMC (Cu_6Sn_5).

5.4.4 Assembly with CuSn Solder Microbump and CuSn Solder Microbump

When assembly with CuSn solder microbump and ENIG pad, both the Si chip and the Si carrier have to be kept at high temperature; therefore, the joining has to be done one by one with chip-to-chip assembly. Now the assembly is done with CuSn solder microbump and CuSn solder microbump. The joining conditions are pressure = 10 MPa; bottom substrate temperature = room temperature; up arm temperature = 350°C; and time = 30 s. This step is only for temporally joining. After that, the assembly is reflowed in a reflow oven with peak temperature of 265°C. Figure 5.45 shows the cross-section of microjoints after reflow. It is found that

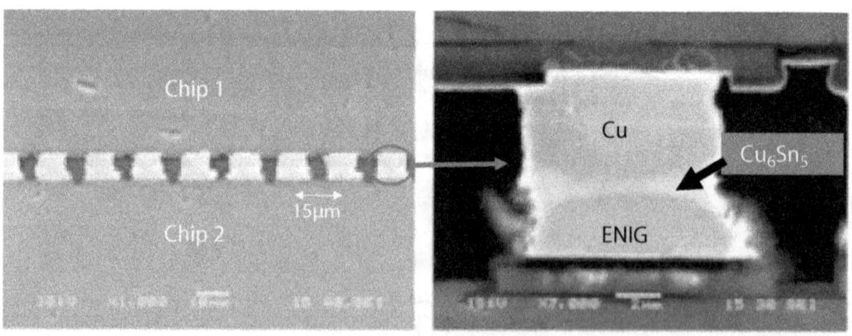

FIGURE 5.44 Cross-section of microjoints with CuSn solder microbumps joined to ENIG pads.

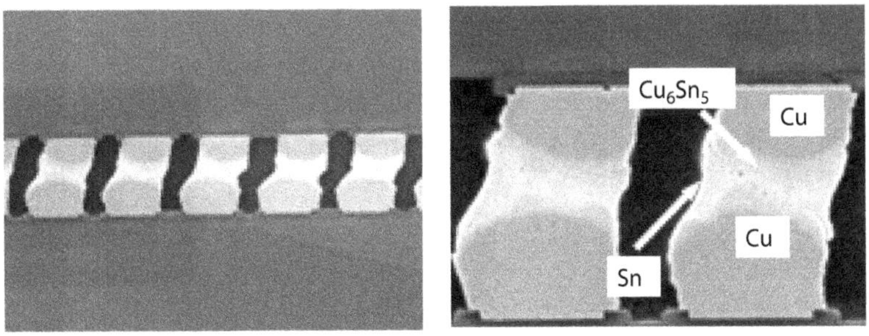

Figure 5.45 Cross-section of microjoints with CuSn solder microbumps joined to CuSn solder microbumps.

there is still some Sn left. In this method, the Si chip 2 is kept at room temperature. Therefore, it is possible to do chip-to-wafer assembly, which means pre-joining the Si chip 1 to wafer of Si chip 2 on the FC150 first, then do batch reflow at the same time in the reflow furnace.

5.4.5 Evaluation of Underfill

As mentioned, the gap between the two jointed Si chips is very narrow; therefore, it is important to evaluate the flow-ability of underfill and the void formation in underfill which contains silica-filler. Considering the fine chip-to-chip gap, fine filler size underfills are used to evaluate the underfill process. Figure 5.46 shows the cross-section image of the underfilled sample. It can be seen that good filling without any void can be achieved with underfill resin containing very fine filler. The average filler size is less than 0.6 μm.

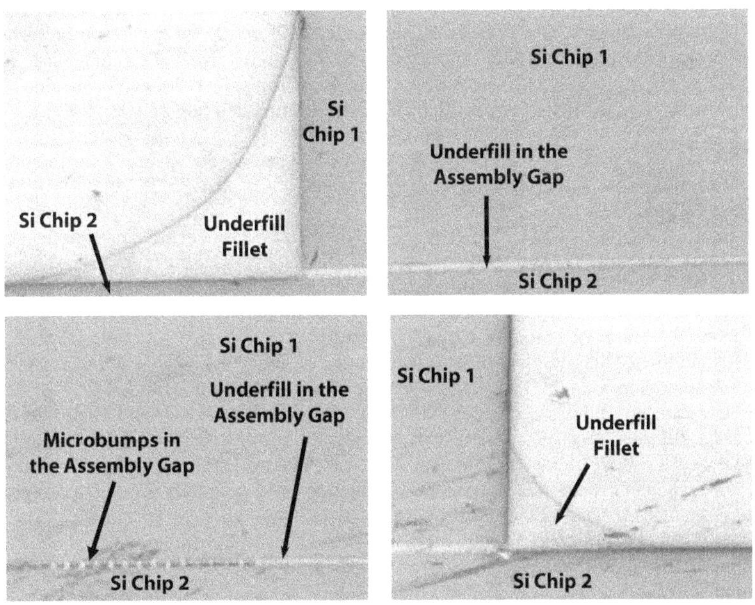

Figure 5.46 Optical photo of cross-section of two chips with underfill in the gap.

5.4.6 Summary and Recommendations

The investigations of solder microbumps in 15 μm pitch and their assembly process have been reported in this section. The fine pitch solder microbumps can be used for stacking of memory chip and microprocessor chip as well as other 3D IC integration systems in package. Some important results and recommendations are summarized as follows [5]:

- For Cu/Sn solder microbumps with 8 μm in diameter and 15 μm in pitch, the thickness of the Sn cap should be thicker than 1.5 μm and less than 4 μm if the total bump height (Cu + Sn) is kept to 10 μm.
- It is found that when the Ti is used as adhesion layer, during the wet etching process, lots of bumps are missing because of undercut. Therefore, the adhesion layer is changed to Ta and is etched by dry etching process.
- With optimized bonding conditions (bottom temperature = 300°C, upper temperature = 350°C, bonding time = 60 s, and bonding pressure = 20 MPa), good joining can be achieved between CuSn microbumps and ENIG UBM pads.
- Bonding of CuSn solder microbumps to CuSn solder microbumps (instead of ENIG UBM) also yields good results. The chips are temporally joining together first on the FC150 bonder (the joining conditions are pressure ~10 MPa; bottom substrate temperature ~25°C; upper arm temperature ~350°C and time ~30 s) and then reflowed in a reflow oven with a peak temperature of 265°C. This approach has the potential to perform chip to wafer assembly.

5.5 References

[1] Yu, A., J. H. Lau, S. Ho, A. Kumar, W. Hnin, W. Lee, M. Jong, et al., "Fabrication of High Aspect Ratio TSV and Assembly with Fine-Pitch Low-Cost Solder Microbump for Si Interposer Technology with High-Density Interconnects," *IEEE Transactions on CPMT*, Vol. 1, No. 9, September 2011, pp. 1336–1344.

[2] Yu, A., J. H. Lau, Ho, S., Kumar, A., Yin, H., Ching, J., Kripesh, V., et al., "Three Dimensional Interconnects with High Aspect Ratio TSVs and Fine Pitch Solder Microbumps," *IEEE Proceedings of ECTC*, San Diego, CA, May 2009, pp. 350–354.

[3] Yu, A., A. Kumar, S. W. Ho, H. W. Yin, J. H. Lau, N. Su, et al., "Development of 25-μm-Pitch Microbumpsfor 3D Chip Stacking," *IEEE Transactions on CPMT*, Vol. 2, No. 11, November 2012, pp. 1777–1785.

[4] Yu, A., A. Kumar, S. Ho, H. Yin, J. H. Lau, J. Ching, V. Kripesh, et al., "Development of Fine Pitch Solder Microbumps for 3D Chip Stacking," *IEEE EPTC Proceedings*, Singapore, December 2008, pp. 387–392.

[5] Yu, A., J. H. Lau, S. W. Ho, A. Kumar, W. Y. Hnin, D.-Q. Yu, M. C. Jong, et al., "Study of 15μm Pitch Solder Microbumps for 3D IC Integration," *Proceedings of IEEE/ECTC*, May 2009, pp. 6–10.

[6] Lim, S., V. Rao, W. Hnin, W. Ching, V. Kripesh, C. Lee, J. H. Lau, et al., "Process Development and Reliability of Microbumps," *IEEE Transactions on CPMT*, Vol. 33, No. 4, December 2010, pp. 747–753.

[7] Selvanayagam, C., J. H. Lau, X. Zhang, S. Seah, K. Vaidyanathan, and T. C. Chai, "Nonlinear Thermal Stress/Strain Analyses of Copper Filled TSV (Through Silicon Via) and Their Flip-Chip Microbumps," *IEEE Transactions on Advanced Packaging*, Vol. 32, No. 4, November 2009, pp. 720–728.

[8] Lee, C. K., T. C. Chang, Y. Huang, H. Fu, J. H. Huang, Z. Hsiao, J. H. Lau, et al., "Characterization and Reliability Assessment of Solder Microbumps and Assembly for 3D IC Integration," *IEEE ECTC Proceedings*, Orlando, FL, June 2011, pp. 1468–1474.

[9] Lee, C. K., T. C. Chang, J. H. Lau, Y. Huang, H. Fu, J. Huang, Z. Hsiao, et al., "Wafer Bumping, Assembly, and Reliability of Fine-Pitch Lead-Free Micro Solder Joints for 3-D IC Integration," *IEEE Transactions on CPMT*, Vol. 2, No. 8, August 2012, pp. 1229–1238.

[10] Lee, C., C. Zhan, J. H. Lau, Y. Huang, H. Fu, J. Huang, Z. Hsiao, et al., "Wafer Bumping, Assembly, and Reliability Assessment of μbumps," *IEEE/ECTC Proceedings*, San Diego, CA, May 2012, pp. 636–640.

[11] Lau, J. H., P.-J. Tzeng, C.-K. Lee, C.-J. Zhan, M.-J. Dai, L. Li, C.-T. Ko, et al., "Wafer Bumping and Characterizations of Fine-Pitch Lead-Free Solder Microbumps on 12" (300mm) wafer for 3D IC Integration," *Proceedings of IMAPS International Conference*, Long Beach, CA, October 2011, pp. 650–656.

[12] Selvanayagam, C., J. H. Lau, X. Zhang, S. Seah, K. Vaidyanathan, and T. Chai, "Nonlinear Thermal Stress/Strain Analysis of Copper Filled TSV (Through Silicon Via) and Their Flip-Chip Microbumps," *IEEE Proceedings of Electronic, Components & Technology Conference*, Orlando, FL, May 27-30, 2008, pp. 1073–1081.

[13] Lim, S., V. Rao, H. Yin, W. Ching, V. Kripesh, C. Lee, J. H. Lau, et al., "Process Development and Reliability of Microbumps," *IEEE/EPTC Proceedings*, Singapore, December 2008, pp. 367–372.

[14] Agarwal, R., W. Zhang, P. Limaye, R. Labie, B. Dimcic, A. Phommahaxay, and P. Soussan, "Cu/Sn Microbumps Interconnect for 3D TSV Chip Stacking," *IEEE Proceedings of ECTC*, June 2010, Las Vegas, NV, pp. 858–863.

[15] Liu, X., Q. Chen, V. Sundaram, M. Simmons-Matthews, K. Wachtler, R. Tummala, and S. Sitaraman, "Thermo-Mechanical Behavior of Through Silicon Vias in a 3D Integrated Package with Inter-Chip Microbumps," *IEEE ECTC Proceedings*, Orlando, FL, June 2011, pp. 1190–1195.

[16] Vos, J., A. Jourdain, M. Erismis, W. Zhang, K. De Munck, A. La Manna, D. Tezcan, et al., "High Density 20μm Pitch CuSn Microbump Process for High-End 3D Applications," *IEEE ECTC Proceedings*, Orlando, FL, June 2011, pp. 27–31.

[17] Selvanayagam, C., J. H. Lau, X. Zhang, S. Seah, K. Vaidyanathan, and T. C. Chai, "Nonlinear Thermal Stress/Strain Analyses of Copper Filled TSV (Through Silicon Via) and Their Flip-Chip Microbumps," *IEEE Transactions on Advanced Packaging*, Vol. 32, No. 4, November 2009, pp. 720–728.

[18] Zhang, X., R. Rajoo, C. S. Selvanayagam, A. Kumar, V. Rao, N. Khan, J. H. Lau, et al., "Application of Piezoresistive Stress Sensor in Wafer Bumping and Drop Impact Test of Embedded Ultrathin Device," *IEEE Transactions on CPMT*, Vol. 2, No. 16, June 2012, pp. 935–943.

[19] Hubbard, R., and B.-S. Lee, "Low Warpage and Improved 2.5/3Dic Process Capability with a Low Stress Polyimide Dielectric Microbump Lithography for 3D Stacking Applications," *Proceedings of International Wafer Level Packaging Conference*, San Jose, November 2013, S12_P2_1–7.

[20] Jaenen, P., J. Slabbekoorn, A. Miller, W. W. Flack, M. Ranjan, G. Kenyon, R. Hsieh, et al., "Microbump Lithography for 3D Stacking Applications," *Proceedings of International Wafer Level Packaging Conference*, San Jose, November 2013, S12_P3_1–8.

[21] Wang, Y., T. Jiang, J. Im, P. S. Ho, S.-H. Chae, R. Dunne, Y. Takahashi, et al., "Effect of Intermetallic Formation on Electromigration Reliability of TSV-Microbump Joints in 3D Interconnect," *Proceedings of IEEE/ECTC*, San Diego, CA, May 2012, pp. 319–325.

[22] Lim, S., L. Ding, M.-B. Yu, M. Z. Ding, S. D. Velez and V. S. Rao, "Process Integration of Solder bumps and Cu Pillar Microbumps on 2.5D Fine Pitch TSV Interposer," *Proceedings of IEEE/EPTC*, Singapore, December 2013, pp. 429–434.

[23] Chen, Z., B. Talebanpour, Z. Huang, P. Kumar and I. Dutta, "Effect of Processing on the Microstructure and Fracture of Solder Microbumps in 3D Packages," *Proceedings of IEEE/EPTC*, Singapore, December 2013, pp. 539–542.

[24] Ito, Y., T. Fukushima, K.-W. Lee, T. Tanaka, and M. Koyanagi, Flux-Assisted Self-Assembly with Microbump Bonding for 3D Heterogeneous Integration," *Proceedings of IEEE/ECTC*, Las Vegas, NV, May 2013, pp. 891–896.

[25] Wang, Y., S.-H. Chae, J. Im, and P. S. Ho, "Kinetics Study of Intermetallic Growth and its Reliability Implications in Pb-Free, Sn-Based Microbumps in 3D Integration," *Proceedings of IEEE/ECTC*, Las Vegas, NV, May 2013, pp. 1953–1958.

[26] Murugesan, M., H. Kobayashi, H. Shimamoto, F. Yamada, T. Fukushima, J. Bea, K. Lee, et al. "Minimizing the Local Deformation Induced Around Cu-TSVs and CuSn/InAu-Microbumps in High-Density 3D-LSIs," *Proceedings of IEEE/IEDM*, San Francisco, December 2012, pp. 28.6.1–28.6.4.

[27] Huffman, A., M. Lueck, C. Bower, and D. Temple, "Effects of Assembly Process Parameters on the Structure and Thermal Stability of Sn-Capped Cu Bump Bonds," *Proceedings of the 57th Electronic Components and Technology Conference*, Reno, NV, May–June 2007, pp. 1589–1596.

[28] Dimcic, B., R. Labie, W. Zhang, I. De Wolf, and B. Verlinden, "Influence of the Processing Method on the Amount and Development of Voids in Miniaturized Interconnections," *Proceedings of the 3rd Electronic System-Integration Technology Conference*, Berlin, Germany, September 2010, pp. 1216–1222.

[29] Wright, S. L., R. Polastre, H. Gan, L. P. Buchwalter, R. Horton, P. S. Andry, E. Sprogis, et al., "Characterization of Micro-Bump C4 Interconnects for Si-Carrier SOP Applications," *Proceedings of the 56th Electronic Components and Technology Conference*, San Diego, CA, May–June 2006, pp. 633–640.

[30] Gan, H., S. L. Wright, R. Polastre, L. P. Buchwalter, R. Horton, P. S. Andry, C. Patel, et al., "Ph-Free Micro Joints (50 Mm Pitch) for the Next Generation Micro-Systems: The Fabrication, Assembly and Characterization," *Proceedinds of the 56th Electronic Components and Technology Conference*, San Diego, CA, May–June 2006, pp. 1210–1215.

[31] Yokomine, K., N. Shimizu, Y. Miyamoto, Y. Iwata, D. Love, and K. Newman, "Development of Electroless Ni/Au Plated Build-Up Filp Chip Package with Highly Reliable Solder Joints," *Proceedings of the 51st Electronic Components and Technology Conference*, Orlando, FL, May–June 2001, pp. 1384–1392.

[32] Iwasaki, T., M. Watanabe, S. Baba, Y. Hatanaka, S. Idaka, Y. Yokoyama, and M. Kimura, "Development of 30 Micron Pitch Bump Interconnections for COC-FCBGA," *Proceedings of the 56th Electronic Components and Technology Conference*, San Diego, CA, May–June 2006, pp. 1216–1222.

[33] Ruhmer, K., E. Laine, K. O'Donnell, J. Kostetsky, K. Hauck, D. Manessis, A. Ostmann, et al., "Alternative UBM for Lead Free Solder Bumping using C4NP," *Proceedings of the 57th Electronic Components and Technology Conference*, Reno, NV, May–June 2007, pp. 15–21.

[34] Khong, C. H., A. Yu, X. Zhang, V. Kripesh, D. Pinjala, D.-L. Kwong, S. Chen, et al., "Sub-Modeling Technique for Thermo-Mechanical Simulation of Solder Microbumps Assembly in 3D Chip Stacking," *Proceedings of the 11th Electronic Packaging Technology Conference*, Singapore, December 2009, pp. 591–595.

[35] Sakuma, K., K. Toriyama, H. Noma, K. Sueoka, N. Unami, J. Mizuno, S. Shoji, et al., "Fluxless Bonding for Fine-Pitch and Low-Volume Solder 3-D Interconnections," *Proceedings of the 61st Electronic and Component Technology Conference*, May–June 2011, pp. 7–13.

[36] Zhan, C.-J., J.-Y. Juang, Y.-M. Lin, Y.-W. Huang, K.-S. Kao, T.-F. Yang, J. H. Lau, et al., "Development of Fluxless Chip-On-Wafer Bonding Process for 3DIC Chip Stacking with 30 μm Pitch Lead-Free Solder Micro Bumps and Reliability Characterization," *Proceedings of the 61st Electronic and Component Technology Conference*, May–June 2011, pp. 14–21.

[37] Hwang, J., J. Kim, W. Kwon, U. Kang, T. Cho, and S. Kang,"Fine Pitch Chip Interconnection Technology For 3D Integration ," *Electronic Components and Technology Conference (ECTC)*, 2010, pp. 1399–1403.

[38] Chen, H. Y., and C. Chen, "Kinetic Study of Eutectic Sn-3.5Ag and Electroplated Ni Metallization in Flip-Chip Solder Joints," Proceedings of *Electronic Materials and Packaging*, 2008, pp. 262–267.

[39] Huang, S., T. Chang, R. Cheng, J. Chang, C. Fan, C. Zhan, J. H. Lau, et al., "Failure Mechanism of 20 μm Pitch Micro Joint Within a Chip Stacking Architecture," *IEEE ECTC Proceedings*, Orlando, FL, June 2011, pp. 886–892.

[40] Sakuma, K., P. S. Andry, B. Dang, J. Maria, C. K. Tsang, C. Patel, S. L. Wright, et al., "3D Chip Stacking Technology with Low-Volume Lead-Free Interconnections," *57th Proceedings of Electronic Components and Technology Conference*, 2007, pp. 627–632.

[41] Juang, J. Y., S. Huang, C. Zhan, Y. Lin, Y. Huang, C. Fan, S. Chung, et al., "Effect of Metal Finishing Fabricated by Electro and Electro-less Plating Process on Reliability Performance of 30μm-Pitch Solder Micro Bump Interconnection," *IEEE/ECTC Proceedings*, Las Vegas, NA, May 2013, pp. 653–659.

[42] Huang, S., C. Zhan, Y. Huang, Y. Lin, C. Fan, S. Chung, J. H. Lau, et al., "Effects of UBM Structure/Material on the Reliability Performance of 3D Chip Stacking with 30μm-pitch Solder Micro Bump Interconnections," *IEEE/ECTC Proceedings*, San Diego, CA, May 2012, pp. 1287–1292.

[43] Lin, Y., C. Zhan, K. Kao, C. Fan, S. Chung, Y. Huang, J. H. Lau, et al., "Low Temperature Bonding using Non-Conductive Adhesive for 3D Chip Stacking with 30μm-Pitch Micro Solder Bump Interconnections," *IEEE/ECTC Proceedings*, San Diego, CA, May 2012, pp. 1656–1661.

[44] Kuo, K. H., J. Lee, S. Chen, F. L. Chien, R. Lee, and J. H. Lau, "Electromigration Performance of Printed Sn0.7Cu Bumps with Immersion Tin Surface Finishing for Flip Chip Applications," *IEEE/ECTC Proceedings*, San Diego, CA, May 2012, pp. 698–702.

[45] Lin, Y., C. Zhan, J. Juang, J. H. Lau, T. Chen, R. Lo, M. Kao, et al., "Electromigration in Ni/Sn Intermetallic Micro Bump Joint for 3D IC Chip Stacking," *IEEE ECTC Proceedings*, Orlando, FL, June 2011, pp. 351–357.

[46] Vempati, S. R., S. Nandar, C. Khong, Y. Lim, K. Vaidyanathan, J. H. Lau, B. P. Liew, et al., "Development of 3–D Silicon Die Stacked Package Using Flip Chip Technology with Micro Bump Interconnects," *IEEE Proceedings of ECTC*, San Diego, CA, May 2009, pp. 980–987.

[47] Yu, D., T. Chai, M. Thew, Y. Ong, S. Vempati, W. Leong, and J. H. Lau, "Electromigration Study of 50 μm Pitch Micro Solder Bumps using Four-Point Kelvin Structure," *IEEE/ECTC Proceedings*, San Diego, CA, May 2009, pp. 930–935.

[48] Choi, K., K. Sung, H. Bae, J. Moon, and Y. Eom, "Bumping and Stacking Processes for 3D IC using Fluxfree Polymer," *IEEE ECTC Proceedings*, Orlando, FL, June 2011, pp. 1746–1751.

[49] Au, K., J. Beleran, Y. Yang, Y. Zhang, S. Kriangsak, P. Wilson, Y. Drake, et al., "Thru Silicon Via Stacking & Numerical Characterization for Multi-Die Interconnections using Full Array & Very Fine Pitch Micro C4 Bumps," *IEEE ECTC Proceedings*, Orlando, FL, June 2011, pp. 296–303.
[50] Lin, T., R. Wang, M. Chen, C. Chiu, S. Chen, T. Yeh, L. Lin, et al., "Electromigration Study of Micro Bumps at Si/Si Interface in 3DIC Package for 28nm Technology and Beyond," *IEEE ECTC Proceedings*, Orlando, FL, June 2011, pp. 346–350.
[51] Wei, C., C. Yu, C. Tung, R. Huang, C. Hsieh, C. Chiu, H. Hsiao, et al., "Comparison of the Electromigration Behaviors Between Micro-bumps and C4 Solder Bumps," *IEEE ECTC Proceedings*, Orlando, FL, June 2011, pp. 706–710.
[52] Meinshausen, L., K. Weide-Zaage, and M. Petzold, "Electro- and Thermomigration in Micro Bump Interconnects for 3D Integration," *IEEE ECTC Proceedings*, Orlando, FL, June 2011, pp. 1444–1451.
[53] Zhang, W., B. Dimcic, P. Limaye, A. Manna, P. Soussan, and E. Beyne, "Ni/Cu/Sn Bumping Scheme for Fine-Pitch Micro-Bump Connections," *IEEE ECTC Proceedings*, Orlando, FL, June 2011, pp. 109–113.
[54] Shigetou, A., T. Itoh, K. Sawada, and T. Suga, "Bumpless Interconnect of 6-um pitch Cu Electrodes at Room Temperature," *IEEE Proceedings of ECTC*, Lake Buena Vista, FL, May 27–30, 2008, pp. 1405–1409.
[55] Shigetou, A., T. Itoh, M. Matsuo, N. Hayasaka, K. Okumura, and T. Suga, "Bumpless Interconnect Through Ultrafine Cu Electrodes by Mans of Surface-Activated Bonding (SAB) Method," *IEEE Transaction on Advanced Packaging*, Vol. 29, No. 2, May 2006, p. 226.
[56] Ko. C., Z. C. Hsiao, P. S. Chen, J. H. Huang, H. C. Fu, Y. J. Huang, C. W. Chiang, et al.—ITRI; Y. J. Chang and K. N. Chen, "Structural Design, Process, and Reliability of a Wafer-Level 3D Integration Scheme with Cu TSVs based on Micro-Bump/Adhesive Hybrid Wafer Bonding," *Proceedings of IEEE/ECTC*, San Diego, CA, May 2012, pp. 1–7
[57] Banijamali, B., S. Ramalingam, H. Liu, and M. Kim, "Outstanding and Innovative Reliability Study of 3D TSV Interposer and Fine Pitch Solder Micro-Bumps," *Proceedings of IEEE/ECTC*, San Diego, CA, May 2012, pp. 309–314.
[58] You, H.-Y., Y. Hwang, J.-W. Pyun, Y.-G. Ryu, and H.-S. Kim, "Chip Package Interaction in Micro Bump and TSV Structure," *Proceedings of IEEE/ECTC*, San Diego, CA, May 2012, pp. 315–318.
[59] Choi, Y., J. Shin, and K.-W. Paik, "3D-TSV Vertical Interconnection Method using Cu/SnAg Double Bumps and B-Stage Non-Conductive Adhesives (NCAs)," *Proceedings of IEEE/ECTC*, San Diego, CA, May 2012, pp. 1077–1080.
[60] Park, Y.-S., J.-W. Shin, Y.-W. Choi, and K.-W. Paik, "Study on the Intermetallic Growth of Fine-Pitch Cu Pillar/SnAg Solder Bump for 3D-TSV Interconnection," *Proceedings of IEEE/ECTC*, San Diego, CA, May 2012, pp. 2053–2056.
[61] Sorono, D. V., S. R. Vempati, L. Bu, S. C. Chong, C. T. W. Liang, and S. W. Wei, "Simultaneous Molding and Under-filling for Void Free Process to Encapsulate Fine Pitch Micro Bump Interconnections of Chip-to-Wafer (C2W) Bonding in Wafer Level Packaging," *Proceedings of IEEE/EPTC*, Singapore, December 2013, pp. 72–77.
[62] Rao, V. S., S. C. Chong, C. Zhaohui, J. L. Aw, E. W. L. Ching, H. Gilho, and D. M. Fernandez, "Development of Bonding Process for High Density Fine Pitch Micro Bump Interconnections with Wafer Level Underfill for 3D Applications," *Proceedings of IEEE/EPTC*, Singapore, December 2013, pp. 548–553.
[63] Son, H.-Y., S.-K. Noh, H.-H. Jung, W.-S. Lee, J.-S. Oh, and N.-S. Kim, "Reliability Studies on Micro-Bumps for 3-D TSV Integration," *Proceedings of IEEE/ECTC*, Las Vegas, NV, May 2013, pp. 29–34.
[64] Yeh, C.-L., Y.-Y. Yeh, J.-C. Kao, T. H. Wang, C.-C. Lee, and H.-M. Tong, " "Micro-Bump Bondability Design Guidelines for High Throughput 2.5D/3D IC Assemblies," *Proceedings of IEEE/ECTC*, Las Vegas, NV, May 2013, pp. 897–903.
[65] Shin, J.-W., Y.-W. Choi, Y. S. Kim, U. B. Kang, Y. K. Jee, and K.-W. Paik, "Effect of NCFs with Zn-Nanoparticles on the Interfacial Reactions of 40 um Pitch Cu Pillar/Sn-Ag Bump for TSV Interconnection," *Proceedings of IEEE/ECTC*, Las Vegas, NV, May 2013, pp. 1024–1030.
[66] De Vos, J., L. Bogaerts, T. Buisson, C. Gerets, G. Jamieson, K. Vandersmissen, A. La Manna, et al., "Key Elements for Sub-50μm Pitch Micro Bump Processes," *Proceedings of IEEE/ECTC*, Las Vegas, NV, May 2013, pp. 1122–1126.
[67] Bertheau, J., P. Bleuet, R. Pantel, J. Charbonnier, F. Hodaj, P. Coudrain, and N. Hotellier, "Microstructural and Morphological Characterization of SnAgCu Micro-Bumps for Integration in 3D Interconnects," *Proceedings of IEEE/ECTC*, Las Vegas, NV, May 2013, pp. 1127–1132.

[68] Shuto, T., K. Iwanabe, L. J. Qiu, and T. Asano, "Room-Temperature High-Density Interconnection Using Ultrasonic Bonding of Cone Bump for Heterogeneous Integration," *Proceedings of IEEE/ECTC*, Las Vegas, NV, May 2013, pp. 1141–1145.
[69] Park, Y.-B., S.-H. Kim, J.-J. Park, J.-B. Kim, H.-Y. Son, K.-W. Han, J.-S. Oh, et al., "Current Density Effects on the Electrical Reliability of Ultra-Fine-Pitch Micro-Bump for TSV Integration," *Proceedings of IEEE/ECTC*, Las Vegas, NV, May 2013, pp. 1988–1993.
[70] Coudrain, P., D. Henry, A. Berthelot, J. Charbonnier, S. Verrun, R. Franiatte, N. Bouzaida, et al., "3D Integration of CMOS Image Sensor with Coprocessor Using TSV Last and Micro-Bumps Technologies," *Proceedings of IEEE/ECTC*, Las Vegas, NV, May 2013, pp. 674–682.
[71] La Manna, A., K. Rebibis, J. De Vos, L. Bogaerts, C. Gerets, and E. Beyne, "Small Pitch Micro-Bumping and Experimental Investigation Under Filling 3D Stacking," *Proceedings of the IMAPS International Symposium on Microelectronics*, San Diego, CA, September 2012, pp. 535–541.
[72] Asgari, R., "Challenges in 3D Inspection of Micro Bumps Used in 3D Packaging," *Proceedings of the IMAPS International Symposium on Microelectronics*, September 2012, San Diego, CA, pp. 542–547.
[73] Lau, J. H., *Through Silicon Via (TSV) for 3D Integration*, McGraw-Hill Book Company, New York, 2013.
[74] Lau, J. H., *Reliability of RoHS Compliant 2D & 3D IC Interconnects*, McGraw-Hill Book Company, New York, 2011.
[75] Lau, J. H., C. K. Lee, C. S. Premachandran, and Yu Aibin, *Advanced MEMS Packaging*, McGraw-Hill Book Company, New York, 2010.
[76] Lau, J. H., *Flip Chip Technology*, McGraw-Hill Book Company, New York, 1995.
[77] Lau, J. H., *Low-Cost Flip Chip Technology for WLCSP*, McGraw-Hill Book Company, New York, 2000.
[78] Lloyd, J. R., N. A. Connelly, X. He, K. J. Ryan, and B.H. Wood, "Fast Diffusers in a Thermal Gradient (Solder Ball)," *Microelectronics Reliability*, Vol. 50, 2010, pp. 1355–1358.
[79] Spivak, A., "Gravity Segregation in Two-Phase Displacement Processes," *SPE Journal*, Vol. 14, No. 6, 1974, pp. 619–632.
[80] Shabestari, S. G., and J. E. Gruzleski, "Gravity Segregation of Complex Intermetallic Compounds in Liquid Al-Si Alloys," *Metallurgical and Materials Transaction A*, Vol. 26, 1995, pp. 999–1006.
[81] Shen, J., Y. C. Chan, and S. Y. Liu, "Growth Mechanism of Ni_3Sn_4 in a Sn/Ni Liquid/Solid Interfacial Reaction," *Acta Materialia*, Vol. 57, 2009, pp. 5196–5206.
[82] Gur, D., and M. Bamberger, "Formation and Growth of Ni_3Sn_4 Intermediate Phase in the Ni-Sn System," *Journal of Materials Science*, Vol. 35, 2000, pp. 4601–4606.

CHAPTER 6

3D Si Integration

6.1 Introduction

As mentioned earlier, 3D integration consists of 3D IC packaging, 3D IC integration, and 3D Si integration. They are different and in general the TSV (through-silicon via) separates 3D IC packaging from 3D IC/Si integrations since the latter two use TSV but 3D IC packaging does not. In this chapter, the origin of 3D integration is presented. Also, the evolution, challenges, and outlook of 3D Si integrations are discussed. The electronic industry will be briefly mentioned first.

3D IC integration will be discussed in Chaps. 7 and 8, while 3D IC packaging will be discussed in Chap. 14. Chapters 9 through 13, respectively discuss the thermal management of 3D IC integration systems, embedded 3D hybrid integration, 3D LED and IC integration, 3D MEMS and IC integration, and 3D CMOS image sensor and IC integration. The key enabling technologies such as TSV, thin-wafer strengthening and handling, package substrates, microsolder bumping, assembly, and reliability have been discussed in Chaps. 2 through 5.

6.2 The Electronic Industry

The electronics industry has been the largest industry since 1996 and may well reach 1.5 trillion dollars (i.e., $\$1.5 \times 10^{12}$) by the end of 2015 [1–3]. The most important invention of the electronics industry is, arguably, the transistor (1947), Fig. 6.1, which earned John Bardeen, Walter Brattain, and William Shockley the 1956 Nobel Prize in Physics. Today, it is not uncommon to find a chip with more than 2.5 billion of transistors.

The invention of IC (integrated circuit) by Jack Kilby in 1958 (which earned him the 2000 Nobel Prize in Physics), and six months later by Robert Noyce (who did not share the Nobel Prize with Jack Kilby because he passed away in 1990) excited the generations of IC integrations [2], Fig. 6.2. Kilby's substrate was made of germanium, and Noyce's substrate was made of silicon. Today, more than 90% of transistors produced are on silicon substrates.

The proposal of doubling the number of transistors on an IC chip (for minimum costs and innovations) every 24 months by Gordon Moore in 1965 (also called Moore's law) [4] has been the most powerful driver for the development of the microelectronics industry in the past ~50 years, Fig. 6.3.

FIGURE 6.1 Bell Labs junction transistor; John Bardeen, Walter Brattain, and William Shockley.

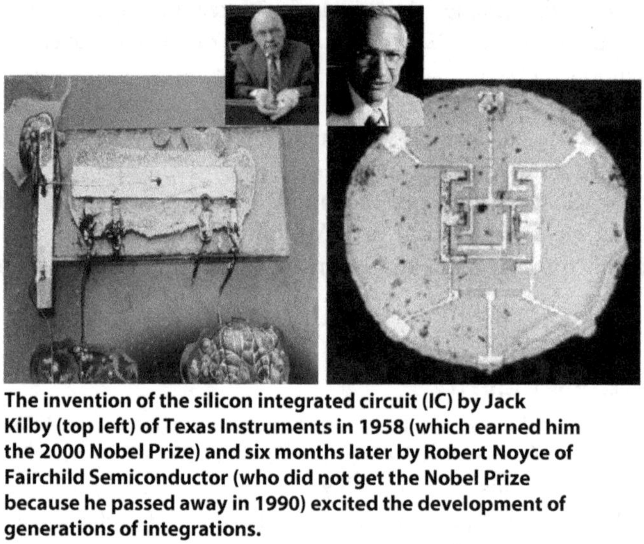

FIGURE 6.2 (Left) Texas Instruments' integrated circuit by Jack Kilby. (Right) Fairchild Semiconductor's integrated circuit by Robert Noyce.

6.3 Moore's Law and More-Than-Moore

Moore's law emphasizes lithography scaling and integration (in 2D) of all functions on a single chip, perhaps through system-on-chip (SoC), as shown in Fig. 6.4. On the other hand, the integration of all these functions can be achieved through 3D integrations [1–3, 5–50] such as 3D IC packaging, 3D IC integration, and 3D Si integration as shown in Figs. 6.4 and 6.5. Since 3D IC packaging is a mature technology and does not use TSVs, it is not included in this chapter. The 3D IC integration and 3D Si integration are shown in the right-hand side of Figs. 6.4 and 6.5 and are some of the more-than-Moore.

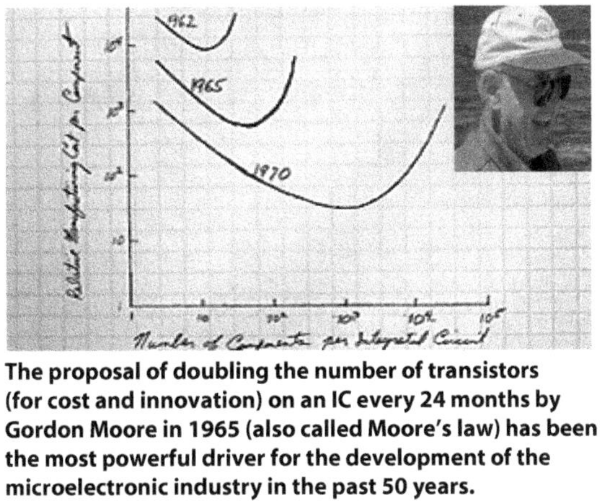

Figure 6.3 Moore's law by Gordon Moore in 1965.

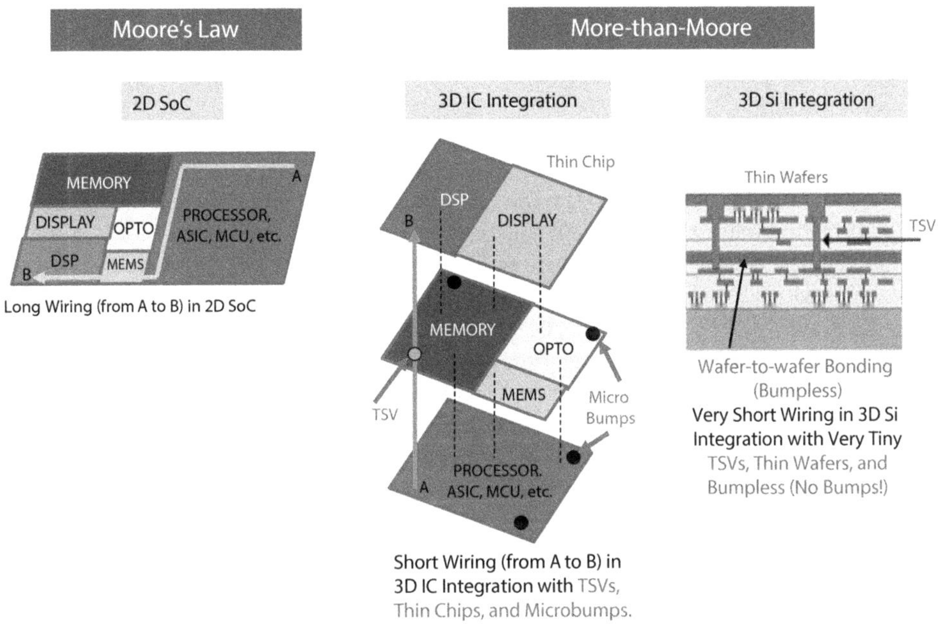

Figure 6.4 Moore's law vs. More-Than-Moore.

6.4 The Origin of 3D Integration

Again, since the focus of this chapter is TSV, thus 3D IC packaging is out of the scope and will only be discussed in Chap. 14. 3D integration is a very old idea [1, 2] which consists of two or more layers of active electronic components that are integrated vertically through TSV (it used to be called vertical interconnection) into a single circuit. It was trigged by the

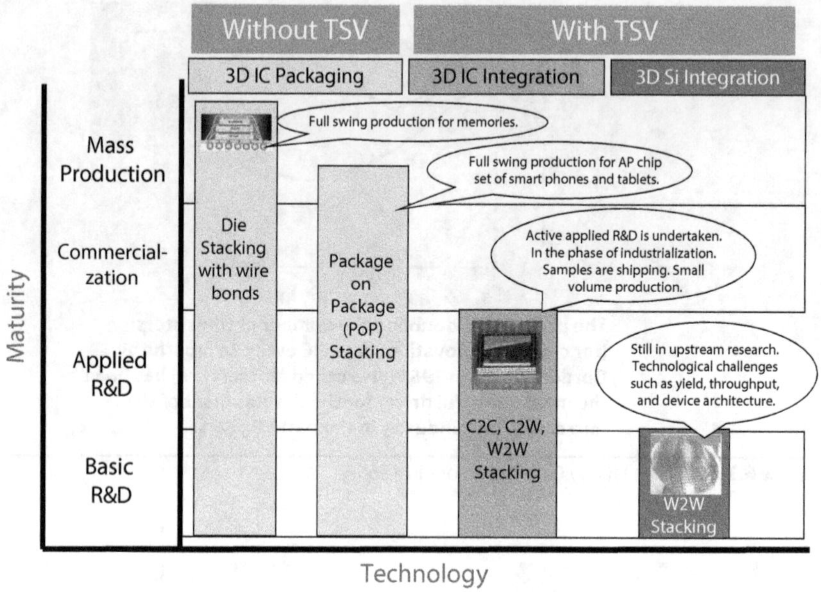

Figure 6.5 Maturity of 3D technology.

advance of the silicon-on-insulator (SOI) technology first reported by Gat and his colleagues more than 30 years ago [51], when semiconductor people thought Moore's law could be hitting the wall by the 1990s. Of course, the fact showed otherwise.

In the early 1980s, there were two schools of thought [27, 28]. One is to stack up the chips with TSV and flip chip solder bumps (3D IC integration), and the other is to stack up the wafers with TSV alone, i.e., bumpless (3D Si integration), as shown in Figs. 6.4 and 6.5. The advantages of 3D Si integration over 3D IC integration are (1) better electrical performance, (2) less power, (3) lower profile, (4) less weight, and (5) higher throughput. In general, at that time, the industry favored the 3D Si integration.

The most powerful pusher on 3D integration is the 1965 Nobel Physics laureate, Richard Feynman. At the Yoshio Nishina Memorial Lecture at Gakushuin University (Tokyo) on August 9, 1985 with the title *Computing Machines in the Future*, he said, "Another direction of improvement (of computing power) is to make physical machines three dimensional instead of all on a surface of a chip. That can be done in stages instead of all at once—you can have several layers and then add many more layers as time goes on." Wow, Feynman not only told us to go for 3D, but also taught us how to do it more than 30 years ago. Even today, many people who are hunting for 3D integration research funding like to cite his 1985 lecture given at Tokyo.

6.5 Overview and Outlook of 3D Si Integration

In the early 1980s, the Japan MITII (Ministry of International Trade and Industry) funded and directed the 3D Research Committee of the 3D Si integration project. Their road map is shown in Fig. 6.6 [27], where it stated that (1) functional models have been fabricated in

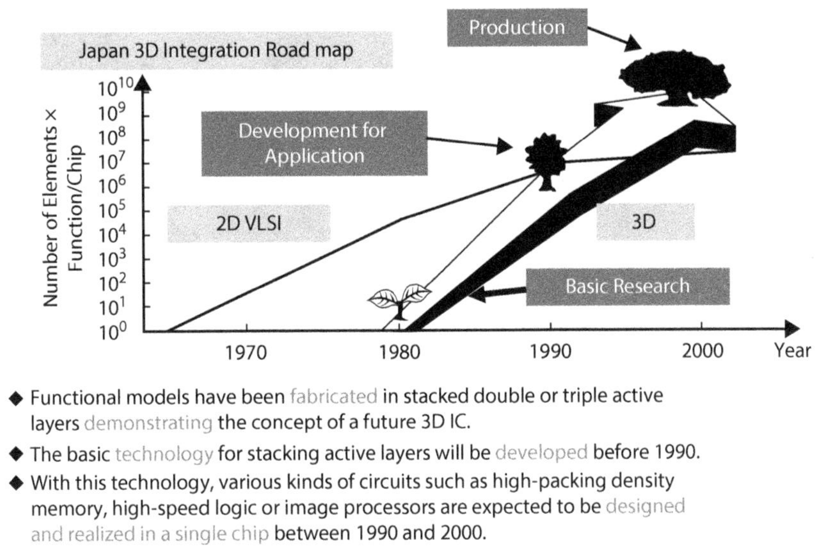

Figure 6.6 Japan MITI 3D Si integration road map.

stacked double or triple active layers demonstrating the concept of a future 3D structure; (2) the basic technology for stacking active layers will be developed before 1990; and (3) with this technology, various kinds of circuits such as high-packing density memory, high-speed logic or image processors are expected to be designed and realized in a 3D single chip between 1990 and 2000. History showed that the technology did not materialize within the timeframe stated by the project.

6.5.1 Bonding Methods for 3D Si Integration

Basically, W2W (wafer-to-wafer) is the most common method to perform the bonding operation for 3D Si integration and yield is a big issue (e.g., some bad chips are forced to bond on the good chips). In addition, the absence of (or an infinitesimal) gap between wafers and thermal management is a huge problem. Furthermore, the requirements of the bonding conditions, such as the surface cleanness, surface flatness, and the class of clean room for 3D Si integration are very high. Some most common W2W bonders are shown in Fig. 6.7.

There are many different W2W bonding methods for 3D Si integration. Basically, there are two groups as shown in Table 6.1, one is with an intermediate layer between the bonding pads/traces and the other is not, i.e., nothing! The focuses of this chapter are on: (a) Cu-to-Cu bonding and (b) oxide-to-oxide bonding, which do not have anything between the bonding pads/traces on the wafers and thus are called direct bonding.

Bonding methods with intermediate layer such as: (a) BCB (Benzocyclobutene)-based polymers; (b) solder IMCs (intermetallic compounds) from TLP (transient liquid phase) bonding; and (c) solder by thermal compression are called indirect bonding and will not be considered in this section. Why? The reasons are (1) both with polymer or solder, they will add lots of process steps and materials and thus increase costs substantially; (2) additional process steps and materials increase reliability problems; (3) they are too "dirty" to be

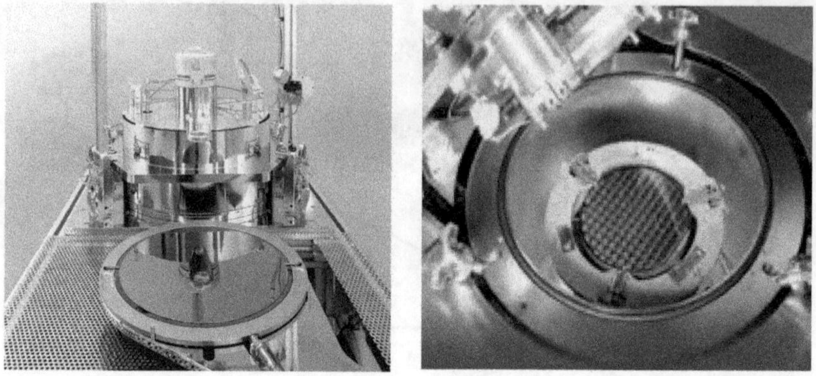

Figure 6.7 (Left) EVG W2W bonder. (Right) SUSS W2W bonder.

operated in semiconductor fabs; and (4) the intermediate polymer adhesive layer has been shown (a) there's a concern on the reliability of the adhesive layer during the post-bond processes and during the device operations, and (b) there's a great challenge for maintaining the alignment accuracy, especially for more than two stacks.

6.5.2 Cu-to-Cu (W2W) Bonding

The advantages of Cu-to-Cu bonding pads are to provide much lower electrical resistivity, very high density, and lower electromigration than any other joints. On the other hand, in

Table 6.1 3D Integration Process Steps

Processes	Methods/Options		
Via forming	Bosch DRIE	Non-Bosch DRIE	Laser
Dielectric deposition	SiO_2	SiN/SiO_2	Polymer
Barrier/Seed layers deposition	Ti (or Ta)/Cu	TiW/Cu W/Cu	W/W
Via filling	Cu	W	Conductive polymer, CNT, solder, etc.
Residue removal	CMP	CMP (Two-step)	
TSV revealing	Wet etch	Dry etch	
TSV process	TSV before bonding, TSV after bonding	Via first, via middle	Via last (front- or back-side)
Thin-wafer handling	Support (Carrier) Wafer	Carrierless	On stacking
Stacking	C2C	C2W	W2W
Micro-interconnect	Solder bump	Cu pillar + solder	Au/Cu studs
Bonding	Nature reflow, thermo compression	Direct bonding	Indirect bonding (intermediate layer)

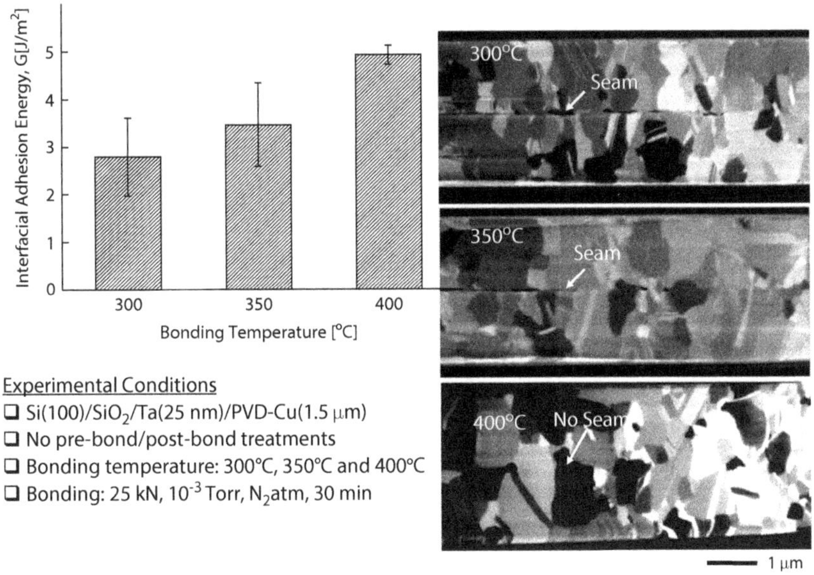

Figure 6.8 Effect of bonding temperature on bonding interface properties: (Left) interfacial adhesion energies and (Right) SEM images of microstructures.

order to reduce the tendency to form native oxides which strongly affect the bonding reliability, Cu-to-Cu bonding usually operates at high temperature (~400°C) and pressure and long process time (60–120 min), which are not good for throughput and the device reliability.

Figure 6.8 shows the effect of bonding temperature on the critical interfacial adhesion energy. (The critical interfacial adhesion energy is also called critical energy release rate at the interface. If the maximum interfacial adhesion energy is larger than the critical interfacial adhesion energy, then the interfacial delamination will take place.) It can be seen that the higher the bonding temperature the higher the critical interfacial adhesion energy (G_c), i.e., the stronger the bond (joint). Also, it is shown in Fig. 6.8 that the higher the temperatures, the less the seams between the interface and the original bond interface tends to disappear due to an activated interdiffusion through the two interfacial layers. That's the major reason why high temperature is needed for Cu to-Cu bonding [26].

Figure 6.9(a) schematically shows the IBM/RPI bonding structure of two device layers bonded in a face-to-face approach, and Figure 6.9(b) shows one with the face-to-back approach [29–31]. A typical Cu-to-Cu interconnect is shown in Fig. 6.10, which shows a high-quality bonding interface. Before bonding, the Cu interconnects (pads) are fabricated

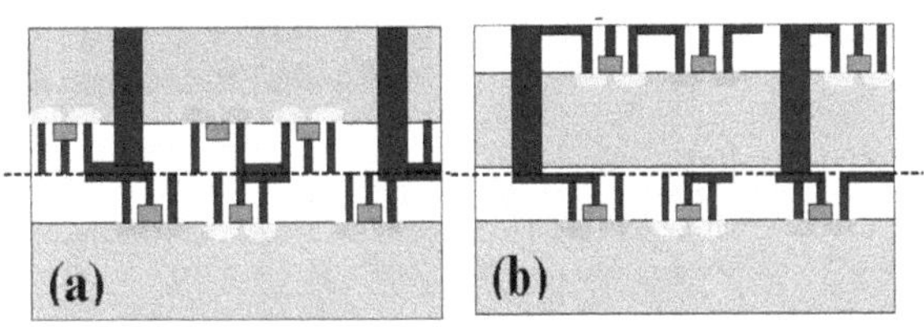

Figure 6.9 Face-to-face and face-to-back W2W bonding.

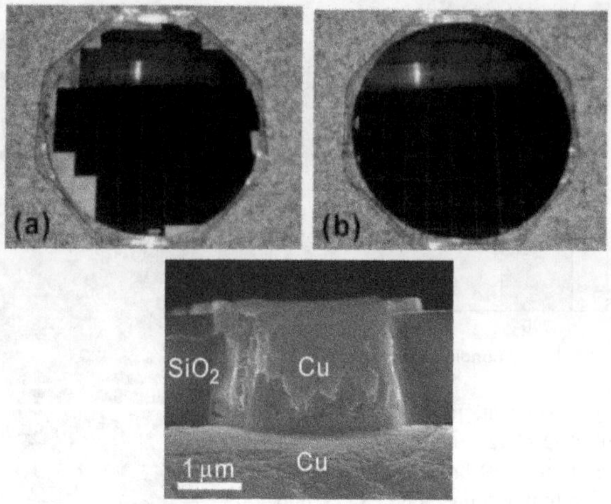

FIGURE 6.10 IBM/RPI's Cu-to-Cu bonding.

with the standard BEOL (back-end-of-line) damascene process, followed by the oxide CMP (chemical-mechanical polishing) processed (oxide touch-up) to recess the oxide level to 40 nm lower than the Cu surface. The bonding temperature is ramped up to 400°C.

6.5.3 Cu-to-Cu (W2W) Bonding with Post-Annealing

One way to reduce the bonding temperature and obtain high quality bonds (joints) is by annealing. Figure 6.11 shows the effects of various annealing temperatures on the critical interfacial adhesion energy, G_c. It can be seen that for bonding temperature at 300°C for 30 min under 25 kN force for a 8" wafer, after annealing temperature at 300°C for 60 min under N_2 atm, the G_c is increased from 2.8 J/m^2 (without annealing) to 12.2 J/m^2. Even for 60 min of annealing temperature at 250°C, the G_c is increased to 8.9 J/m^2. However, too low an annealing temperature won't help, e.g., 200°C, as shown in Fig. 6.11.

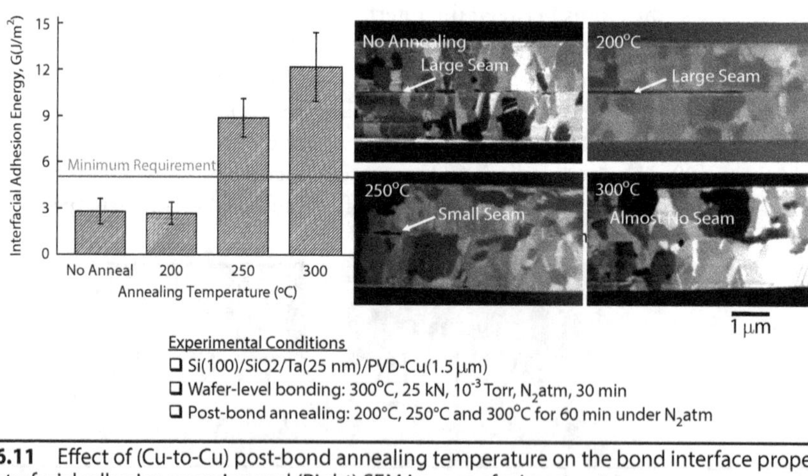

FIGURE 6.11 Effect of (Cu-to-Cu) post-bond annealing temperature on the bond interface properties: (Left) interfacial adhesion energies, and (Right) SEM images of microstructures.

6.5.4 Cu-to-Cu (W2W) Bonding at Room Temperature

W2W (Cu-to-Cu) bonding at room temperature leads to the highest throughput and least device reliability concerns as well as very low costs. However, the drawbacks of room temperature bonding are the stringent requirements on: (1) pad/trace/wafer planarization and (2) surface treatment to ensure smooth hydrophilic surfaces for high quality bonding. Figure 6.12 (left-hand side) schematically shows the NIMS/AIST/Toshiba/University of Tokyo bonding structure of two device layers bonded at room temperature [32–38]. A typical cross-sectional scanning electron microscope (SEM) image of the interface between the bumpless electrodes (pads) after a high-temperature storage test is shown in the right-hand side of Fig. 6.12. It can be seen that a tight adhesion between surfaces is maintained even after exposed to 150°C for 1000 h [32–38].

6.5.5 SiO_2-to-SiO_2 (W2W) Bonding

W2W (SiO_2-to-SiO_2) bonding usually takes three steps: pre-bonding, bonding, and post-bonding. The pre-bonding is operated at room temperature, which eliminates run-out errors in wafer alignment, and thus leads to higher post-bond alignment accuracy. In order to achieve covalent bonds (joints), the bonding temperature is very high (~400°C). In order to achieve strong chemical bonds (joints) with lower annealing (post-bond) temperature (200°C to 400°C), the surface chemistry must be modified by plasma activation. Figure 6.13 shows the effects of annealing temperature on the critical bonding energy. As expected, the higher the annealing temperatures the stronger the critical bonding energy. Unfortunately, due to the maximum allowable temperature of most devices, 400°C is the most used. Figure 6.14 shows the effect of temperature annealing time on the critical surface energy at 300°C annealing temperature. It can be seen that (1) the longer the annealing time the larger the critical surface energy, (2) one hour annealing time is more than enough, and (3) the plasma activation on the surface chemistry before bonding has a great impact on the critical surface energy.

Figure 6.15 schematically shows MIT's oxide-to-oxide bonding structure of three device layers bonded at 275°C [39–45]. It can be seen that (a) two completed circuit wafers (Tier 1 and Tier 2) are planarized, aligned, and bonded face-to-face together; (b) a wet etch of the handle silicon exposes the buried oxide (BOX) of the upper wafer; (c) 3D vias

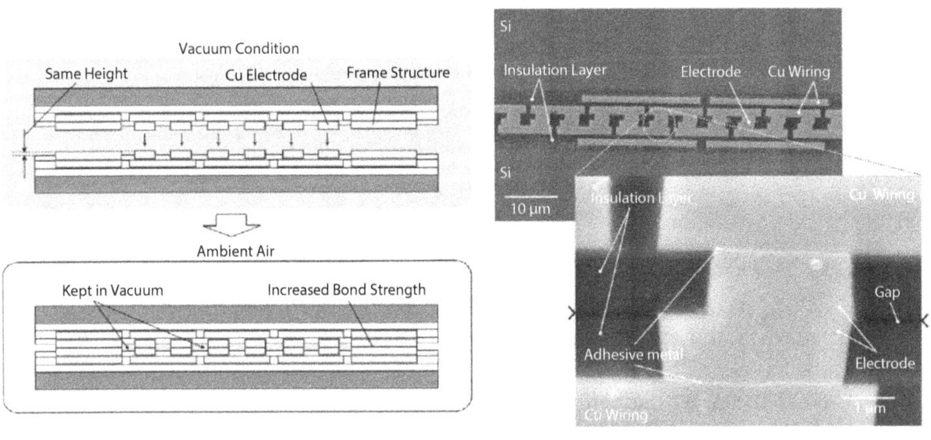

FIGURE 6.12 NIMS/AIST/Toshiba/University of Tokyo's Cu-to-Cu bonding.

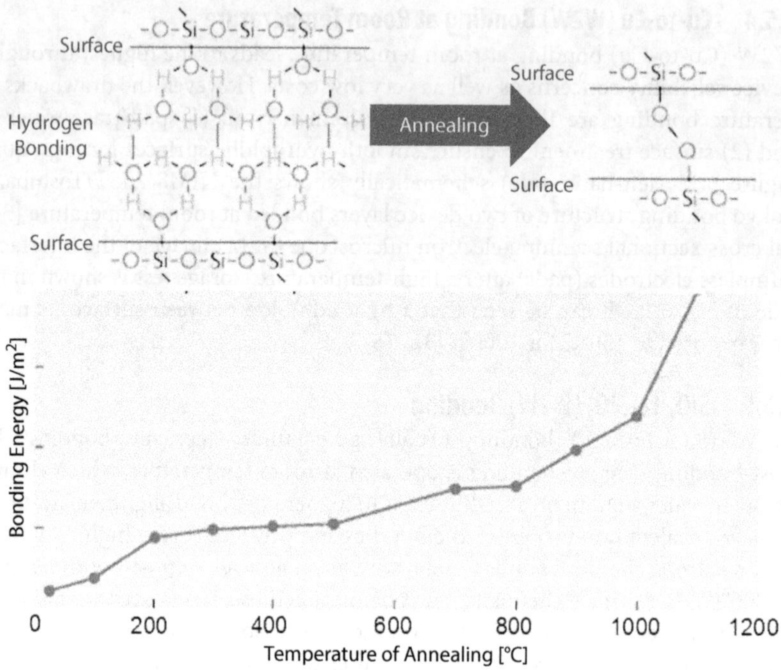

Figure 6.13 Bonding energy (SiO_2-to-SiO_2) as a function of temperature of annealing.

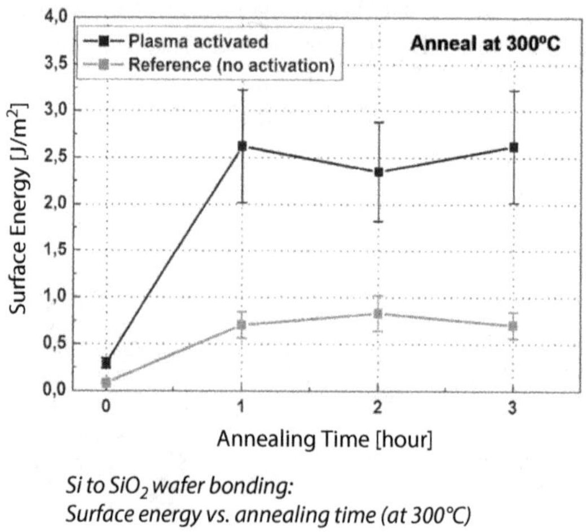

Figure 6.14 Surface energy as a function of thermal annealing time at 300°C (SiO_2-to-SiO_2).

are patterned and etched through the BOX and deposited oxides expose metal contacts in both layers; (d) a Ti/TiN liner and 1 mm of tungsten (W) are deposited to fill the 3D vias (with the larger diameter equal to 1.5 μm) and electrically connect the two layers; and (e) a third layer can then be bonded with its face to the BOX (back) of Tier (layer) 2 and 3D vias are formed. Figure 6.16 shows a typical cross section of the three-layer 3D (ring oscillator) structure. It can be seen that (1) the layers are bonded and interconnected with

MIT's Assembly Process for a 3D Si integration:

1. Two completed circuit wafers are planarized, aligned, and bonded face to face.
2. The handle silicon is removed.
3. 3-D vias are etched through the deposited buried oxide (BOX) and the field oxides.
4. Tungsten plugs are formed to connect circuits in both tiers.
5. After tier 3 is transferred, bond pads are etched through the BOX for testing and packaging.

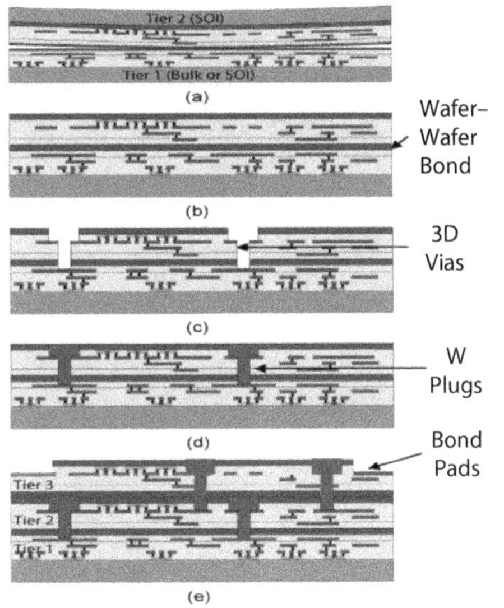

Figure 6.15 MIT's assembly process for SiO_2-to-SiO_2.

W-plugs, (2) the conventional interlevel connections are in the bottom two layers, and (3) the 3D vias are located in the isolation (field) region between transistors. A few functional 3D structures/circuits have been created and demonstrated [39–45].

Figure 6.17 schematically shows Leti/Freescale/STMicroelectronics' dielectric-to-dielectric bonding structure of two device layers bonded at <400°C [46–48]. It can be seen that (a) first, a metal level is formed on a 200 mm bulk wafer and SOI wafer; next, these wafers are bonded face to face, and then the bulk silicon of the SOI wafer is removed down to the BOX layer; (b) the interstrata vias (ISVs) are formed, which make contact from upper strata to lower strata; and (c) a metal layer is formed at the top of the back side of the SOI wafer. The typical cross-section of an ISV is shown in Fig. 6.17 [46–48], and it can be seen that this ISV (~1.5 μm) makes good contact.

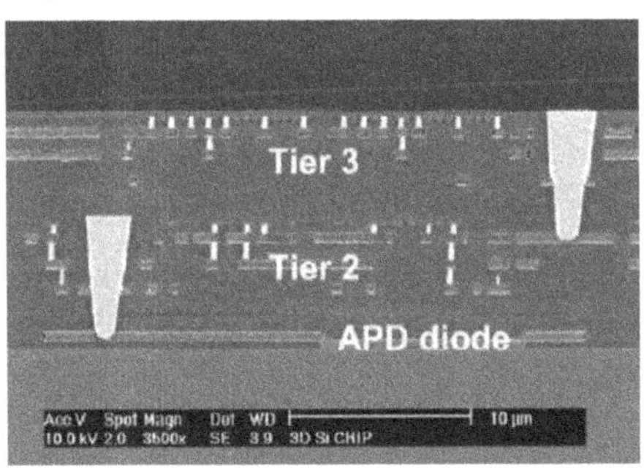

The tiers are bonded and interconnected with tungsten plugs; the conventional interlevel connections are seen in tiers 2 and 3, which are FDSOI tiers. Note that the 3D vias are located in the isolation (field) region between transistors

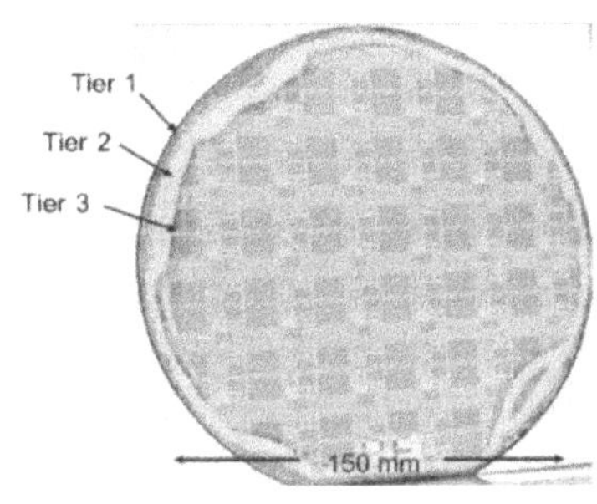

The metal pattern of tier 3 is viewed through the BOX. Tiers 2 and 1 are visible at the edge of the wafer where the tiers did not bond because of nonplanar surfaces

Figure 6.16 MIT's SiO_2-to-SiO_2 bonding results.

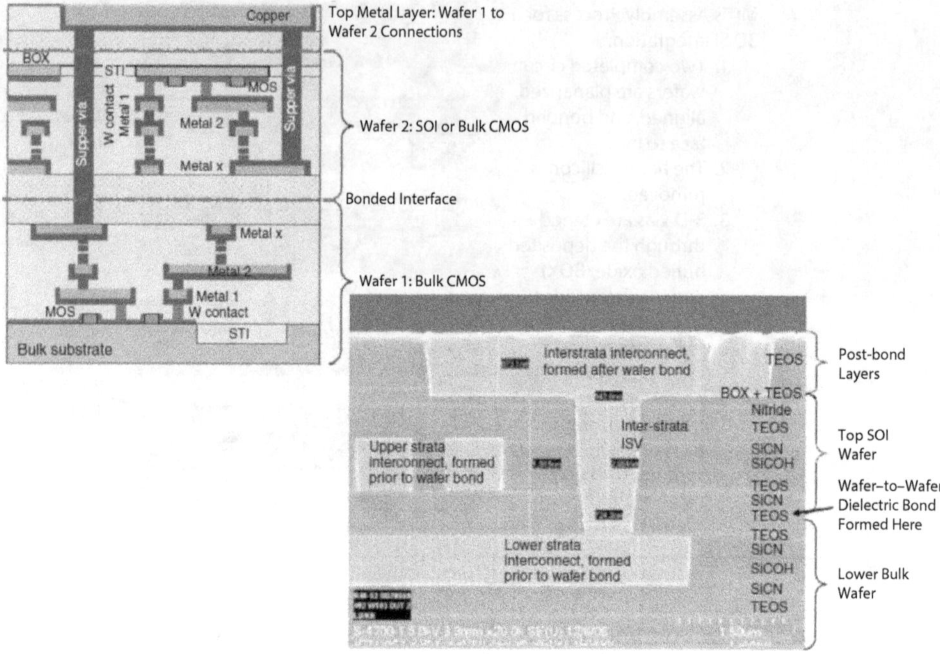

FIGURE 6.17 Leti/Freescale/STMicroelectronics' SiO_2-to-SiO_2 bonding.

6.5.6 A Few Notes on W2W Bonding

It should be noted and emphasized that for 3D Si integration in this book: (1) it is bumpless/adhesiveless, (2) fabricate TSVs before Cu-to-Cu bonding, (3) fabricate TSVs after SiO_2-to-SiO_2 bonding, (4) the TSV diameter is very small (usually $\leqq 2$ μm) and thus the aspect ratios are very high, and (5) the TSVs are usually filled with Tungsten by chemical vapor deposition. Also, for heterogeneous systems, due to different chip size and pin-out, W2W bonding is very difficult. In this case, C2W is one way to go and improves yield.

Recently, W2W hybrid bonding is mentioned in the 3D integration community. Basically, W2W hybrid bonding is performing the metal (e.g., Cu-to-Cu) and dielectric (e.g., SiO_2-to-SiO_2) bonding at the same time.

6.6 3D Si Integration Technology Challenges

Tremendous of work still need to be done before products can be manufactured using the 3D Si integration technology. More research and development efforts should also be placed on areas such as:

1. Cost reduction
2. Thermal management
3. Vias formation
4. Thin-wafer handling
5. Design

6. Process parameter optimization
7. Bonding environment
8. W2W bonding alignment
9. Wafer distortion
10. Wafer bow (warpage)
11. Inspection and testing
12. Contact performance
13. Contact integrity
14. Contact reliability
15. Manufacturing yield
16. Manufacturing throughput
17. Packaging the 3D Si integration module cost-effectively, systematically and reliably to the next level of interconnect.

6.7 3D Si Integration EDA Challenges

Besides technology issues, the EDA (electronic design automation), which is the soul of 3D Si integration [49], is far from ready. Urgently, the industry needs an ecosystem (e.g., standard and infrastructure) for 3D Si integration. Then, the EDA can write the design, simulation, analysis and verification, manufacturing preparation, and test software with the following guidelines:

1. Design automation from high level description to layout generation/optimization.
2. Verification all dedicated and tuned to 3D integration.
3. Addressing the 3rd dimension not like a packaging bumping (bumpless).
4. Addressing the true 3rd dimension, with partitioning, floor planning, automatic placing and routing.
5. Full extraction with the 3rd dimension, full 3D DRC (design rule checks), 3D LVS (layout vs. schematic) with all tiers together in a same database.
6. The 3D Si integration has then to be seen as a whole system distributed in several tiers, and not just a stack of predefined chips.

In the next 10 years, the industry will be hard-pressed for making products at high volume with the 3D Si integration technology, except for very niche applications. However, it should be noted and emphasized that 3D Si integration is the right way to go and compete with Moore's law. The industry should *strive* to make this happens!

6.8 Summary and Recommendations

Some important results and recommendations are summarized as follows:

- 3D Si integration is the right way to go and compete with Moore's law for the semiconductor industry.
- A set of technology challenges (for R&D) for 3D Si integration has been provided.

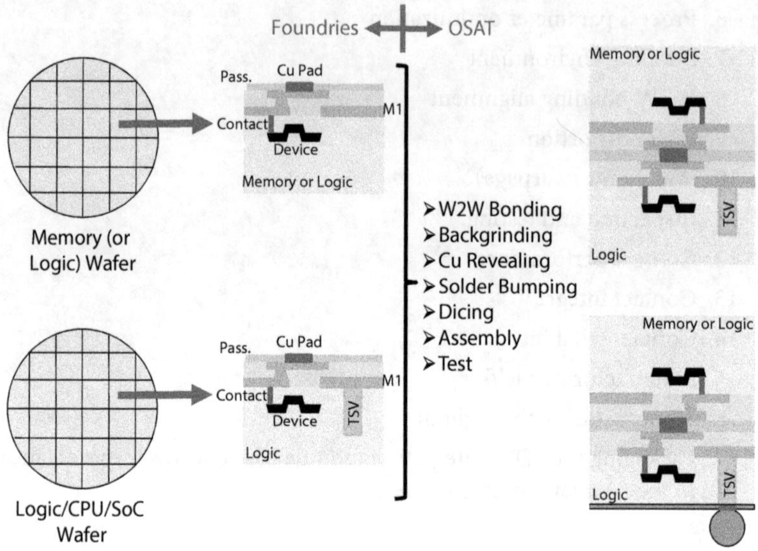

FIGURE 6.18 3D Si integration assembly process of wide I/O memory (or logic-to-logic).

- A simple guideline on developing the EDA has been presented for 3D Si integration.
- The semiconductor industry should strive to make 3D Si integration at high volume happen! Figure 6.18 schematically shows the assembly process for the wide I/O memory (or logic-on-logic) shown in Fig. 1.19. The memory wafer (without TSV)

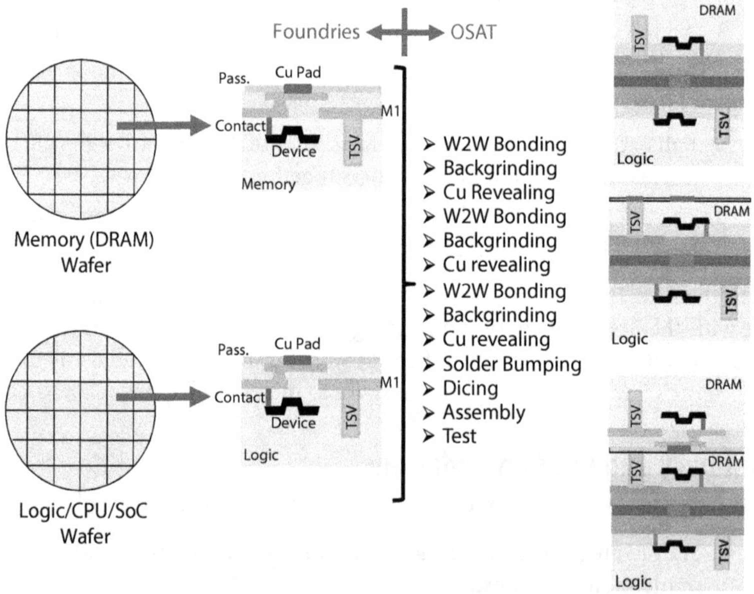

FIGURE 6.19 3D Si integration assembly process of wide I/O DRAM (or HMC).

is face-to-face and Cu-to-Cu bonded to the logic wafer with TSVs. Then, backgrind the logic wafer, Si dry etch, low temperature SiN/SiO_2, and CMP (Cu revealing). It is followed by solder wafer bumping and dicing and then putting on a package substrate. Figure 6.19 schematically shows the assembly process for wide I/O DRAM or HMC shown in Fig. 1.20. First, Cu-to-Cu bond the DRAM wafer to the logic wafer face-to-face, then backgrind the DRAM wafer and Cu reveal. It is followed by Cu-to-Cu bonding another DRAM wafer (face-to-back) on top of the DRAM wafer. Then, backgrinding and Cu revealing. Repeat these processes until the last DRAM wafer (without TSV) is stacked. It is followed by backgrinding and Cu revealing of the bottom logic wafer, solder wafer bumping, dicing and attaching the individual stacked module (wide I/O DRAM or HMC) to the package substrate. For both cases, the devices, metal contacts, metal layers, and TSVs should be fabricated by the foundries. However, the W2W bonding, backgrinding, Cu revealing, solder bumping, dicing, assembly, and testing should be done by the OSATs.

6.9 References

[1] Lau, J. H., *Through-Silicon Via for 3D Integration*, McGraw-Hill, NY, 2013.
[2] Lau, J. H., *Reliability of RoHS Compliant 2D & 3D IC Interconnects*, McGraw-Hill, NY, 2011.
[3] Lau, J. H., C. K. Lee, C. S. Premachandran, and A. Yu, *Advanced MEMS Packaging*, McGraw-Hill, NY, 2010.
[4] Gordon, M., "Cramming More Components Onto Integrated Circuits," *Electronics*, Vol. 38, No. 8, April 19, 1965.
[5] Lau, J. H., "Overview and Outlook of TSV and 3D Integrations," *Journal of Microelectronics International*, Vol. 28, No. 2, 2011, pp. 8–22.
[6] Lau, J. H., "Overview and Outlook of 3D IC Packaging, 3D IC Integration, and 3D Si Integration," *ASME Transactions, Journal of Electronic Packaging*, Vol. 136, No. 4, December 2014, pp. 1–15.
[7] Lau, J. H., C. Lee, C. Zhan, S. Wu, Y. Chao, M. Dai, R. Tain, et al., "Low-Cost Through-Silicon Hole Interposers for 3D IC Integration," *IEEE Transactions on CPMT*, Vol. 4, No. 9, September 2014, pp. 1407–1419.
[8] Lau, J. H., P. Tzeng, C. Lee, C. Zhan, M. Li, J. Cline, K. Saito, et al., "Redistribution Layers (RDLs) for 2.5D/3D IC Integration," *IMAPS Transactions, Journal of Microelectronic Packaging*, Vol. 11, No. 1, First Quarter 2014, pp. 16–24.
[9] Lau, J. H., "The Future of Interposer for Semiconductor IC Packaging," *Chip Scale Review*, Vol. 18, No. 1, January–February, 2014, pp. 32–36.
[10] Hsieh, M. C., S. T. Wu, C. J. Wu, and J. H. Lau, "Energy Release Rate Estimation for Through Silicon Vias in 3D Integration," *IEEE Transactions on CPMT*, Vol. 4, No. 1, January 2014, pp. 57–65.
[11] Lau, J. H., "Supply Chains for High-Volume Manufacturing of 3D IC Integration," *Chip Scale Review*, Vol. 17, No. 1, January–February 2013, pp. 33–39.
[12] Khan, N., H. Li, S. Tan, S. Ho, V. Kripesh, D. Pinjala, J. H. Lau, et al., "3D Packaging With Through-Silicon Via (TSV) for Electrical and Fluidic Interconnections," *IEEE Transactions on CPMT*, Vol. 3, No. 2, February 2013, pp. 221–228.
[13] Lau, J. H., M. S. Zhang, and S. W. R. Lee, "Embedded 3D Hybrid IC Integration System-in-Package (SiP) for Opto-Electronic Interconnects in Organic Substrates," *ASME Transactions, Journal of Electronic Packaging*, Vol. 133, September 2011, pp. 1–7.
[14] Lau, J. H., "Critical Issues of 3D IC Integrations," *IMAPS Transactions, Journal of Microelectronics and Electronic Packaging*, First Quarter Issue, 2010, pp. 35–43.
[15] Lau, J. H., "Design and Process of 3D MEMS Packaging," *IMAPS Transactions, Journal of Microelectronics and Electronic Packaging*, First Quarter Issue, 2010, pp. 10–15.
[16] Lau, J. H., Lee, R., Yuen, M., and Chan, P., "3D LED and IC Wafer Level Packaging," *Journal of Microelectronics International*, Vol. 27, No. 2, 2010, pp. 98–105.
[17] Lau, J. H., "State-of-the-art and Trends in 3D Integration," *Chip Scale Review*, March/April, 2010, pp. 22–28.

[18] Lau, J. H., "TSV Manufacturing Yield and Hidden Costs for 3D IC Integration," *IEEE Proceedings of ECTC*, Las Vegas, NV, June 2010, pp. 1031–1041.
[19] Lau, J. H., Y. S. Chan, and R. S. W. Lee, "3D IC Integration with TSV Interposers for High-Performance Applications," *Chip Scale Review*, September/October, 2010, pp. 26–29.
[20] Yu, A., N. Khan, G. Archit, D. Pinjalal, K. Toh, V. Kripesh, S. Yoon, et al., "Development of Silicon Carriers with Embedded Thermal Solutions for High Power 3D Package," *IEEE Transactions on Components and Packaging Technology*, Vol. 32, No. 3, September 2009, pp. 566–571.
[21] Tang, G., O. Navas, D. Pinjala, J. H. Lau, A. Yu, and V. Kripesh, "Integrated Liquid Cooling Systems for 3D Stacked TSV Modules," *IEEE Transactions on Components and Packaging Technologies*, Vol. 33, No. 1, 2010, pp. 184–195.
[22] Khan, N., V. Rao, S. Lim, H. We, V. Lee, X. Zhang, E. Liao, et al., "Development of 3D Silicon Module With TSV for System in Packaging," *IEEE Transactions on CPMT*, Vol. 33, No. 1, March 2010, pp. 3–9.
[23] Lim, S., V. Rao, W. Hnin, W. Ching, V. Kripesh, C. Lee, J. H. Lau, et al., "Process Development and Reliability of Microbumps," *IEEE Transactions on CPMT*, Vol. 33, No. 4, December 2010, pp. 747–753.
[24] Selvanayagam, C., J. H. Lau, X. Zhang, S. Seah, K. Vaidyanathan, and T. Chai, "Nonlinear Thermal Stress/Strain Analysis of Copper Filled TSV (Through Silicon Via) and Their Flip-Chip Microbumps," *IEEE Transactions on Advanced Packaging*, Vol. 32, No. 4, November 2009, pp. 720–728.
[25] Lau, J. H., and Tang, G., "Thermal Management of 3D IC Integration with TSV (Through Silicon Via)," *IEEE Proceedings of ECTC*, San Diego, May 2009, pp. 635–640.
[26] Kim, B., T. Matthias, M. Wimplinger, P. Kettner, and P. Lindner, "Comparison of Enabling Wafer Bonding Techniques for TSV Integration," *ASME Paper No. IMECE2010-400002.*
[27] Akasaka, Y., "Three-dimensional IC Trends," *Proceedings of the IEEE*, Vol. 74, No. 12, December 1986, pp. 1703–1714.
[28] Akasaka, Y., and Nishimura, T., "Concept and Basic Technologies for 3D IC Structure," *IEEE Proceedings of International Electron Devices Meetings*, Vol. 32, 1986, pp. 488–491.
[29] Chen, K., S. Lee, P. Andry, C. Tsang, A. Topop, Y. Lin, J. Lu, et al., "Structure, Design and Process Control for Cu Bonded Interconnects in 3D Integrated Circuits," *IEEE Proceedings of International Electron Devices Meeting*, (IEDM 2006), San Francisco, CA, December 11–13, 2006, pp. 367–370.
[30] Liu, F., R. Yu, A. Young, J. Doyle, X. Wang, L. Shi, K. Chen, et al., "A 300-Wafer-Level Three-Dimensional Integration Scheme Using Tungsten Through-Silicon Via and Hyprid Cu-Adhesive Bonding," *IEEE Proceedings of IEDM*, December 2008, pp. 1–4.
[31] Yu, R., F. Liu, R. Polastre, K. Chen, X. Liu, L. Shi, E. Perfecto, et al., "Reliability of a 300-mm Compatible 3DI Technology Base on Hybrid Cu-adhesive Wafer Bonding," *Proceedings of Symposium on VLSI Technology Digest of Technical Papers*, 2009, pp. 170–171.
[32] Shigetou, A., T. Itoh, K. Sawada, and T. Suga, "Bumpless Interconnect of 6–um pitch Cu Electrodes at Room Temperature," In *IEEE Proceedings of ECTC*, Lake Buena Vista, FL, May 27–30, 2008, pp. 1405–1409.
[33] Tsukamoto, K., E. Higurashi, and T. Suga, "Evaluation of Surface Microroughness for Surface Activated Bonding," *Proceedings of IEEE CPMT Symposium Japan*, August 2010, pp. 147–150.
[34] Kondou, R., C. Wang, and T. Suga, "Room-temperature Si-Si and Si-SiN wafer bonding," *Proceedings of IEEE CPMT Symposium Japan*, August 2010, pp. 161–164.
[35] Shigetou, A., T. Itoh, M. Matsuo, N. Hayasaka, K. Okumura, and T. Suga, "Bumpless Interconnect Through Ultrafine Cu Electrodes by Mans of Surface-Activated Bonding (SAB) Method," *IEEE Transaction on Advanced Packaging*, Vol. 29, No. 2, May 2006, p. 226.
[36] Wang, C., and T. Suga, "A Novel Moire Fringe Assisted Method for Nanoprecision Alignment in Wafer Bonding," In *IEEE Proceedings of ECTC*, San Diego, CA, May 25–29, 2009, pp. 872–878.
[37] Wang, C., and T. Suga, "Moire Method for Nanoprecision Wafer-to-Wafer Alignment: Theory, Simulation and Application," *IEEE Proceedings of Int. Conference on Electronic Packaging Technology & High Density Packaging*, August 2009, pp. 219–224.
[38] Higurashi, E., D. Chino, T. Suga, and R. Sawada, "Au-Au Surface-Activated Bonding and Its Application to Optical Microsensors with 3D Structure," *IEEE Journal of Selected Topic in Quantum Electronics*, Vol. 15, No. 5, September/October 2009, pp. 1500–1505.
[39] Burns, J., B. Aull, C. Chen, C. Chen, C. Keast, J. Knecht, V. Suntharalingam, et al., "A Wafer-Scale 3D Circuit Integration Technology," *IEEE Transactions on Electron Devices*, Vol. 53, No. 10, October 2006, pp. 2507–2516.
[40] Chen, C., K. Warner, D. Yost, J. Knecht, V. Suntharalingam, C. Chen, J. Burns, et al., "Sealing Three-Dimensional SOI Integrated-Circuit Technology," *IEEE Proceedings of Int. SOI Conference*, 2007, pp. 87–88.

[41] Chen, C., C. Chen, D. Yost, J. Knecht, P. Wyatt, J. Burns, K. Warner, et al., "Three-dimensional integration of silicon-on-insulator RF amplifier," *Electronics Letters*, Vol. 44, No. 12, June 2008, pp. 1–2.

[42] Chen, C., C. Chen, D. Yost, J. Knecht, P. Wyatt, J. Burns, K. Warner, et al., "Wafer-Scale 3D Integration of Silicon-on-Insulator RF Amplifiers," *IEEE Proceedings of Silicon Monolithic IC in RF Systems*, 2009, pp. 1–4.

[43] Chen, C., C. Chen, P. Wyatt, P. Gouker, J. Burns, J. Knecht, D. Yost, et al., "Effects of Through-BOX Vias on SOI MOSFETs," *IEEE Proceedings of VLSI Technology, Systems and Applications*, 2008, pp. 1–2.

[44] Chen, C., C. Chen, J. Burns, D. Yost, K. Warner, J. Knecht, D. Shibles, et al., "Thermal Effects of Three Dimensional Integrated Circuit Stacks," *IEEE Proceedings of Int. SOI Conference*, 2007, pp. 91–92.

[45] Aull, B., J. Burns, C. Chen, B. Felton, H. Hanson, C. Keast, J. Knecht, et al., "Laser Radar Imager Based on 3D Integration of Geiger-Mode Avalanche Photodiodes with Two SOI Timing Circuit Layers," *IEEE Proceedings of Int. Solid-State Circuits Conference*, 2006, p. 16.9.

[46] Chatterjee, R., M. Fayolle, P. Leduc, S. Pozder, B. Jones, E. Acosta, B. Charlet, et al., "Three dimensional chip stacking using a wafer-to-wfer integration," *IEEE Proceedings of IITC*, 2007, pp. 81–83.

[47] Ledus, P., F. Crecy, M. Fayolle, M. Fayolle, B. Charlet, T. Enot, M. Zussy, et al., "Challenges for 3D IC integration: bonding quality and thermal management," *IEEE Proceedings of IITC*, 2007, pp. 210–212.

[48] Poupon, G., N. Sillon, D. Henry, C. Gillot, A. Mathewson, L. Cioccio, B. Charlet, et al., "System on Wafer: A New Silicon Concept in Sip," *Proceedings of the IEEE*, Vol. 97, No. 1, January 2009, pp. 60–69.

[49] Lau, J. H., "Heart and Soul of 3D IC Integration," posted at 3D InCites on June 29, 2010, http://www.semineedle.com/posting/34277.

[50] Lau, J. H., "Who Invented the TSV and When?" posted at 3D InCites on April 24, 2010, http://www.semineedle.com/posting/31171.

[51] Gat, A., L. Gerzberg, J. Gibbons, T. Mages, J. Peng, and J. Hong, "CW Laser of Polyerystalline Silicon: Crystalline Structure and Electrical Properties," *Applied Physics Letter*, Vol. 33, No. 8, October 1978, pp. 775–778.

CHAPTER 7
2.5D/3D IC Integration

7.1 Introduction

Unlike 3D Si integration (as shown in the left-hand side of Fig. 7.1), 3D IC Integration (as shown in the right-hand side of Fig. 7.1) is defined as stacking up whatever Moore law's IC thin chips in the 3rd dimension with TSVs (through-silicon vias) and microbumps [1–63] to achieve high performance and density, low consumption power, wide bandwidth, small form factor, and light weight. For 3D IC integration, microsolder bumps (discussed in Chap. 5) are everywhere inside the integrated structure!

Today, unlike 35 years ago, most people favor 3D IC integration because it is much easier to make than 3D Si integration. (However, it should be emphasized again that 3D Si

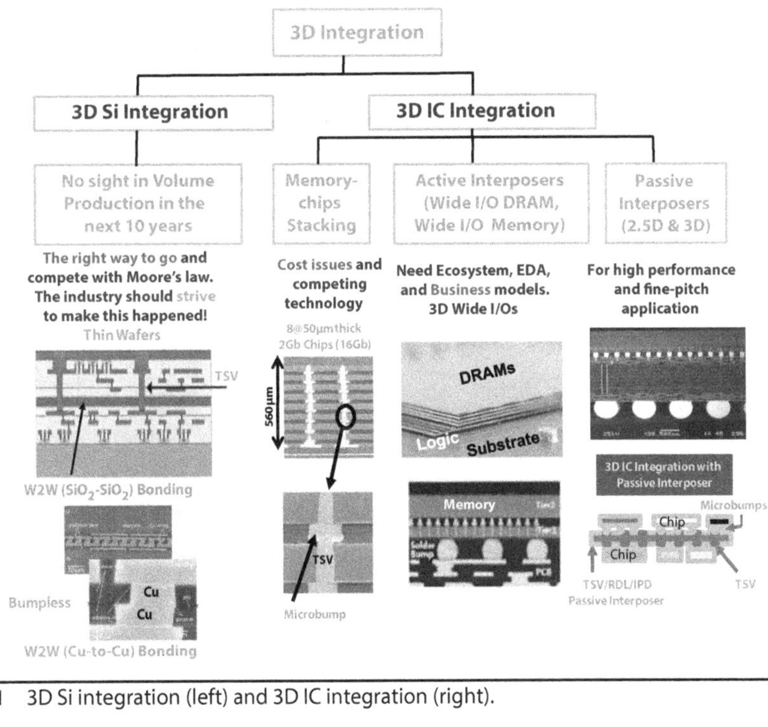

Figure 7.1 3D Si integration (left) and 3D IC integration (right).

integration is the right way to go and compete with Moore's law. The industry should strive to develop the technology mentioned in Sec. 6.6, the EDA mentioned in Sec. 6.7, and the assembly process in Sec. 6.8.)

There are two different kinds of interposers, as shown in Fig. 7.1. One is with active devices such as transistors, TSVs, and RDLs and is called active interposer—for example, the SoC/logic (with TSVs) to support/control the wide I/O memory and wide I/O DRAM (dynamic random-access memory). The other one is with TSVs and RDLs (redistribution layers) but without active devices and is called passive interposer (also called 2.5D IC integration)—for example, a piece of dummy silicon with TSVs and RDLs to support the chips (without TSVs). Both of them could have IPD (integrated passive devices) to enhance their electrical performance. In this chapter, 3D IC integration and 2.5D IC integration will be discussed.

7.2 TSV Process for 3D IC Integration

As mentioned earlier, for 3D Si integration, the TSV is made before Cu-to-Cu bonding and after SiO_2-to-SiO_2 bonding. For 3D IC integration, the TSV can be fabricated with at least four different processes. (The definition is based on the device fabrication. For more information, please see Sec. 1.7.) First, tiny vias inside a chip will be briefly mentioned and then the four processes will be discussed.

7.2.1 Tiny Vias on a Chip

Figure 7.2 shows the cross-section of a 9 Cu-low-k layers chip (made in 2006) with the 65 nm process technology by Chartered Semiconductor, which is called GlobalFoundries

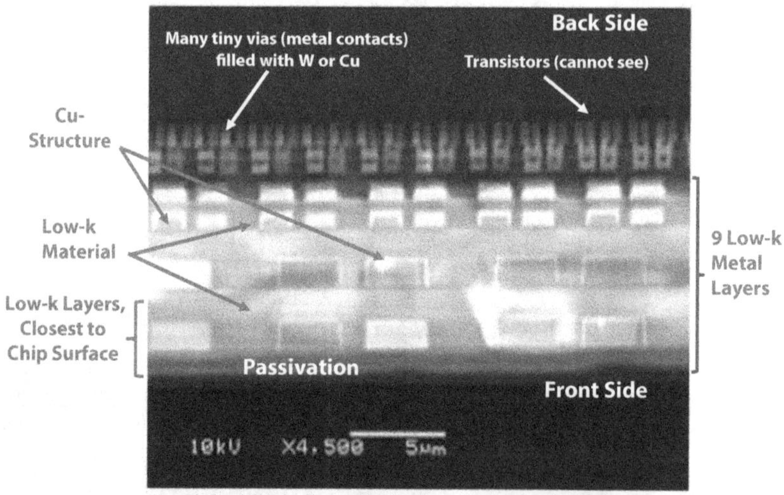

65 nm Cu/low-k Chip by *Chartered Semiconductor (2006) to show the tiny vias.*

Besides the tiny devices such as transistors (which cannot be seen in this SEM image), there are many tiny vias (metal contacts) on a chip. They are connected to devices (e.g., four vias for each transistor) to build the first metal (M1) layer. Today, the number of these tiny vias, for many chips, already exceeds the world population of over 7 billion . One of the core competences of foundries is to make these tiny vias. These vias are not the same TSVs for 3D IC integration.

Figure 7.2 The tiny vias (metal contacts) in a chip (not the TSVs).

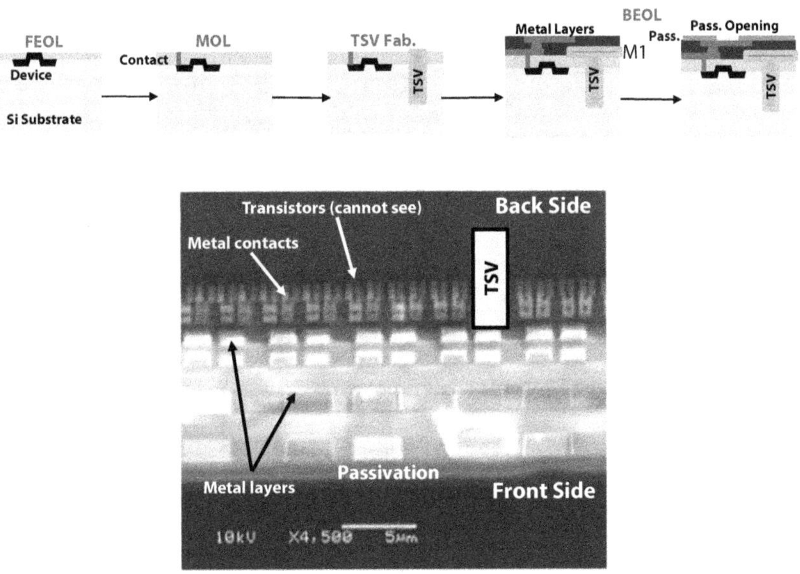

Figure 7.3 TSV via-middle process.

today. Besides the tiny devices such as transistors (which cannot be seen in this SEM image), there are many tiny vias (metal contacts) on the chip. They are connected to devices (e.g., four vias for each transistor) to build the first metal (M1) layer. Today, the number of these tiny vias, for many chips, already exceeds the world population of over 7 billion. One of the core competences and major businesses of foundries is to make these tiny vias (in addition to the transistors). These vias are not the same TSVs for 3D IC integration. They are much smaller and their number is many times more than that of the TSVs.

7.2.2 Via-First Process

For via-first process, the TSV is fabricated (on a bare wafer) before the implantation of the semiconductor devices such as the transistors.

7.2.3 Via-Middle Process

For via-middle process, the TSV is made after the fabrication of the devices and metal contacts, and before the metal layers, as shown in Fig. 7.3. Figure 7.4 shows a SEM image of the cross-section of TSVs made by the via-middle process [40].

7.2.4 Via-Last from the Front-Side Process

For via-last process from the front-side, the TSV is fabricated after the making of the devices and metal contacts, all the metal layers, and passivation. The TSV is made from the front-side of the wafer.

7.2.5 Via-Last from the Back-Side Process

For via-last process from the back side, the TSV is fabricated after the making of the devices and metal contacts, all the metal layers, and passivation. The TSV is made from the back side of the wafer (Fig. 7.5). Figure 7.6 shows a SEM image of the cross-section of TSVs made by the via-last from the back-side process [41].

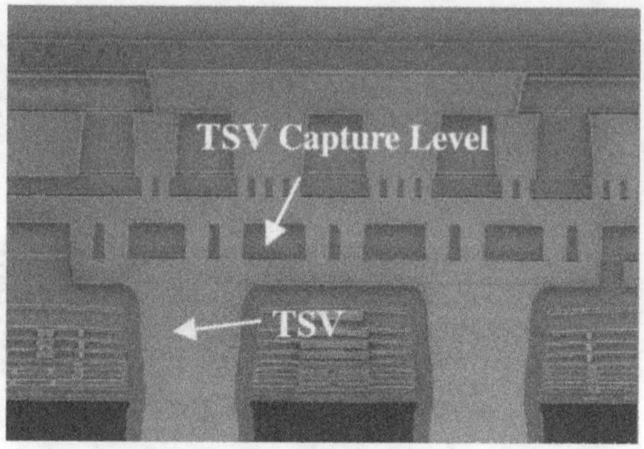

Figure 7.4 An example of TSV via-middle process.

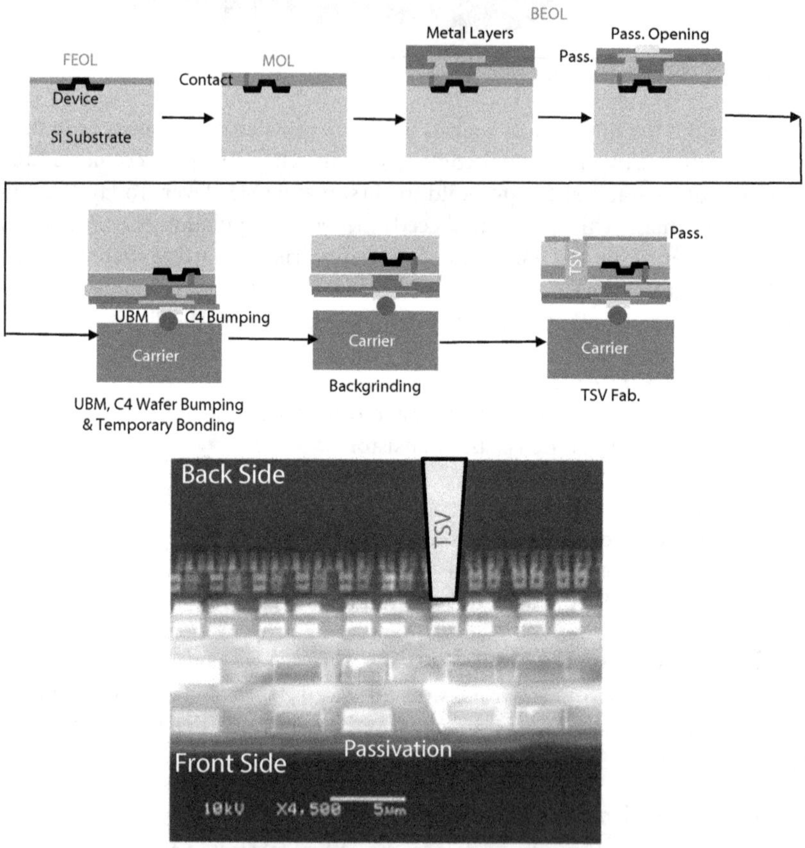

Figure 7.5 TSV via-last (from the back side) process.

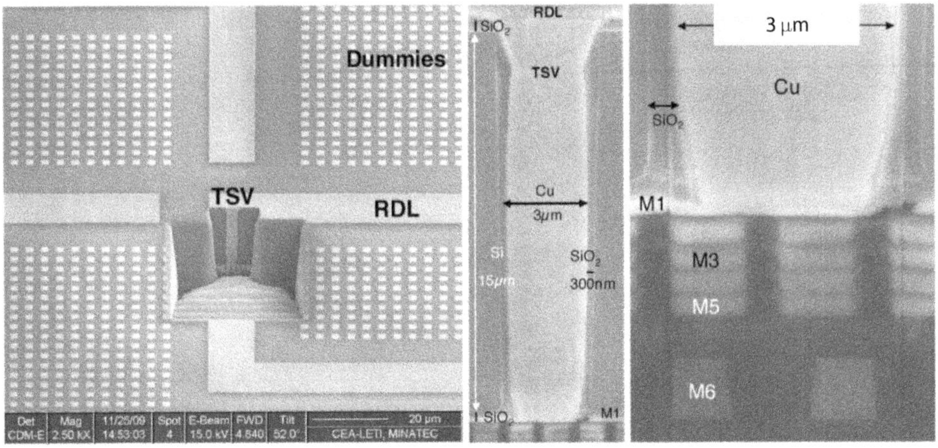

FIGURE 7.6 An example of TSV via-last (from the back side) process.

7.2.6 Summary and Recommendations

As discussed in Sec. 1.7, some important results and recommendations are summarized as follows:

- For device wafers and high volume production, TSVs should be made by the via-middle process and by the foundries.
- For device wafers and high volume production, the cost of making the TSVs should not be more than 4% of the cost in fabricating the ($\leq$32 nm) device wafers.
- For device-less wafers, just make the TSVs on the dummy Si wafers.

7.3 The Potential Application of 3D IC Integration

The right-hand side of Fig. 7.1 shows that 3D IC integration basically consists of three groups: (a) the memory-chip stacking, (b) the active interposer, and (c) the passive interposer. The potential applications of 3D IC integration are shown in Fig. 7.7. Basically, there are four groups: (a) memory-chip stacking, (b) wide I/O memory or logic-on-logic, (c) wide I/O DRAM or hybrid memory cube (HMC), and (d) wide I/O interface (interposer or 2.5D IC integration). These will be the focus of this chapter. Wide I/O 2 and high bandwidth memory (HBM) will also be briefly mentioned.

7.4 Memory-Chip Stacking

7.4.1 The Chips

The far-left of Figs. 7.7 and 7.8 show the simplest example of memory-chip stacking. It can be seen that even with eight chips (stacked by Samsung in 2006), their total thickness (560 µm) is still less than that of an ordinary chip. These chips could be DRAM or NADA Flash with less than 100 I/Os (78 to be exact).

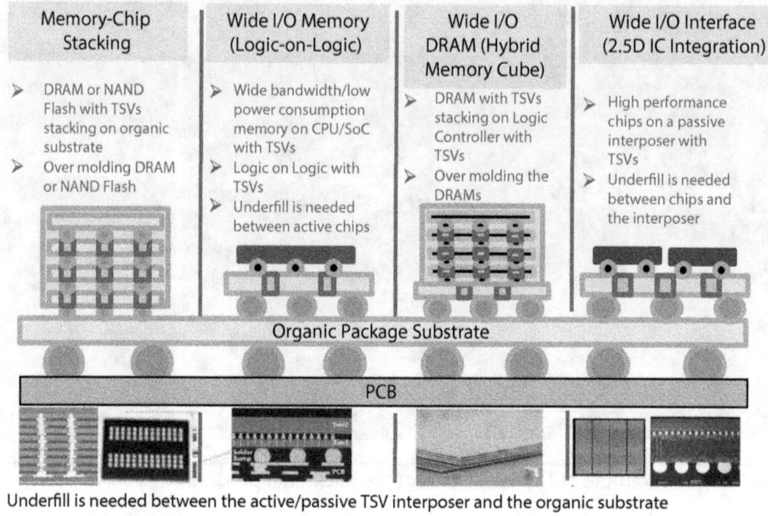

Figure 7.7 Potential applications of 3D IC integration.

7.4.2 The Potential Products

With this technology, Samsung mass-produced (August 2014) industry's first TSV-based 64GB DDR4 (double data rate type 4) DRAM module which consists of 36 DDR4 DRAM chips, each of which consists of four 4-gigabit (GB) DDR4 DRAM dies (Fig. 7.9). The new 64GB RDIMM (Registered dual in-line memory module) with TSVs and microbumps performs twice as fast as a module that uses wire-bonding packaging, while consuming approximately half the power. The module is for next-generation server application.

According to JEDEC's road map, DDR3 now comes in two speed grades: 1333 and 1600 MHz. A 1866-MHz version is also on the road map. Following DDR3, Samsung, Hynix, and Micron are developing monolithic parts based on the next-generation interface—DDR4.

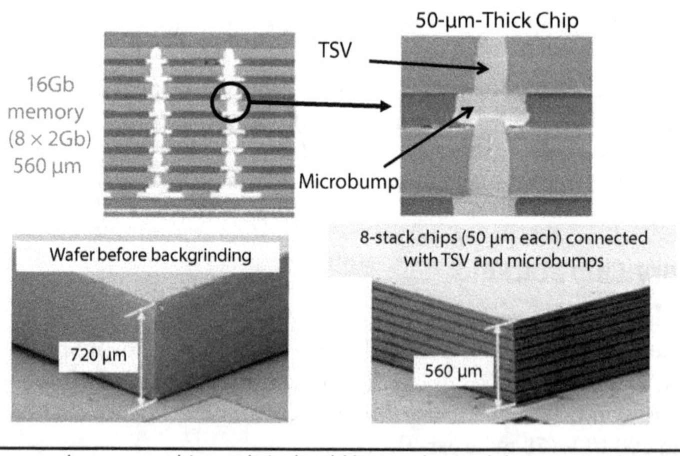

Figure 7.8 Samsung's memory-chip stack. It should be emphasized that even with 8 chips, the total thickness (560 μm) is less than that of an ordinary chip (720 μm).

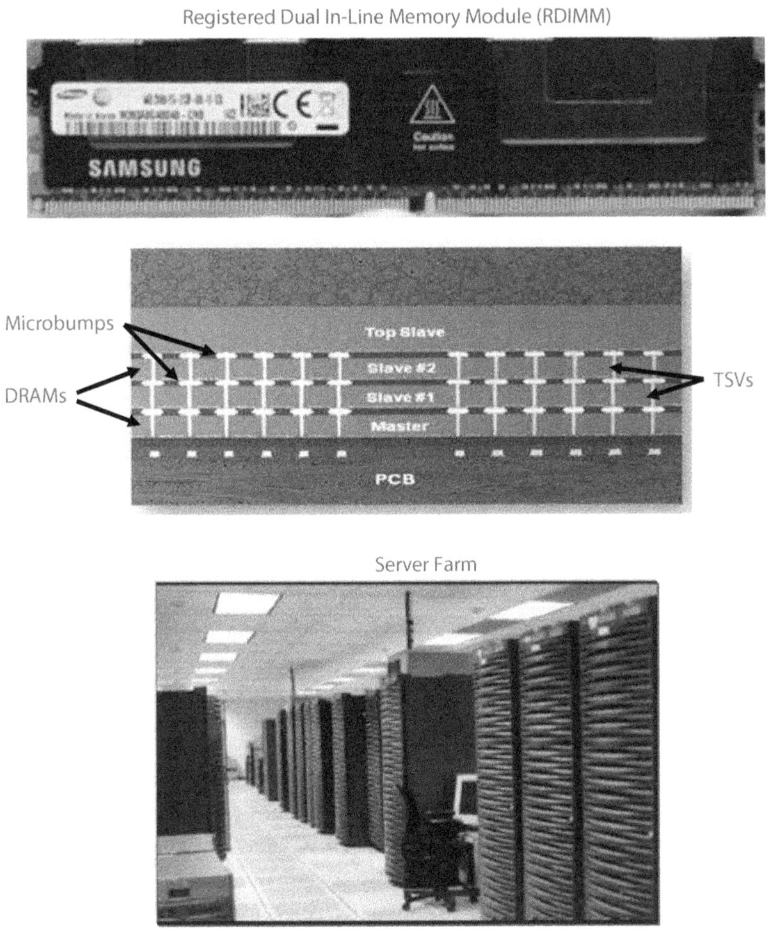

Figure 7.9 Samsung's TSV-based DDR4 RDIMM for server application.

The first DDR4 DRAMs will come in 4GB densities and 2xnm processes. The devices will operate at 2133 and 2400 MHz, which are slated for delivery in 2015, respectively, according to the JEDEC road map. The 3200-MHz and 16-GB versions are due out in 2020. Based on a 1.2-V technology and a 16-bank architecture, DDR4 is expected to be about 30% to 40% lower power than DDR3. DDR4 also features VDDQ DQ termination, a 500,000 page size for x4 devices, new RAS features and error correction on the SODIMM using a common connector.

Why apply TSV to next-generation servers? Microprocessors used to be the power hog in a server farm system, as shown in Fig. 7.9. However, the DRAM shares the blame now, as memory takes up more than 30% of the power consumption in a server. TSV can reduce the power consumption, save energy, and provide a green environment.

Why not use TSV for memory-chip stacking in high volume consumer products such as the smartphones and tablets today? Unfortunately, it is due to cost issues and competing technologies, e.g., wirebonding, as shown in Fig. 1.4.

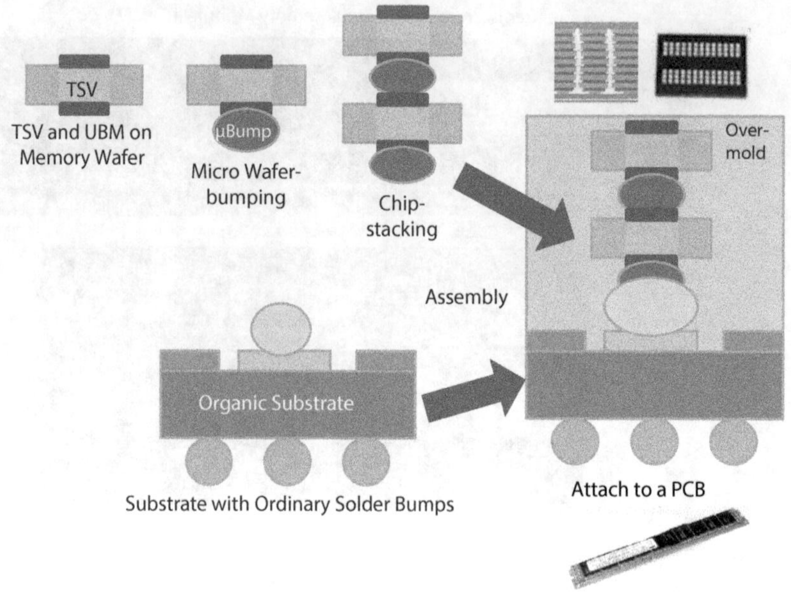

FIGURE 7.10 Assembly process for memory-chip stacking.

7.4.3 Assembly Process

The assembly process of memory-chip stacking is shown in Fig. 7.10. After fabrication of the TSVs (e.g., with the technologies mentioned in Chap. 2 of [1]), perform microsolder wafer bumping (e.g., with the technologies mentioned in Chap. 5) and thin-wafer handling and dicing (e.g., with the technologies mentioned in Chap. 5 of [1]). Then after stacking of the memory-chips (e.g., with the technologies shown in Chap. 8 of [1]), assemble them on an organic substrate. Now after overmolding of the memory-chips, perform solder ball mounting on the bottom-side of the substrate. And finally, the memory-stack module with solder ball is SMT (surface mount technology) assembly on the PCB.

7.5 Wide I/O Memory or Logic-on-Logic

7.5.1 The Chips

The second-left side of Fig. 7.7 shows a wide I/O memory. It consists of a low-power and wide-bandwidth memory usually with more than thousands of interface pins. This memory is supported by a CPU/Logic or SoC (system-on-chip) with TSVs. It should be noted that this memory chip is very different from those in Sec. 7.4 and usually have more than thousand I/Os.

7.5.2 The Potential Products

Figure 7.11 shows one of Intel's technology road maps [64], which shows that memory chip directly attached on top of the CPU/Logic/SoC leads to the optimal bandwidth and best electrical performance. Figure 7.12 schematically shows one possible structure [22]. Since mobile products want it, samples with TSVs have been made and published by, e.g., Samsung as shown in Fig. 7.13 by the via-middle process. Comparing with the 3D IC

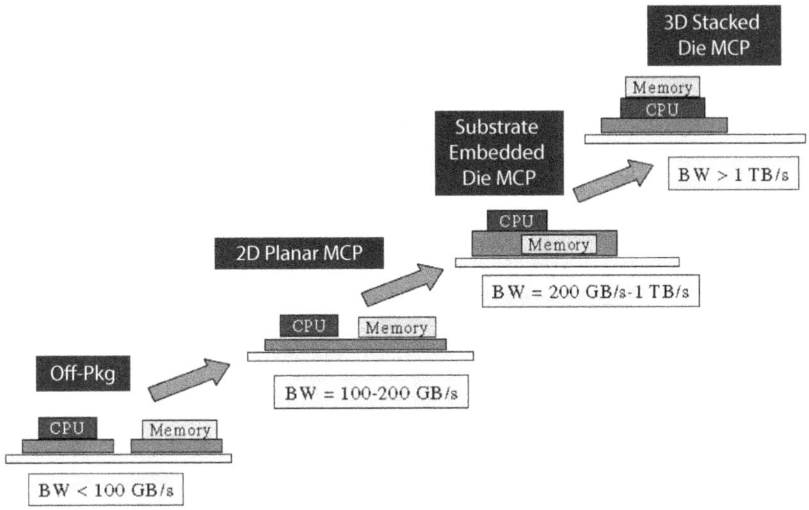

Figure 7.11 Intel's high-density, high-performance, and wide I/O packaging configurations.

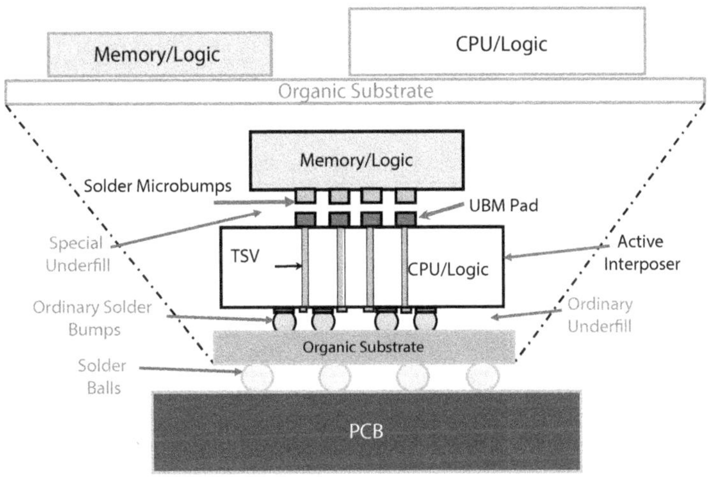

Figure 7.12 The low-power consumption and wider bandwidth memory (with thousands of interface pins) is right on top of the SoC/logic with TSVs (an example of wide I/O memory).

FC-PoP (flip chip package-on-package) with LPDDR2 memory technique (without TSVs), it can be seen from Fig. 7.13 that the memory I/O power consumption of the one with TSVs and wide I/O memory (44 mW) is only 1/4 of that (176 mW) of FC-PoP without TSVs.

Figure 7.14 shows a similar structure presented by IBM [40] for high-end server applications. It can be seen that IBM's chip-to-chip interconnect is face-to-face and the powers, grounds, and signals are coming out from the back side of the bottom chip with TSVs. The key dimensions for the TSVs are diameter = 15 μm, depth = 100 μm, pitch = 50 μm, and annular copper ring = 4–5 μm. The TSVs are fabricated by the via-middle process.

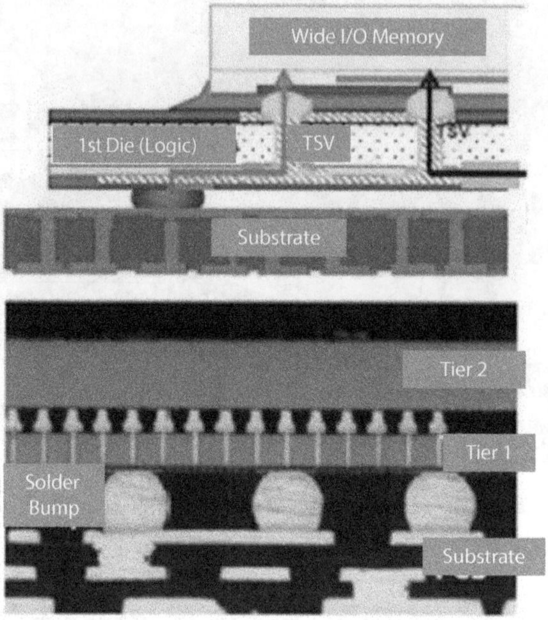

	Conventional 3D IC Packaging (FC-PoP) with LPDDR2	TSV-SiP with Wide I/O Memory
Memory I/O Power Consumption	176 mW	44 mW

Figure 7.13 Samsung's wide I/O memory. The power consumption is only 1/4 of that of a PoP.

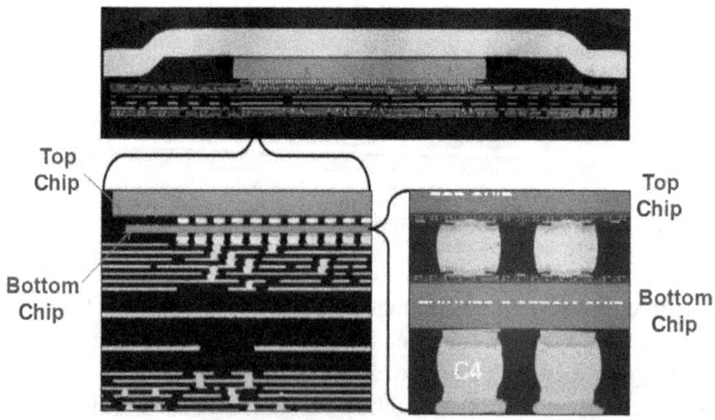

Figure 7.14 IBM's wide I/O made by TSV via-middle process [45].

Is wide I/O memory in volume production today? No! The readiness of the supply chain infrastructures, including industry standards, supply chain business models, and competitive pricing are not ready. Also, much cheaper solutions such as those face-to-face and chip-to-chip interconnects (without using TSVs) are available, even their performances are not as good as the ones with TSVs. Please read Chap. 14 for more information.

7.5.3 Assembly Process

Figure 7.15 shows the assembly process of the Samsung structure shown in Fig. 7.13. Basically, there are three key tasks: (1) fab process, e.g., TSV fabrication, metallization and wafer bumping the C4 (controlled-collapse chip connection) bumps, and temporary bonding of a carrier wafer; (2) post fab process, e.g., TSV Cu revealing and microbumping; and (3) packaging assembly, e.g., wafer bumping the microbumps on the memory chip, C2W (chip-to-wafer) bonding, underfilling, de-bonding, dicing, and flip chip assembly twice (one with microbumps between the memory chip and the SoC and the other with C4 bumps between the SoC and the organic substrate.) It is interesting to note that Samsung's SoC is not (like IBM's) face-to-face with the memory chip.

Similar to Samsung's via-middle process, after FEOL, IBM made the TSVs by DRIE [40]. Then, PECVD for conformal oxide layer, PVD for barrier and seed layers, electroplating Cu for filling the vias, and CMP for removing the excess copper. A dielectric for the contact level metal is put down and then the top metal levels are conventionally defined. IBM used a node-agnostic (via-middle) TSV, which made contact to the upper metal layers in

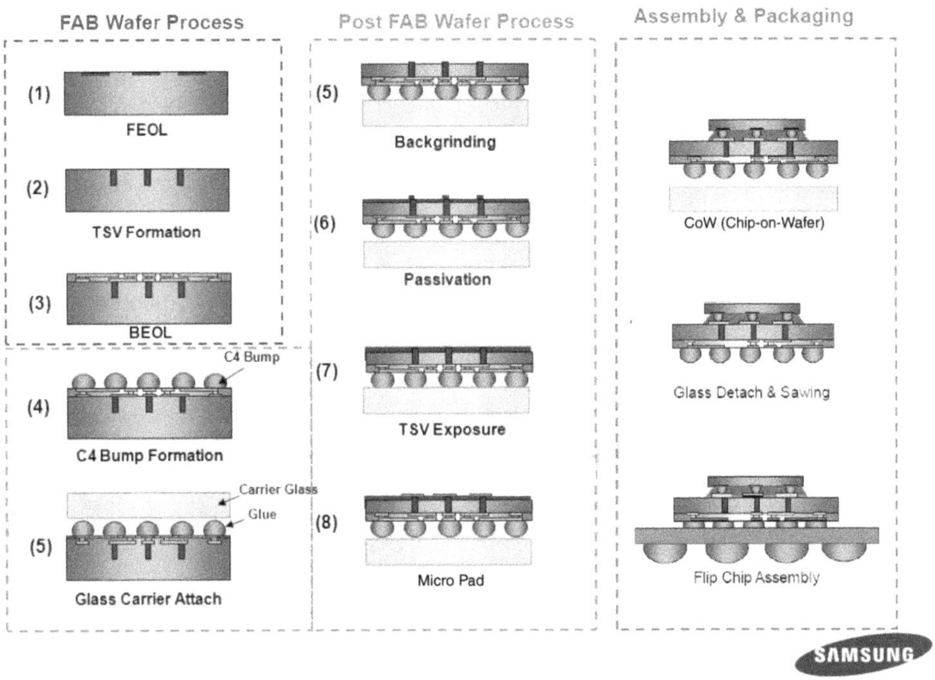

FIGURE 7.15 Process steps in making Samsung's wide I/O memory shown in Figure 7.13.

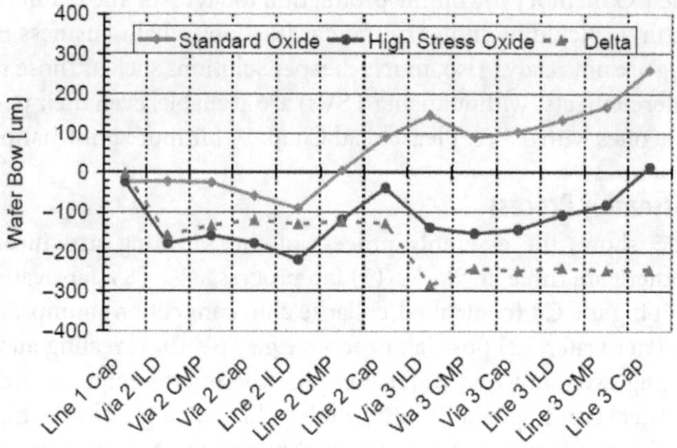

FIGURE 7.16 Introduction of a high compressive stress silane oxide (−280 MPa vs. −80 MPa for standard TEOS) at the via 2 and 3 levels can result in a 250 μm bow reduction [45].

the device chip (fabricated with HKMG processes ranging from 90 to 32 nm). By doing so, the TSVs have anything from 3 to 9 metal layers below the contact level and must cope with dielectric k-values ranging from 4.1 down to 2.4, and bulk and SOI wafers. Warpage is always a big problem for making wafer with TSVs. For IBM, the warpage reached 250 when they just completed M3, Fig. 7.16. IBM's solution is to use a compressive oxide to pull the wafer flat during via 2 and via 3 levels. IBM used glass as a supporting wafer for their thin-wafer handling. After Cu revealing, IBM deposited and patterned a protective oxide/nitride before forming and defining the RDLs. They had done some thermal cycling test and the results are so far so good [40]. Finally, the solder ball is attached to the bottom of the organics substrate and then SMT (surface mount technology) assembly on the PCB, as shown in Fig. 7.17.

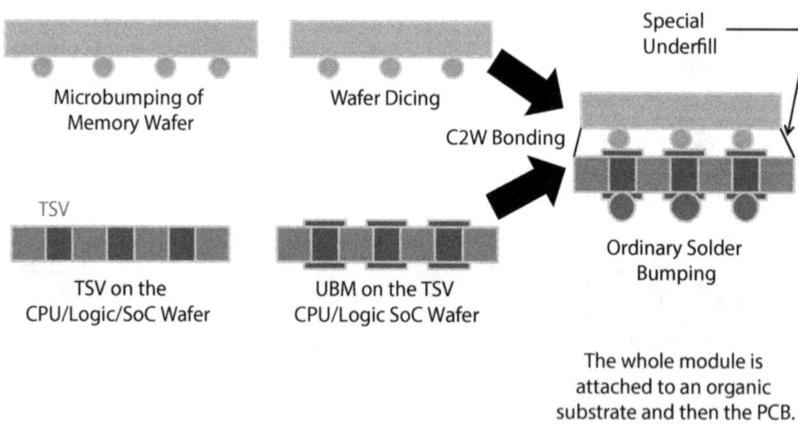

FIGURE 7.17 Assembly process of a typical wide I/O memory.

7.6 Wide I/O DRAM or Hybrid Memory Cube

7.6.1 The Chips

The second-right side of Fig. 7.7 shows a wide I/O DRAM. Samsung has been publishing many papers on this topic for at least 5 years, e.g., [48] and finally at the IEEE ISSCC in San Francisco, February 2011, they showed the very first sample of 2 DRAMs on a master logic chip with TSVs (Figs. 7.18 to 7.20) [49]. For this DRAM, the number of interface pins is about 1000. JEDEC (JESD229) put it to 1200 as shown in Fig. 7.21. Please note that the TSV cage area of Samsung's sample is almost 15% of the DRAM chip size! Comparing the TSV technology with the PoP technology (without TSVs), the size can be reduced by 35%, the power can be reduced by 50%, and the bandwidth can be increased by 8 times, as shown in Fig. 7.22. Again, the standard of the interface pins for these DRAMs is 1200, which is many times more than those mentioned in the memory-chip stacking, Sec. 7.4.

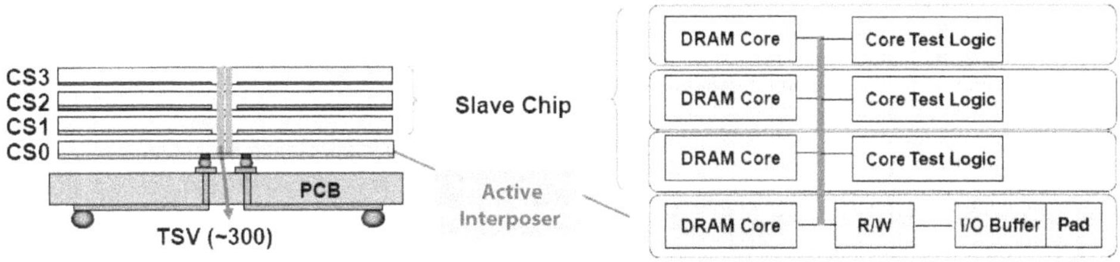

FIGURE 7.18 Samsung's wide I/O DRAM with the memory cube and a SoC/logic master chip with TSVs.

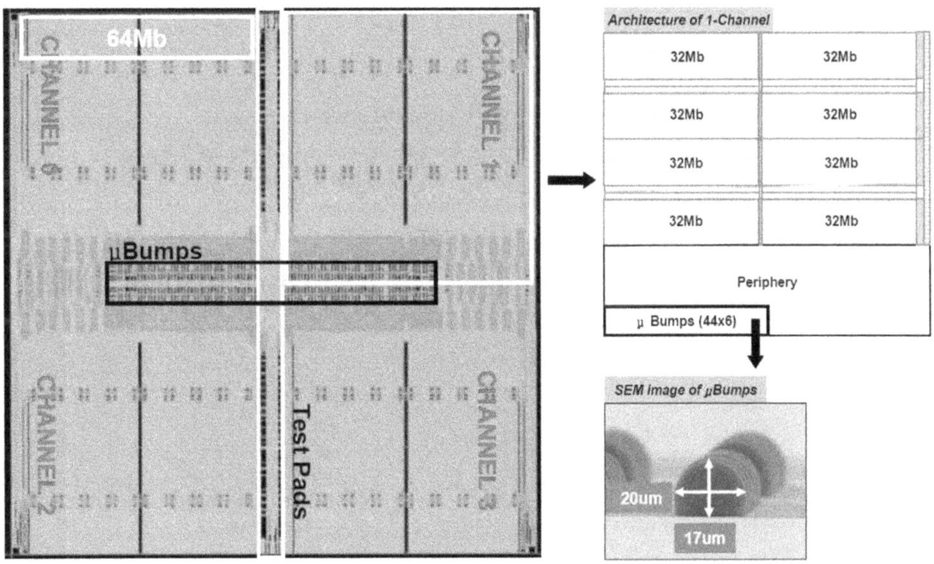

FIGURE 7.19 Samsung's wide I/O DRAM showing the area of TSVs and microbumps.

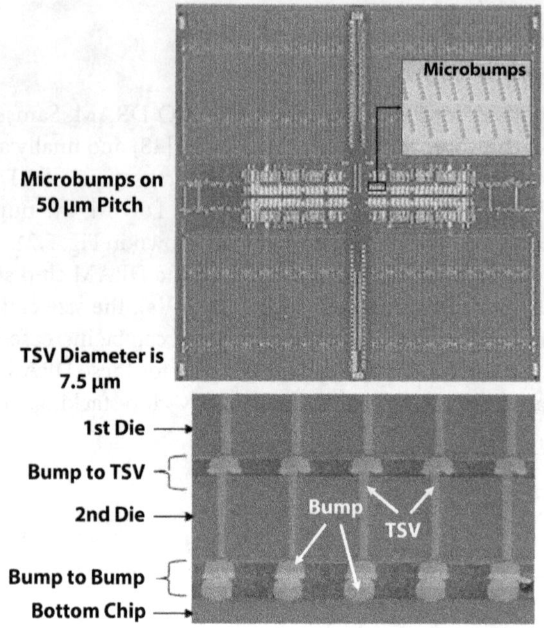

FIGURE 7.20 Samsung's wide I/O DRAM sample with 2 DRAMs on top of the logic chip with TSVs.

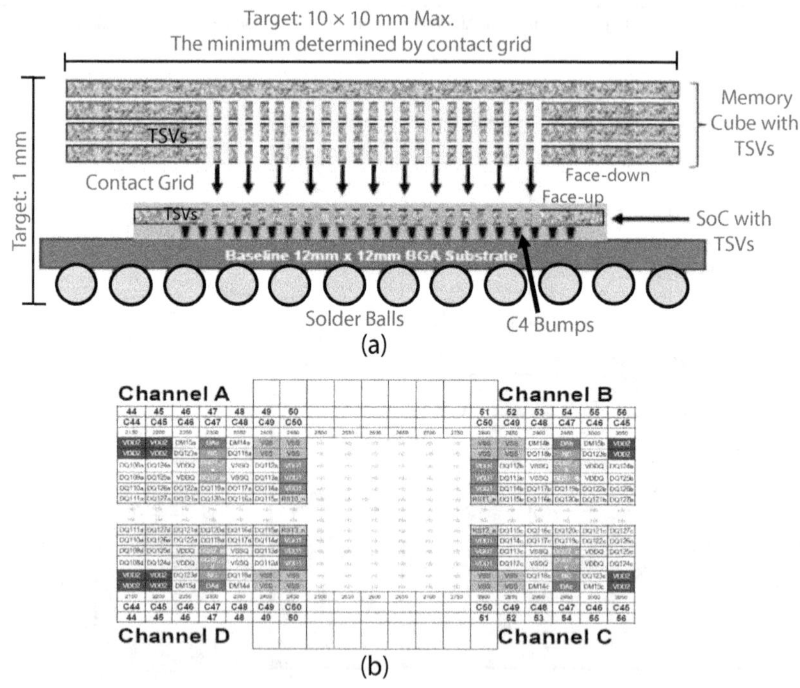

FIGURE 7.21 JEDEC/Micron wide I/O DRAM.

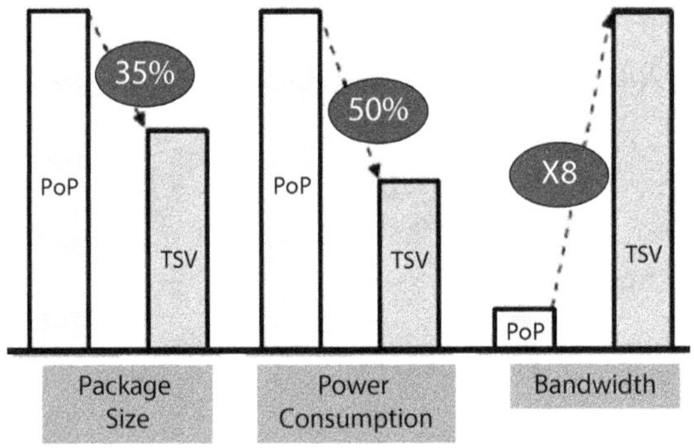

FIGURE 7.22 Samsung's wide I/O (comparing with a PoP) leads to 35% reduction in size and 50% in power, and eight times increase in bandwidth.

7.6.2 The Potential Products

Wide I/O DRAM with TSVs is very suitable for mobile products such as smartphones and tablets as well as for games, because of their hunger for bandwidth and density. As showed in Sec. 1.5.4 of [1] that if we have TSVs run through a 4(DDR3-1600)-DRAM stacking with x512 bit data rate, we could have a total memory bandwidth of 102GB.

People who have been working (via simulation) on using the wide I/O memory for their mobile products are rethinking and doing the trend-off between the wide I/O DRAM and wide I/O memory (Sec. 7.5) since the latter can only stack one large memory chip (without TSVs) on top of the SoC, i.e., the capacity is limited. Unlike wide I/O DRAM, which is growth architecture, i.e., by principle, you can stack up as many DRAMs as you please!

Is it in high volume production today? No! However, the good news are (1) Micron, Samsung, Altera, ARM, IBM, Open-Silicon, SK Hynix, and Xilinx formed a HMC consortium; (2) the HMC specification (with more than 120 adopters) was published on April 2, 2013 and revised (HMC specification 2.0) on November 26, 2014; and (3) Micron/IBM (Micron is responsible for the memory cube and IBM is responsible for the logic, Fig. 7.23) shipped the very first sample (Fig. 7.23) in the last week of September 2013. It is a HMC which consists of four DRAMs each with >2000 TSVs stacking on top of a logic controller with TSVs. The HMC is then attached to an organic package substrate. The microbumps are Cu pillars (20 μm tall) with solder caps.

At the 2014 International Supercomputing Conference, Intel announced that "Knights Landing" processor unit would debut in 2015. It will support for up to 384GB of on board DDR4 RAM and 16GB of Micron HMC stacked DRAM on-package, providing up to 500GB/s of memory bandwidth (Fig. 7.24). Micron reports that having such HMC in the CPU (central processor unit) package is expected to deliver 5× the sustained memory bandwidth versus GDDR5 (graphics double data rate type 5) with one-third the energy per bit in half the footprint. Unfortunately, the HMC is not for consumer products such as smartphones and tablets, rather it is for high performance computing, networking, energy, wireless communications, transportation, security, and high-end servers.

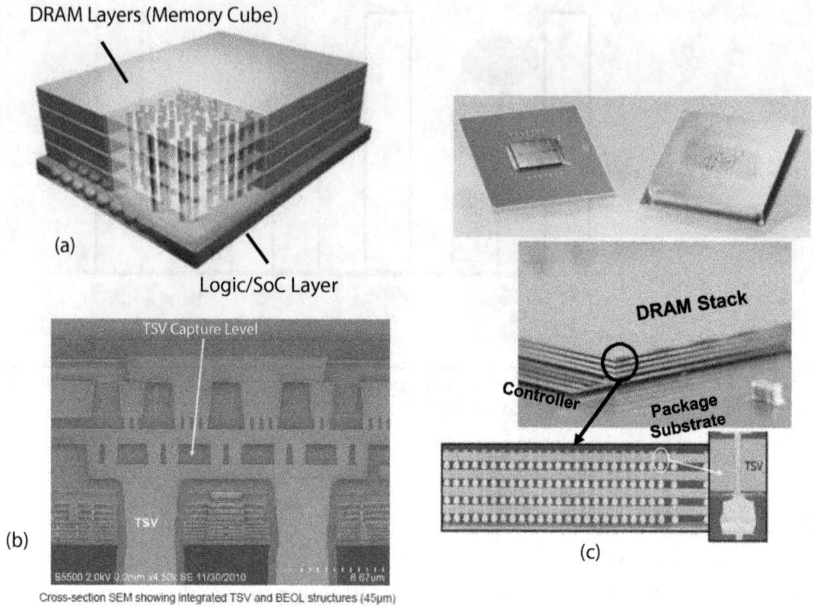

FIGURE 7.23 (a) Hybrid memory cube. (b) IBM is going to fabricate the SoC/logic chip. (c) Micron is going to fabricate the memory cube.

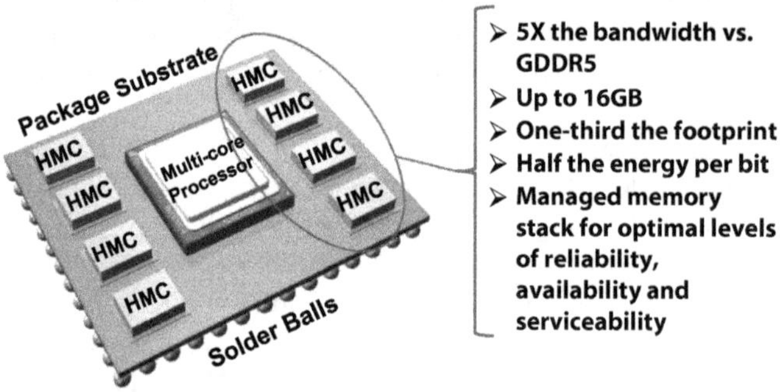

FIGURE 7.24 Intel's Knights Landing processor unit with Micron's HMCs.

7.6.3 Assembly Process

The assembly process (Fig. 7.25) of wide I/O DRAM is very similar to those shown (conceptually and schematically) in Figs. 7.10 and 7.17. All the key enabling technologies such as via formation, dielectric deposition, barrier and seed layers disposition, via filling, CMP, TSV revealing, thin-wafer handling, metallization, UBM (under bump metallurgy), microsolder wafer bumping and assembly, C2C (chip-to-chip) and C2W bonding, thermal management, etc. presented in Chaps. 2 through 9 of [1] can be used to fabricate the wide I/O DRAM or HMC products.

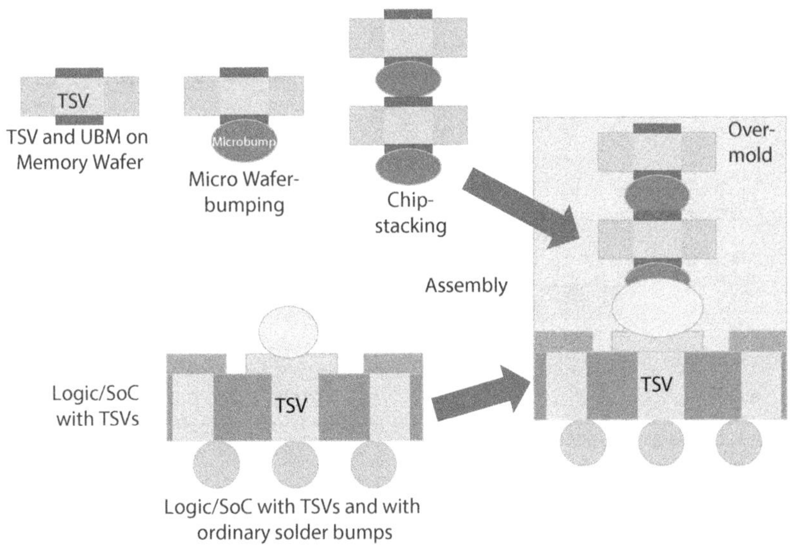

Figure 7.25 Assembly process of a typical wide I/O DRAM.

7.7 Wide I/O 2 and High Bandwidth Memory

Figure 7.26 schematically shows a wide I/O 2 provided by JEDEC standard (JESD229-2). It can be seen that the microbumps on the DRAMs are divided into four quadrants with signal assignments mirrored both horizontally and vertically as shown in Fig. 7.26, where the bump pitch (40 μm) of the area array is also shown. The dimensions of each quadrant are 2880 × 200 μm. There will be a space between quadrants in the x-direction (1000 μm) and in the y-direction (120 μm).

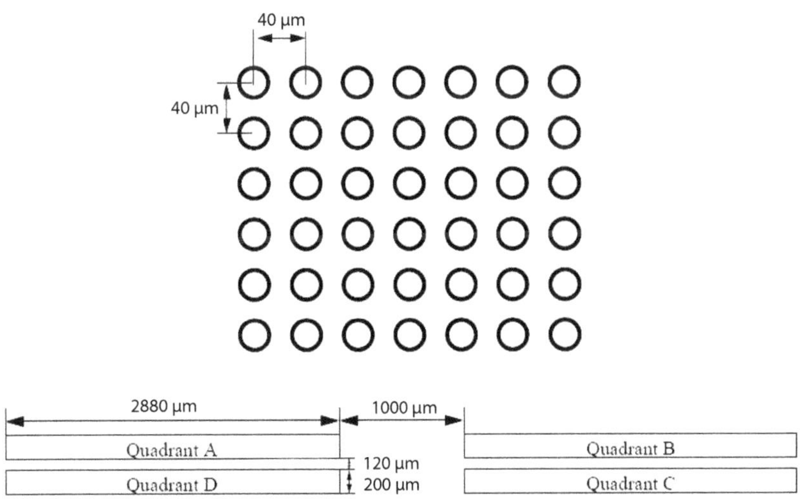

Figure 7.26 Wide I/O 2. The pitch of the TSVs in the x- and y-direction is 40 μm.

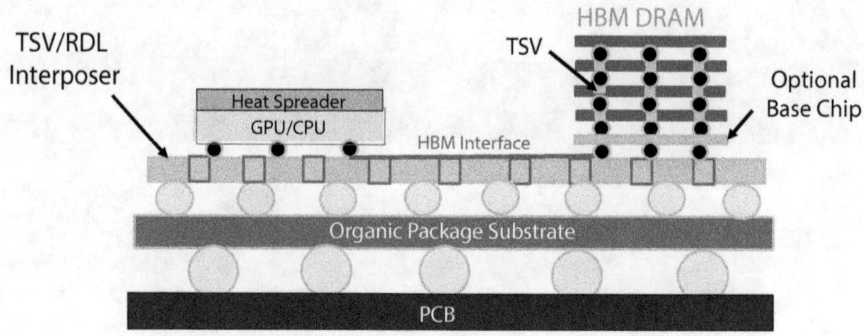

FIGURE 7.27 High bandwidth memory. The DRAM memory cube is connected to the CPU or GPU via the HBM interface.

Figure 7.27 schematically shows a HBM system developed by Hynix/AMD, which is based on JEDEC standard (JESD235). It is meant for graphics applications supporting bandwidth from 128GB/s to 256GB/s. A TSV/RDL interposer is used to support/connect mainly the lateral communication (HBM interface) between the HBM DRAM memory cube with TSVs and the SoC such as graphic processor unit or CPU without TSVs. The optional base chip is used for buffering and signal rerouting of the HBM DRAM cube.

7.8 Wide I/O Interface (2.5D IC Integration)

One of the potential applications of 3D IC integration is wide I/O interfaces [1], which consists of a piece of device-less silicon with TSVs and high-performance, high-density IC chips without TSV. This piece of device-less silicon (also called a passive interposer) is used to support the chips and has RDLs (mainly) for lateral communication between the chips as shown in Fig. 7.28 and the far-right side of Fig. 7.7. It is also called 2.5D IC integration.

As mentioned in [36], one of the key reasons to have the intermediate substrate (passive interposer) is created by the ever increasing IC density and pin-out, and shrinking IC pad-pitch and size, the conventional package substrate cannot support these IC requirements. A few examples are shown in Fig. 7.29 (Xilinx/TSMC) [50–52], Fig. 7.30 (Altera/TSMC) [53–54], and Fig. 7.31 (Xilinx/UMC/SPIL) [55]. The samples in Figs. 7.29 and 7.30 are fabricated/assembled by a turn-key (vertically integrated) manufacturing process called CoWoS (chip on [interposer] wafer on [package] substrate) developed and operated by TSMC. Xilinx and Altera designed the structures. On the other hand, the sample shown in Fig. 7.31 is made by the collaboration of Xilinx (design), UMC (TSV/RDL interposer), and SPIL (MEOL). For the definition of MEOL, please see Sec. 1.8.

7.8.1 Real Applications of TSV/RDL Passive Interposers

In all these samples, it can be seen from Figs. 7.29 through 7.31 that even with 12 build-up layers (6-2-6) on the package substrate, it is still not enough to support the four 28-nm FPGA (field-programmable gate array) chips. In addition, a TSV (10-μm-diameter) silicon interposer (100 μm deep) with four top RDLs (three Cu damascene layers and one aluminum

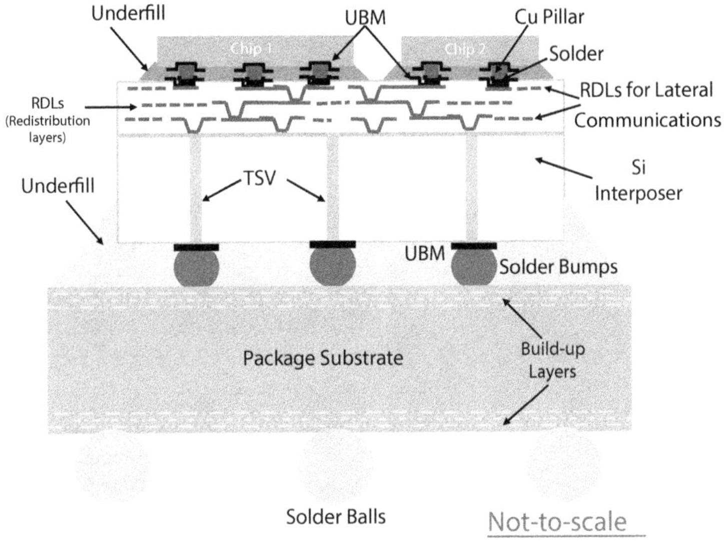

FIGURE 7.28 TSV/RDL passive interposer on a package substrate.

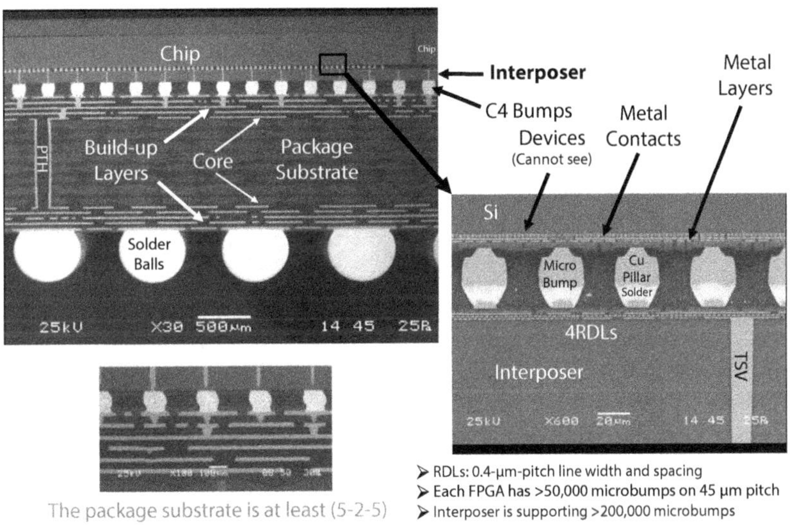

FIGURE 7.29 Xilinx/TSMC's CoWoS.

layer) is needed. Why? As shown in Fig. 7.32, in order for better device manufacturing yield (to save cost), a very large SoC has been sliced into four smaller FPGA chips made by TSMC's 28-nm process technology. The 10,000+ of lateral interconnections between FPGA chips are connected mainly by the 0.4-μm-pitch (minimum) RDLs of the interposer. The minimum thickness of the RDLs and passivation is ~1 μm. Each FPGA has more than 50,000 Cu-pillars with solder cap microbumps (200,000+ microbumps on the interposer) at 45 μm pitch as

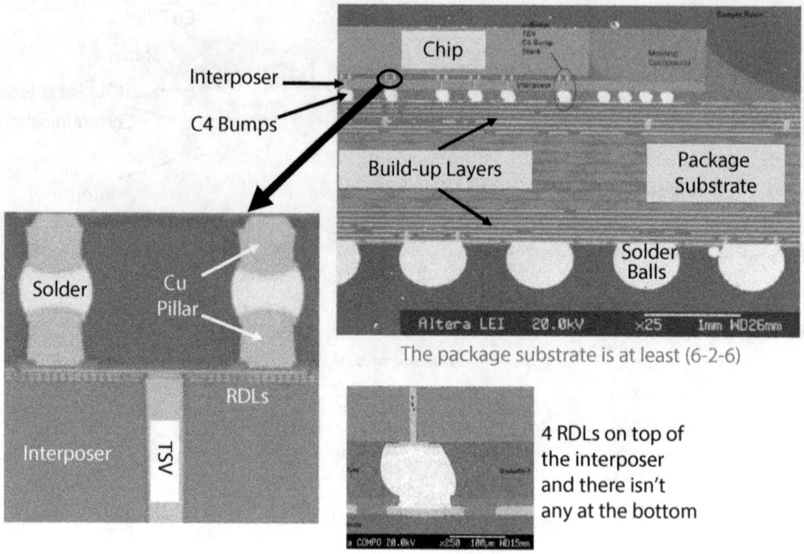

Figure 7.30 Altera/TSMC's CoWoS.

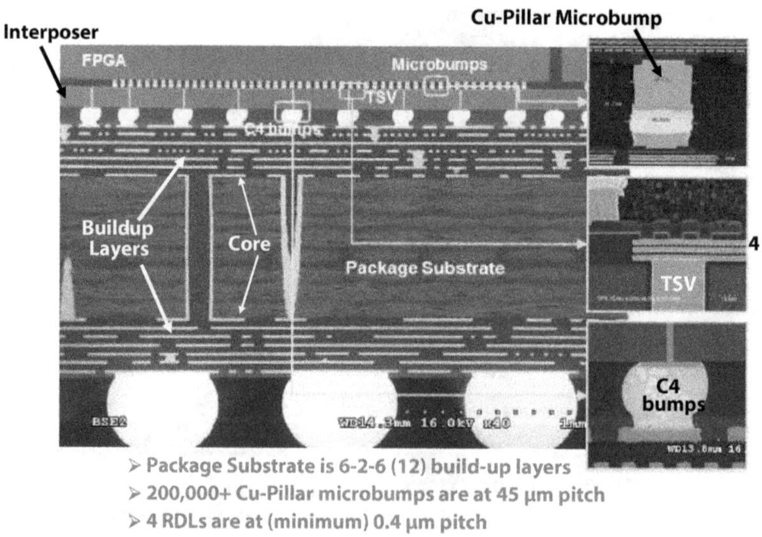

Figure 7.31 Xilinx/UMC/SPIL's 2.5D IC integration.

shown in Figs. 7.29 through 7.31. Thus, passive TSV/RDL interposers are for extremely fine-pitch, high-I/O, high-performance, and high-density semiconductor IC applications.

7.8.2 Fabrication of Interposers

The fabrication of interposers consists of two key tasks, namely, the fabrication of TSV and the fabrication of RDLs.

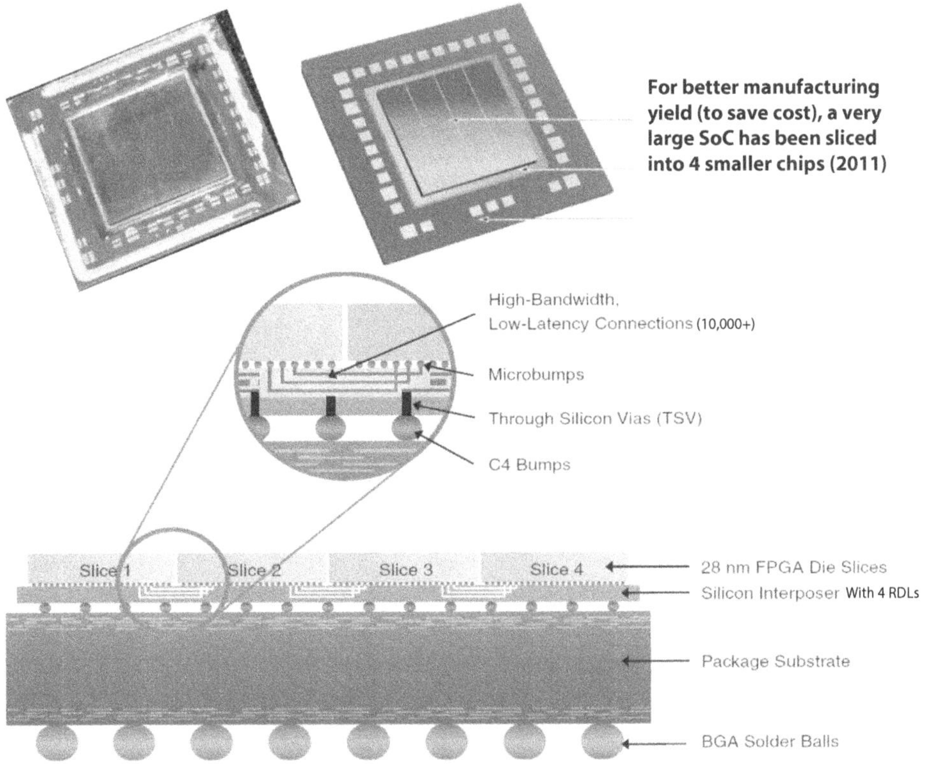

Figure 7.32 A very large SoC has been sliced into four smaller FPGA chips made by TSMC's 28nm process technology. The lateral communications of the chips are mainly performed by the 4 RDLs on top of the interposer.

7.8.3 Fabrication of TSVs

The fabrication process of TSVs is shown in Fig. 7.33. The process starts with an SiN_x/SiO_x insulation layer by either thermal oxidation or PECVD (plasma enhanced chemical vapor deposition) as shown in Fig. 7.33. After photoresist and TSV lithography, the TSV is etched into the Si substrate by Bosch-type DRIE (deep reactive ion etch) [56] to form a high aspect ratio (10.5) via structure. The etched TSV structure is then processed with an SiO_x liner by SACVD (subatmosphere chemical vapor deposition), a Ta barrier layer, and a Cu seed layer by PVD (physical vapor deposition) [57]. Cu electroplating is used to fill the TSV structure. The final blind TSV has a top opening of approximately 10 μm in diameter and a depth of about 105 μm, which gives an aspect ratio of 10.5. In such a high aspect ratio via structure, a bottom-up plating mechanism is applied to ensure a seamless TSV with a reasonably low Cu thickness in the field.

The SEM (scanning-electron microscopy) cross-sectional images are shown in Fig. 7.34. It can be seen that the diameter of the TSV is slightly decreased at the bottom, which is expected from the etching process. The Cu thickness at the field is <5 μm. The postplating anneal is at 400°C for 30 min. To complete the TSV process, excess Cu in the field is removed by CMP (chemical-mechanical polishing) [58].

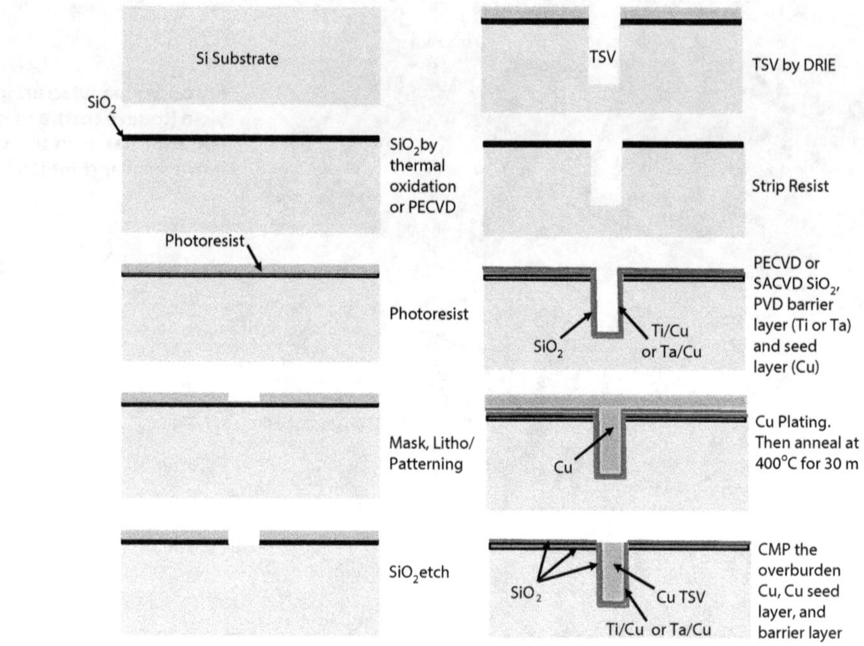

FIGURE 7.33 TSV fabrication process flow.

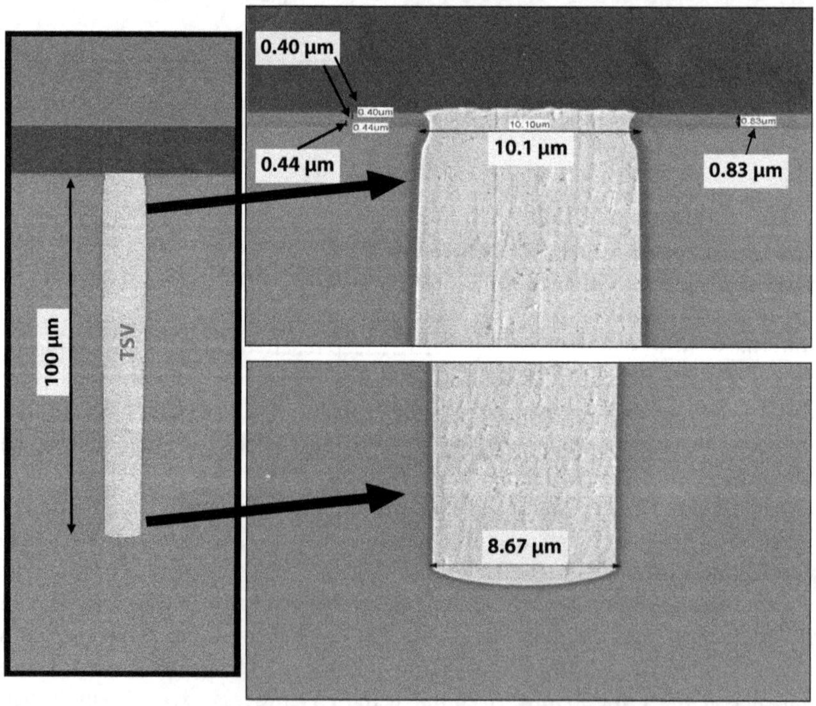

FIGURE 7.34 SEM images of TSV cross sections.

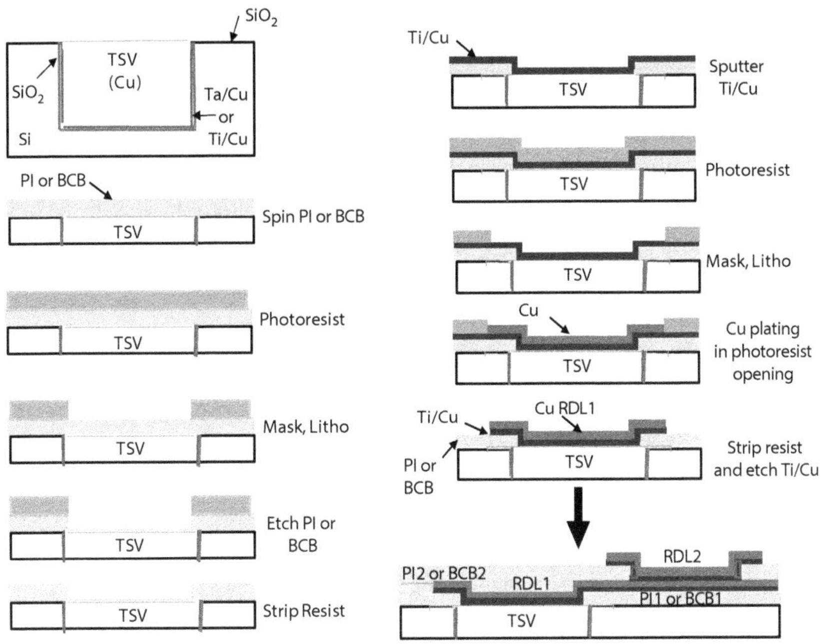

FIGURE 7.35 RDL fabrication process with polymers as passivation and Cu plating as metal layers.

7.8.4 Fabrication of RDLs

There are at least two ways to fabricate RDLs [4]. The first method is by using polymers, such as polyimide (PI) PWDC 1000 (Dow Corning), benzocyclobutene (BCB) cyclotene 4024-40 (Dow Chemical), polybenzo-bisoxazole (PBO) HD-8930 (HD Micro Systems), and the fluorinated aromatic AL-X 2010 (Asahi Glass Corporation), to make the passivation layer and electroplating (such as Cu) to make the metal layers. This method has been used by the OSAT (outsourced semiconductor assembly and test) to fabricate RDLs (without using semiconductor equipment) for wafer-level (fan-in) chip scale package [65, 66], embedded wafer-level (fan-out) ball grid array package [67], and (fan-out) redistribution chip package [68]. The second method is the Cu damascene method, which is primarily modified from the conventional semiconductor back-end-of-line to make the Cu metal RDLs such as those shown in Figs. 7.29 through 7.31. In general, much thinner structures (both dielectric layers and Cu RDLs), finer pitches, smaller line-widths, and spacing can be obtained with the Cu damascene method. The polymer/Cu-plating method will be mentioned first and then the Cu damascene method.

7.8.5 Fabrication of RDLs—Polymer/Cu-Plating Method

Continuing with the wafer from Fig. 7.33, the fabrication process for the RDLs using polymers is shown in Fig. 7.35, and the steps are listed below (UBM is included as well):

Step 1. Spin the polymers such as PI or BCB on the wafer and cure for 1 h. This will form a 4- to 7-μm-thick layer.

Step 2. Apply photoresist and mask, and then use photolithography techniques (align and expose) to open vias on the PI or BCB.

Step 3. Etch the PI or BCB.

Step 4. Strip off the photoresist.

Step 5. Sputter Ti and Cu over the entire wafer.

Step 6. Apply a photoresist and mask, and then use photolithography techniques to open the redistribution-trace locations.

Step 7. Electroplate Cu in photoresist openings.

Step 8. Strip off the photoresist.

Step 9. Etch off the Ti/Cu and RDL1 is completed.

Step 10. Repeat steps 1 to 9 for RDL2, and so forth.

Step 11. Repeat step 1 (for UBM).

Step 12. Apply photoresist and mask, and then use photolithography techniques (align and expose) to open vias on the PI or BCB for the desired bump pads and to cover the redistribution traces.

Step 13. Etch the desired vias on the PI or BCB.

Step 14. Strip off the photoresist.

Step 15. Sputter Ti and Cu over the entire wafer.

Step 16. Apply photoresist and mask, and then use photolithography techniques to open the vias on the bump pads to expose the areas with UBM.

Step 17. Electroplate the Cu core.

Step 18. Strip off the photoresist.

Step 19. Etch off the Ti/Cu.

Step 20. Electroless Ni and immersion Au. UBM is completed.

A typical cross-section of the RDLs with polymers (e.g., BCB) as passivation and Cu plating as metal layers is shown in Fig. 7.36 [59]. It can be seen that the thickness of the passivation layers, BCB1 and BCB2, is about 6 to 7 μm, and the RDL is about 4 μm.

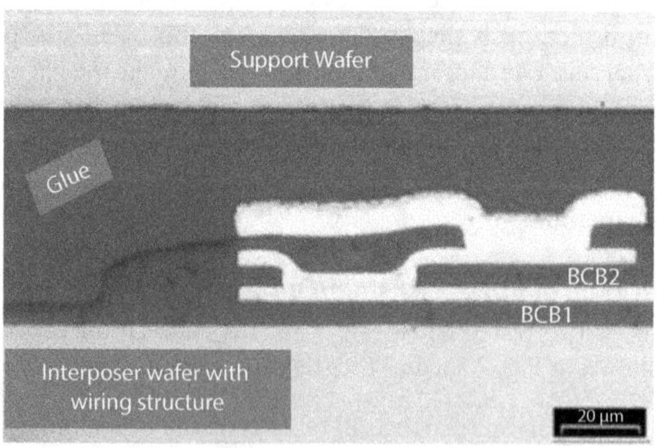

FIGURE 7.36 IZM's RDLs with BCB polymer.

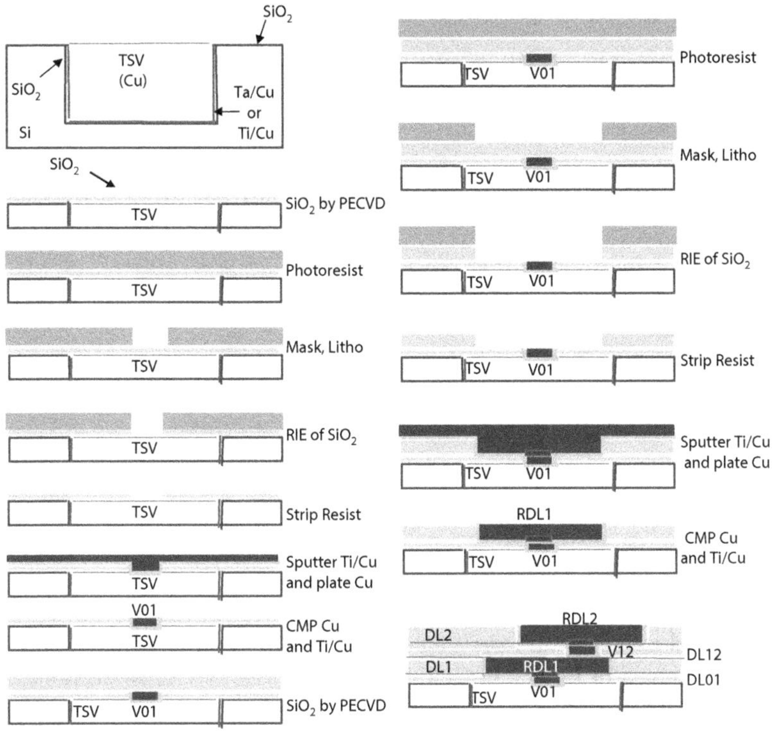

Figure 7.37 Process flow of RDLs fabricated by Cu Damascene method.

7.8.6 Fabrication of RDLs—Cu Damascene Method

Another method of RDL fabrication is by a Cu damascene process. If starting with the wafer from Fig. 7.33, the fabrication process of RDLs with a Cu damascene technique is primarily based on the semiconductor back-end-of-line process. The details are shown in Fig. 7.37, and the steps are listed below [4]:

Step 1. SiO_2 layer by PECVD.

Step 2. Apply photoresist and mask, then use photolithography techniques (align and expose) to open vias on the SiO_2.

Step 3. RIE (reactive ion etch) of SiO_2.

Step 4. Strip off the photoresist.

Step 5. Sputter Ti and Cu and electroplate Cu over the entire wafer.

Step 6. CMP the Cu and Ti/Cu. V01 (the via connecting the TSV to RDL1) is completed.

Step 7. Repeat step 1.

Step 8. Apply photoresist and mask, and then use photolithography techniques to open the redistribution trace locations.

Step 9. Repeat step 3.

Step 10. Repeat step 4.

Step 11. Repeat step 5.

Step 12. CMP the Cu and Ti/Cu. RDL1 is completed.

Step 13. Repeat steps 1 through 6 to complete V12 (the via connecting the RDL1 to RDL2).

Step 14. Repeat steps 7 through 12 to complete RDL2 and any additional layers.

Step 15. (For UBM) Repeat step 1.

Step 16. Apply photoresist and mask, and then use photolithography techniques (align and expose) to open vias on the SiO_2 for the desired bump pads and cover the redistribution traces.

Step 17. Etch the desired vias on the SiO_2.

Step 18. Strip off the photoresist.

Step 19. Sputter Ti and Cu over the entire wafer.

Step 20. Apply photoresist and mask, and then use photolithography techniques to open the vias on the bump pads to expose the areas with UBM.

Step 21. Electroplate Cu core.

Step 22. Strip off the photoresist.

Step 23. Etch off the Ti/Cu.

Step 24. Electroless Ni and immersion Au. UBM is completed.

It should be noted that the RDLs can also be fabricated by the dual Cu damascene method as shown in Fig. 7.38 [4]. SEM images of the RDL cross-sections fabricated by a Cu damascene technique are shown in Figs. 7.39 and 7.40. The minimum RDL line width is

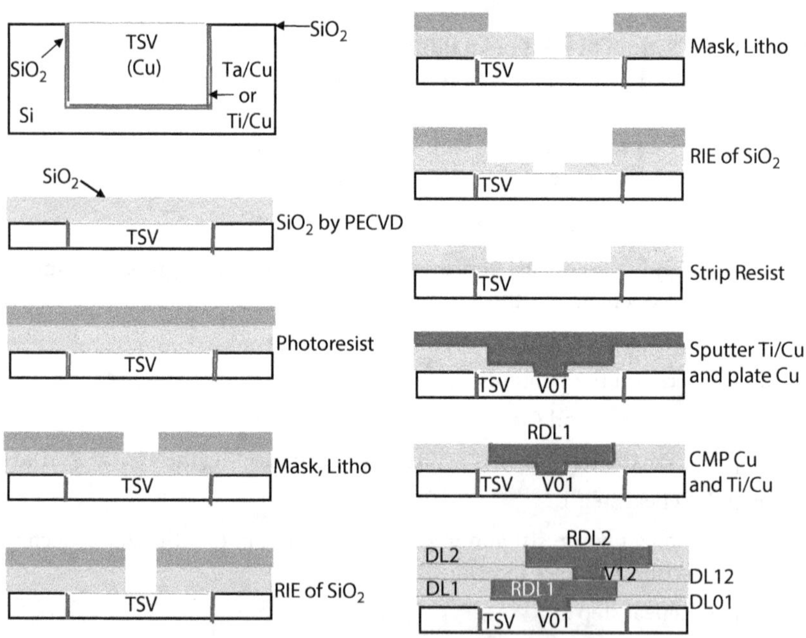

FIGURE 7.38 Process flow of RDLs fabricated by dual Cu Damascene method.

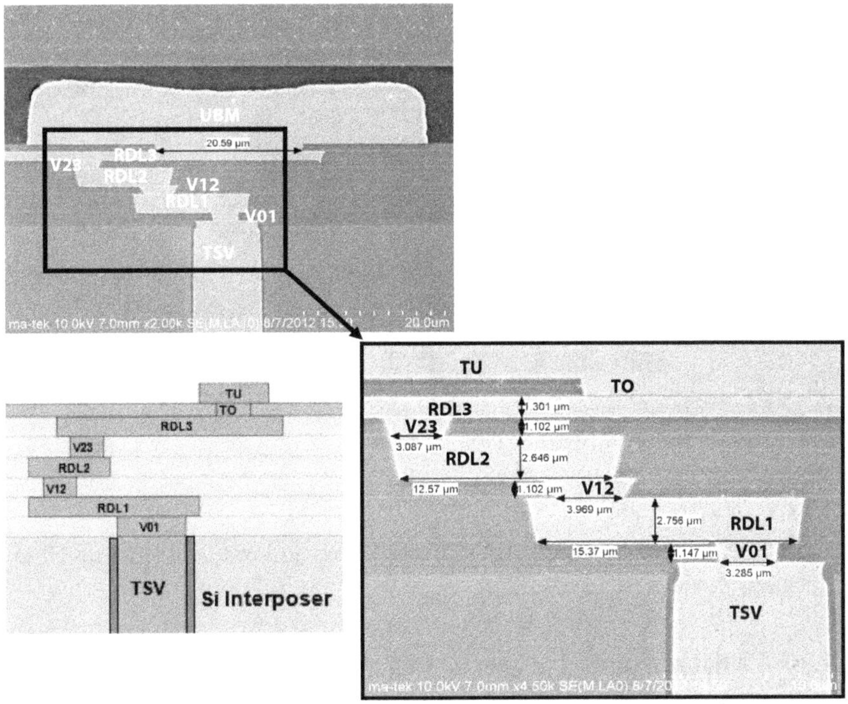

Figure 7.39 SEM images of cross-sections of RDLs fabricated by Cu damascene method.

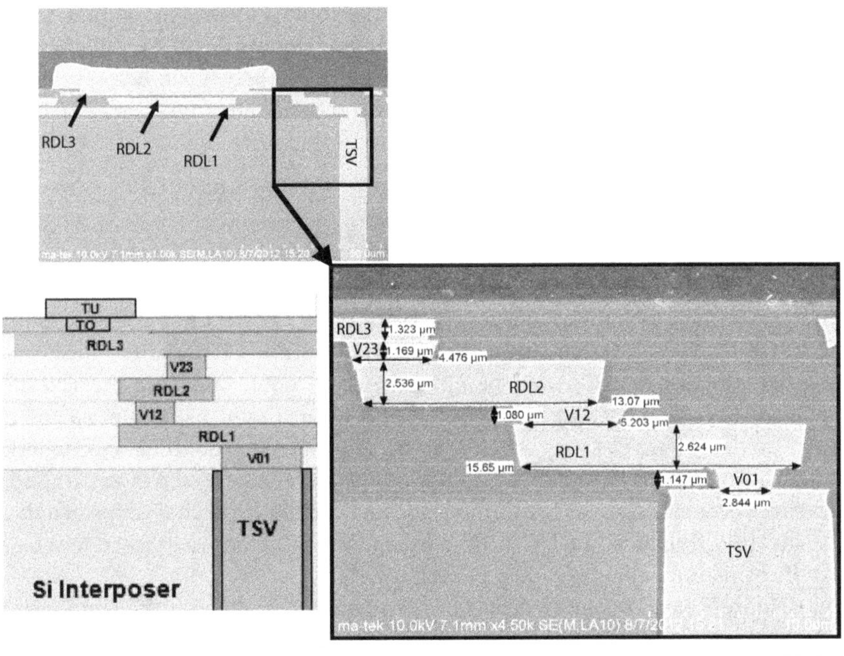

Figure 7.40 SEM images of cross-sections of RDLs fabricated by Cu damascene method.

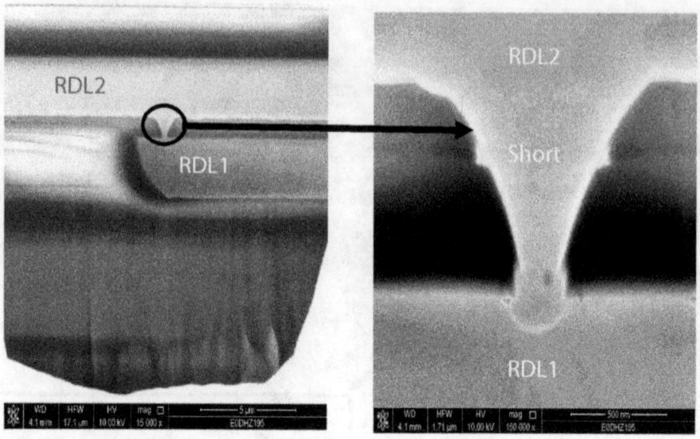

Figure 7.41 SEM/FIB showing the short between RDL1 and RDL2. The thickness of the passivation layer between RDL1 and RDL2 is <1 μm.

3 μm. The thickness of RDL1 and RDL2 is 2.6 μm and of RDL3 is 1.3 μm. The passivation thickness between RDLs is 1 μm.

7.8.7 A Note on Contact Aligner for Cu Damascene Method

Lithography using a contact aligner provides a low-cost process as compared with a stepper/scanner under the same resolution requirements. Since the minimum line width is 3 μm in this case, the mask had to be placed very close to the (photoresist) surface of the 300 mm wafer. In a few cases, particles on the contact aligner mask punched holes on the photoresist. In this case, shorts may happen as shown in Fig. 7.41, which happened while fabricating the V12 (via connecting RDL1 and RDL2). This can be prevented by cleaning the mask between exposures. Alternately, if cost is not an issue, using a stepper/scanner is another solution.

7.8.8 Back-Side Processing and Assembly

The process flow of back side and assembly [47] is shown in Fig. 7.42. It can be seen that after the fabrication of TSV, RDLs, passivation, and UBM, the topside of the interposer wafer is temporary bonded to a carrier by adhesive. The next step is backgrinding the interposer wafer, Si etching, low temperature passivation, and Cu revealing. Then, back-side RDL (optional), UBM, and C4 wafer bumping are carried out. After that, the next step is to temporarily bond another carrier wafer to the back side (with solder bumps) and de-bond the first carrier wafer. This step is followed by chip-on-wafer bonding and underfilling. After the whole (chip-on) interposer wafer is completed, the next step is to de-bond the second carrier wafer and transfer the thin interposer wafer with attached chips to a dicing tape for singulation. The individual TSV/RDL interposer with chips is attached to the package substrate by natural reflow and then underfilled.

Figure 7.43 shows more details on Cu revealing. Right after the temporary bonding of the support carrier, backgrinding the wafer to a few microns to the TSV, Si dry etching (by RIE) to a few microns below the TSV, and low-temperature passiving the SiN/SiO_2 are

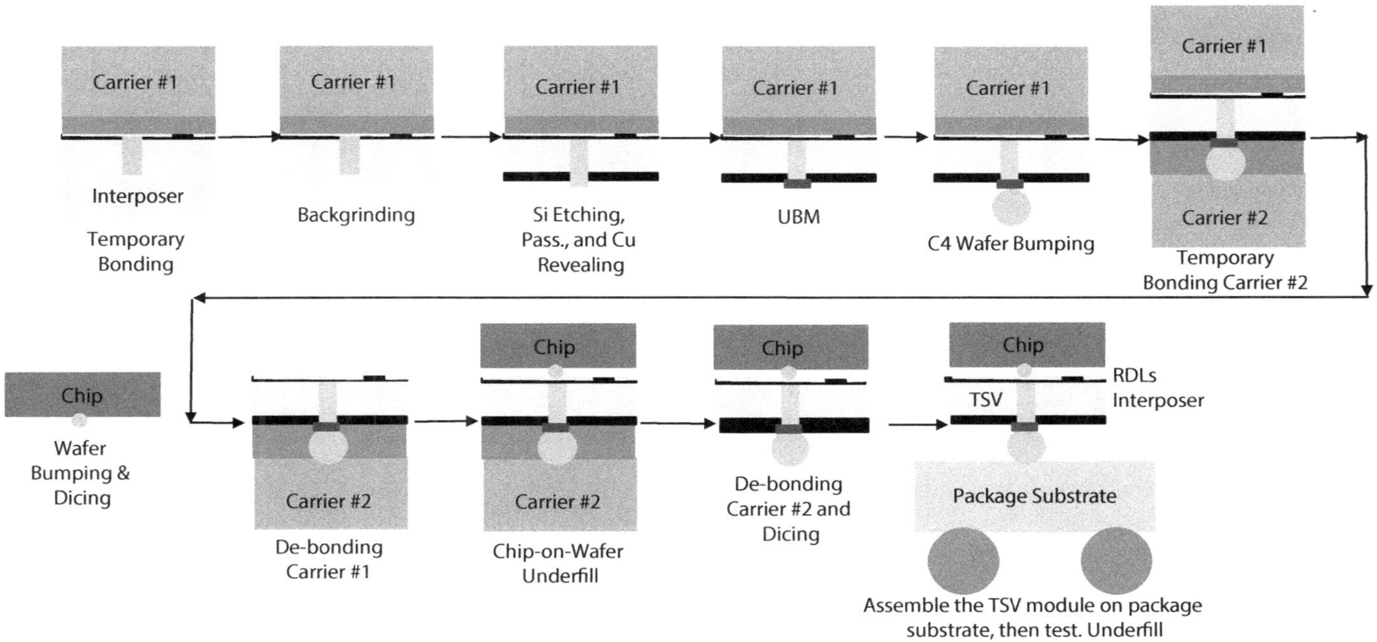

Figure 7.42 Conventional process flow for 2.5D/3D IC integration (chip on interposer wafer on package substrate).

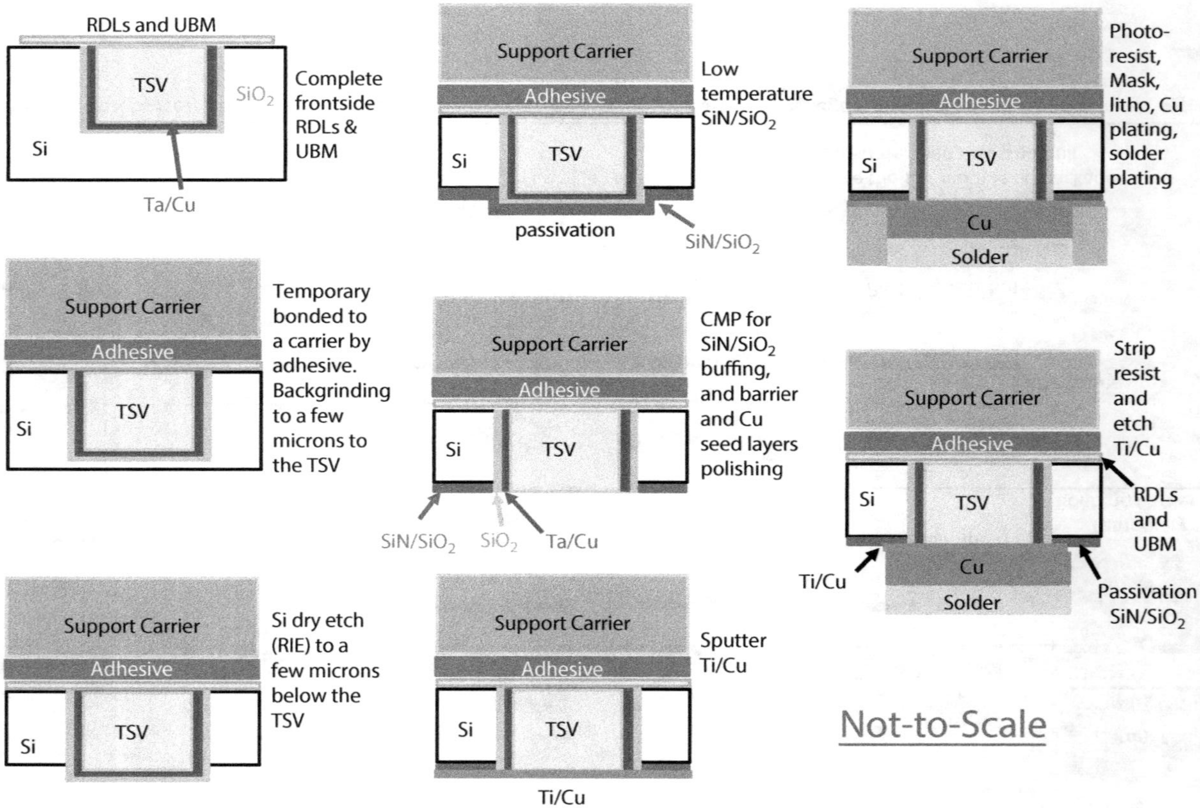

Figure 7.43 Back-side Cu reveal and UBM/solder plating process flow.

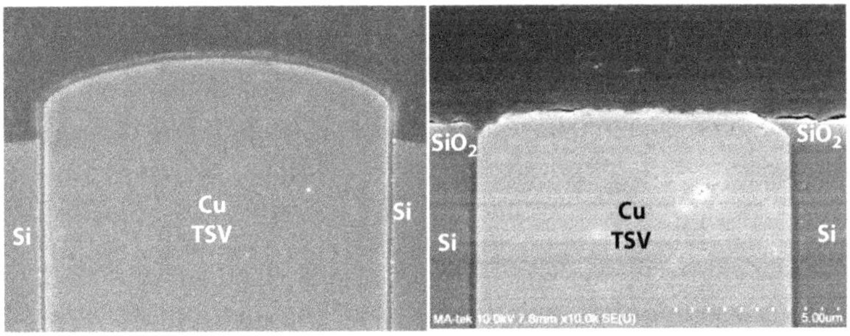

Figure 7.44 TSV Cu revealing. (Left) Before dry etch of Si. (Right) After Si dry etching, low-temperature SiN/SiO_2, and removal (CMP) of the isolation, barrier, and seed layer.

performed. Then, CMP for SiN/SiO_2 buffing and barrier and Cu seed layers polishing are carried out. Cu revealing is completed and shown in Fig. 7.44 [4].

7.8.9 Summary and Recommendations

Materials and processes for the fabrication of RDLs with two different methods, namely (1) polymer (e.g., BCB) to make the passivation and Cu plating to make the metal layers, and (2) semiconductor back-end-of-line Cu damascene, have been presented. The fabrications of TSVs, UBMs, and Cu reveal of the interposer have also been described. Some important results and recommendations are summarized as follows:

- The RDLs fabricated by the polymer method are usually thicker. The thickness of the passivation is 6 to 7 μm and the Cu metal layer is 3 to 4 μm.
- The RDLs fabricated by the Cu damascene method are thinner. The thickness of the passivation and the Cu metal layer is ~1 μm.
- Lithography using a stepper/scanner can go down to submicron line width and line spacing. However, for a few microns' applications, a contact aligner will provide a low-cost process under the same resolution requirements. In order to prevent shorts (not to allow holes in the photoresist and thus the passivation during the fabrication of the vias) between the RDLs, all the particles must be cleaned off from the mask.

7.9 Thin-Wafer Handling

One of the key reasons why 3D IC integration is so attractive to semiconductor industry today is because it uses thin chips/interposers. For example, 50 μm or less for memory chips, 100 μm or less for passive interposers, and 200 μm or less for active interposers.

Making the wafer thin is not a big problem because most backgrinding machine can grind wafers to as thin as 5 μm. However, handling thin TSV/RDL wafers through all the semiconductor fabrication and packaging assembly processes is difficult. Usually, the TSV/RDL chip/interposer wafer is temporarily bonded with an adhesive on a support (carrier) wafer before backgrinding to expose the Cu TSVs (Cu revealing) and make the wafer thin. Then it goes through the fabrication processes, such as metallization, passivation, UBM, and wafer bumping. After all these are done, removing the thin TSV/RDL wafer from the support wafer (de-bonding) poses another big challenge such as broken wafers and adhesive residues [1–25], which translate to higher yield-lost.

In this section, a different route is taken that uses a heat-spreader wafer as the supporting carrier. When the manufacturing process is completed, the heat-spreader remains on the thin TSV/RDL interposer. In this case, the temporary bonding and de-bonding processes are eliminated and a simple and low-cost thin-wafer handling method is achieved. TI and TSMC's thin-wafer handling method with a polymer or molding compound [60–62] will also be presented. The conventional thin-wafer handling method will be discussed first.

7.9.1 Conventional Thin-Wafer Handling Method

As mentioned earlier, Fig. 7.42 shows a conventional thin-wafer handling method. After the fabrication of TSV, RDLs, passivation, and UBM, the TSV/RDL (active or passive) interposer wafer is temporarily bonded to a supporting carrier by adhesive. Then, backgrind the TSV/RDL interposer wafer, Si etch, low temperature passivation, and Cu reveal. It is followed by back-side RDL (optional), UBM, and ordinary C4 wafer bumping. Then, temporarily bond another carrier wafer to the back side (with solder bumps) and de-bond the first carrier wafer. It is followed by chip-on-wafer bonding (attaching the microbumped chip on the interposer wafer) and underfilling. After the whole interposer wafer is completed, then de-bond the second carrier wafer and transfer the thin TSV/RDL interposer wafer with attached chips to a dicing tape for singulation.

Today, the available adhesive materials are (1) the LC-series and LTHC (light-to-heat convention) layer provided by 3M, (2) HT-10.10 and ZoneBOND by Brewer Science, (3) TZNR-A series and WSP series by TOK, (4) XP-BCB by DOW, (5) WL-40XX and WL-30XX by Dow Corning, (6) HD-3007 and HD-7010 by Du Pont, and (7) TMAT by Thin Materials. The supporting carrier wafer materials are (1) glass and (2) silicon. The temporary bonding and de-bonding equipments are (1) TWM12000 for bonding and TWR12000 for de-bonding provided by TOK, (2) 850TB for temporary bonding and 850DB for de-bonding by EVG, (3) XBS300 for both temporary bonding and de-bonding provided by SUSS, and (4) WSS by 3M. The de-bonding methods are (1) mechanical peel-off, (2) solvent release, (3) a combination of mechanical peel-off and solvent release, (4) thermal-sliding, and (5) laser irradiation.

7.9.2 TI's TSV-WCSP Integration Process

Figure 7.45 shows TI's [60] TSV-WCSP (wafer-level chip scale package) integration process flow. Up to Cu reveal and UBM, the processes are exactly the same as those (Fig. 7.42) in the conventional process flow. Then, instead of wafer bumping, TI performs the chip-on-wafer bonding and underfilling. It is followed by overmolding the whole wafer and de-bonding the carrier. Then, UBM, wafer bumping, and dicing are carried out. The individual TSV module is attached on the package substrate and underfilled. In this case, the second set of temporary bonding and de-bonding, as shown in Fig. 7.42, is eliminated.

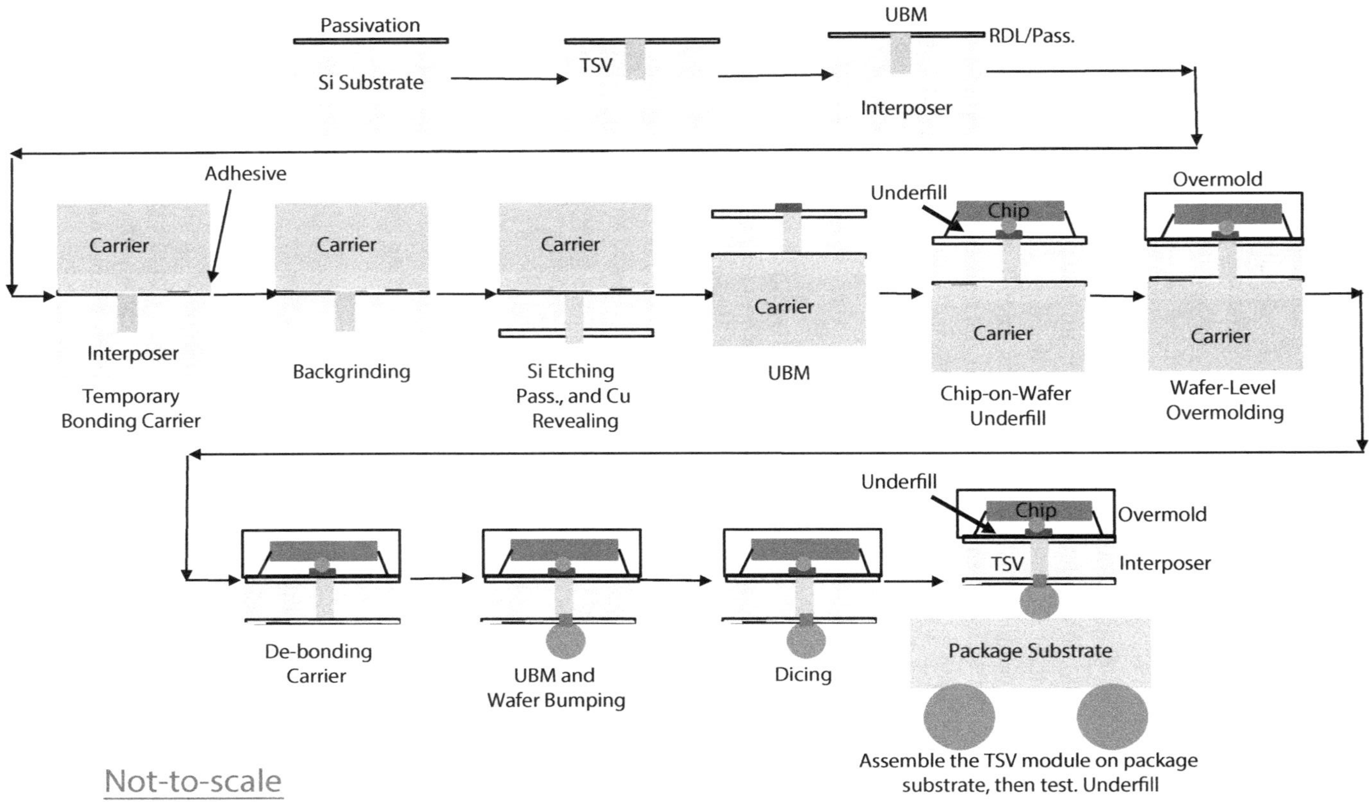

FIGURE 7.45 TI's TSV-WCSP integration process flow.

7.9.3 TSMC's Thin-Wafer Handling with Polymer

In contradiction to TI, TSMC perform the chip-on-wafer bonding [61] right after the fabrication of TSVs, top RDLs, and UBM as shown in Fig. 7.46. It is followed by underfilling, overmolding the whole wafer, removing the portions of molding over chips, temporary bonding a carrier, backgrinding, low-temperature passivation, Cu revealing, UBM, and wafer bumping. Then, the carrier is de-bonded and the TSV module is transferred to a dicing tape for singulation. The individual TSV module is attached on the package substrate and underfilled. Again, the second set of temporary bonding and de-bonding, as shown in Fig. 7.42, is eliminated.

7.9.4 TSMC's Thin-Wafer Handling without Temporary Bonding and De-Bonding

In contradict to the conventional thin-wafer handling method as shown in Fig. 7.42, Fig. 7.47 shows TSMC's thin-wafer handling method with molding chips on interposer wafer [62]. After the fabrication of TSV, RDLs, passivation, and UBM, the microbumped chips are bonded on the TSV/RDL full-thickness (>700 μm) interposer wafer and then underfilled. Then, overmold the chips on top of the interposer wafer with a molding compound and a stress relief feature to reduce the wafer warpage while backgrinding to reveal the Cu TSV. It is followed by UBM, C4 wafer bumping, dicing, and attaching the individual TSV/RDL interposer with chips and molding compound on a package substrate by natural reflow and then underfilling.

Figure 7.48 shows more details on TSMC's method. In Fig. 7.48(a), the microbumped chips are bonded on the full-thickness TSV/RDL interposer wafer. Underfill is applied to cement the chips to the interposer to ensure the microbump reliability. Stress relief "dams" (walls) are formed between the chips on the topside of the interposer. The material of the stress relief walls should have a high CTE (coefficient of thermal expansion) and a low modulus. In Fig. 7.48(b), compression molding is performed and mold compound now covers the chips and the stress relief features. Then, the mold compound and the stress relief features may be cured. Figure 7.48(c) shows the top-grinding of the mold compound to expose the back side of the chips and backgrinding of the silicon substrate to expose (reveal) the Cu TSV. Figure 7.48(d) shows the C4 wafer bumping. In TSMC's manufacturing process, temporary bonding and de-bonding processes are eliminated. Instead, similar to embedded wafer-level ball grid array package (eWLP) [47] by using molding compound as a stiffer, in addition they also use the stress relief features to increase the stiffness of the molding compound to resist the warpage.

7.9.5 Thin-Wafer Handling with a Heat-Spreader Wafer

Figure 7.49 shows the process flow of a proposed thin-wafer handling with a low CTE and high thermal conductivity heat-spreader wafer [63]. After the fabrication of TSV, RDLs, passivation, and UBM, the microbumped chips are bonded on the TSV/RDL full-thickness (>700 μm) interposer wafer and then underfilled as shown in Fig. 7.50(a). Then, the heat-spreader wafer is attached to the back side of all the chips with a high-temperature stable TIM (thermal interface material), Fig. 7.50(b). It is followed by backgrinding the TSV/RDL interposer and Cu revealing to expose the Cu TSV, Fig. 7.50(c). Then, UBM and C4 wafer bumping, as shown in Fig. 7.50(d), and singulation are carried out. The individual thin TSV/RDL interposer with chips and heat-spreader is then attached to the package substrate by natural reflow and then underfilled.

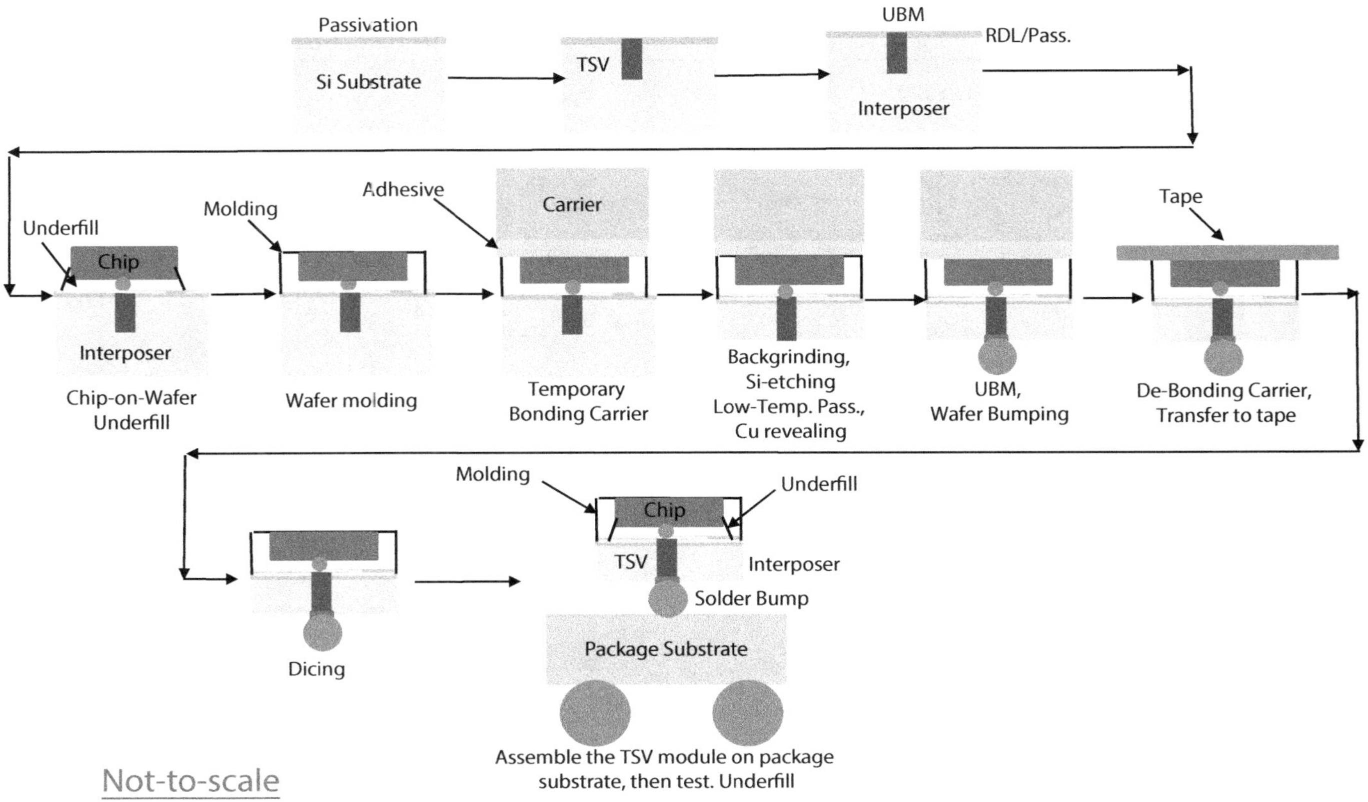

Figure 7.46 TSMC's chip-on-wafer integration process flow.

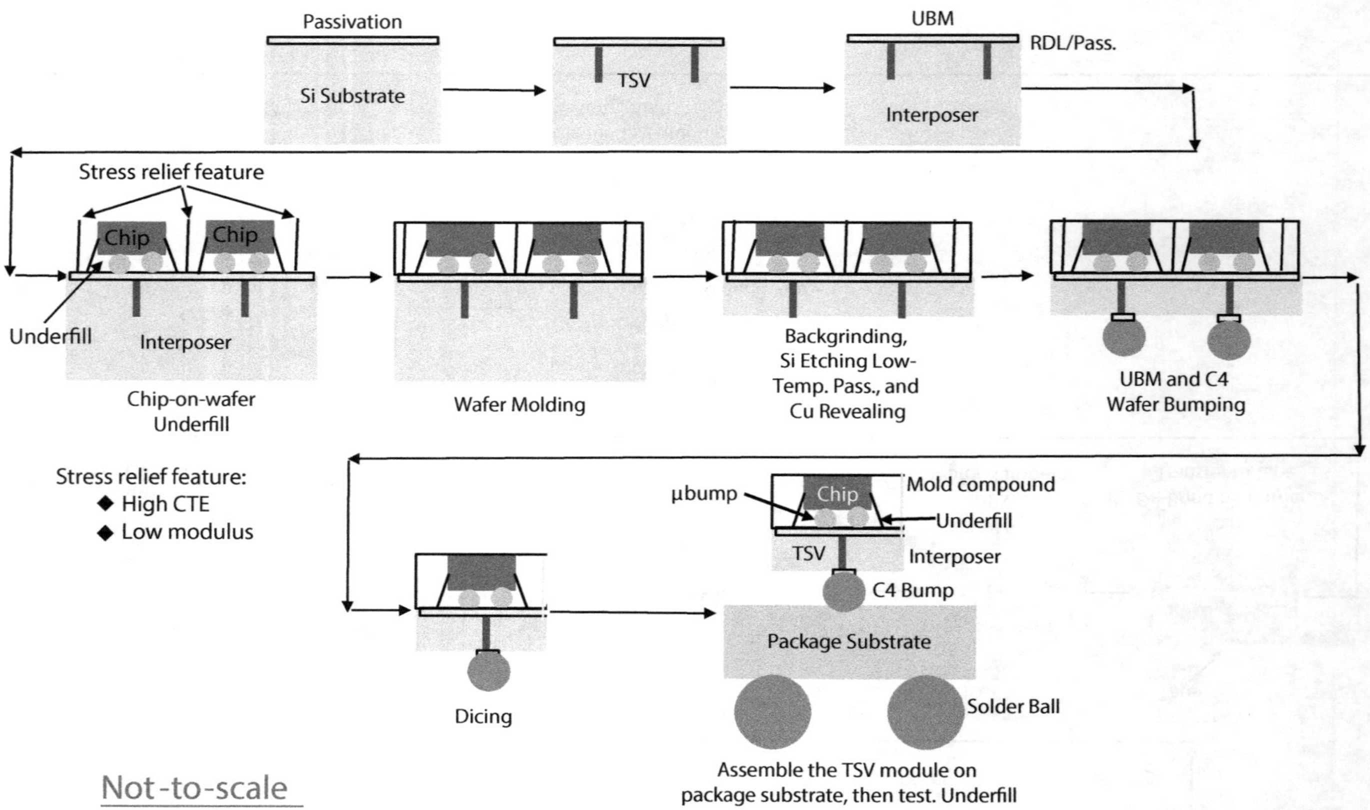

Figure 7.47 TSMC's chip-on-wafer integration process flow (without temporary bonding and de-bonding).

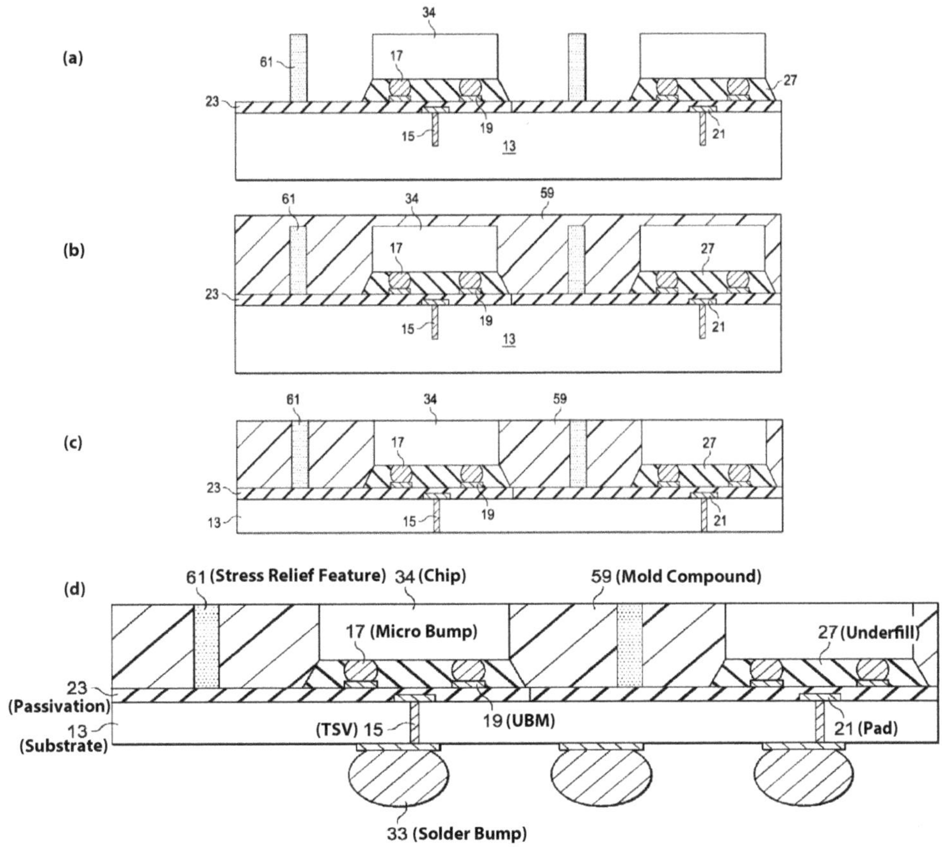

FIGURE 7.48 TSMC's molding chips on interposer wafer (without temporary bonding and de-bonding).

7.9.6 Summary and Recommendations

Conventional thin-wafer handling such as adhesive materials, supporting carriers, temporary bonding and de-bonding equipments, and de-bonding methods has been discussed. TI and TSMC's think-wafer handling methods have also been presented and discussed. A new thin-wafer handling with a heat-spreader wafer is proposed and the process flow has been presented. Some important results are summarized as follows:

- With the proposal of bonding a heat-spreader wafer to the back side of chips on top of a TSV/RDL interposer wafer for thin-wafer handling, the conventional temporary bonding and de-bonding processes can be eliminated.
- With the new proposal, much less process steps are used (save cost) and much less chances to have yield-lost (save cost).
- One stone two birds! The heat-spreader wafer can be considered (used) as the supporting carrier during manufacturing and as the thermal management tool (heat-spreader) of the chips during operations.

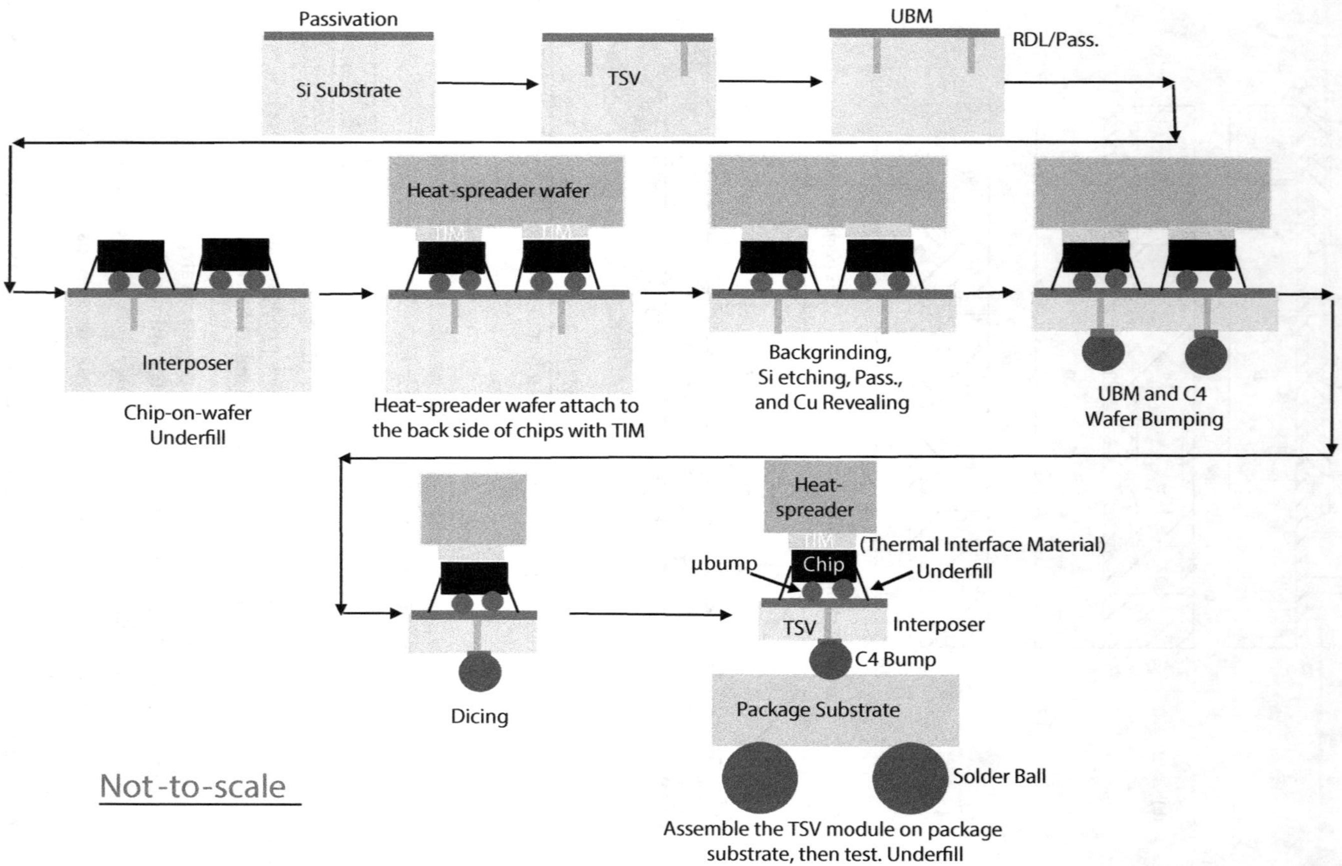

Figure 7.49 Process flow of thin-wafer handling of 2.5D/3D integration with a heat-spreader wafer.

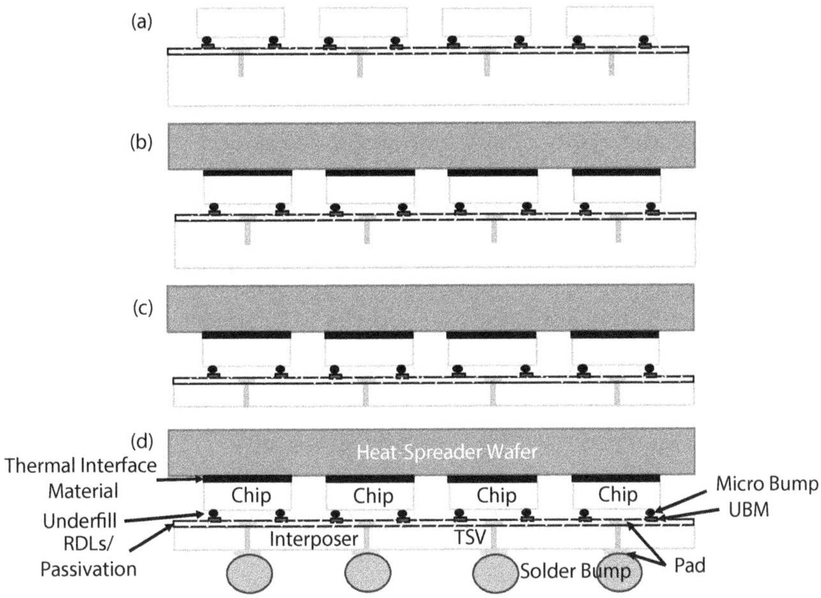

FIGURE 7.50 Thin-wafer handling with a heat-spreader wafer.

7.10 References

[1] Lau, J. H., *Through Silicon Via (TSV) for 3D Integration*, McGraw-Hill, New York, 2013.

[2] Lau, J. H., *Reliability of RoHS compliant 2D & 3D IC Interconnects*, McGraw-Hill, New York, 2011.

[3] Lau, J. H., C. K. Lee, C. S. Premachandran, and Y. Aibin, *Advanced MEMS Packaging*, McGraw-Hill, New York, 2010.

[4] Lau, J. H., P. Tzeng, C. Lee, C. Zhan, M. Li, J. Cline, K. Saito, et al., "Redistribution Layers (RDLs) for 2.5D/3D IC Integration," *IMAPS Transactions, Journal of Microelectronic Packaging*, Vol. 11, No. 1, First Quarter 2014, pp. 16–24.

[5] Lau, J. H., "The Future of Interposer for Semiconductor IC Packaging," *Chip Scale Review*, Vol. 18, No. 1, January–February, 2014, pp. 32–36.

[6] Hsieh, M. C., S. T. Wu, C. J. Wu, and J. H. Lau, "Energy Release Rate Estimation for Through Silicon Vias in 3-D Integration," *IEEE Transactions on CPMT*, Vol. 4, No. 1, January 2014, pp. 57–65.

[7] Lau, J. H., "Supply Chains for High-Volume Manufacturing of 3D IC Integration," *Chip Scale Review*, Vol. 17, No. 1, January–February 2013, pp. 33–39.

[8] Khan, N., H. Li, S. Tan, S. Ho, V. Kripesh, D. Pinjala, J. H. Lau, et al., "3-D Packaging With Through-Silicon Via (TSV) for Electrical and Fluidic Interconnections," *IEEE Transactions on CPMT*, Vol. 3, No. 2, February 2013, pp. 221–228.

[9] Lau, J. H., and G. Y. Tang, "Effects of TSVs (through-silicon vias) on thermal performances of 3D IC integration system-in-package (SiP)," *Journal of Microelectronics Reliability*, Vol. 52, No. 11, November 2012, pp. 2660–2669.

[10] Lau, J. H., "Recent Advances and New Trends in Nanotechnology and 3D Integration for Semiconductor Industry," *The Electrochemical Society, ECS Transactions*, Vol. 44, No. 1, 2012, pp. 805–825.

[11] Chien, H. C., J. H. Lau, Y. Chao, R. Tain, M. Dai, S. T. Wu, W. Lo, et al., "Thermal Performance of 3D IC Integration with Through-Silicon Via (TSV)," *IMAPS Transactions, Journal of Microelectronic Packaging*, Vol. 9, 2012, pp. 97–103.

[12] Chen, J. C., J. H. Lau, P. J. Tzeng, S. Chen, C. Wu, C. Chen, H. Yu, et al., "Effects of Slurry in Cu Chemical Mechanical Polishing (CMP) of TSVs for 3-D IC Integration," *IEEE Transactions on CPMT*, Vol. 2, No. 6, June 2012, pp. 956–963.

[13] Lee, C. K., T. C. Chang, J. H. Lau, Y. Huang, H. Fu, J. Huang, Z. Hsiao, et al., "Wafer Bumping, Assembly, and Reliability of Fine-Pitch Lead-Free Micro Solder Joints for 3-D IC Integration," *IEEE Transactions on CPMT*, Vol. 2, No. 8, August 2012, pp. 1229–1238.

[14] Sekhar, V. N., L. Shen, A. Kumar, T. C. Chai, X. Zhang, C. S. Premchandran, J. H. Lau, et al., "Study on the Effect of Wafer Back Grinding Process on Nanomechanical Behavior of Multilayered Low-k Stack," *IEEE Transactions on CPMT*, Vol. 2, No. 1, January 2012, pp. 3–12.

[15] Zhang, X., R. Rajoo, C. S. Selvanayagam, A. Kumar, V. Rao, N. Khan, J. H. Lau, et al., "Application of Piezoresistive Stress Sensor in Wafer Bumping and Drop Impact Test of Embedded Ultrathin Device," *IEEE Transactions on CPMT*, Vol. 2, No. 16, June 2012, pp. 935–943.

[16] Wu, C., S. Chen, P. Tzeng, J. H. Lau, Y. Hsu, J. Chen, Y. Hsin, et al., "Oxide Liner, Barrier and Seed Layers, and Cu-Plating of Blind Through Silicon Vias (TSVs) on 300 mm Wafers for 3D IC Integration," *IMAPS Transactions, Journal of Microelectronic Packaging*, Vol. 9, No. 1, First Quarter 2012, pp. 31–36.

[17] Lau, J. H., M. S. Zhang, and S. W. R. Lee, "Embedded 3D Hybrid IC Integration System-in-Package (SiP) for Opto-Electronic Interconnects in Organic Substrates," *ASME Transactions, Journal of Electronic Packaging*, Vol. 133, September 2011, pp. 1–7.

[18] Chai, T. C., X. Zhang, J. H. Lau, C. S. Selvanayagam, D. Pinjala, Y. Hoe, Y. Ong, et al., "Development of Large Die Fine-Pitch Cu/low-*k* FCBGA Package with through Silicon via (TSV) Interposer," *IEEE Transactions on CPMT*, Vol. 1, No. 5, May 2011, pp. 660–672.

[19] Lau, J. H., "TSV Interposers: The Most Cost-Effective Integrator for 3D IC Integration," *Chip Scale Review*, Vol. 15, No. 5, September/October 2011, pp. 23–27.

[20] Sharma, G., V. Rao, A. Kumar, Y. Lim, K. Houe, S. Lim, J. H. Lau, et al., "Design and Development of Multi-Die Laterally Placed and Vertically Stacked Embedded Micro-Wafer-Level Packages," *IEEE Transactions on CPMT*, Vol. 1, No. 5, May 2011, pp. 52–59.

[21] Kumar, A., X. Zhang, Q. Zhang, M. Jong, G. Huang, V. Lee, J. H. Lau, et al., "Residual Stress Analysis in Thin Device Wafer Using Piezoresistive Stress Sensor," *IEEE Transactions on CPMT*, Vol. 1, No. 6, June 2011, pp. 841–851.

[22] Yu, A., J. H. Lau, S. Ho, A. Kumar, W. Hnin, W. Lee, M. Jong, et al., "Fabrication of High Aspect Ratio TSV and Assembly with Fine-Pitch Low-Cost Solder Microbump for Si Interposer Technology with High-Density Interconnects," *IEEE Transactions on CPMT*, Vol. 1, No. 9, September 2011, pp. 1336–1344.

[23] Lau, J. H., "Overview and Outlook of TSV and 3D Integrations," *Journal of Microelectronics International*, V.28, No. 2, 2011, pp. 8–22.

[24] Lau, J. H., C.-J. Zhan, P.-J. Tzeng, C.-K. Lee, M.-J. Dai, H.-C. Chien, Y.-L. Chao, W. Li, et al., "Feasibility Study of a 3D IC Integration System-in-Packaging (SiP) from a 300mm Multi-Project Wafer (MPW)," *IMAPS Transactions, Journal of Microelectronic Packaging*, Vol. 8, No. 4, Fourth Quarter 2011, pp. 171–178.

[25] Sheu, S., Z. Lin, J. Hung, J. H. Lau, P. Chen, S. Wu, K. Su, et al., "An Electrical Testing Method for Blind Through Silicon Vias (TSVs) for 3D IC Integration," *IMAPS Transactions, Journal of Microelectronic Packaging*, Vol. 8, No. 4, Fourth Quarter 2011, pp. 140–145.

[26] Lau, J. H., "Critical Issues of 3D IC Integrations," IMAPS Transactions, *Journal of Microelectronics and Electronic Packaging*, First Quarter Issue, 2010, pp. 35–43.

[27] Lau, J. H., Y. S. Chan, and R. S. W. Lee, "3D IC Integration with TSV Interposers for High-Performance Applications," Chip Scale Review, Vol. 14, No. 5, September/October 2010, pp. 26–29.

[28] Lau, J. H., "Design and Process of 3D MEMS Packaging," IMAPS Transactions, *Journal of Microelectronics and Electronic Packaging*, First Quarter Issue, 2010, pp. 10–15.

[29] Lau, J. H., Lee, R., Yuen, M., and Chan, P., "3D LED and IC Wafer Level Packaging," *Journal of Microelectronics International*, Vol. 27, Issue 2, 2010, pp. 98–105.

[30] Lau, J. H., "State-of-the-art and Trends in 3D Integration," *Chip Scale Review*, Vol. 14, No. 2, March/April, 2010, pp. 22–28.

[31] Tang, G. Y., S. Tan, N. Khan, D. Pinjala, J. H. Lau, A. Yu, V. Kripesh, et al., "Integrated Liquid Cooling Systems for 3-D Stacked TSV Modules," *IEEE Transactions on CPMT*, Vol. 33, No. 1, March 2010, pp. 184–195.

[32] Khan, N., V. Rao, S. Lim, H. We, V. Lee, X. Zhang, J. H. Lau, et al., "Development of 3-D Silicon Module With TSV for System in Packaging," *IEEE Transactions on CPMT*, Vol. 33, No. 1, March 2010, pp. 3–9.

[33] Lim, S., V. Rao, W. Hnin, W. Ching, V. Kripesh, C. Lee, J. H. Lau, et al., "Process Development and Reliability of Microbumps," *IEEE Transactions on CPMT*, Vol. 33, No. 4, December 2010, pp. 747–753.

[34] Lee, C., A. Yu, L. Yan, H. Wang, J. Han, Q. Zhang, et al., "Characterization of Intermediate In/Ag Layers of Low Temperature Fluxless Solder Based Wafer Bonding for MEMS Packaging," *Journal of Sensors and Actuators A*, Vol. 154, 2009, pp. 85–91.

[35] Yu, A., N. Khan, G. Archit, D. Pinjala, K. Toh, V. Kripesh, J. H. Lau, et al., "Fabrication of Silicon Carriers With TSV Electrical Interconnections and Embedded Thermal Solutions for High Power 3-D Packages," *IEEE Transactions on CPMT*, Vol. 32, No. 3, September 2009, pp. 566–571.

[36] Selvanayagam, C., J. H. Lau, X. Zhang, S. Seah, K. Vaidyanathan, and T. C. Chai, "Nonlinear Thermal Stress/Strain Analyses of Copper Filled TSV (Through Silicon Via) and Their Flip-Chip Microbumps," *IEEE Transactions on Advanced Packaging*, Vol. 32, No. 4, November 2009, pp. 720–728.

[37] Zhang, X., A. Kumar, Q. X. Zhang, Y. Y. Ong, S. W. Ho, C. H. Khong, J. H. Lau, et al., "Application of Piezoresistive Stress Sensors in Ultra-Thin Device Handling and Characterization," *Journal of Sensors & Actuators: A. Physical*, Vol. 156, November 2009, pp. 2–7.

[38] Lau, J. H., "Overview and Outlook of 3D IC Packaging, 3D IC Integration, and 3D Si Integration," *ASME Transactions, Journal of Electronic Packaging*, December 2014, Vol. 136, No. 4, pp. 1–15.

[39] Lau, J. H., C. Lee, C. Zhan, S. Wu, Y. Chao, M. Dai, R. Tain, et al., "Low-Cost Through-Silicon Hole Interposers for 3D IC Integration," *IEEE Transactions on CPMT*, Vol. 4, No. 9, September 2014, pp. 1407–1419.

[40] Farooq, M. G., T. L. Graves-Abe, W. F. Landers, C. Kothandaraman, B. A. Himmel, P. S. Andry, C. K. Tsang, et al., "3D Copper TSV Integration, Testing and Reliability," *Proceedings of IEEE/IEDM*, Washington, DC, December 2011, pp. 7.1.1–7.1.4.

[41] Chaabouni, H., M. Rousseau, P. Ldeus, A. Farcy, R. El Farhane, A. Thuaire, G. Haury, et al., "Investigation on TSV impact on 65nm CMOS devices and circuits," *Proceedings of IEEE/IEDM*, San Francisco, CA, December 2010, pp. 35.1.1–35.1.4.

[42] Xu, C., and K. Banerjee, "Compact Capacitance and Capacitive Coupling-Noise Modeling of Through-Oxide Vias in FDSOI Based Ultra-High Density 3-D ICs," *Proceedings of IEEE/IEDM*, Washington, DC, December 2011, pp. 34.8.1–34.8.4.

[43] Arnaud, F., S. Colquhoun, A. L. Mareau, S. Kohler, S. Jeannot, F. Hasbani, R. Paulin, et al., "Technology-Circuit Convergence for Full-SOC Platform in 28 nm and Beyond," *Proceedings of IEEE/IEDM*, Washington, DC, December 2011, pp. 15.7.1–15.7.4.

[44] Batude, P., M. Vinet, B. Previtali, C. Tabone, C. Xu, J. Mazurier, O. Weber, et al., "Advances, Challenges and Opportunities in 3D CMOS Sequential Integration," *Proceedings of IEEE/IEDM*, Washington, DC, December 2011, pp. 7.3.1–7.3.4.

[45] Lee, K.-W., Y. Ohara, K. Kiyoyama, S. Konno, Y. Sato, S. Watanabe, A. Yabata, et al., "Characterization of Chip-Level Hetero-integration Technology for High-Speed, Highly Parallel 3D-stacked Image Processing System," *Proceedings of IEEE/IEDM*, San Francisco, CA, December 2012, pp. 33.2.1–33.2.4.

[46] Xu, Z., and J. J.-Q. Lu, "Hybrid Modeling and Analysis of Different Through-Silicon-Via (TSV)-Based 3D Power Distribution Networks," *Proceedings of IEEE/IEDM*, San Francisco, CA, December 2012, pp. 30.6.1–30.6.4.

[47] Tzeng, P. J., J. H. Lau, C. Zhan, Y. Hsin, P. Chang, Y. Chang, J. Chen, et al., "Process Integration of 3D Si Interposer with Double-Sided Active Chip Attachments," *IEEE/ECTC Proceedings*, Las Vegas, NA, May 2013, pp. 86–93.

[48] Kang, U., H. Chung, S. Heo, D. Park, H. Lee, J. Kim, S. Ahn, et al., "8 Gb 3-D DDR3 DRAM Using Through-Silicon-Via Technology," *IEEE Journal of Solid-State Circuits*, Vol. 45, No. 1, January 2010, pp. 111–119.

[49] Kim, J., C. Oh, H. Lee, D. Lee, H. Hwang, S. Hwang, B. Na, et al., "A 1.2V 12.8GB/s 2Gb Mobile Wide-I/O DRAM with 4x128 I/Os Using TSV-Based Stacking," *Proceedings of IEEE/ISSCC*, San Francisco, CA, February 2011, pp. 496–498.

[50] Banijamali, B., S. Ramalingam, K. Nagarajan, and R. Chaware, "Advanced Reliability Study of TSV Interposers and Interconnects for the 28nm Technology FPGA," *Proceedings of IEEE/ECTC*, Orlando, FL, June 2011, pp. 285–290.

[51] Chaware, R., K. Nagarajan, and S. Ramalingam, "Assembly and Reliability Challenges in 3D Integration of 28nm FPGA Die on a Large High Density 65nm Passive Interposer," *Proceedings of IEEE/ECTC*, San Diego, CA, May 2012, pp. 279–283.

[52] Banijamali, B., S. Ramalingam, H. Liu, and M. Kim, "Outstanding and Innovative Reliability Study of 3D TSV Interposer and Fine Pitch Solder Micro-bumps," *Proceedings of IEEE/ECTC*, San Diego, CA, May 2012, pp. 309–314.

[53] Xie, J., H. Shi, Y. Li, Z. Li, A. Rahman, K. Chandrasekar, D. Ratakonda, et al., "Enabling the 2.5D Integration," *Proceedings of IMAPS International Symposium on Microelectronics*, San Diego, CA, September 2012, pp. 254–267.

[54] Li, Z., H. Shi, J. Xie, and A. Rahman, "Development of an Optimized Power Delivery System for 3D IC Integration with TSV Silicon Interposer," *Proceedings of IEEE/ECTC*, San Diego, CA, May 2012, pp. 678–682.

[55] Kwon, W., M. Kim, J. Chang, S. Ramalingam, L. Madden, G. Tsai, S. Tseng, et al., "Enabling a Manufacturable 3D Technologies and Ecosystem using 28nm FPGA with Stack Silicon Interconnect Technology," *IMAPS Proceedings of International Symposium on Microelectronics*, Orlando, FL, October 2013, pp. 217–222.

[56] Hsin, Y. C., C. Chen, J. H. Lau, P. Tzeng, S. Shen, Y. Hsu, S. Chen, et al., "Effects of Etch Rate on Scallop of Through-Silicon Vias (TSVs) in 200mm and 300mm Wafers," *Proceedings of IEEE/ECTC*, Orlando, FL, May 2011, pp. 1130–1135.

[57] Wu, C., S. Chen, P. Tzeng, J. H. Lau, Y. Hsu, J. Chen, Y. Hsin, et al., "Oxide Liner, Barrier and Seed Layers, and Cu-Plating of Blind Through Silicon Vias (TSVs) on 300mm Wafers for 3D IC Integration," *IMAPS Transactions, Journal of Microelectronic Packaging*, Vol. 9, No. 1, First Quarter 2012, pp. 31–36.

[58] Chen, J. C., J. H. Lau, P. J. Tzeng, S. Chen, C. Wu, C. Chen, H. Yu, et al., "Effects of Slurry in Cu Chemical Mechanical Polishing (CMP) of TSVs for 3-D IC Integration," *IEEE Transactions on CPMT*, Vol. 2, No. 6, June 2012, pp. 956–963.

[59] Zoschke, K., J. Wolf, C. Lopper, I. Kuna, N. Jürgensen, V. Glaw, K. Samulewicz, et al., "TSV based Silicon Interposer Technology for Wafer Level Fabrication of 3D SiP Modules," *Proceedings of IEEE/ECTC*, Orlando, Florida, May 2011, pp. 836–842.

[60] Dunne, R., Y. Takahashi, K. Mawatari, M. Matsuura, T. Bonifield, P. Steinmann, and D. Stepniak, "Development of a Stacked WCSP Package Platform using TSV (Through Silicon Via) Technology," *IEEE proceedings of ECTC*, San Diego, CA, May 2012, pp. 1062–1067.

[61] Lin, J. C., H. Chang, and S. T. Lin, "Chip-on-Wafer Structures and Method for Forming the Same," *US Patent Publication No. 2013/0134559 A1*, Publication Date: May 30, 2013.

[62] Wang, C., C. Wu, S. Lu, and J. Lin, "Apparatus and Methods for Molding Die on Wafer Interposers," *US Patent Publication No. 2013/0075937 A1*, Publication Date: March 28, 2013.

[63] Lau, J. H., H. C. Chien, S. Wu, Y. Chao, W. Lo, and M. Kao, "Thin-Wafer Handling with a Heat-Spreader Wafer," *Proceedings of IMAPS International Symposium on Microelectronics*, Orlando, FL, September 2013, pp. 389–396.

[64] Hu, G., H. Kalyanam, S. Krishnamoorthy, and L. Polka, "Package Technology to Address the Memory Bandwidth Challenge for Tera-Scale Computing," *INTEL Technology Journal*, Vol. 11, 2007, pp. 197–206.

[65] Lau, J. H., C. Ouyang, and R. Lee, "A Novel and Reliable Wafer-Level Chip Scale Package (WLCSP)," *Proceedings of Chip Scale International Conference*, San Jose, CA, September 1999, pp. H1–H9.

[66] Lau, J. H., R. Lee, C. Chang, and C. Chen, "Solder Joint Reliability of Wafer Level Chip Scale Packages (WLCSP): A Time-Temperature-Dependent Creep Analysis," *ASME Paper No. 99-IMECE/EEP-5*, November 1999.

[67] Brunnbauer, M., E. Furgut, G. Beer, T. Meyer, H. Hedler, J. Belonio, E. Nomura, et al., "An Embedded Device Technology Based on a Molded Reconfigured Wafer," *Proceedings of IEEE/ECTC*, San Diego, CA, May 2006, pp. 547–551.

[68] Keser, B., C. Amrine, T. Duong, O. Fay, S. Hayes, G. Leal, W. Lytle, et al., "The Redistributed Chip Package: A Breakthrough for Advanced Packaging," *Proceedings of IEEE/ECTC*, Reno, NV, May 2007, pp. 286–291.

CHAPTER 8

3D IC Integration with Passive Interposer

8.1 Introduction

2.5D IC integration with an interposer has been discussed in Chap. 7. 3D IC integration with an interposer will be the focus of this chapter. Emphasis is placed on (1) interposer with double-sided chip attachments, (2) interposer with chips on both sides, and (3) low-cost interposer for 3D IC integration. The advantages of 3D IC integration with a TSV/RDL passive interposer will be briefly mentioned first.

8.2 3D IC Integration with TSV/RDL Interposer

One of the applications of 3D IC integration is the passive interposer, also called 2.5D IC integration, as mentioned in Chap. 7. In general, it consists of a piece of device-less silicon with TSVs (through-silicon vias), RDLs (redistribution layers), and/or IPDs (integrated passive devices) supporting one or more high-performance, high-density, and fine-pitch chips without TSVs [1–12]. This is schematically shown in the top drawing of Fig. 8.1. It can be seen that the TSV/RDL interposer is supporting Chip-1 and Chip-2 side-by-side on its top surface. Another design is shown in the bottom drawing of Fig. 8.1. It can be seen that the interposer is supporting these two chips on its top and bottom side. In this case, the size of the interposer can be smaller (or more chips can be placed on the same size of interposer), and the electrical performance can be better because the chip-to-chip interconnects are face-to-face instead of side-to-side [13–26, 33–35]. Also, it is truly a 3D IC integration with a passive interposer, which will be the focus of this chapter.

8.3 TSV/RDL Interposer with Double-Sided Chip Attachments

8.3.1 The Structure

Figure 8.2 shows the schematics of the double-sided interposer for 3D IC integration under consideration. Figure 8.3 shows the layout of the interposer. It can be seen that (a) on the left side, there are pads for the top two chips and (b) on the right side; there are pads for the bottom chip and the organic package substrate. The size of the package substrate is 35 mm × 35 mm × 970 μm. Underfills are used between the chips and interposer and the interposer

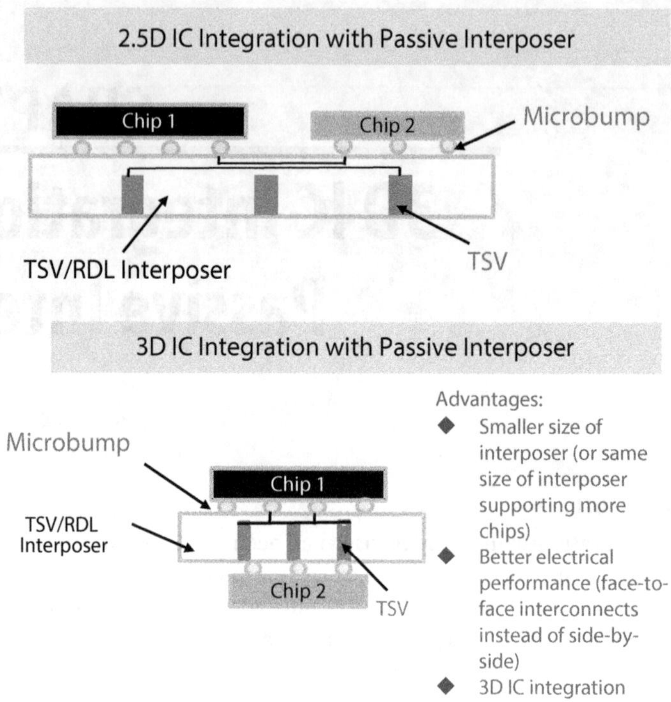

Figure 8.1 3D IC integration with a passive interposer.

and package substrate. The TSVs' diameter is 10 μm and on 150 μm pitch. The diameter of the solder bumps between the chips and interposer and between the interposer and package substrate is 90 μm and on 125 μm pitch. The diameter of the solder balls between the package substrate and the PCB is 600 μm and on 1000 μm pitch. Table 8.1 summarizes the geometry of the 3D IC integration with a TSV interposer supporting chips on both sides [16–20].

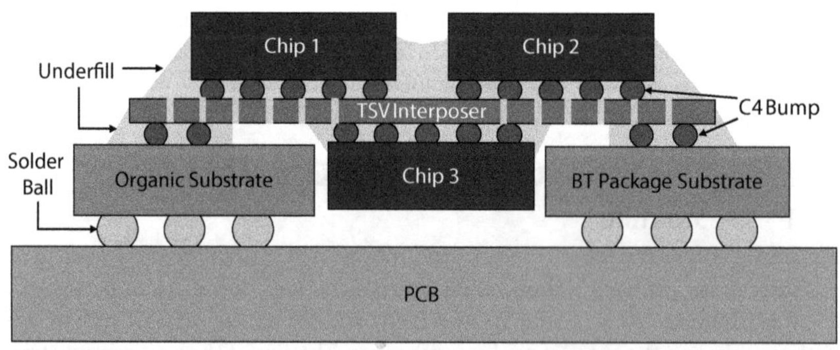

Figure 8.2 Schematics of the interposer (with two chips on its top side and one chip on its bottom side) for 3D IC integration.

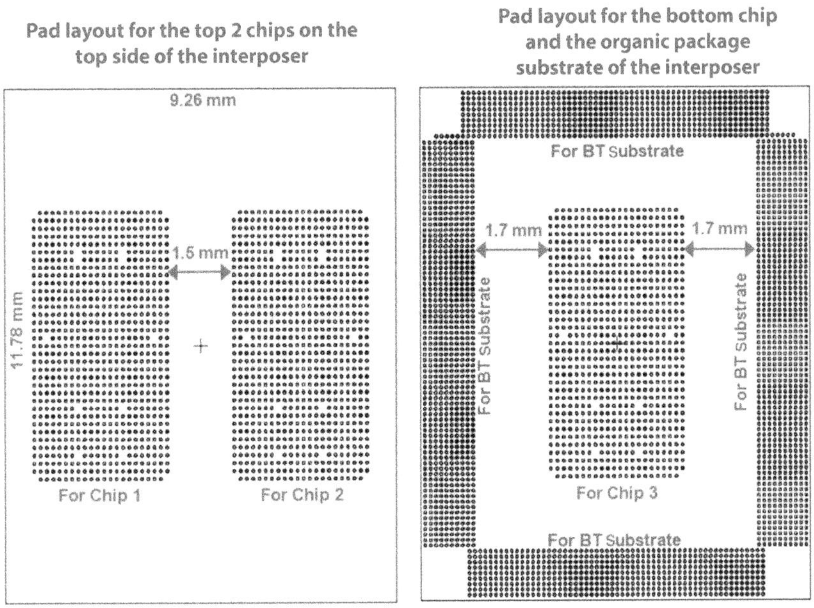

FIGURE 8.3 (Left) Pad layout for the top two chips on the top side of the interposer. (Right) Pad layout for the bottom chip and the organic package substrate of the interposer.

TABLE 8.1 Geometry Parameter of the 3D IC Integration Module

	Length	Width	Height	Diameter	Pitch	Note
Chip#1-3	6.5 mm	3.8 mm	850 μm	—	—	The three chips are in the same sizes
Interposer	11.9 mm	9.4 mm	100 μm	—	—	
Organic substrate	35 mm	35 mm	970 μm	—	—	
Cavity	5.74 mm	8.46 mm	—	—	—	The cavity is in the organic substrate
PCB	60 mm	60 mm	1600 μm	—	—	
Heat spreader	3.35 mm	3.35 mm	500 μm	—	—	
C4 bump	—	—	—	90 μm	125 μm	Between chip and interposer
C4 bump	—	—	—	75 μm	125 μm	Between interposer and organic substrate
Solder ball	—	—	—	600 μm	1000 μm	Between organic substrate and PCB
TSV	—	—	—	10 μm	150 μm	Regular distribution in the interposer

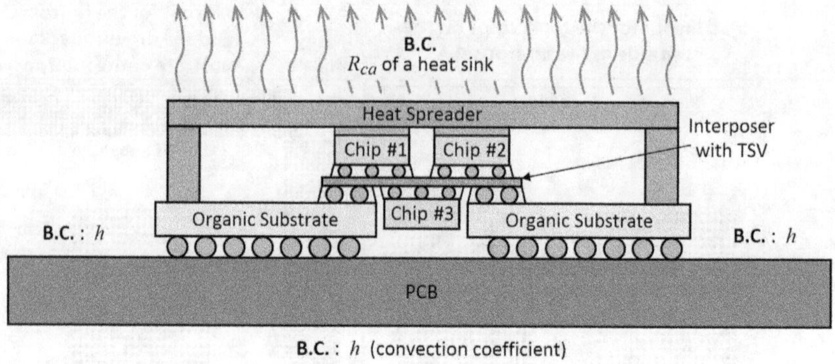

Figure 8.4 Schematic of the thermal analysis model and the boundary conditions.

8.3.2 Thermal Analysis—Boundary Conditions

Boundary conditions for the thermal analyses of the 3D IC integration module are shown schematically in Fig. 8.4; the thermal loading conditions are included in Tables 8.1 and 8.2 along with the boundary conditions. As for the part's geometries, the dimensions are listed in Tables 8.1 and 8.2. It can be seen that (a) the power dissipation of each chip soldered on the interposer is 5 W; (b) the thermal resistance, R_{ca}, to imitate the cooling capability of a suppositional heat sink that is attached on the heat spreader is from 0.1 to 4.0°C/W; and (c) the heat transfer coefficient (h) of both sides of the PCB is 20 W/m²°C.

8.3.3 Thermal Analysis—TSV Equivalent Model

The equivalent Eqs. (2.19) and (2.20) in Chap. 2 can be used for the design and analysis of 2.5D/3D IC integration by establishing an equivalent model. This equivalent model is used to replace the real detail TSV model for simplifying the thermal simulations. In the conversion, the arrays of TSV, solder bump, and solder ball can be replaced by many coupled equivalent zones with their equivalent thermal conductivities (Fig. 8.5). Particularly, the SiO_2 layer on interposer has to be retained in the equivalent model for accurateness. For example, a 2 × 2 TSV array with 10 μm in diameter, 0.2 μm in thickness of sidewall SiO_2, 100 μm in height (thickness), and 50 μm in pitch, could be converted to an equivalent zone with 100 × 100 μm in area and having 144.56 W/m·K of k_{xy} and 156.34 W/m·K of k_z.

8.3.4 Thermal Analysis—Solder Bump/Underfill Equivalent Model

The equivalent thermal conductivities of solder bump and solder ball should be determined by thermal resistance parallel-series correlation. Because of disconnection and large pitch, the equivalent thermal conductivity of a bump array that is with underfill material should

Table 8.2 Thermal Loading and Boundary Conditions used in the Analyses

Thermal loading	5 W per chip (#1, #2, #3)
Boundary conditions	The thermal resistance (R_{ca}) of a heat sink: tunable for parameter study The convection coefficient (h): 20 W/m².K on PCB both sides Other surfaces are regarded as adiabatic

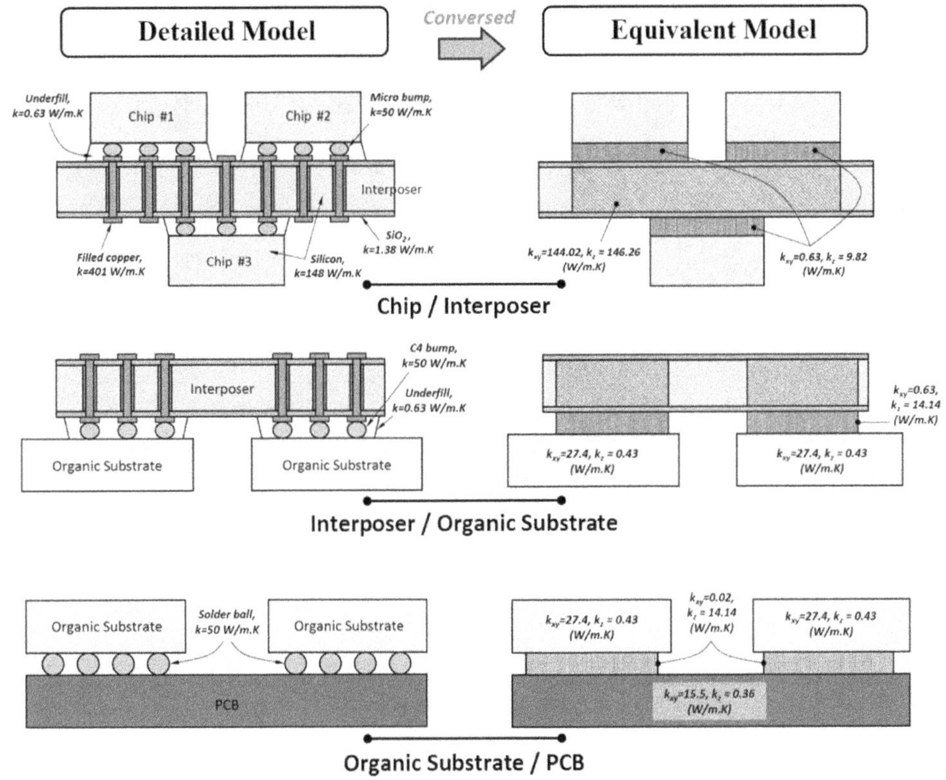

FIGURE 8.5 Model conversion rules and the calculated thermal conductivities of each equivalent zone.

approach to that of the underfill in the x-y direction. For the same reason, a solder ball array without underfill, its equivalent k_{xy} should be very small and is nearly equal to zero. Figure 8.5 shows the model conversion rules and the calculated thermal conductivities of each equivalent zone.

8.3.5 Thermal Analysis—Results

Figure 8.6 shows the top view temperature contours of each chip (applied boundary condition: R_{ca} = 1.0°C/W on heat spreader), and Fig. 8.7 shows the 3D temperature distribution of the 3D IC integration module. From these temperature distribution contours, it can be seen that the temperature distribution of chip #1 and chip #2 are rather nonuniform, but that of chip #3 is not. The maximum temperature difference of chips #1 and #2 is about 4.7°C (63.2°C to 58.5°C), and that of chip #3 is just about 0.4°C (70.6°C to 70.2°C). Therefore, chips #1 and #2 have a more severe thermal issue about temperature nonuniformity than that of chip #3; the issue can damage the chips' quality and reliability performance. Also, from the same temperature distribution contours, it can be seen that a heat sink is necessary for cooling the 3D IC integration module, because chips #1 and #2, and chip #3 can reach to very high temperatures, 410°C and 417°C, respectively, under a natural convection condition and without heat spreader attachment.

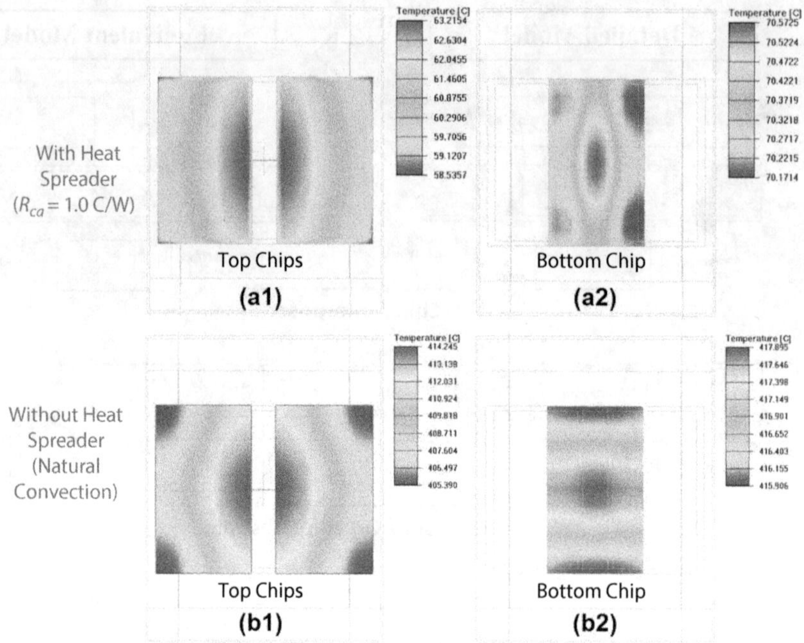

FIGURE 8.6 Temperature distribution of (a1) chips #1 and #2 and (a2) chip #3 (Rcs = 1.0C/W); and (b1) Chips #1 and #2 and (b2) chip #3 (without heat spreader, natural convention). The four temperature bars are not in the same scale.

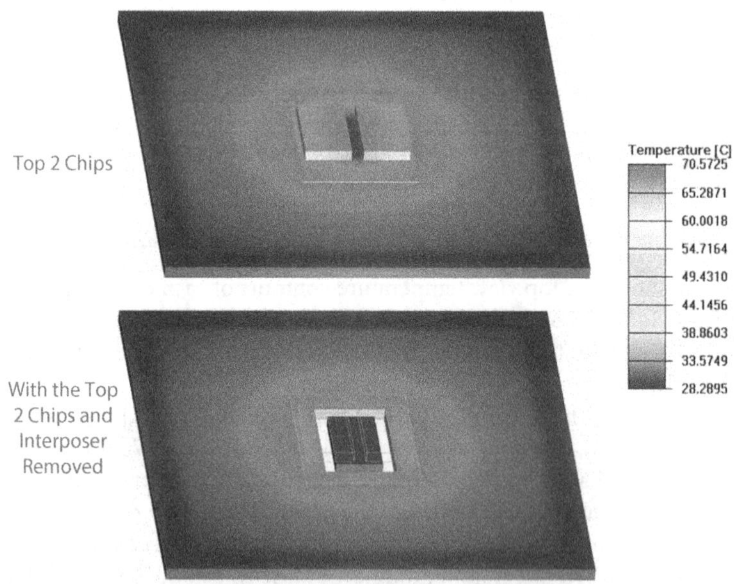

FIGURE 8.7 The temperature contour of the integration of organic substrate, interposer, and chips #1, #2 and #3. (Top) chip #3 is shielded. (Bottom) Interposer and chips #1 and #2 are invisible.

The relationships between chip's average temperature and the applied cooling capability (R_{ca}) to the heat spreader/sink are shown in Fig. 8.8. The figure can help us select an appropriate heat sink that is attached on the heat spreader. For instance, if the temperature specification of the chips is 100°C, a heat sink with cooling capability, R_{ca}, less than 3.7°C/W, is essential to cool the 3D IC integration module. On the other hand, if the chips' temperatures must be under 60°C, then we have to choose a very powerful heat sink that has an R_{ca} less than 0.3°C/W.

Figure 8.8 shows a severe thermal issue for the 3D IC integration structure. The issue is that the chip planted on the bottom side of the interposer is always hotter than the chips planted on the top side of the interposer. This is because the chips on top side of interposer can be in direct contact with the main cooling mechanism—heat spreader and heat sink—but the chip on the bottom side of interposer cannot. The difference in the chips' average temperature is equal to 9.0°C. The higher temperature of chip #3 is inevitable because the heat spreader plays a major role to dissipate the chips' generated heat, and the temperatures of chips #1 and #2 barricade the heat dissipation from the chip #3 to the heat spreader. It is very hard to solve the severe issue without any structural change in the 3D IC integration module. Thickening of the heat spreader could reduce the chips' temperatures, but the enhancement is limited. A useful and simple technique has been proposed in [24], which is to insert a metallic heat slug from the PCB side and to directly contact the back side of chip #3. The inserted slug can effectively drain out the heat of chip #3 and also help to reduce the temperature of chips #1 and #2 noticeably (please see Sec. 9.6 of Chap. 9).

8.3.6 Thermomechanical Analysis—Boundary Conditions

For thermomechanical simulations, nonlinear finite element modeling and analysis are conducted to realize the thermal stress/strain distributions of the structure. At the same time,

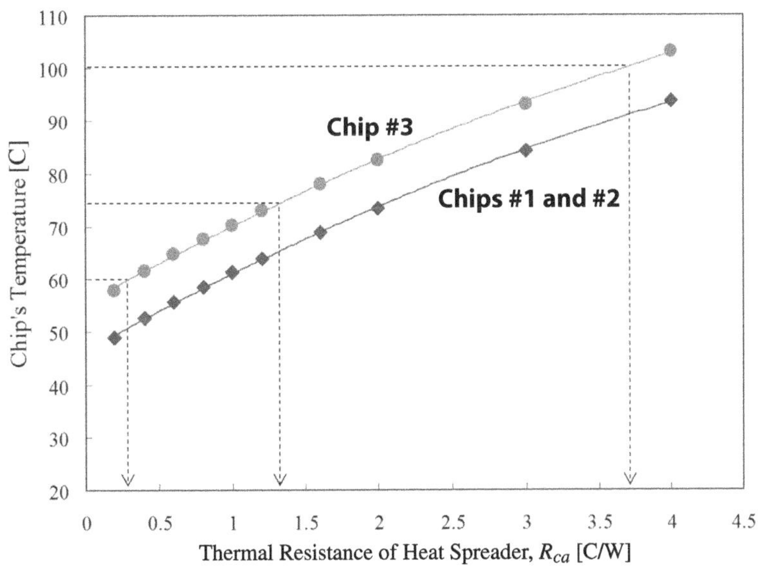

FIGURE 8.8 Correlation between the chips' average temperatures and the applied cooling capability on the heat spreader.

the reliability estimation of solder joints in the architecture is examined. In order to realize the nonlinear temperature and time-dependent creep behavior of solder joints, the structure is simulated under a thermal cycling condition of −25°C to 125°C for 5 cycles.

8.3.7 Thermomechanical Analysis—Material Properties

All the thermomechanical material properties such as the Young's modulus, Poisson's ratio, and coefficient of thermal expansion for the material used in the structure are shown in Table 8.3. Since lead-free solders are temperature and time dependent, the following nonlinear temperature and time-dependent constitutive equations are used [2]:

$$\frac{d\varepsilon}{dt} = C_1[\sinh(C_2\sigma)]^{C_3} \exp\left(-\frac{C_4}{T}\right)$$

$$E(GPa) = 49 - 0.07 \times T(°C)$$

$$\alpha(ppm/°C) = 21.301 + 0.017 \times T(°C)$$

where
C1 = 50000 (s^{-1})
C2 = 0.01 (MPa^{-1})
C3 = 5
C4 = 0.5 (K)

In these equations, E is the Young's modulus, α is the coefficient of thermal expansion, ε is the creep strain, K is in degree Kevin, and t is time.

Only 2D finite element modeling is performed in the thermomechanical analyses. Since the structure is symmetry and thus only half of the structure is modeled. The symmetry axis

Table 8.3 Material Properties for Finite Element Analyses

	Young's modulus (GPa)	Poisson's ratio	CTE (ppm/K)
Silicon	130	0.28	2.8
BT substrate	X: 26 Y: 11	0.39	X: 15 Y: 52
FR4	X: 22 Y: 10	0.28	X: 18 Y: 70
Underfill	9.07	0.3	40.75
Solder alloy	Temperature dependent	0.35	Temperature dependent
Electroplated Cu	70	0.34	18
Cu low-k pad	8	0.3	10
SiO_2	70	0.16	0.6
IMC (Cu_6Sn_5)	125	0.3	18.2
Nickel	131	0.3	13.4

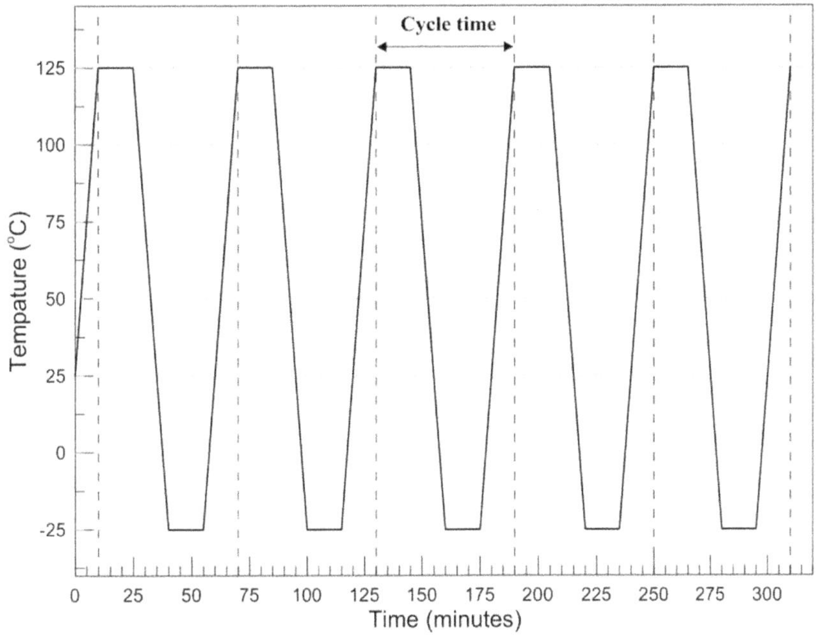

FIGURE 8.9 Thermal cycling boundary condition (−25°C ⇆ 125°C).

is the y-axis and proper displacement and rotational boundary conditions are applied. The temperature cycling condition is shown in Fig. 8.9, which is −25°C ⇆ 125°C on a 60-min cycle. Stress-free is at 25°C.

8.3.8 Thermomechanical Analysis—Results

The 2D element (PLANE182) of ANSYS Mechanical R14 is used to construct the half finite element model as shown in Figs. 8.10 and 8.11. It can be seen that the symmetry axis is the y-axis. Detailed modeling of the TSVs in the interposer and the surrounding areas is shown in Fig. 8.10. Detailed modeling of the solder bumps between the chips and interposer and the interposer and package substrate, and the solder balls between the package substrate and PCB is shown in Fig. 8.11. The whole model is subjected to the thermal cycling loading shown in Fig. 8.9 and the results are shown in the following section.

Figure 8.12 shows the von Mises stress contours in the corner TSVs at 125°C (stress-free is at 25°C). It can be seen that the maximum stress is ~135 MPa and the critical locations are near the interface between the filled Cu and SiO_2 of the TSV and underfill.

It is important to study the creep responses for multiple cycles by observing when the hysteresis loops become stabilized. Figure 8.13 shows the shear stress and shear creep strain hysteresis loops for multiple cycles at the corner solder bumps between Chip-2 and the interposer. It can be seen that the shear creep strain vs. shear stress loop is quite stabilized after the third cycle. As a matter of fact, for those solder bumps their hysteresis loops are stabilized after the first cycle. Figure 8.13 shows the creep strain energy density history for the corner solder bumps between Chip-2 and the interposer. It can be seen that the creep strain energy density per cycle of the corner solder bumps between Chip-2 and the interposer is

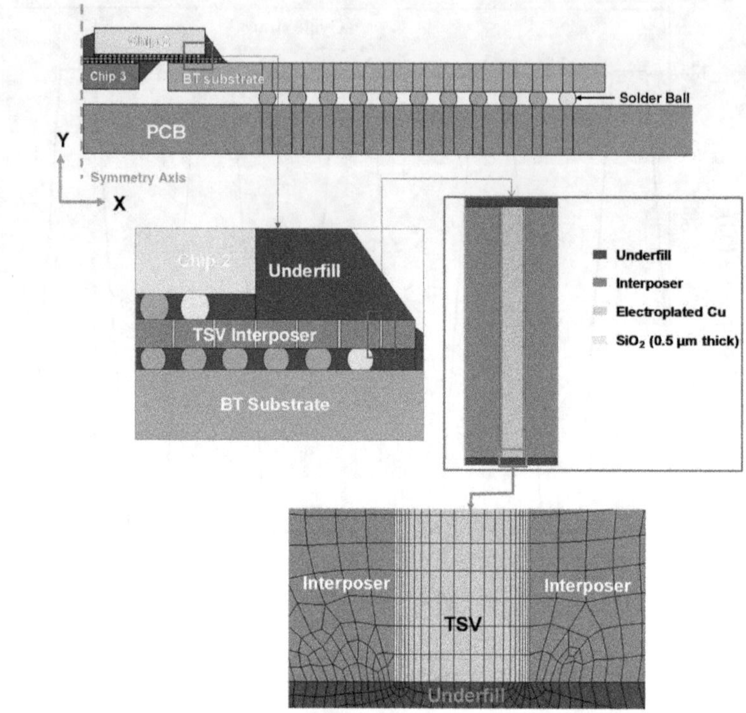

FIGURE 8.10 Finite element analysis model of half of the 3D IC integration module (showing the details of TSV and surrounding areas).

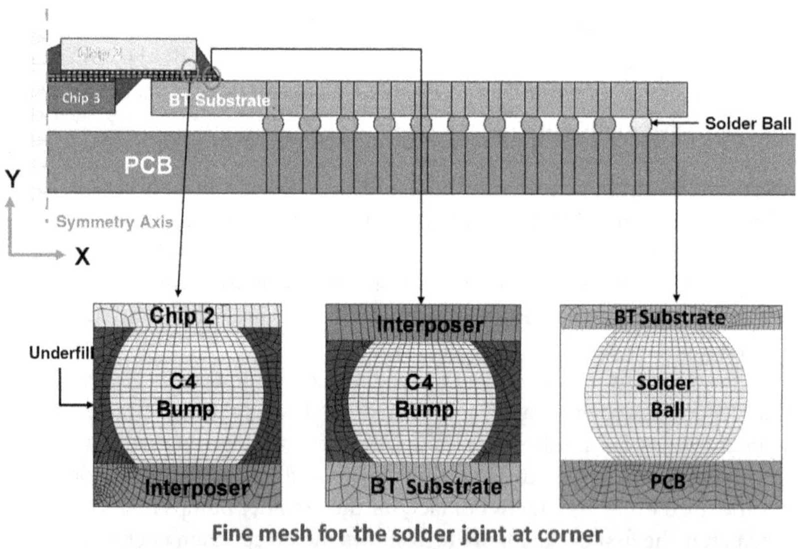

FIGURE 8.11 Finite element analysis model of half of the 3D IC integration module (showing the details of solder bumps between chip #2 and interposer and interposer and package substrate, and solder ball between the package substrate and PCB (not-to-scale).

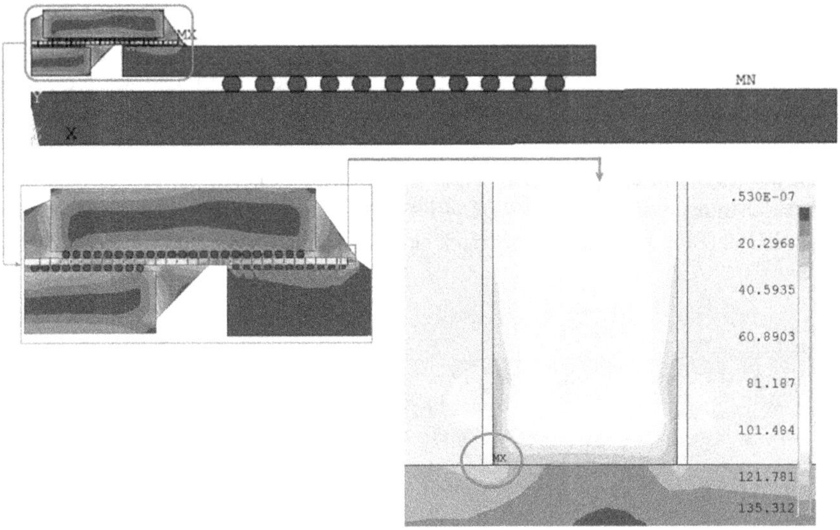

Figure 8.12 Maximum stress acting at the TSV at 125°C (stress free is at 25°C).

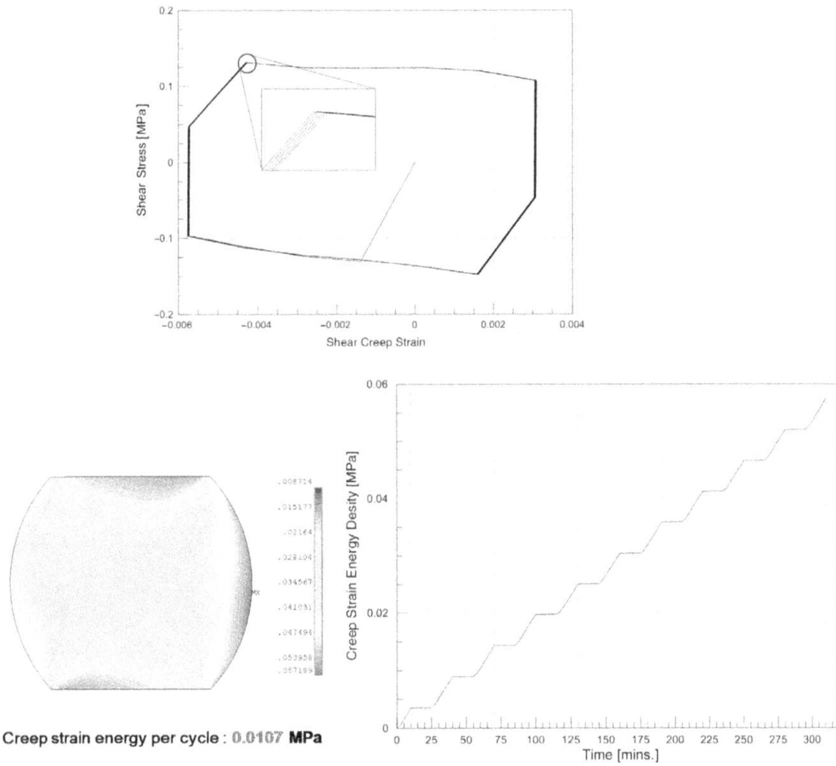

Figure 8.13 (Top) Hysteresis loop (five cycles) at the corner solder bump between chip #2 and interposer. (Bottom) Creep strain energy density vs. time at the corner solder bump between chip #2 and interposer.

0.0107 MPa. This magnitude is too small to create solder joint thermal-fatigue reliability problem [2] under the environmental condition: −25°C ⇆ 125°C on a 60-min cycle. Underfill helps! For the creep responses at other solder joint locations, pleased see [18].

8.3.9 Fabrication of the TSV

The schematic of the present TSV/RDL Si interposer is shown in Fig. 8.14. It can be seen that the interposer has three RDLs (TR1, TR2, and TR3) on its front side (top side) and two RDLs (BR1 and BR2) on its back side, which are fabricated on a 300 mm Si wafer. All

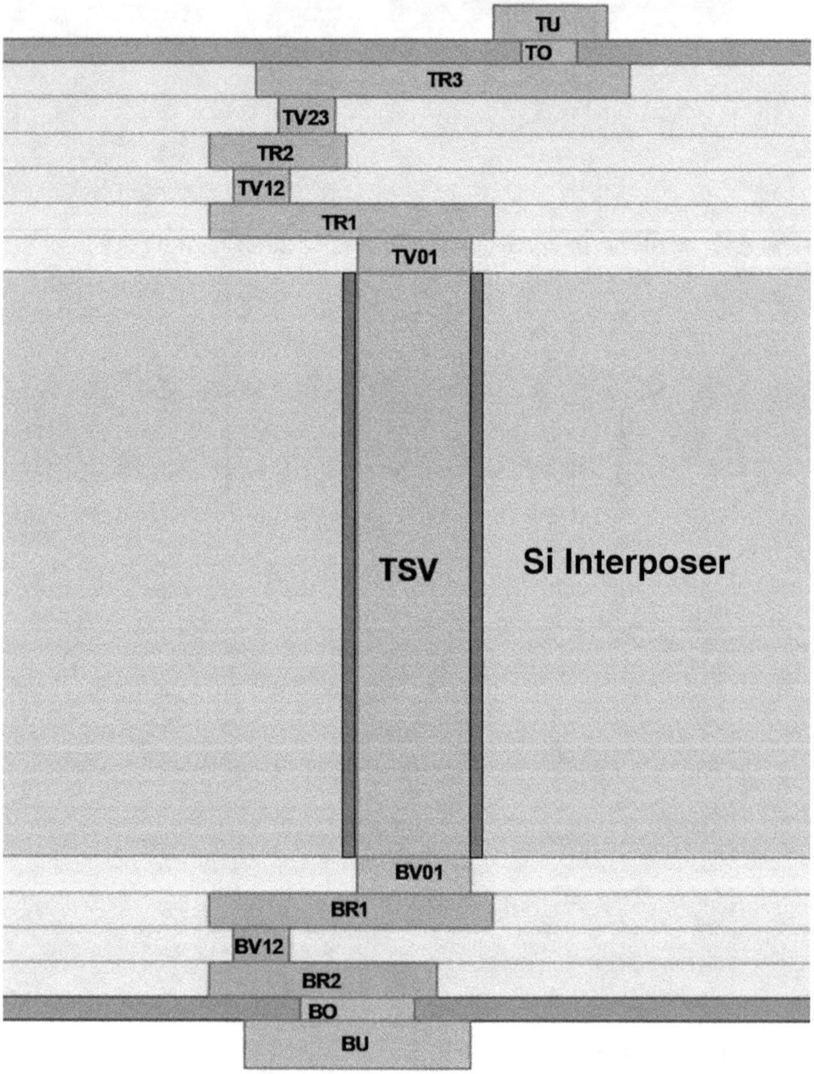

Figure 8.14 Schematic of the cross section of the TSVRDL interposer with three RDLs on top and two RDLs at the bottom.

the RDLs are formed by a Cu damascene technique. Lithography using a contact aligner provides a low-cost process as compared to a stepper/scanner under the same resolution requirements. The five layers pose challenges for manufacturing the back-side RDL using a Cu damascene process. The process flow of this TSV/RDL interposer is listed in Table 8.4.

Table 8.4 Fabrication Process Flow of TSV/RDL Interposer

Process stage	
Inter-layer dielectric (ILD)	
TSV	• TSV diameter = 10 μm • TSV depth = 105 μm • TSV SiO liner (SACVD) = 0.5 μm @ top • Ta barrier = 0.08 μm @ top
Three front side RLDs	• Cu damascene • Minimum line-width = 3 μm • Minimum via diameter = 3 μm • Metal thickness = 2 μm and 1.5 μm • Ta barrier = 0.05 μm
Top passivation opening	• Opening diameter = 60 μm
Top UBM	• UBM size = 75 μm • Cu thickness = 5 μm • NiPdAu = 2 μm
Front-side temporary bonding to Si carrier	• Glue thickness = 25 μm
Back-side thinning	• Grinding + CMP
Si recess & TSV Back-side reveal	• Dry etching
Back-side isolation	• PECVD SiO • Temperature < 200°C • Thickness = 1 μm
CMP for TSV Cu exposion	• Back-side isolation remained > 0.8 μm on the field area
Two Back-side RDLs	• Cu dual-damascene • Minimum line-width = 5 μm • Minimum via diameter = 5 μm • Metal thickness = 2 μm • Ta barrier = 0.05 μm
Bottom passivation opening	• Opening diameter = 60 μm
Bottom UBM	• UBM size = 75 μm • Cu thickness = 5 μm • NiPdAu = 2 μm

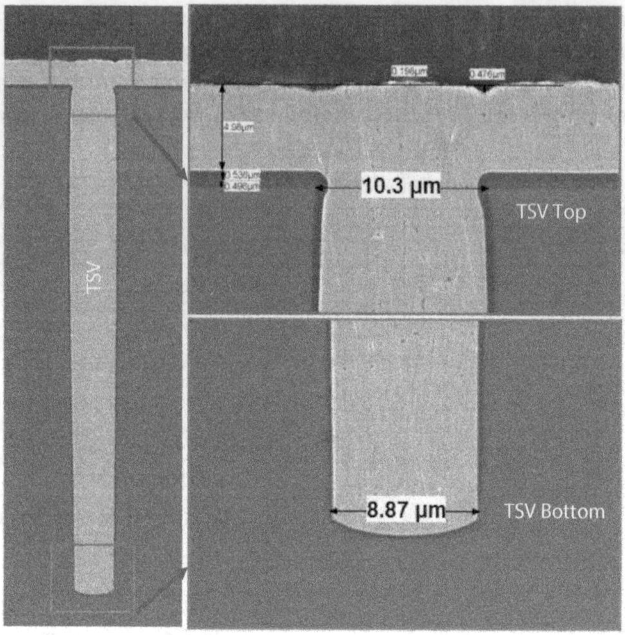

Figure 8.15 SEM cross-section images of the TSV.

The process starts with a SiN_x/SiO_x insulation layer by PECVD (plasma-enhance chemical vapor deposition). After TSV lithography, the TSV is etched into the Si substrate by Bosch-type DRIE (deep reactive ion etch) to form a high-aspect ratio via structure. The etched TSV structure is then processed with a procedure of SiO_x liner by SACVD (Sub-atmospheric chemical vapor deposition), Ta barrier and Cu seed layers by PVD (physical vapor deposition). Cu electroplating is used to fill the TSV structure. The final blind TSV has a top opening of approximately 10 μm in diameter and a depth about 105 μm, which give an aspect ratio of 10.5. In such a high-aspect ratio via structure, a bottom-up plating mechanism is applied to ensure a seamless TSV with a reasonably low Cu thickness at the field. The SEM cross-section images are shown in Fig. 8.15. It can be seen that the diameter of the TSV is slightly decreased at the bottom, which is expected from the etching process point of view. The Cu thickness at the field is less than 5 μm. After annealing (400°C for 30 min), to complete the TSV process, excess Cu at the field is removed by CMP (chemical and mechanical polishing).

8.3.10 Fabrication of the Interposer with Top-Side RDLs

After the TSV formation, the Si interposer wafer goes through three RDL processes on the front side, using a Cu damascene technology, primarily modified from the conventional Cu BEOL (back-end-of-line) process. The CD (critical dimension) of these RDL layers is 3 μm, which is limited by the resolution of the contact aligner. Three RDL layers are built sequentially, followed by the front side UBM pads for flip chip solder bumps attachment. The UBM consists of Cu-UBM pad and NiPdAu finishing, which gives room for underfill between the active chip and the interposer and also lowers the possibility of bonding failure. Figure 8.16 shows the SEM cross-sectional images of the front-side layers (TSV, three front side RDLs,

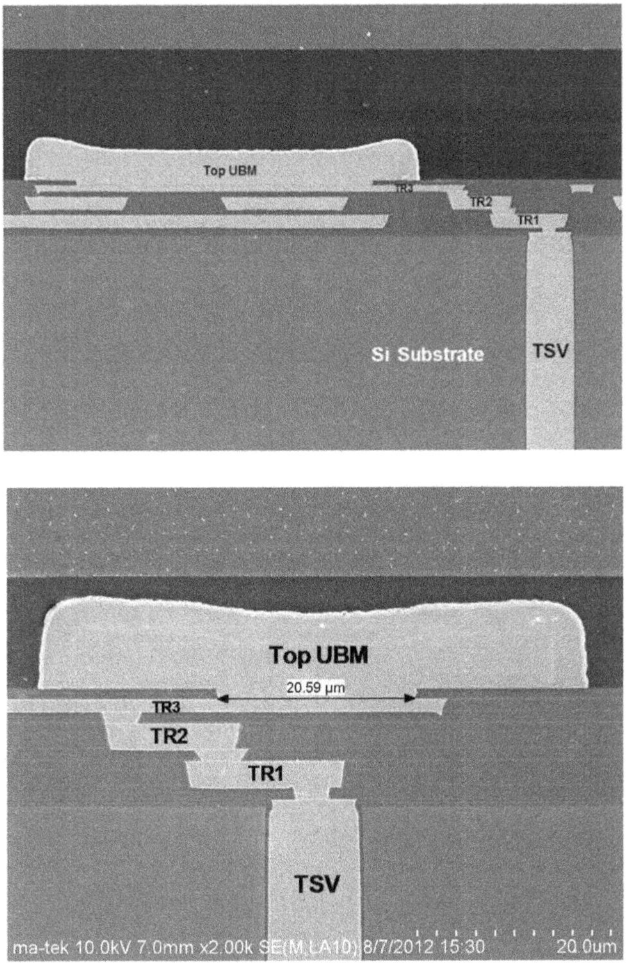

FIGURE 8.16 SEM cross-sectional images of the top side of the interposer (TSV, RDLs: TR1, TR2, TR3, and the top UBM).

and top UBM). Figure 8.17 shows the top-view of the interposer. As shown in the top-view photo, interconnecting region refers to the area for the two active die attachments, while the rest of the area is designed for passive test structures.

8.3.11 TSV Reveal of the Cu-Filled Interposer with Top-Side RDLs

After the front-side process is done, wafers thinning starts with the temporary bonding process. The Si interposer wafer is temporarily bonded to a Si carrier before being grounded and polished at the back side of the Si interposer wafer, which stops before the TSV is revealed. The thickness distribution across the wafer for the remaining Si-substrate after the back side grinding/polishing process is very uniform. The profile distribution is shown in Fig. 8.18. The wafer back side was then dry-etched to reveal the TSV and also to recess the Si surface. The isolation dielectrics are deposited before CMP to expose the TSV back side Cu surface for connecting to back side RDL. Due to the maximum working

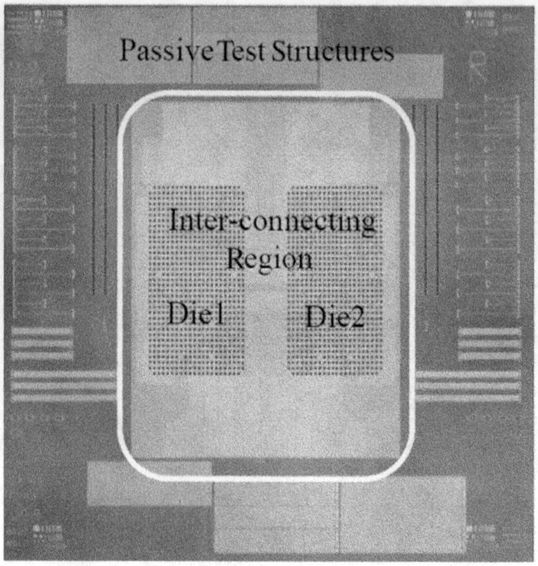

Figure 8.17 Top-view of the TSV/RDL interposer.

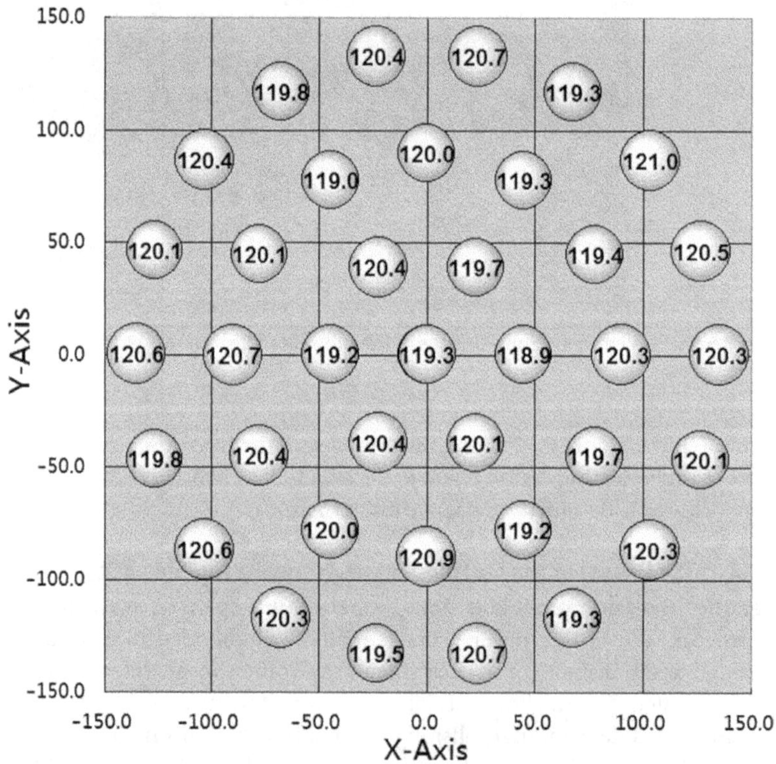

Figure 8.18 Mapping of the Si-substrate remained thickness after back side grinding.

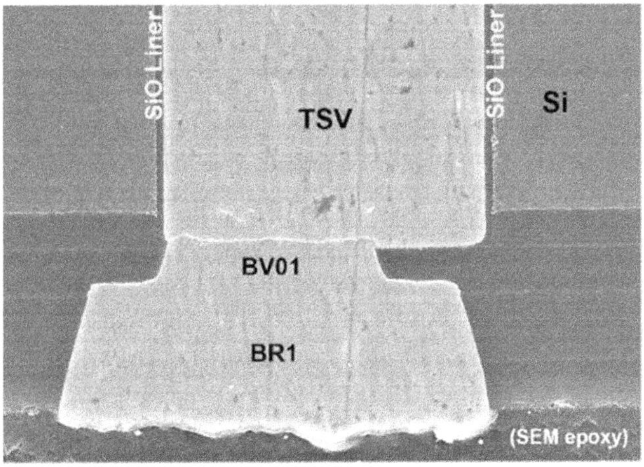

FIGURE 8.19 SEM cross-sectional image of Cu dual-damascened first back side RDL (BR1 + BV01) connecting to the bottom of TSV.

temperature of the temporary bonding adhesive, the temperature of the back-side processes is restricted to less than 200°C.

8.3.12 Fabrication of the Interposer with Bottom-Side RDLs

Due to the concerns of mask alignment capabilities with the contact aligner and the wafer warpage from both front-side processing and the temporary bonding process, the minimum CD of the back-side RDL is relaxed to 5 μm. Also, to reduce process-induced cracking or chipping on the thinned wafer, the back-side RDL process utilizes Cu dual-damascene technique (RDL + via formation with one ECD and CMP). The SEM cross-section image of the first back-side RDL/via Cu dual-damascene metallization (connecting to the TSV backside) is shown in Fig. 8.19. Similar to the front-side process, two back-side RDLs were done sequentially, followed by the back-side UBM pads for the solder bumps for the back-side flip chip attachment and for the organic substrate, respectively.

Figure 8.20 shows the SEM cross-section of all back-side layers (TSV, 2 back-side RDLs, and bottom UBM), and Fig. 8.21 shows the bottom-view of the interposer. To complete the Si interposer process, de-bonding of the thinned Si interposer wafer from the Si carrier is performed before the thinned Si interposer wafer is diced. The diced interposer die is ready for SiP (system-in-package) assembly and also for passive electrical characterization on the interposer and package.

8.3.13 Passive Electrical Characterization of the Interposer

Measurement and modeling results for various micro-striplines, striplines, and coplanar waveguide lines of the fabricated interposer have been reported in [20]. Figure 8.22 shows the electrical performance (insertion loss and return loss) for three different cases. There is discrepancy shown between the measured and the 3D field solver modeling results when nominal design dimensions are used. However, when actual manufactured dimensions are used (from Fig. 8.23), the correlation between the model and measured results is very good.

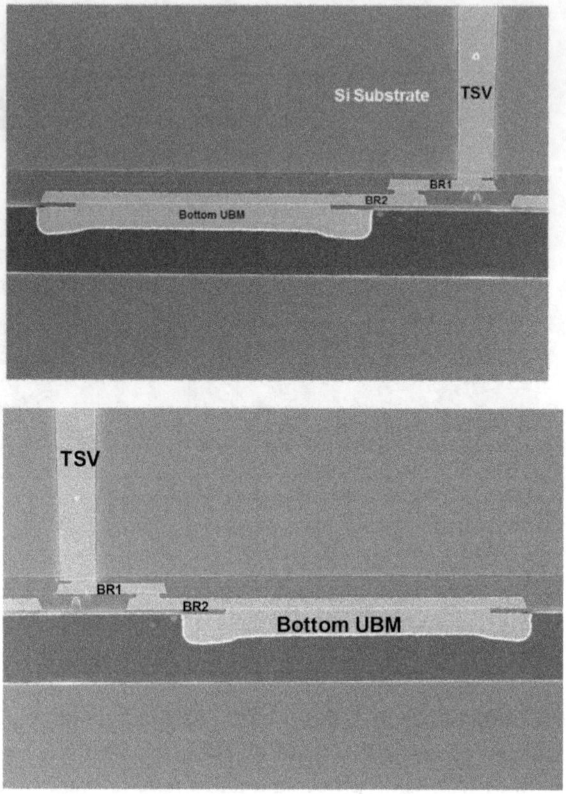

Figure 8.20 SEM cross-sectional images of the back-side of the interposer (TSV, RDLs: BR1, BR2, and bottom UBM).

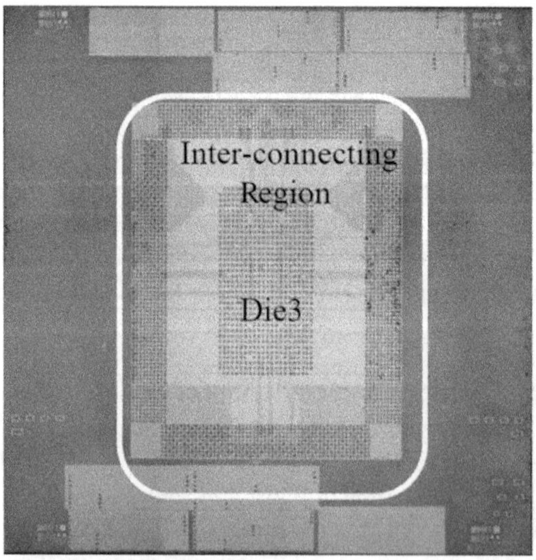

Figure 8.21 Bottom-view of the TSV/RDL interposer.

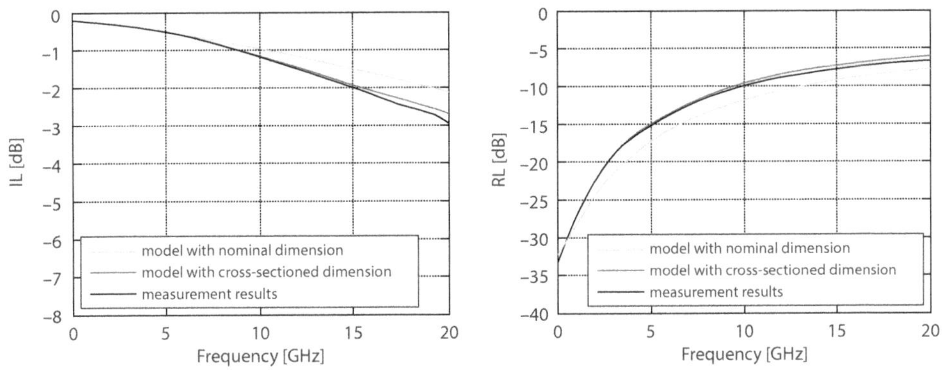

FIGURE 8.22 Simulation and measurement correlation results.

8.3.14 Final Assembly

Figure 8.24 schematically shows the process flow of the final module assembly. First, attach the back-side chip on the interposer, then naturally reflow the solder, apply and cure the underfill. Attach the interposer on the top side of the organic package substrate and naturally reflow. Then apply the underfill and cure. Attach the two top-side chips on the top side of the interposer, naturally reflow the solder, then apply and cure the underfill. Finally, solder ball mounting on the bottom side of the package substrate and reflow as shown in steps 7 and 8 of Fig. 8.24.

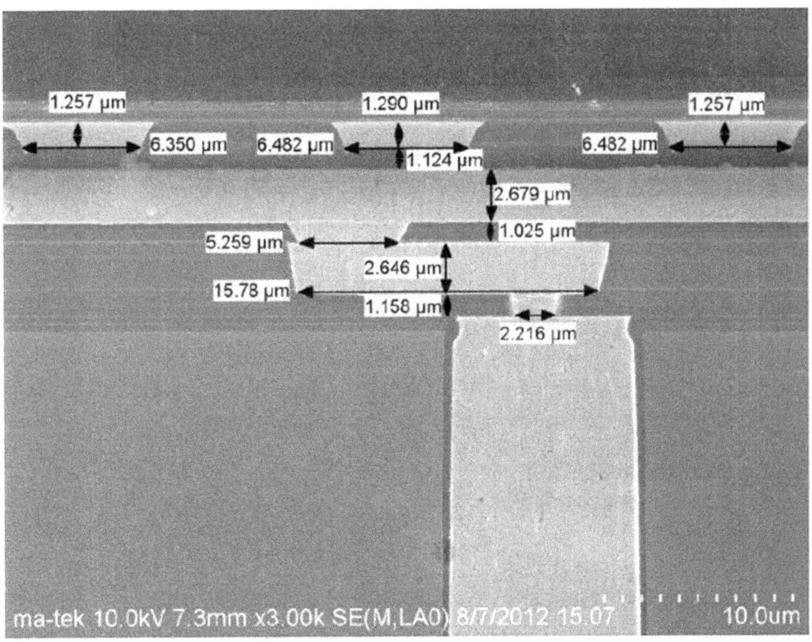

FIGURE 8.23 SEM cross-sectional image of the interposer for simulation and measurement.

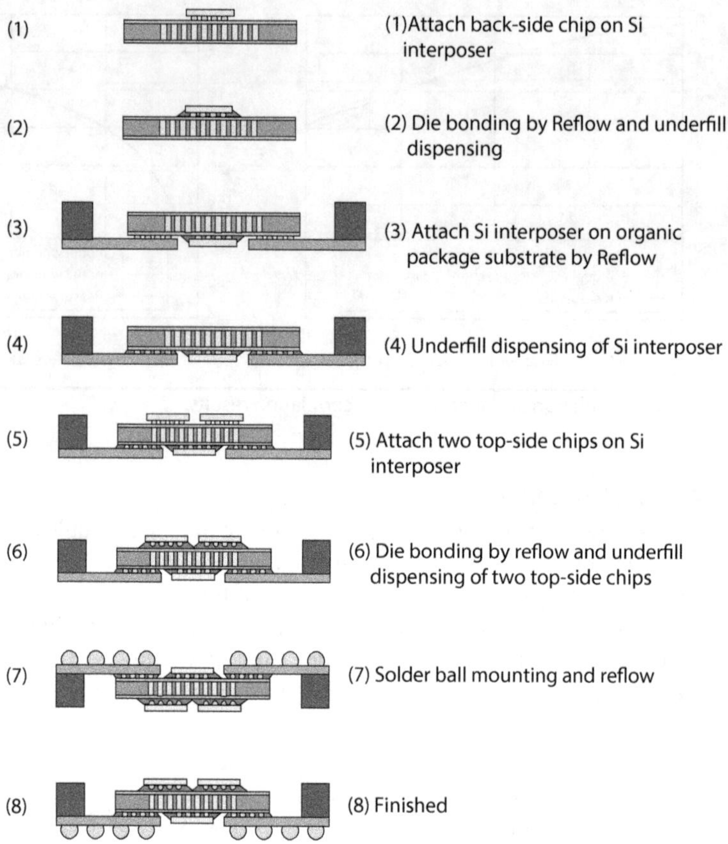

Figure 8.24 Final TSV/RDL interposer and chips assembly process flow.

Figure 8.25 shows an assembled package, which consists of a passive TSV/RDL interposer supporting two active chips on the top side (with underfill) (left-hand side of Fig. 8.25) and one active chip on the bottom side (with underfill) (right-hand side of Fig. 8.25). This interposer with the chip set is soldered to the top side of a package substrate with cavity.

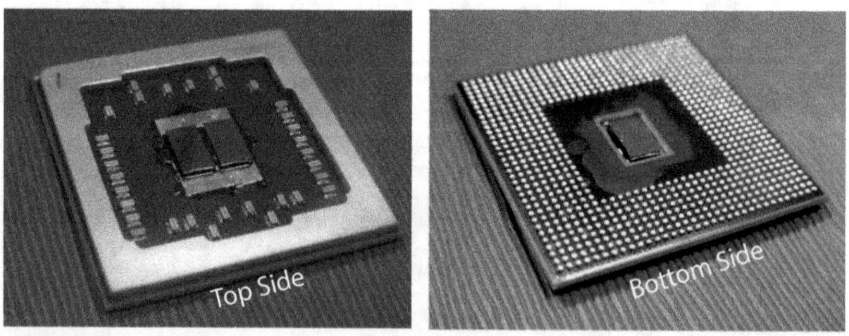

Figure 8.25 Fully assembled 3D IC integration module. (Left) Top side showing the two chips. (Right) Bottom side showing the one chip.

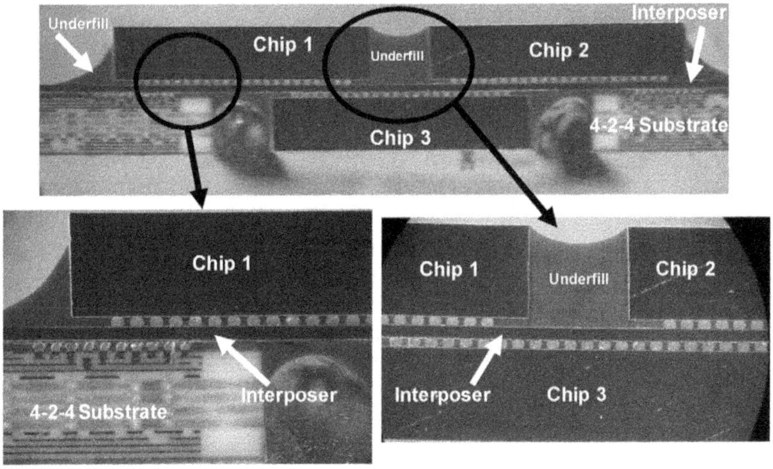

Figure 8.26 Cross-section of the fully assembled module (three chips, interposer, package substrate, and underfill).

Figure 8.26 shows the cross-sections of the fully assembly module. It can be seen that the interposer is properly supporting the three chips with underfill. This interposer is soldered (with underfill) to a 4-2-4 package substrate. Figure 8.27 shows the X-ray image from the top-view of the fully assembly module, demonstrating the assembly is done properly. It can be seen that the two dies (un-thinned) at the front side and one die at the back side are perfectly aligned, with solder bumps (connecting the interposer to the package BT-substrate) at the peripheral regions on the interposer back side.

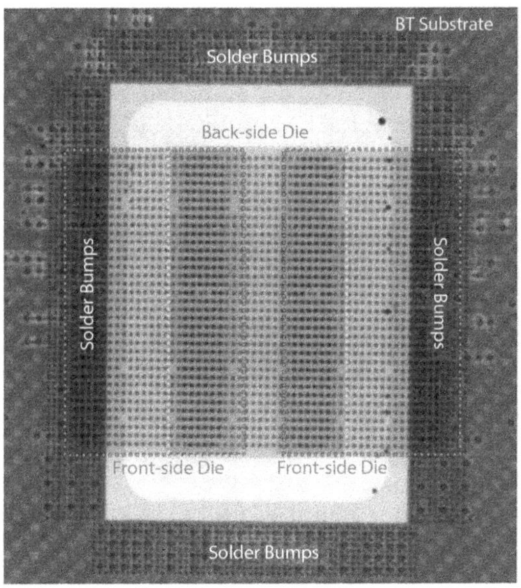

Figure 8.27 X-ray image from the top-view of the fully assembled SiP.

8.3.15 Summary and Recommendations

Some important results and recommendations are summarized as follows [16–20]:

- Equivalent thermal conductivities of TSV with SiO_2 and Cu-filling have been provided.
- Examples in using the equivalent thermal conductivities for the present 3D IC integration module have been presented and discussed.
- The relationships between chips' average temperature and the applied cooling capability (R_{ca}) to the heat spreader/sink of the 3D IC integration module have been provided. Examples in using these relationships have also been given.
- The maximum stress (135 MPa) at 125°C (stress-free is at 25°C) in the TSVs occurred at the corner TSV and the location is at the interface between the SiO_2, Cu, and underfill.
- The maximum creep strain energy density per cycle occurred at the corner solder bumps between Chip-2 and interposer, Chip-3 and interposer, and interposer and package substrate and the magnitudes are respectively 0.0107, 0.0117, and 0.0125 MPa. These values are too small to have solder joint thermal-fatigue problem. Underfill helps!
- The maximum creep strain energy density per cycle of the solder balls between the package substrate and PCB occurred at the corner solder ball and the magnitude is 0.071 MPa. Again, this value is too small to have solder ball thermal-fatigue problem.
- There are five RDLs on the passive interposer; three RDL on its front side, and two RDL at its back side.
- Contact aligner has been used for all lithography processes for the interest on process cost. However, the wafer warpage induced lithography difficulty using contact aligner is a concern.
- The RDL process on both sides of the interposer using a contact aligner has been modified from the typical CMOS BEOL Cu-damascene process technique.
- Minimum CD is relaxed for the back-side process based on the concerns of the thinned-wafer warpage induced by front-side process and temporary bonding.
- SEM cross-section images showed that all the double-sided wiring layers (RDLs) and vias have been properly fabricated. Also, void-less Cu-filled TSVs have been achieved.
- For the final assembly, the chips have been successfully attached to the interposer and underfilled. The multichip interposer module has also been attached onto an organic package substrate. The cross-section and X-ray results indicated that the fully assembly module with the active dies on the double-sided TSV/RDL interposer is done properly.
- The electrical VNA testing of passive test structures corresponded well with 3D field solver modeling results.
- Reliability thermal cycling testing should be the next phase of this project.

8.4 TSV Interposer with Chips on Both Sides

8.4.1 The Structure

The 3D IC SiP (system-in-package) has an interposer that supported one CPU (central processor unit) or ASIC (application specific IC) chip on its top side and two memory chips on its bottom side, as shown in Fig. 8.28 [21–26]. TSVs are embedded in the interposer uniformly and the sizes are 25 μm in diameter, 150 μm in depth, and 150 μm in pitch; the thickness of dielectric layer (SiO_2) on TSV sidewall is 0.5 μm. The interposer is attached to a BT substrate with ordinary bumps and the BT substrate is attached to a PCB with solder balls. There is a copper heat spreader to attach to the back side of the CPU/ASIC with a thin layer of thermal grease for serving as a cooling module.

Figure 8.29 shows the detail analyzed structure and Table 8.5 lists the dimensional properties of the structure. It can be seen that the interposer is 28 mm × 28 mm × 150 μm, the CPU/ASIC is 22 mm × 18 mm × 400 μm, and the DRAM is 10 mm × 10 mm × 100 μm.

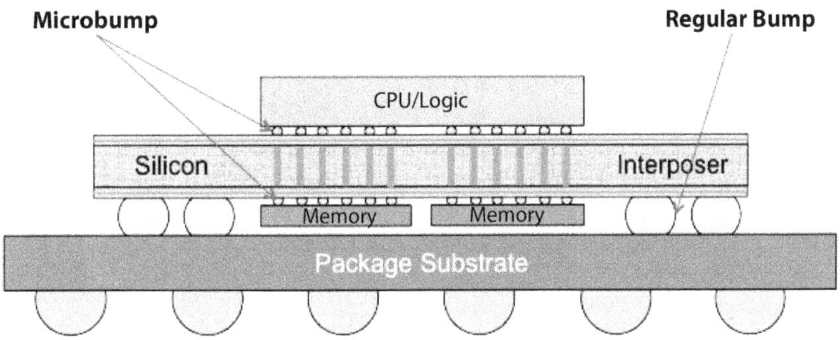

FIGURE 8.28 Schematic of the 3D IC integration with a TSV/RDL passive interposer.

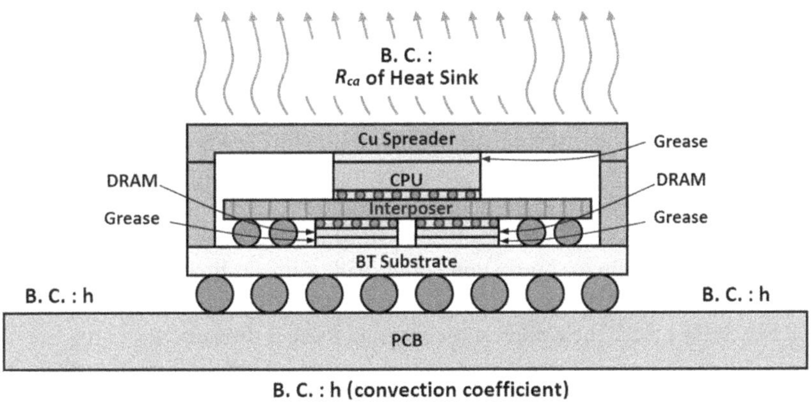

FIGURE 8.29 Schematic of the structure for analysis and boundary conditions.

Table 8.5 Dimensions of the Analyzed 3D IC Integration Structure

Chips and substrates					
	Quantity	Thickness (μm)	Length (mm)	Width (mm)	Power consumption (W)
CPU	1	400	22	18	75
DRAM	2	100	10	10	8 (= 4+4)
Interposer	1	150	28	28	0
BT substrate	1	950	40	40	0
PCB	1	1600	114.3	101.6	0
TSVs, bumps, and solder balls					
	Diameter (μm)	**Pitch (μm)**	**Distribution**		**Notes**
TSV	25	150	uniform		in interposer
Microbump	12	250	uniform		between CPU and interposer
Microbump	12	250	uniform		between DRAM and interposer
Ordinary bump	150	250	2 groups (28mm × 6.7mm) on chip both ends		between interposer and BT substrate
Solder ball	600	1000	uniform		between BT substrate and PCB

The package substrate is 40 mm × 40 mm × 950 μm and the PCB is 114.3 mm × 101.6 mm × 1600 μm. The diameter of microsolder bumps between all the chips and the interposer is 25 μm and on 250 μm pitch. The diameter of the ordinary solder bumps is 150 μm and on 250 μm pitch. The diameter of the solder balls is 600 μm and on 1000 μm pitch.

8.4.2 Thermal Analysis—Material Properties

The materials thermal properties are listed in Table 8.6. For grease, silicon, and cooper, the thermal conductivity in cross-plane direction (k_z) and in-plane direction (k_{xy}) are equal due to homogeneous material. For BT substrate and PCB, they are assumed as laminated materials in which there are several high conductive (Cu) layers inside the organic material, so their k_{xy} is obviously higher than k_z. As for the interposer, microbump, ordinary bump and solder ball, the anisotropic thermal conductivities could be calculated by using Eqs. (2.19) and (2.20) for interposer zone and using resistance series/parallel correlation for the zones of microbump, ordinary bump and solder ball as those shown in Secs. 8.3.3 and 8.3.4. The equivalent k_{xy} and k_z of microbump zone are nearly the same and equal to the value of underfill material because their pitch is too large and size is too small.

8.4.3 Thermal Analysis—Boundary Conditions

The analyzed 3D IC SiP is designed for a network system and thus it is supposed to be placed in a cabinet; therefore, we assumed the ambient temperature is 55°C. The boundary

TABLE 8.6 Thermal Material Properties of the Analyzed 3D IC Integration Structure

	k_{xy}(W/m·K)	k_z	Thickness (μm)	Notes
Grease	2.5	2.5	38	Between DRAM & BT substrate
Grease	2.5	2.5	12.5	Between CPU & heat spreader
Silicon	148	148	—	
Cooper	401	401	—	
Interposer	142.8	151.2	150	Calculated equi. properties
BT substrate	13.4	0.21	950	
PCB	56.9	0.36	1600	
Microbump	0.63	0.63	12	Calculated equi. properties
Ordinary bump	0.63	14.12	150	Calculated equi. properties
Solder ball	0.05	14.12	600	Calculated equi. properties

condition on PCB surfaces varies, and it should be $h = 0$, 10, and 50 W/m^{2}K for imitating the PCB under adiabatic (the worst case), mixed natural convection (weak air flow), and force convection (strong air flow) conditions, respectively. For heat spreader, we use the thermal resistance, R_{ca}, instead of a cooling module for simplifying the simulations. The R_{ca} is varied from 0.001 to 1.0°C/W to imitate the heat spreader that is cooled by different cooling modules: an extra-low R_{ca} ($\leq$0.3°C/W) denotes a liquid cooling module of two-phase flow or single phase flow; a medium R_{ca} (about 0.2 to 0.8°C/W) denotes a heat sink with a working fan; and a larger R_{ca} ($\geq$0.6°C/W) means a heat sink without fan. The power generation of ASIC is 75 W; that of each memory is 4 W; and the total power consumption of the SiP is 83 W. In this section, Icepak 12.1.6 is used as the simulation tool to solve the heat conduction problems using finite volume method.

8.4.4 Thermal Analysis—Result and Discussions

Figure 8.30 shows a cross-section temperature distribution of the SiP under the boundary conditions of $R_{ca} = 0.6$°C/W on heat spreader top side and $h = 10$ W/m^{2}K on PCB surfaces. The condition of $R_{ca} = 0.6$°C/W means there is a heat sink with a coupled fan to cool the spreader; the $h = 10$ W/m^{2}K implies there is a weak air flow fans the PCB surfaces. Power generations of ASIC and each DRAM are 75 W and 4 W that yield the temperatures of 118°C and 122°C, respectively, under ambient temperature 55°C.

Figure 8.31 shows the chip's temperature distributions of ASIC and DRAMs, the SiP is under the boundary conditions of $R_{ca} = 0.4$°C/W on heat spreader top side and $h = 10$ W/m^{2}K on PCB both surfaces. It can be seen that the temperature distribution is not uniform for ASIC chip and also for DRAM chips: the maximum and minimum temperatures of ASIC are 107.5°C and 89.5°C and that of DRAM are 107.5°C and 99.5°C. The both chips' average temperatures are 101°C for ASIC and 104°C for DRAMs.

To investigate the thermal effects caused by various capability of cooling modules, the thermal resistance, R_{ca}, applied on heat spreader is varied from 0.001 to 1.0°C/W, the boundary condition of PCB surfaces is $h = 10$ W/m^{2}K. Figure 8.32 shows the appearance

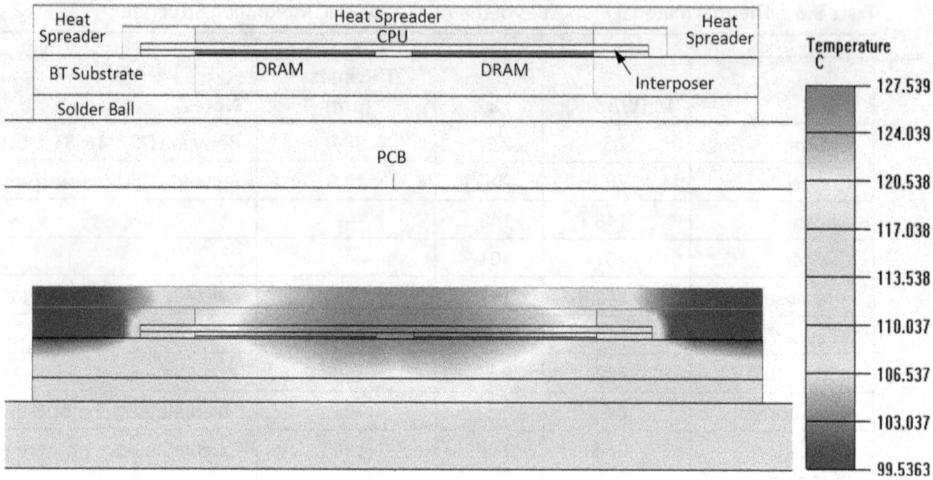

FIGURE 8.30 A typical cross-section temperature distribution of the analyzed 3D IC integration with a passive interposer.

temperature distribution of the SiP: a higher thermal resistance R_{ca} results in a higher system temperature. Figure 8.33 is the correlation between R_{ca} and chips' average temperatures. By using the correlation, it can be seen that if the operating temperature of ASIC or DRAM has to be under 100°C, the thermal module attached on the spreader must have a cooling capability less than 0.3°C/W. Also, if the chips' temperature must be under 80°C, we should choose a thermal module with cooling capability less than 0.1°C/W.

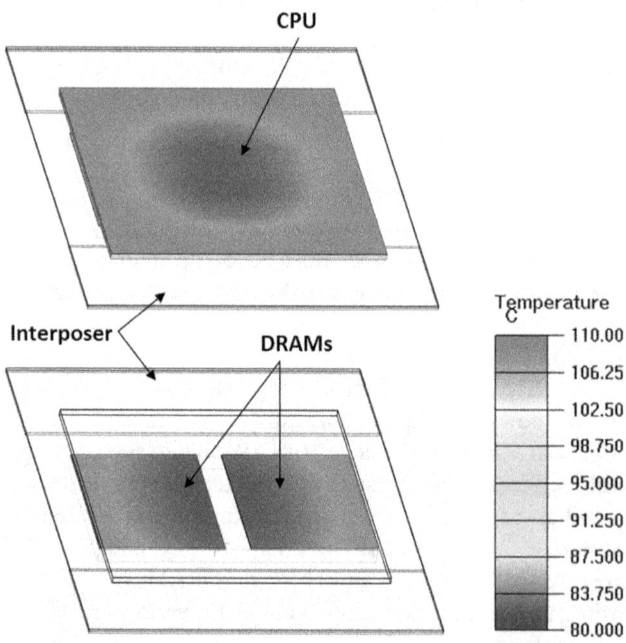

FIGURE 8.31 Temperature distribution of CPU/ASIC and DRAM memories.

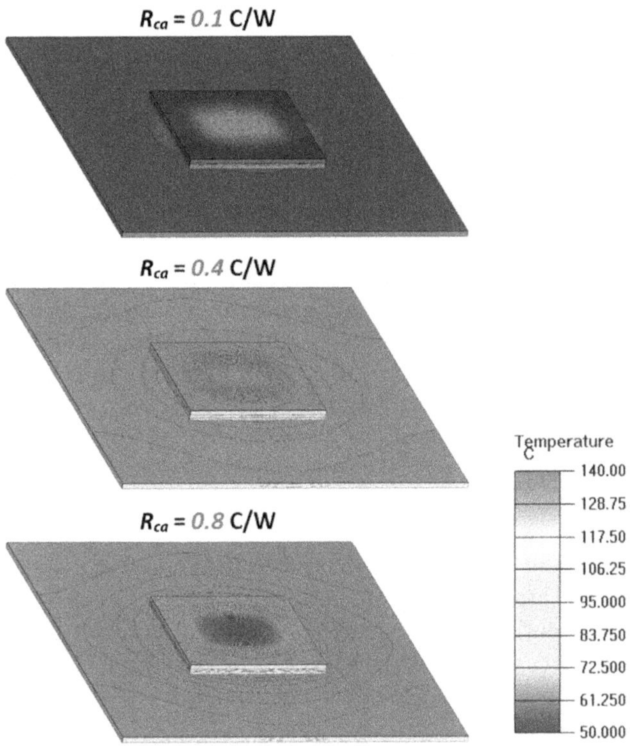

Figure 8.32 Appearance temperature distribution of the analyzed SiP under various boundary conditions.

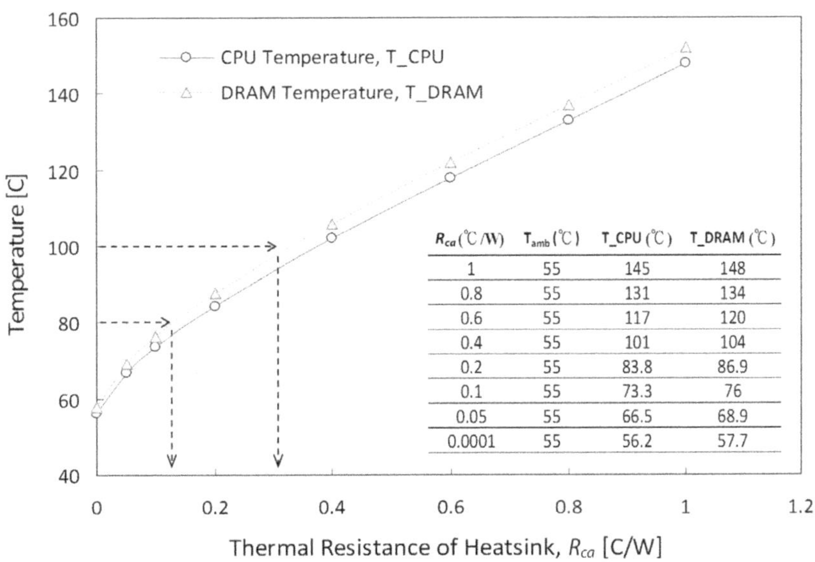

R_{ca}(℃/W)	T_{amb}(℃)	T_CPU (℃)	T_DRAM (℃)
1	55	145	148
0.8	55	131	134
0.6	55	117	120
0.4	55	101	104
0.2	55	83.8	86.9
0.1	55	73.3	76
0.05	55	66.5	68.9
0.0001	55	56.2	57.7

Figure 8.33 Correlation between Rca and chips' average temperatures.

Figure 8.33 shows that even having much less heat generation, temperature of the memory is little higher than the ASIC temperature. That is a big issue for a 3D IC SiP thermal management, and means (1) the temperature of memory is closely related to that of ASIC, and (2) memory is more vulnerable than ASIC because it can bear lower operating temperature and there is no easy solution for reducing its temperature.

A simulation is proceeded to clarify the thermal connection between memories and ASIC in such a 3D IC SiP. Here, we suppose there is no heat generation produced from the both memories; the ASIC still generates a heat of 75 W; and the boundary conditions are $R_{ca} = 0.4°C/W$ and $h = 10\ W/m^2K$. The result shows ASIC average temperature decreases from original 101°C to 97.8°C, and memories average temperatures are still high and approach to the ASIC temperature 97.6°C. Due to temperature superposition effect, we know the total temperature rise of the memory is to sum up its own temperature rise and those induced from the heating ASIC chip. For the reason, unless the memories are far-off to the ASIC or the heat dissipating path from memory to PCB is smooth enough, the memories are inevitable to have a high and irreducible temperature. To attach a Cu slug to the back of the memory chips could help, which is shown in Sec. 9.6 of Chap. 9.

8.4.5 Thermomechanical Analysis—Material Properties

All the material properties such as the Young's modules, Poisson's ratio, and CTE (coefficient of thermal expansion) for the material used in the structure are shown in Table 8.3. The material properties of Cu_6Sn_5 are from [27], of Ni are from [28], and of Cu-low-k are from [29]. The lead-free solder equations are the same as those in Sec. 8.3.7.

8.4.6 Thermomechanical Analysis—Boundary Conditions

There is a symmetry for the structure and thus only one-half of the structure is modeled as shown in Figs. 8.34 and 8.35. The symmetry axis is the y-axis and proper displacement boundary conditions have been applied. The temperature cycling condition is shown in Fig. 8.9, which is −40°C ⇆ 125°C on a 40-min cycle. Stress-free is at 25°C.

8.4.7 Thermomechanical Analysis—Results and Discussions

The ANSYS Multi Physics, v. 13.1 has been used to construct the half finite element model as shown in Fig. 8.35. It can be seen that the symmetry axis is the y-axis. The interposer, ordinary solder bumps, and organic (BT) substrate are shown in the right-hand side of Fig. 8.35. The ASIC chip, Cu-low-k pad, microsolder bump (Cu_6Sn_5), interposer, and DRAM chip are shown in Fig. 8.35, where the cross-section and meshing are detailed. The whole model is subjected to the thermal loading as shown in Fig. 8.9 and the results are presented in the following section.

Figure 8.36 shows the von Mises stress contours in the 3D IC integration SiP at 125°C (stress-free is at 25°C). It can be seen that the maximum stress occurs at the interface between the Si interposer, the Cu UBM (under bump metallurgy), and the Cu TSV.

Figure 8.37(a) highlights the key elements of the interface between the Cu TSV, the Cu UBM on the interposer, and the Si interposer. Figure 8.37(b) shows the von Mises stress contours and the maximum stress (281 MPa) at 125°C (stress-free = 25°C). The location of the maximum stress occurs at the interface between the Si interposer, Cu-filled TSV, and Cu UBM for the microbumps of the ASIC chip (the top contours of Fig. 8.37(b)). The other maximum stress (271 MPa) occurs at the interface between the Cu-filled TSV, Si interposer, and

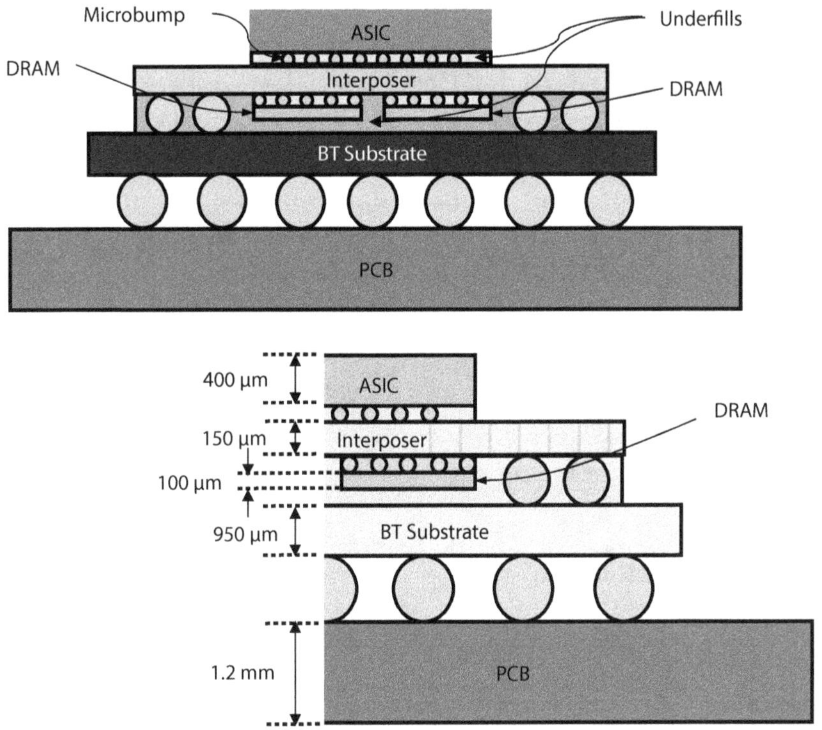

FIGURE 8.34 Schematic of the 3D IC integration SiP.

the Cu UBM for the microbumps of the DRAM chip (the bottom contours of Fig. 8.37[b]). These interfaces are potential failures (such as delaminations and cracks) locations.

Figure 8.38 shows the von Mises stress contours for the microsolder joint (Cu_6Sn_5) of the ASIC chip and DRAM chip at 125°C (stress-free temperature = 25°C). It can be seen that the maximum stress location for both case occurs at the interface between the Cu_6Sn_5 and the Ni and the Cu_6Sn_5 and the Cu UBM. Their magnitudes are, respectively, 102 MPa (for ASIC) and 125 MPa (for DRAM). The major driving force for these stresses is from the thermal expansion mismatch between the Cu-filled TSV, Si interposer, and the Cu TSV.

The maximum creep strain energy density time-history acting at the corner ordinary solder joint between the Si interposer and the organic (BT) substrate subjected to temperature cycling (−40 to 125°C) is shown in Fig. 8.38, where the von Mises stress contours are also provided. The shear stress vs. creep shear strain hysteresis loops are also shown in Fig. 8.38, which demonstrated that the creep responses are stabilized after the second cycle. It can be seen that that the creep strain energy density per cycle is 0.018 MPa which is much less than 0.1 MPa and thus it should be reliable for most of the environmental condition. Underfill helps! For stress and creep strain energy densities at other locations please see [23]. Table 8.7 summarizes the maximum stress and creep strain energy density per cycle of the key elements of the 3D IC integration SiP.

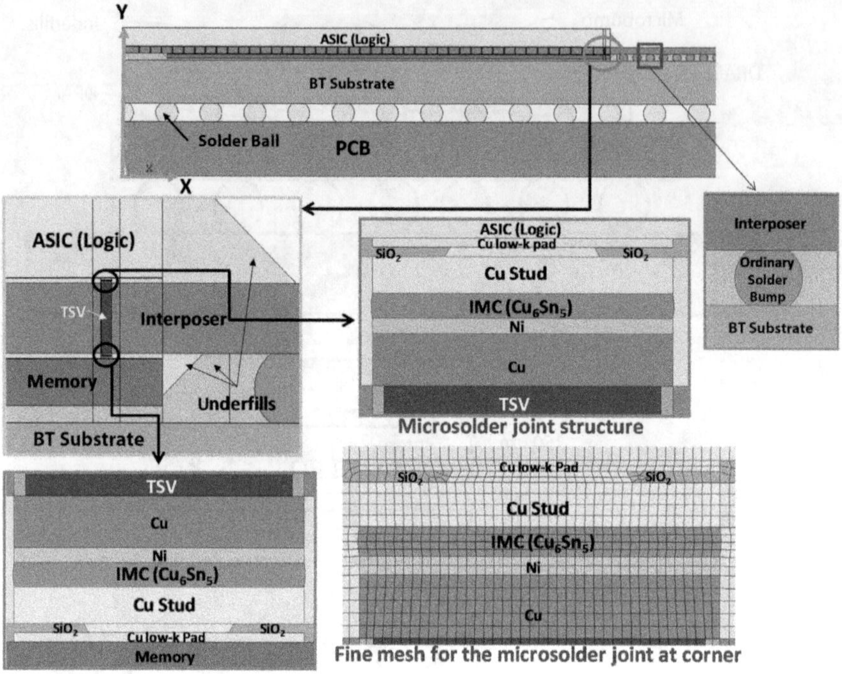

Figure 8.35 Finite element model. (Left) The area with interposer, ordinary solder bumps, and organic substrate. (Right) The area with ASIC, Cu-low-k pads, Cu studs, Cu_6Sn_5, IMC, Ni, Cu-filled TSV, the bottom memory chips, and organic substrate.

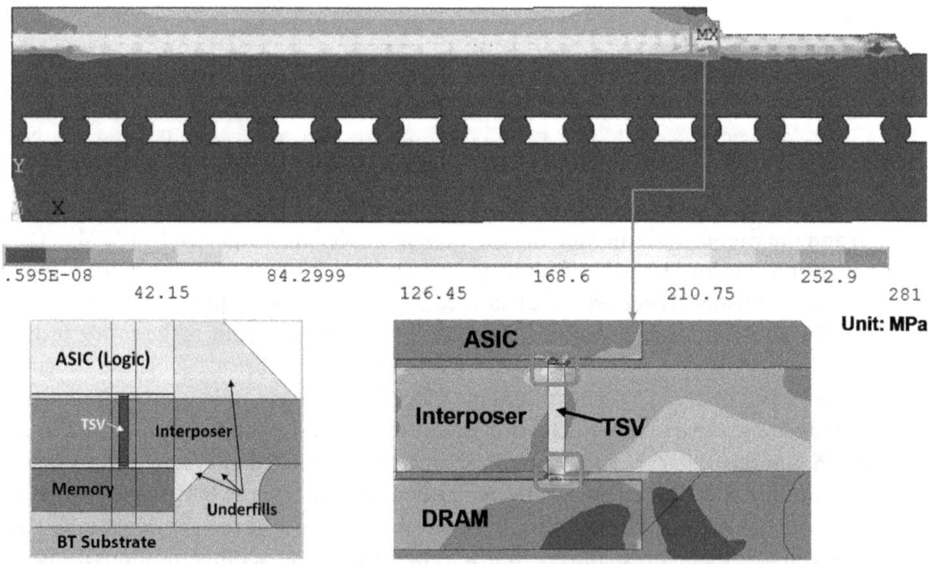

Figure 8.36 Von Mises stress contour at 125°C (reference temperature = 25°C). The maximum stress occurs at the interface between the Si interposer, the Cu UBM on the interposer, and the Cu-filled TSV.

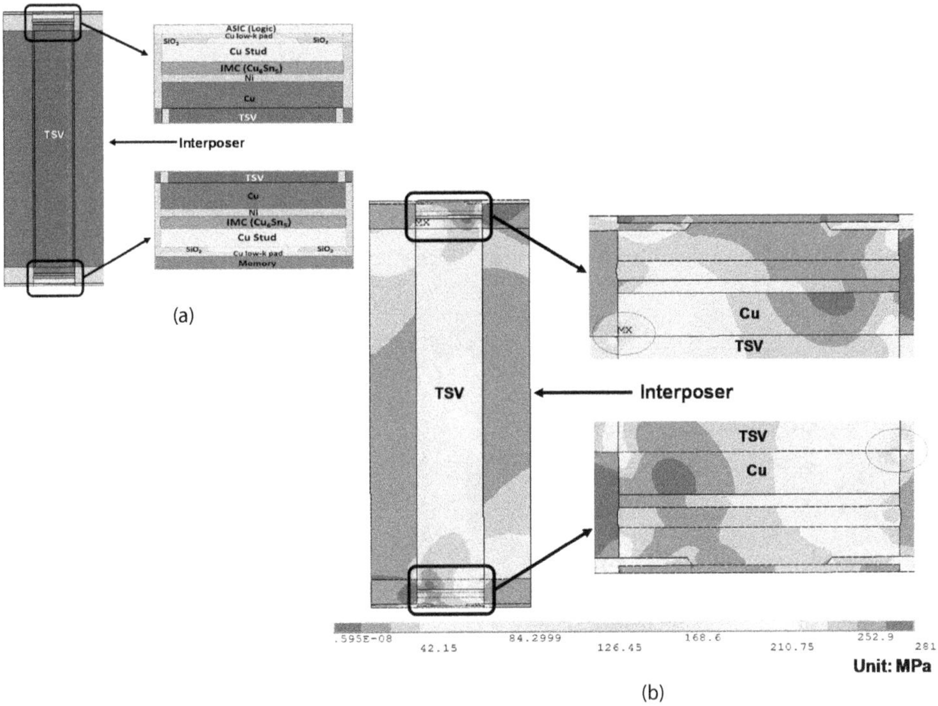

Figure 8.37 Von Mises stress contour at 125°C (reference temperature = 25°C). The maximum stress (281 MPa) occurs at the interface between the Si interposer, the Cu-filled TSV, the Cu UBM on the interposer for the ASIC chip, and the Cu-filled TSV. The second maximum stress (271 MPa) occurs at the interface for the DRAM chip.

8.4.8 Interposer Fabrication

The interposer has been fabricated from a 12" (300 mm) wafer as shown in Fig. 8.39. The fabrication steps in making the interposer with double-sided wiring layers are as follows. The vias are formed by DRIE [30] and the SiO_2 is fabricated by SACVD [31]. The barrier layer (Ta) and Cu seed layer are made by PVD [31]. The vias are filled by electroplating the Cu [31] and the overburden Cu is polished by CMP [32]. Figure 8.40 shows the SEM images of the cross-section of the Cu-filled TSV with SiO_2 layer. It can be seen that there is no void in the Cu-filled TSV. Also, the thickness of the SiO_2 liner is 0.883 μm on the top surface of the wafer, 0.814 μm at the corner near the top surface, 0.504 μm at the middle sidewall of the TSV, 0.46 μm at the upper-corner of the TSV bottom, 0.437 μm at the lower-corner of the TSV bottom, and 0.405 μm at the TSV bottom.

The double-sided wiring layers and the connecting vias are fabricated by the Cu damascene process. The size of the top-side UBM (TU) is 25 μm. It consists of Ti barrier = 0.1 μm, Cu = 5 μm, and NiPdAu = 2 μm. Figure 8.41 shows the SEM images of the cross-section of the TSV interposer (with front-side metallization and UBM) near the microbump area. The size of the TU is 25 μm.

After the front-side metallization, the TSV/RDL interposer is temporarily bonded to a Si carrier with 30 μm of adhesive. Then, backgrind + CMP the TSV wafer, dry Si etch, low

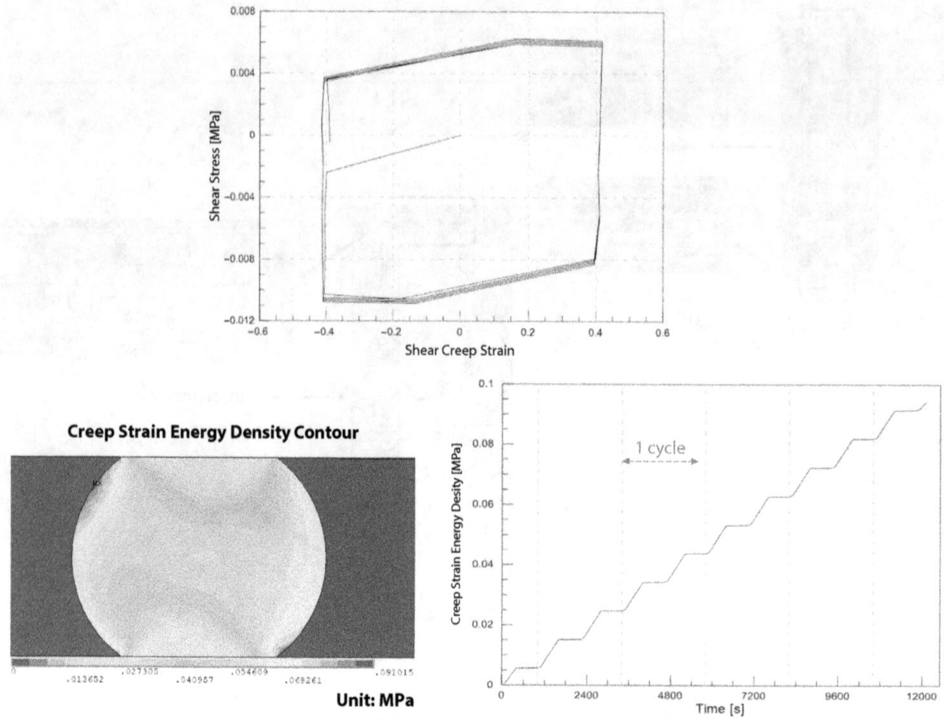

FIGURE 8.38 Shear stress vs. creep shear strain hysteresis loops. Creep strain energy density contour of the corner ordinary solder joint subjected to temperature cycling (−25°C ⇆ 125°C). Creep strain energy density vs. time history.

TABLE 8.7 Summary of Maximum Stress and Creep Strain Energy of Some Key Elements of the 3D IC Integration Structure

	Max. von Mises stress@ 125°C (MPa)	Creep strain energy density per cycle (MPa)
Microbump between ASIC and TSV interposer	At Cu low-k pad: 60	
	At Cu_6Sn_5: 102	
	At the interface of μbump, Cu-filled TSV and Si interposer: 281	
Microbump between memory and TSV interposer	At Cu low-k pad: 74	
	At Cu_6Sn_5: 125	
	At the interface of μbump, Cu-filled TSV and Si interposer: 271	
Ordinary solder bump	0.3	0.018
Solder ball	0.25	0.007

FIGURE 8.39 Fabricated TSV interposer wafer with double sided wiring (RDL) layers.

temperature (<200°C) SiN PECVD (1 μm), and CMP to reveal the Cu TSV [32]. The back-side passivation opening and via are also fabricated by the Cu damascene process. The size of the back side under bump metallurgy (BU) is 25 μm with a Ti barrier = 0.1 μm, Cu = 5 μm, and NiPdAu = 2 μm.

8.4.9 Microbump Wafer Bumping

The materials and process of wafer bumping of the microbumps on the chips of a 300 mm wafer are shown in Table 8.8. It can be seen that there are three key tasks: (a) alignment marks, (b) passivation reopening, and (c) wafer bumping.

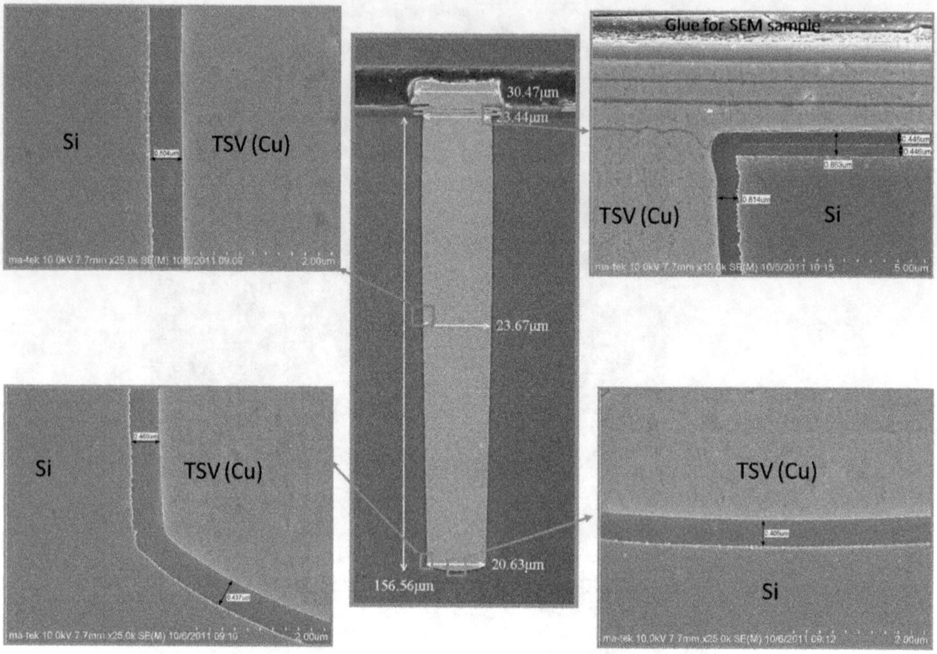

Figure 8.40 SEM cross-section images of TSV with filled Cu and SiO_2 liner.

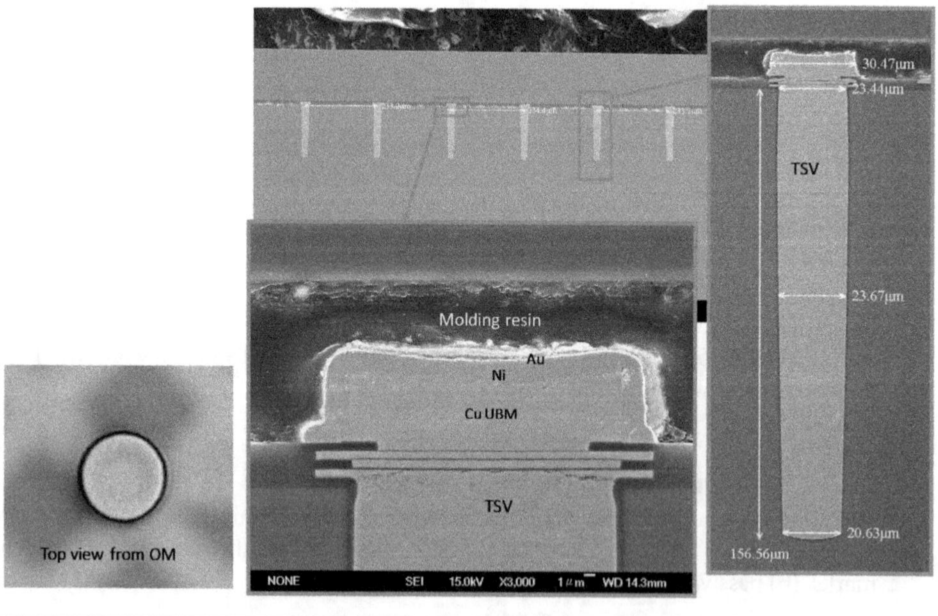

Figure 8.41 Front-side metallization (near microbump area). TU size = 25 µm.

TABLE 8.8 Material and Process for Wafer Bumping of Microbumps

1. For Alignment Mark a. PECVD oxide (2000 Å) b. Sputter Ti (0.1 μm)/Cu (0.3 μm) c. Spin coating photo resist d. Lithography e. Developer f. Cu plating for alignment mark g. Strip h. Seed layer etching (Ti & Cu)
2. For Passivation Reopening a. PECVD oxide (3000 Å) b. Spin coating photo resist c. Lithography (TO = 20 μm) d. Developer e. Oxide etching f. Strip
3. For Bumping a. Sputter Ti (0.1 μm)/Cu (0.3 μm) b. Spin coating photo resist c. Lithography (TU = 25 μm) d. Developer e. Cu (4 μm)/Sn (5 μm) plating f. Strip g. Seed layer etching (Ti & Cu)

The alignment marks are very important for flip chip final assembly on the interposer. It starts with PECVD to grow an oxide (2000 Å) on the 300 mm wafer. Then sputter 0.1 μm of Ti and 0.3 μm of Cu on the whole wafer. Follow by spin coating, photoresist, lithography, developing, and Cu plating for the alignment marks. Then strip off the resist and etch off the Cu and Ti seed layers.

The original passivation opening is too big for microsolder bumps, which is targeted for 20 μm passivation opening for a 25 μm UBM; thus, the passivation needs to be redefined as shown in Figs. 8.42 and 8.43. This can be done by PECVD to grow an oxide layer (3000 Å). Then, spin coating photoresist, lithography (opening = 20 μm), developing, oxide etching, and striping are performed.

The microwafer bumping process starts off by sputtering Ti (0.1 μm) and Cu (0.3 μm). Then, spin coating photoresist, lithography (opening for UBM = 25 μm), developing, Cu plating (4 μm), Sn plating (5 μm), striping off the resist, and etching off the Cu and Ti seed layers are performed. Figure 8.44 shows a typical 3D view of the electroplated Cu-Sn microbump. The average (from 80 samples) shearing force per bump is 8.1 g. Due to the very small size of the microbumps, the variation of the shearing force is large.

8.4.10 Final Assembly

Figure 8.45 schematically shows the final module assembly. First, perform the larger chip TCB (thermal compression bonding) on the top side of the interposer, then apply an

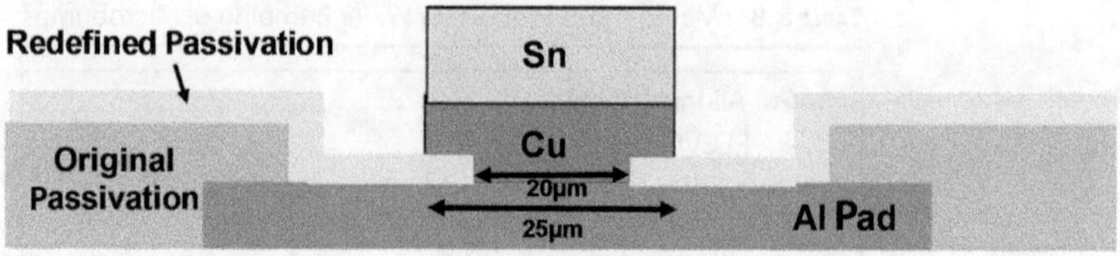

Figure 8.42 Schematic of the original passivation opening and the targeted redefined passivation opening.

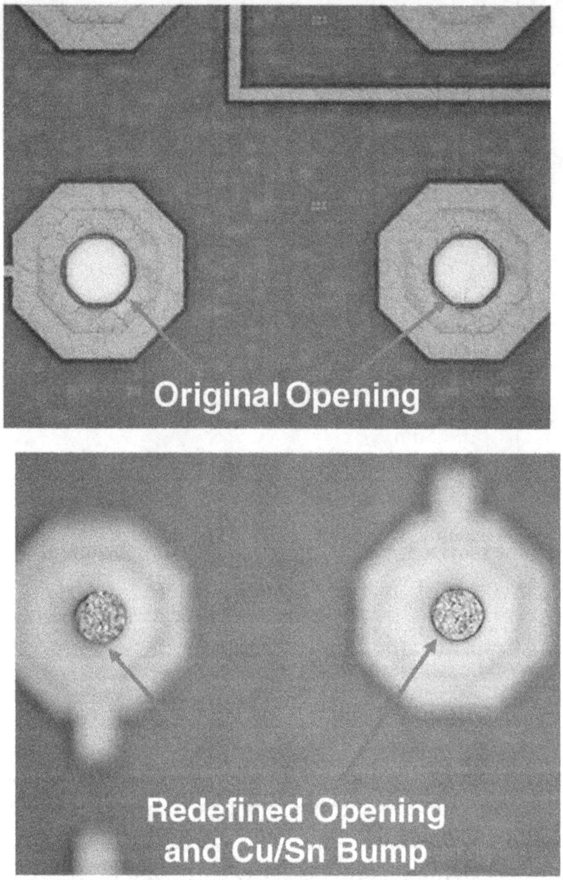

Figure 8.43 (Top) Original passivation openings. (Bottom) Redefined opening and Cu-Sn bumps.

underfill between them. Then, solder printing with a mini stencil on the back side of the interposer and reflowing the solder into ordinary solder bumps. TCB of one of the smaller chips on the back side of the interposer and apply underfill between them. Also, TCB of the other smaller chip on the back side of the interposer and apply underfill between them. Finally, flip the module and reflow the ordinary solder bumps on an organic substrate as shown in Fig. 8.45. Figure 8.46 shows an assembled package, which consists of a passive

Shear test results			
Sample-1		Sample-2	
Bump #	Shear force (g)	Bump #	Shear force (g)
1	4.99	1	7.15
2	3.69	2	6.9
3	3.36	3	7.82
4	3.44	4	9.6
5	11.25	5	9.6
6	7.5	6	8.41
7	6.17	7	9.22
8	10.45	8	6.9
9	2.94	9	8.81
10	10.25	10	7.14
11	12.13	11	8.45
12	5.74	12	10.28
13	7.14	13	9.32
14	10.75	14	2.99
15	4.32	15	10.21
16	4.42	16	9.23
17	7.98	17	9.43
18	12.17	18	10.28
19	8.92	19	9.95
20	10.79	20	7.66
21	7.49	21	8.92
22	6.09	22	4.49
23	10.63	23	4.63
24	9.33	24	9.17
25	10.82	25	9.11
26	10.92	26	5.97
27	10.62	27	8.43
28	11.29	28	6.48
29	10.68	29	10.37
30	5.64	30	10.01
Ave.	8.1	Ave.	8.2
Ave.	8.1		

Figure 8.44 A typical 3D-view of electroplated Cu-Sn bump. Shear test results.

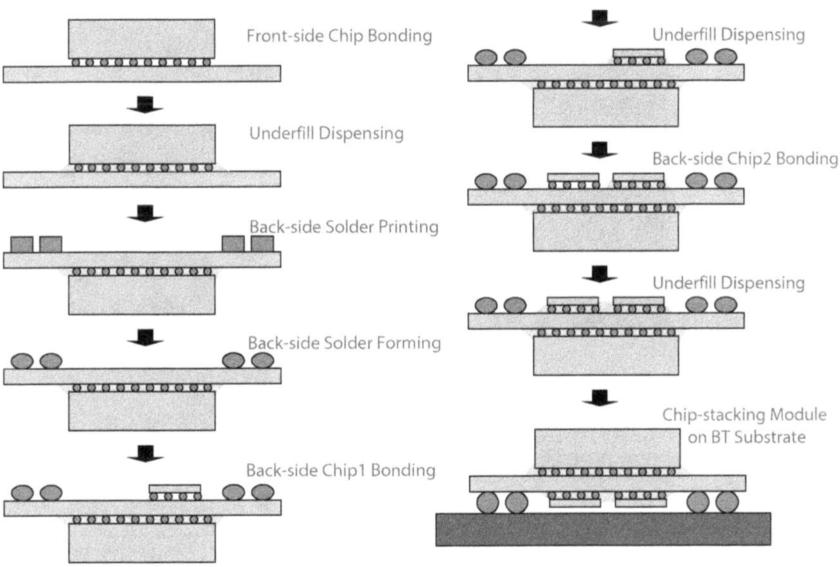

Figure 8.45 Final module assembly process flow.

Figure 8.46 A final assembly. The larger chip on the top side of the interposer (with underfill) and two smaller chips at its bottom side (with underfill). The module is on an organic substrate.

interposer supporting a large chip on its top side (with underfill) and two smaller chips on its bottom side (with underfill). This interposer with the chip set is attached on an organic substrate.

The SEM image of a cross-section of the 3D IC integration SiP assembly is shown in Fig. 8.47. It can be seen that there is a larger and thicker chip on the top side of the passive interposer and a smaller and thinner chip on its bottom side. The microsolder joint between the top chip and the interposer is also enlarged and shown in Fig. 8.47. It can be seen that

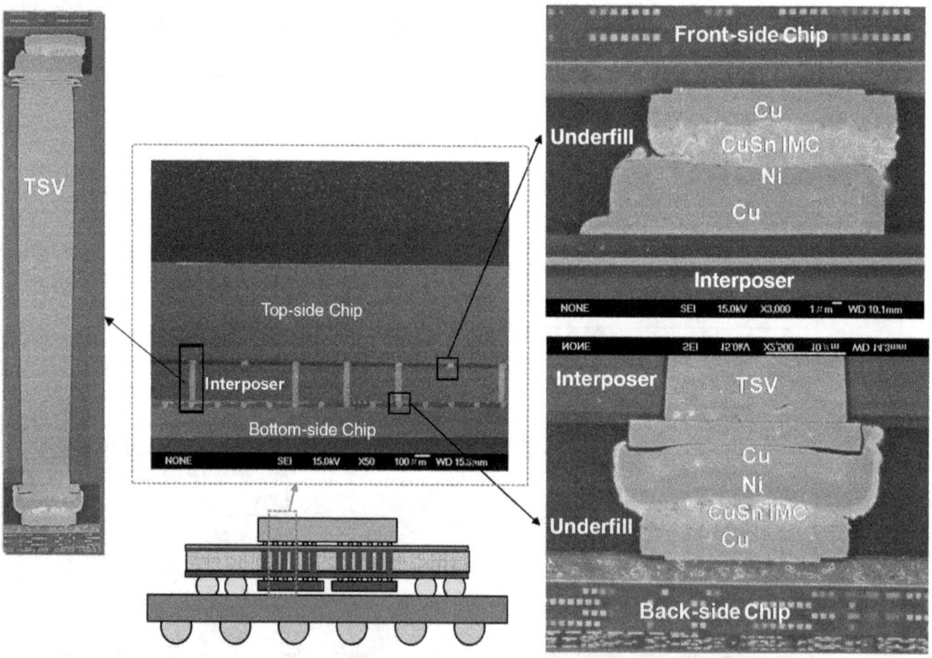

Figure 8.47 SEM cross section images of an interposer supporting a larger chip on its top side (with underfill) and two smaller chips on its bottom side (with underfill). The microsolder joints between the top chip and the interposer and the bottom chips and the interposer.

(a) the UBM on the interposer is Cu and Ni; (b) the solder becomes the IMC (intermetallic compound), Cu_6Sn_5; and (c) the Cu UBM from the larger chip. Similarly, the microsolder joint between the interposer with a TSV and the smaller chip is enlarged and shown in Fig. 8.47.

The 3D X-ray image of the 3D IC integration module is shown in Fig. 8.48. It can be seen that the TSV and microsolder joints between (a) the larger chip (U1), the smaller chip (U2), and the interposer; (b) the interposer and the larger chip (U1); and (c) the smaller chip (U2), larger chip (U1), and the interposer.

8.4.11 Summary and Recommendations

The feasibility of a 3D IC integration SiP has been demonstrated in this section. The heart of this SiP is a TSV interposer with double-sided wiring layers. This interposer is used to

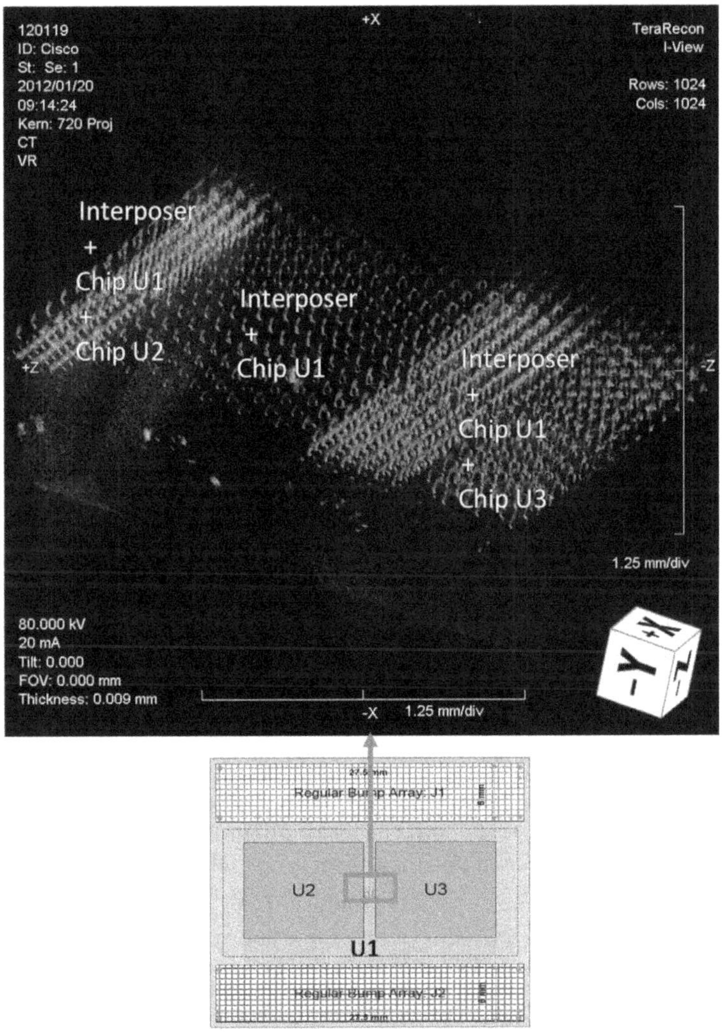

Figure 8.48 3D X-ray image of the 3D IC integration SiP with a passive interposer.

support a very large chip on its top side and two smaller chips at its bottom side. Some important results and recommendations are summarized as follows [21–26]:

- By using an equivalent model, we can rapidly and efficiently obtain a large quantity of simulation data to verify a concept, analyze a structure and modify a thermal design.
- In this study, the reliable anisotropic thermal conductivities for SiO_2-filled Cu TSV interposer have been provided and used for a 3D IC integration SiP.
- In a 3D IC SiP, an add-on cooling module attached on package top side could dominate the temperature of the chips upon an interposer, but influence less on that of the chips under the interposer. Because the add-on cooling module is usually the main solution for the SiP package cooling; therefore, the chips with lager power consumption are suggested to be placed on an interposer top side for solving their thermal problems.
- The chips underneath an interposer in a SiP have a severe thermal issue. Namely, they have a quite high and irreducible temperature because temperature superposition effect and having a bad heat dissipation path from the chips to ambient via PCB. Generally, only <10% of generated heat by the chips under an interposer could be dissipated via this path. A feasible design is to attach a cooling module to the back side of the top chip and a heat slug to the back side of the bottom chip as demonstrated in [24]. In this case, the organic substrate will have a cavity in its center.
- The driving force for the stress is due to the thermal expansion mismatch between the Cu-filled TSV, Si interposer, and the Cu UBM. The maximum stress (271 MPa) occurs at the interface between the Cu-filled TSV, Si interposer, and the Cu UBM for the microbumps of the ASIC chip. The second maximum stress (271 MPa) occurs at the interface between the Cu-filled TSV, Si interposer, and the Cu UBM for the microbumps of the DRAM chip. These interfaces are the potential failure locations.
- With underfills, the maximum stress at the Cu-low-k pad of the ASIC chip is 60 MPa and of the DRAM chip is 74 MPa. These are acceptable.
- With underfill, the maximum stress at the microsolder joint (Cu_6Sn_5) is 102 MPa for the ASIC chip and 125 MPa for the DRAM chip. Again, these are acceptable.
- With underfill, the creep strain energy density per cycle of the ordinary solder joint (between the Si interposer and the organic substrate) is only 0.018 MPa. Thus, it is reliable for most environmental condition.
- The creep strain energy density per cycle of the solder (ball) joint (between the organic substrate and the PCB) is 0.007 MPa. Thus, it is reliable for most environmental condition even there is no underfill.
- For the interposer, voidless Cu-filled TSVs have been achieved. The minimum thickness of the SiO_2 liner occurs at the bottom of the TSV and is equal to 0.405 μm. All the double-sided wiring layers and vias have also been properly fabricated.
- For microbumps on the chips of wafer, the alignment marks, passivation reopening, and wafer bumping have been properly fabricated. The average shearing force of the microsolder bump has been determined to be 81 g per bump.

- For final package assembly, the chips have been successfully TCB to the interposer and underfilled. The multichip interposer module has also been attached on an organic substrate.
- Reliability testing should be the next phase of this project.

8.5 Low-Cost TSH Interposer for 3D IC Integration

TSV is the heart and most important key enabling technology of 3D IC integration. Usually, there are six key steps in making a TSV, namely: (a) via formation by either DRIE or laser drilling, (b) dielectric layer by PECVD, (c) barrier and seed layers by PVD, (d) via Cu-filling by electroplating, (e) CMP to remove the overburden Cu, and (f) TSV Cu reveal by back-grinding, Si dry-etching, low-temperature passivation, and CMP. Thus, how to make low-cost TSVs is one of the important research topics for 3D IC integration. In this section, a class of very low-cost interposer with through-silicon holes (TSHs) and with chips on its both sides (a real 3D IC integration) is proposed and developed [33–35].

8.5.1 The New Design

Figure 8.49 schematically shows a SiP with a TSH interposer supporting a few chips on its top and bottom side. The key feature of TSH interposers is that there is no metallization in the holes. Thus, dielectric layer, barrier and seed layers, via filling, CMP for removing over-burden copper, and Cu revealing are not necessary. Comparing to the TSV interposers, TSH interposers only need to make holes (by either laser or DRIE) on a piece of silicon wafer. Just like the TSV interposers, RDLs are needed by the TSH interposers.

The TSH interposers can be used to support the chips on its top side as well as bottom-side. The holes can let the signals of the chips on the bottom side transmit to the chip on the

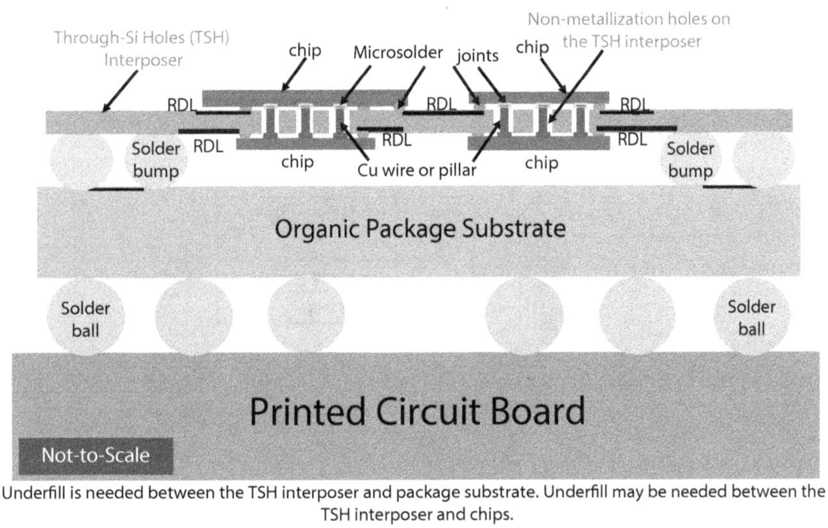

FIGURE 8.49 A SiP which consists of a TSH interposer supporting chips with Cu pillars on its top side and chips with solder bumps on its bottom side.

top side (or vice versa) through the Cu pillars and solders. The chips on the same side can communicate to each other with the RDLs of the TSH interposer. Physically, the top chips and bottom chips are connected through Cu pillars and microsolder joints. Also, the peripherals of all the chips are soldered to the TSH interposer for structural integrity to resist shock and thermal conditions. In addition, the peripherals of the bottom side of the TSH interposer have ordinary solder bumps which are attached to a package substrate.

In this section, a very simple test vehicle is fabricated to demonstrate the feasibility of this SiP with TSH interposer technology. The design, materials, and process of the top chip with Cu pillars, bottom-chip with solder bumps, and TSH interposer will be presented. The final assembly of the SiP test vehicle which consists of the chips, interposer, package substrate, and PCB will also be provided. Shock and thermal cycling tests will be performed to demonstrate the integrity of the SiP structure. Electrical simulations of the proposed structure and the results compared with the structure with TSVs will be briefly mentioned first.

8.5.2 Electrical Simulation

Figure 8.50 shows the material, geometry, and dimension of the TSV and TSH under simulations. It can be seen that 3D axial-symmetry structure are used and the finite element code is ANSYS-HFSS and simulation frequency is up to 20 GHz. The thickness of both interposers is the same (100 μm). Two different pitches are considered, one is 100 μm and the other is 200 μm. The diameter of the TSV is 15 μm and two different SiO_2 thicknesses are considered, 0.2 and 0.5 μm. Two different sizes of the air-hole are considered, 7.5 and 15 μm.

The simulation results are shown in Figs. 8.51 and 8.52 for via pitches equal to 100 and 200 μm, respectively. It can be seen that (1) for both pitches, there is very small difference in

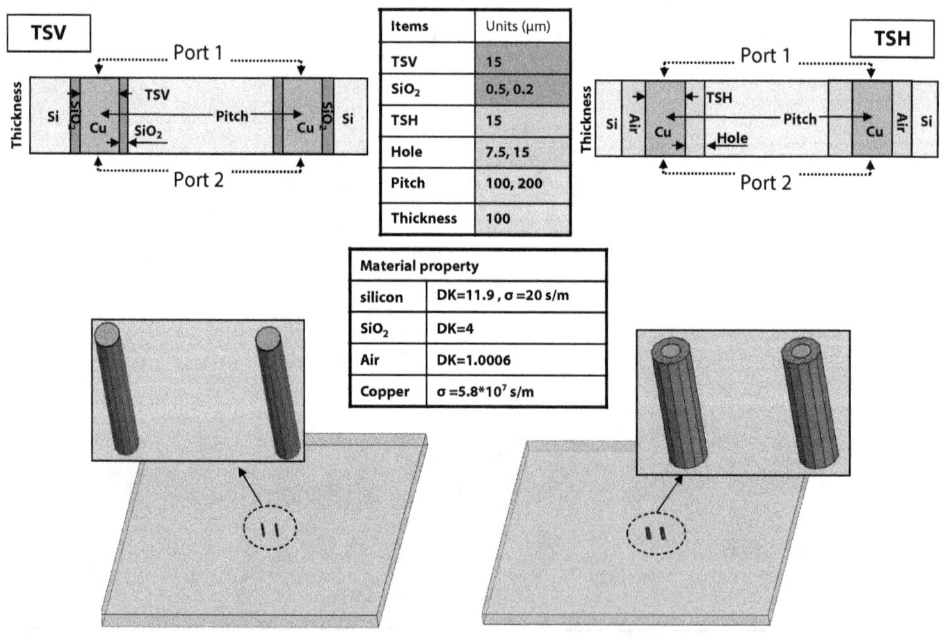

FIGURE 8.50 Electrical simulation structure of the TSV and TSH interposers.

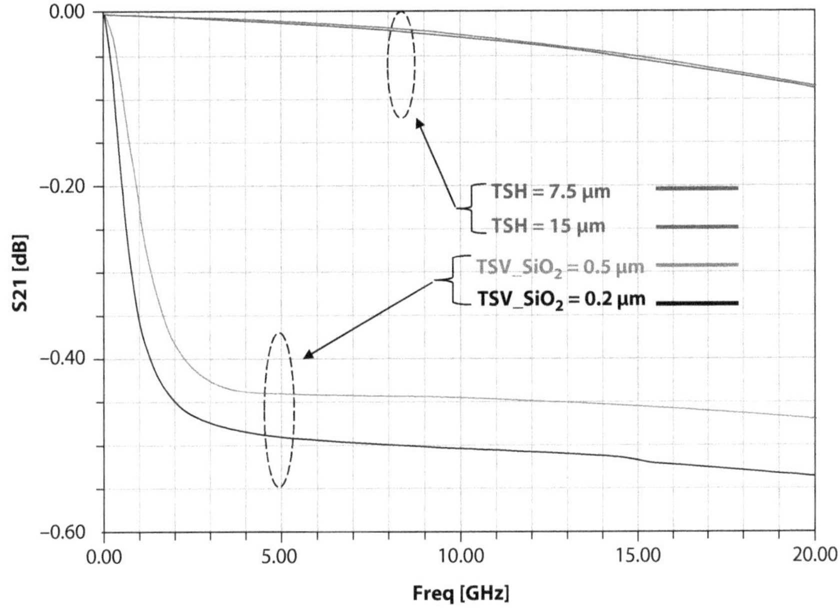

Figure 8.51 S21 for TSH and TSV interposers (pitch = 100 μm).

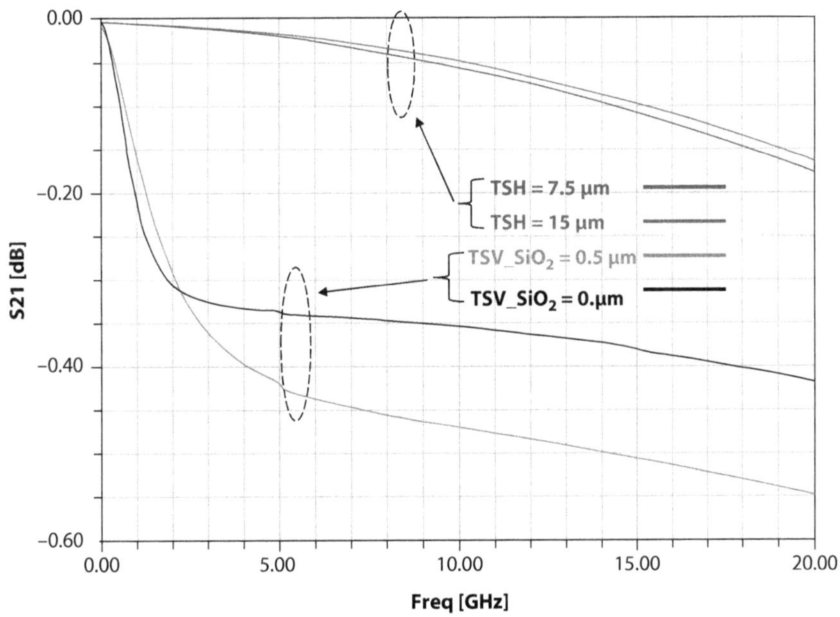

Figure 8.52 S21 for TSH and TSV interposers (pitch = 200 μm).

S21 for the two sizes of air-hole considered in TSH; this gives more freedom for mechanical design without affecting the electrical performance, and (2) for both pitches, the S21 of the TSH is much better than TSV, which means the TSH has much smaller insertion loss than TSV for high frequency signal transit. Thus, as expected, the electrical performance of the new design TSH is better that the conventional TSV.

8.5.3 Test Vehicle

The test vehicle is shown in Fig. 8.53. It can be seen that it consists a TSH interposer, which is supporting a top chip with Cu pillars and a bottom chip with UBM and solder. The interposer module is connected to a package substrate and then attached to a PCB.

The dimensions of the top chip are 5 mm × 5 mm × 725 μm. The chip has (16 × 16 = 256) Cu pillars at its central portion and two-row (176) of Cu UBM/pads at its peripherals. The diameter of the Cu pillars is 50 μm. They are 100 μm tall and on 200 μm pitch as shown in Fig. 8.54 and Table 8.9. The thickness of the peripheral Cu UBM/pads is 9 μm and with electroless (2 μm) Ni and immersion (0.05 μm) Au (ENIG). The dimensions of the bottom chip are also 5 mm × 5 mm × 725 μm. The chip has 432 Cu UBM/pads (4 μm) and coated with Sn solders (5 μm). The central 256 are for the interconnections of the Cu pillars of the top chip.

The dimensions of the TSH interposer are 10 mm × 10 mm × 70 μm (Figs. 8.54 and 8.55). It has 256 holes at its central portion to let the Cu pillars to pass through. The diameter of the holes is 100 μm and the pitch of the holes is 200 μm. There are two rows (176) of peripheral Cu UBM/pads (4 μm) coated with Sn solders (5 μm) on the top side of the TSH interposer for the interconnections of the top chip. On the other hand, there are two rows (176) of peripheral Cu UBM/pads (9 μm) with ENIG (2 μm) on the bottom side of the TSH interposer for the interconnections of the bottom chip.

The dimensions of the organic package substrate are 15 × 15 × 1.6 mm. There is a cavity (6 × 6 × 0.5 mm) on the top side of the substrate for the bottom-chip. The dimensions of the PCB (printed circuit board) are 132 × 77 × 1.6 mm, which is a standard size of the JEDEC (JESD22-B111) specification [36].

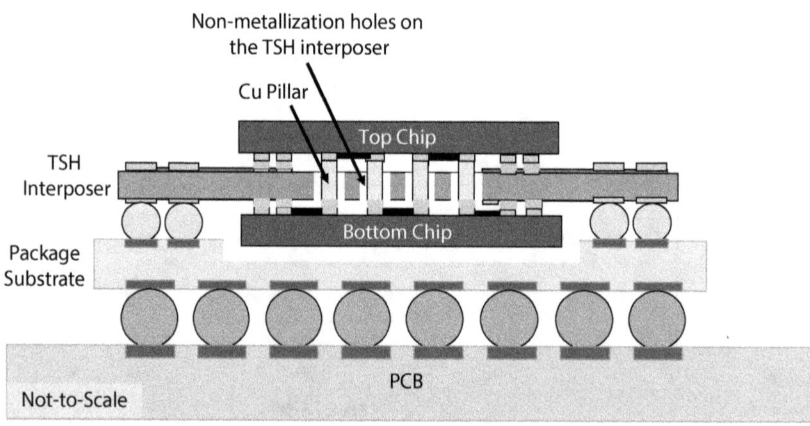

FIGURE 8.53 SiP test vehicle which consists of a top chip with Cu pillars and a bottom chip with solder bumps on a TSH interposer.

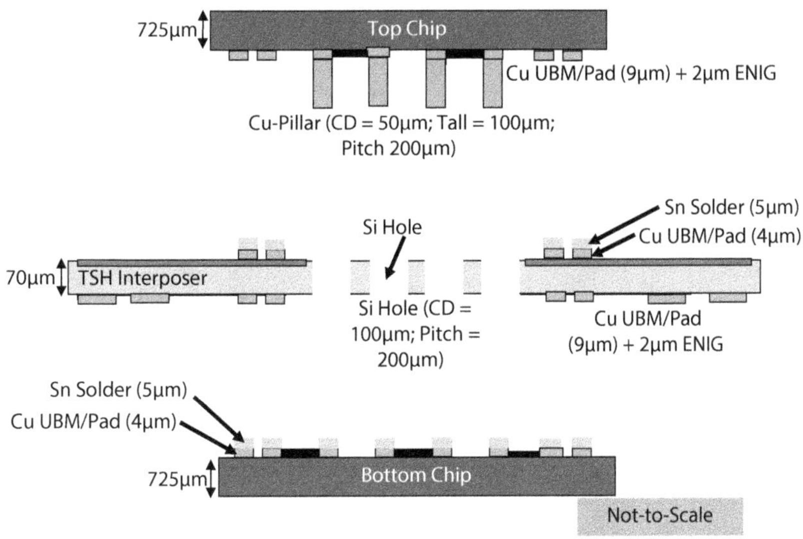

Figure 8.54 Geometry, dimension, and interconnects of the top chip, bottom chip, and TSH interposer.

Table 8.9 Key Elements of the 3D IC Integration Structure and their Dimensions

Top chip with Cu-pillar	
Dimension	5 mm × 5 mm × 725 µm
Cu-Pillar	CD = 50 µm; Tall = 100 µm; Pitch = 200 µm
Cu UBM/Pad	9 µm + 2 µm (ENIG)
Bottom chip without Cu-pillar	
Dimension	5 mm × 5 mm × 725 µm
Cu UBM/Pad	4 µm
Sn Solder	5 µm
TSH interposer	
Dimension	10 mm × 10 mm × 70 µm
TSH	CD = 100 µm; Pitch = 200 µm; Depth = 70 µm
Top side: Cu UBM/Pad and Sn Solder	4 µm (Cu) and 5 µm (Sn)
Bottom side: Cu UBM/Pad	9 µm + 2 µm (ENIG)
Package substrate (Cavity)	15 × 15 × 1.6 mm (6 × 6 × .9 mm)
PCB	132 × 77 × 1.6 mm

8.5.4 Top Chip with UBM/Pad and Cu Pillar

The process flow of fabricating the top chip with Cu-pillar is shown in Fig. 8.56. Since this is not a device chip but a piece of silicon, thus the daisy chain (RDL) will be fabricated first. SiO_2 is deposited on a Si wafer with PECVD provided by Novellus (now Lam Research) at 200°C. Then, (0.1 µm) Ti and (0.3 µm) Cu are sputtered by the MRC machine. It is followed

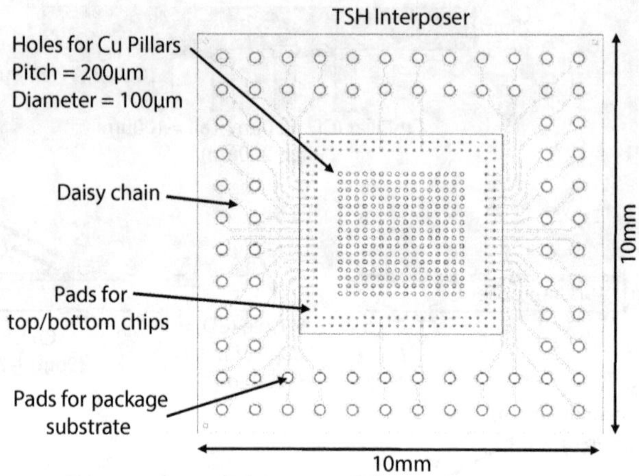

Figure 8.55 Layout of the TSH interposer.

by spin coating a photoresist and patterning with a mask using a photolithography technique (align and expose). Electroplate the Cu (2 μm) RDL. Then, strip off the photoresist and etch off the TiCu. SiO_2 (0.5 μm) is deposited on top of the whole wafer with PECVD. Again, photoresist and patterning are carried out. Then, dry etch the SiO_2 by RIE (reactive ion etch). It is followed by stripping off the photoresist and now it is ready for wafer bumping.

First, sputter the seed layer: (0.1 μm) Ti and (0.3 μm) Cu. Then, photoresist (9.5 μm) and patterning of all 432 pads are carried out. It is followed by electroplating the Cu

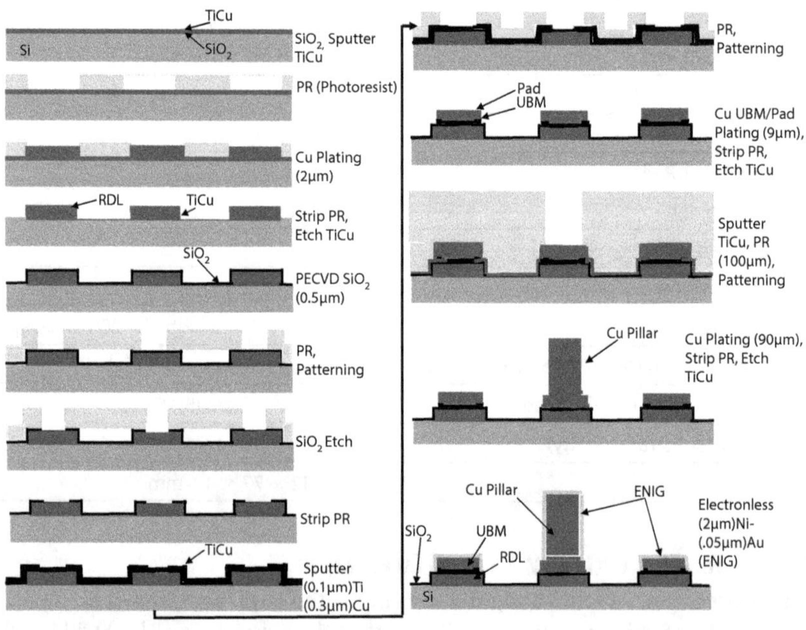

Figure 8.56 Process flow in fabricating the top chip with RDL, Cu UBM/pad, and Cu pillars.

UBM/pad (9 μm). Strip off the photoresist and etch off the TiCu seed layer. Then, a positive type of photoresist (100 μm) provided by HD Microsystems is laminated hard pressed on the whole wafer and it is followed by patterning only the central 256 pads, and Cu plating (90 μm) with Semitool (now Applied Materials) at room temperature. Then, the top surface of the wafer is flattened by DISCO's fly cut. It is followed by stripping off the photoresist and etching off the seed layer TiCu. Finally, electroless Ni (2 μm) and immersion Au (0.05 μm). Figure 8.57 shows the SEM (scanning electron microscopy) images of the Cu pillars, Cu UBM/pads, and Cu RDLs (daisy chains) on the top chip. The smaller diameter (45 μm) on top of the Cu pillars is because of the smaller opening on top of the dry film photoresist.

8.5.5 Bottom Chip with UBM/Pad/Solder

The process flow in fabricating the bottom chip with UBM/pad and solder is shown in Fig. 8.58. It can be seen that the process in making the Cu RDL is the same as those of the top chip. For most of the wafer bumping processes, they are the same except the photoresist thickness and solder. After photoresist (9.5 μm) and patterning of all (432) pads, it is followed by electroplating the Cu UBM/pad (4 μm) and electroplating the Sn solder (5 μm). Strip off the photoresist and etch off the TiCu seed layer. A photo image of the RDL, Cu UBM/pad, and Sn solder cap on the bottom chip is shown in Fig. 8.58.

8.5.6 TSH Interposer Fabrication

The process flow in fabricating the TSH interposer with Cu UBM/pad + Sn solder on its top side and Cu UBM/pad on its bottom side is shown in Fig. 8.59. It can be seen that the RDL and Cu UBM/pad on the bottom side of the interposer will be fabricated first and the

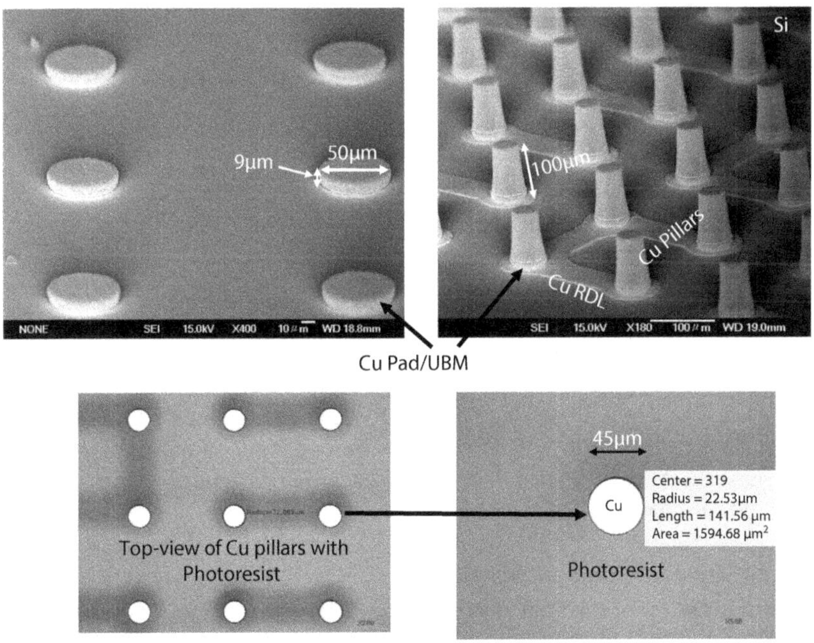

Figure 8.57 SEM images of the Cu UBM/pads and Cu pillars on the to-chip.

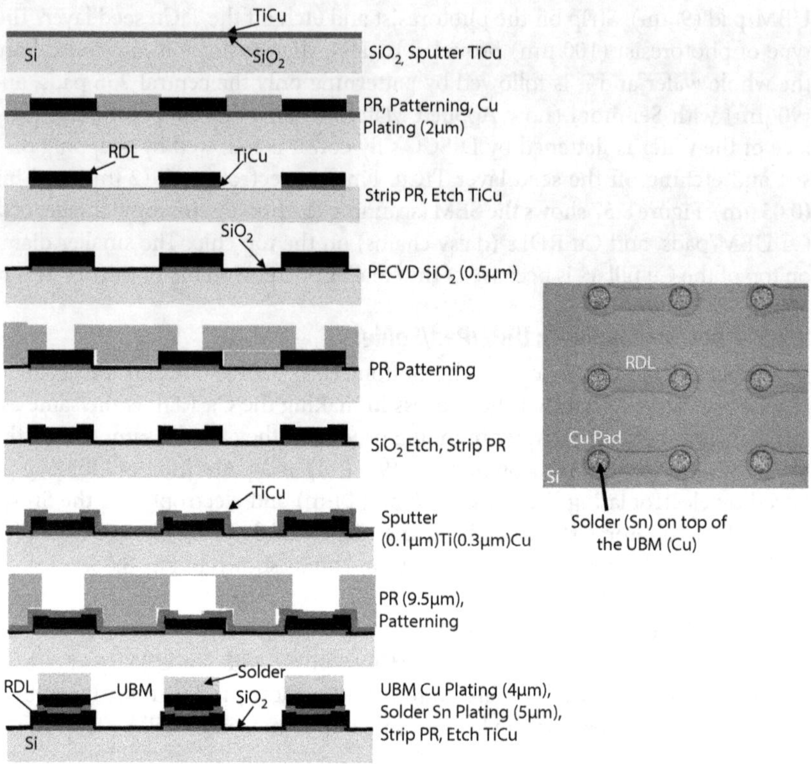

Figure 8.58 Process flow in fabricating the bottom chip with RDL, and Cu UBM/pad + Sn solder.

processes are basically the same as those of the top chip. Except after the strip off of the photoresist and etch off of the seed layer (TiCu) of the 9-μm-thick UBM/pad, it is followed by ENIG (2 μmNi–0.05 μmAu). Then, the bottom side of the interposer wafer with UBM/pad is temporarily bonded with an adhesive to a 750-μm-thick silicon supporting wafer (carrier). It is followed by thinning the top side of the interposer wafer down to 70 μm. Then, repeat all the process steps in fabricating the UBM/pad + Sn solder of the bottom chip mentioned earlier. Finally, de-bond the carrier wafer from the interposer wafer at room temperature. At this stage, the interposer wafer with 176 peripheral UBM/pads + Sn solders on its top side and 176 peripheral UBM/pads at its bottom side, as shown in Figs. 8.54 and 8.55, is achieved. The 256 holes are fabricated with UV laser drilling by the Siemens MicroBeam 3205. The power is 3400 mW. Figure 8.60 shows the photo images of the 70-μm-thick TSH interposer wafer with RDLs, pads for the package substrate, peripheral pads for the chips, and 100-μm-diameter holes on 200 μm pitch.

8.5.7 Final Assembly

The process flow of the final assembly of the SiP with TSH interposer test vehicle is shown in Fig. 8.61. First, the top chip with the 176 peripheral UBM/pads is TC (thermocompression) bonded (with the SuSS FC-150 bonder) to the peripheral UBM/pads + Sn solders

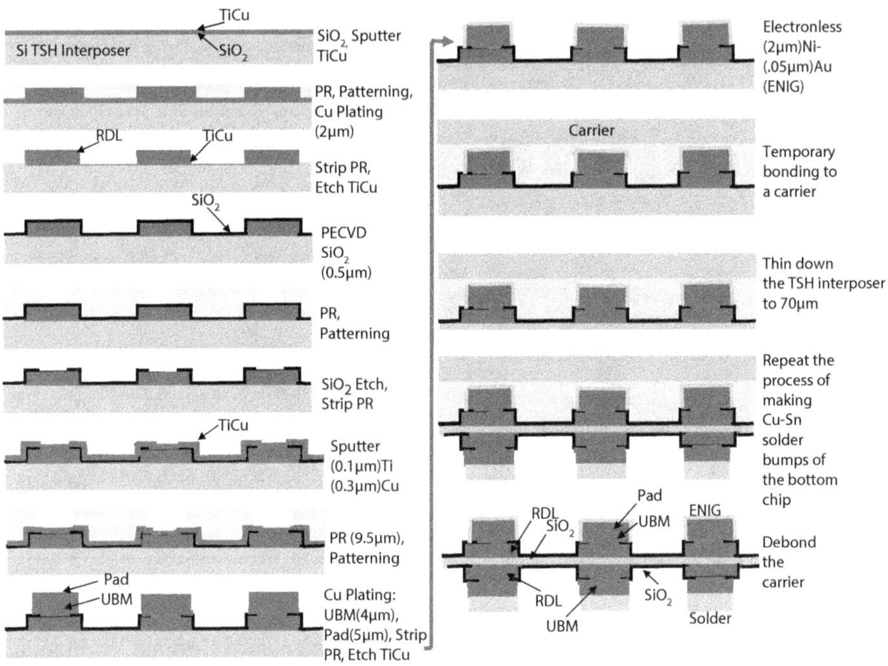

FIGURE 8.59 Process flow in fabricating the TSH interposer with Cu UBM/pad + Sn solder on its top side and Cu UBM/pad on its bottom side.

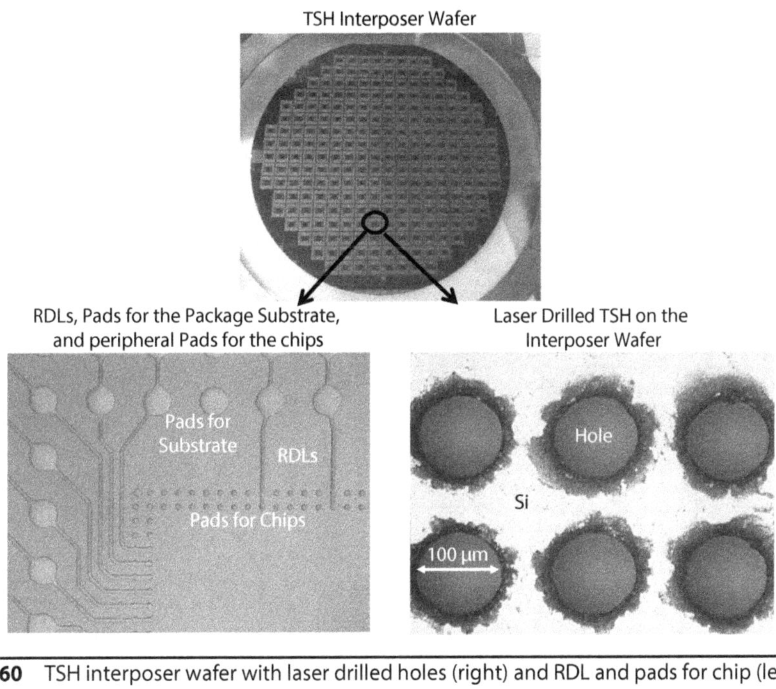

FIGURE 8.60 TSH interposer wafer with laser drilled holes (right) and RDL and pads for chip (left).

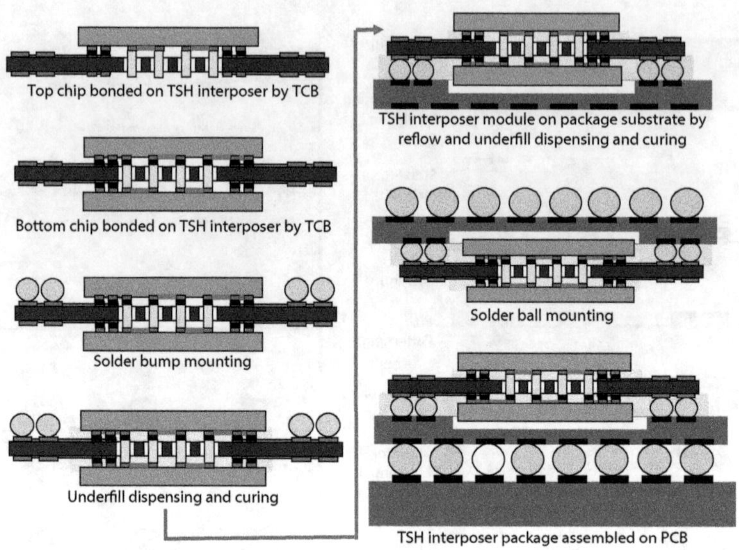

Figure 8.61 Process flow in the final assembly of SiP.

on the top side of the TSH interposer. (The Cu pillars passed through the holes of the TSH interposer.) The TC bonding conditions are shown in Fig. 8.62. It can be seen that (a) the maximum bonding force is 1600 g, (b) the maximum temperature of the chuck is 150°C, (c) the maximum temperature of the head is 250°C, and (d) the cycle time is 120 s.

Then, all the 432 UBM/pads + Sn solders on the bottom chip are TC bonded to the tip of the 256 central Cu pillars and 176 peripheral UBM/pads on the bottom side of the TSH interposer. The bonding conditions are basically the same as those of the top chip except the bonding force is reduced to 800 g. It is followed by solder (Sn3wt%Ag0.5wt%Cu) bump (350 μm diameter) mounting on the bottom side of the TSH interposer. Then, a capillary-type underfill with 50% filler contents (average filler size = 0.3 μm and maximum filler size = 1 μm) is dispensed along two adjacent sides of the top chip. After the underfill (a) fills the gap between the top chip and the TSH interposer, (b) flows through the holes of the TSH interposer, and (c) fills the gap between the bottom chip and the TSH interposer, then it is cured at 150°C for 30 min.

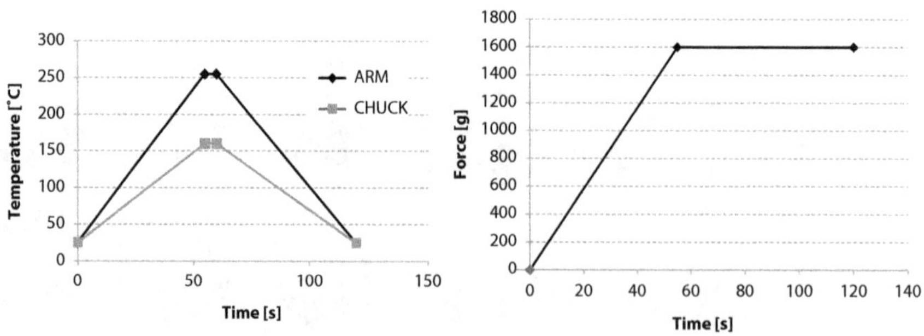

Figure 8.62 TCB conditions in bonding the top chip to the top side of TSH interposer: temperature (left) and force (right).

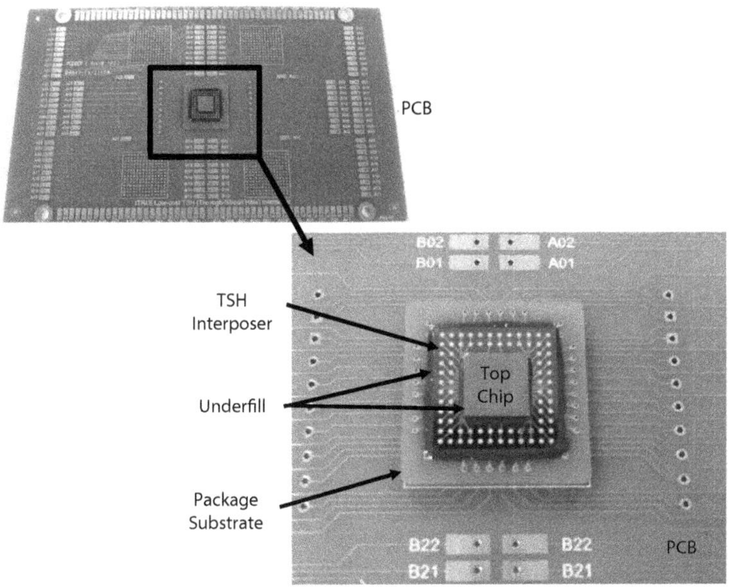

FIGURE 8.63 The final assembled SiP.

The whole TSH interposer module is solder reflowed on the package substrate by a standard lead-free temperature profile with a maximum at 240°C. In order to enhance the solder joint reliability, the underfill is applied between the TSH interposer and the organic package substrate. It is followed by solder (Sn3wt%Ag0.5wt%Cu) ball (450 μm diameter) mounting on the bottom side of the package substrate. Finally, the whole SiP package is solder reflowed on the PCB with the same lead-free reflow temperature profile just mentioned. The final assembled SiP test vehicle is shown in Fig. 8.63. It can be seen that the PCB is supporting the package substrate, which is supporting the TSH interposer, which is supporting the top chip. The bottom chip is blocked by the TSH interposer and cannot be seen.

Figure 8.64 shows the X-ray images of the final assembled SiP. It can be seen that (a) the Cu pillars are not contacting the sidewall of the TSH, and (b) the Cu pillars are almost at the center of the TSH. Figure 8.65 shows the SEM image of a cross-section of the SiP, which includes all the key elements such as the top chip, TSH interposer, bottom chip, package substrate, PCB, microbumps, solder bumps, solder ball, TSH, and Cu pillars. It can be seen through the X-ray and SEM images that the key elements of SiP structure are properly fabricated.

8.5.8 Reliability Assessments

In this section, the reliability assessments to verify the structural and thermal integrity of the assembled SiPs are drop test and thermal cycling test.

Shock (Drop) Test and Result

The drop test board and setup are based on JESD-B111 [36]. The SiPs are facing upward during the test. Four standoffs on the fixture provide the support and the spacing for deflections of the PCB during impacts as shown in Fig. 8.66. The drop height is 460 mm, which leads to the acceleration = 1500 g as shown in Fig. 8.67. After 10 drops, there is no failure, i.e., no resistance change of the daisy chains and there is no obvious failure.

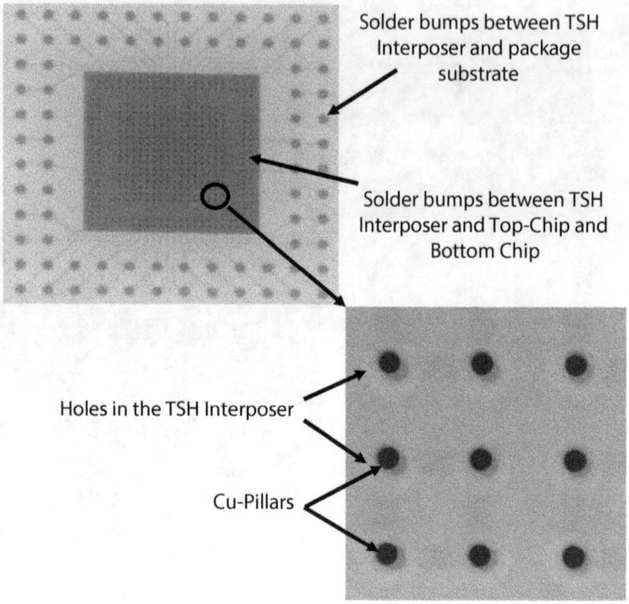

Figure 8.64 X-ray images showing the location of Cu pillars and TSHs of the SiP which consists of the top chip, TSH interposer, bottom chip, package substrate, and PCB.

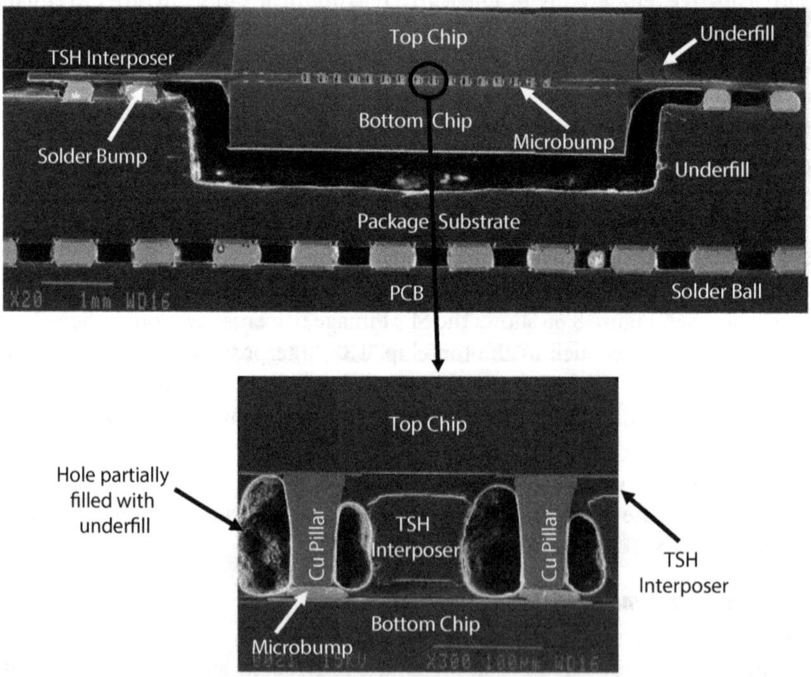

Figure 8.65 SEM image showing a cross-section of the SiP which consists of the top chip, TSH interposer, bottom chip, package substrate, and PCB.

FIGURE 8.66 Drop test of the SiPs set-up.

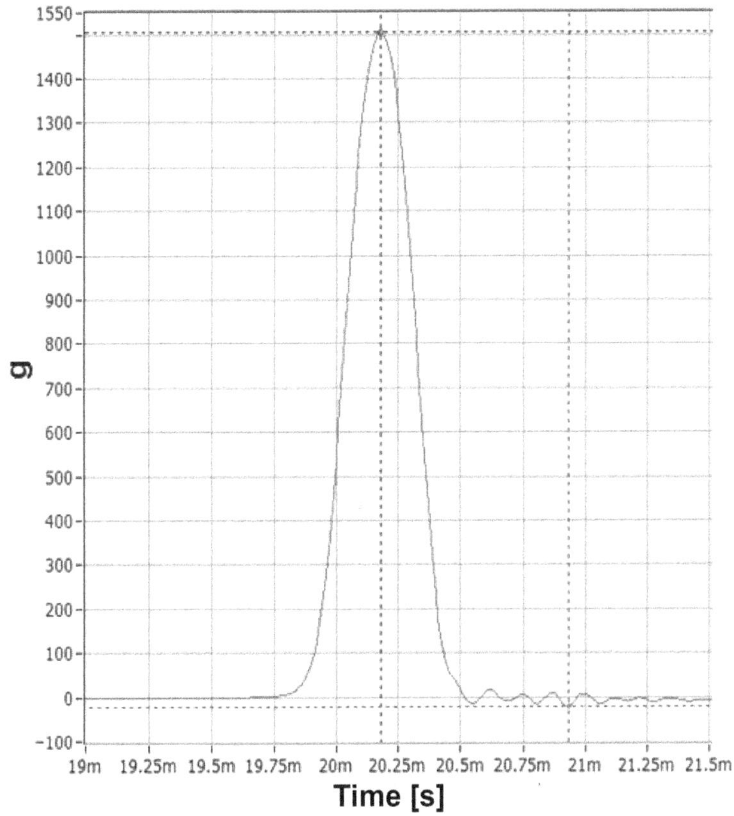

FIGURE 8.67 Drop test profile according to JESD22-B111.

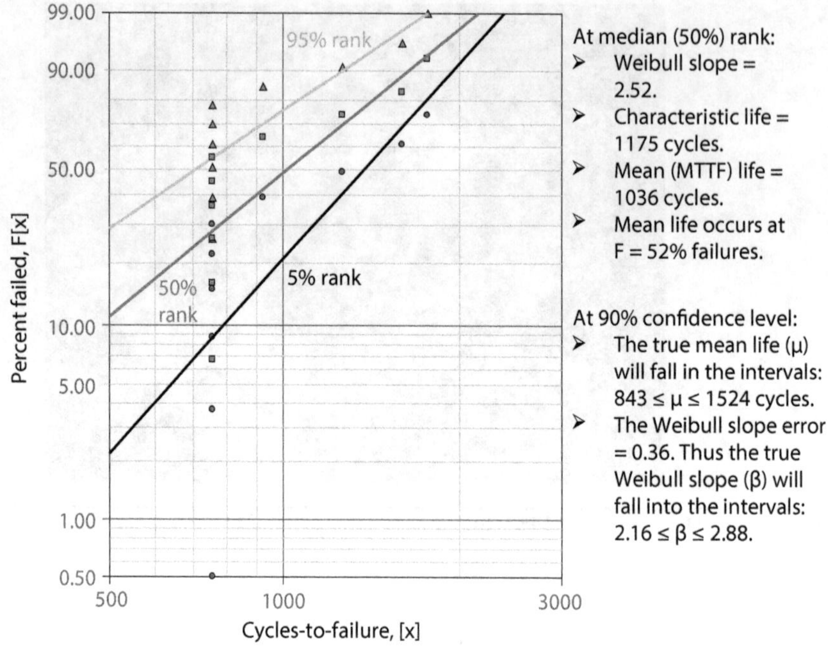

Figure 8.68 Weibull plot of daisy-chained solder joint under thermal cycling testing, −55°C ↔ 125°C, and one hour cycle (15 min ramp up and ramp down and 15 min at dwells). The required confidence level was 90%.

Thermal Cycling Test and Results

The thermal cycling test conditions are: $-55°C \leftrightarrow 125°C$, and 1 hour cycle (15 min ramp up and ramp down and 15 min at dwells). Figure 8.68 shows a Weibull plot of the test results and it can be seen that for the median (50%) rank, the Weibull slop is 2.52, and the sample characteristic life (at 63.2% failures) is 1175 cycles. (For the given test conditions, if the characteristic life is larger than 1000 cycles, then it is considered acceptable.) The sample mean life of the daisy chained solder joints is defined as the mean time to failure (MTTF) $= 1175\Gamma(1 + 1/2.52) = 1036$ cycles, where Γ is the Grammar function. This mean life occurs at $F(1036) = 1 - \exp[-(1036/1175)2.52] = 0.52$, i.e., 52% failures.

Figure 8.68 also shows the test results at 90% confidence level, i.e., in 9 of 10 cases we would like to find out the intervals for the true Weibull slope and the true mean life of the daisy chained solder joints. It can be seen that the true mean life of the solder joints (in 90 of 100 cases, the other 10 cases, no one knows) will be no less than 843 cycles but no larger than 1524 cycles. Also, the true Weibull slope (β) falls into the intervals $2.16 \leq \beta \leq 2.88$.

One of the typical failure modes is shown in Fig. 8.69. It can be seen that there is a crack in the solder bump between the TSH interposer and the pad of the organic package substrate. The failure (crack) location is in the solder bump near the interface between the solder and the UBM of the TSH interposer. It failed at 1764 cycles and the failure criterion is infinitive resistance changed.

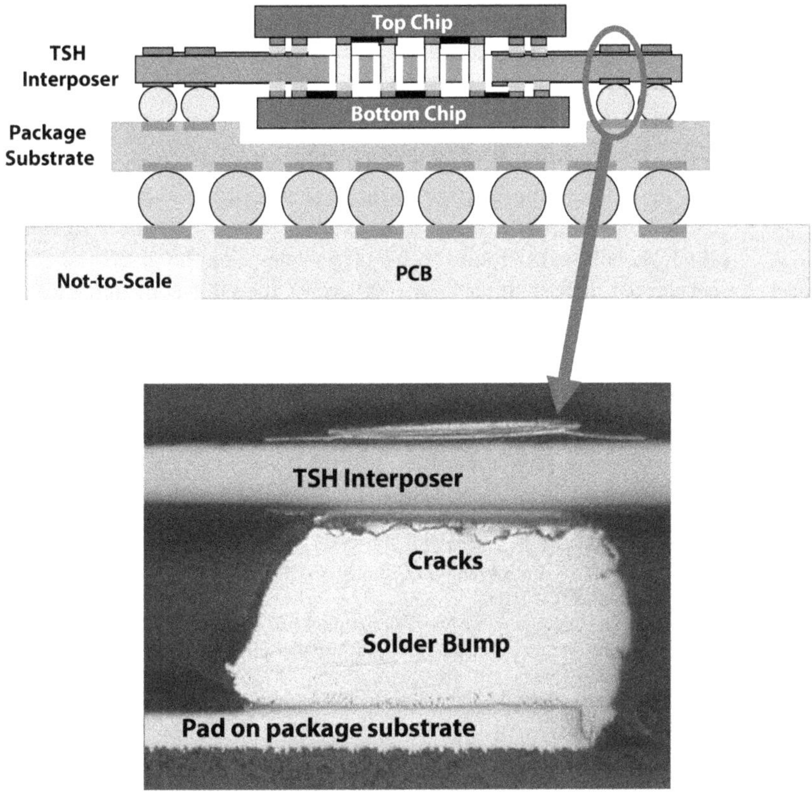

Figure 8.69 Failure (solder joint cracking) modes due to thermal cycling test.

8.5.9 Summary and Recommendations

A SiP which consists of a TSH interposer with chips on its top and bottom side has been developed. Drop and thermal cycling tests have been performed to demonstrate the integrity of the SiP structure. Some important results and recommendations are summarized as follows [33–35]:

- The electrical performance of the TSH interposer is better than that of the TSV interposer.
- The RDL (daisy-chain) top-chip with 256 100 μm-Cu-pillar + 2 μm-ENIG at its central portion and 176 9 μm-UBM/pad + 2 μm-ENIG around the peripherals has been properly fabricated.
- The RDL (daisy-chain) bottom-Chip with 432 4 μm-UBM/pad + 5 μm-Sn-solder has been properly fabricated.
- The RDL (daisy-chain) TSH interposer with 256 central holes (drilled by laser), 176 4 μm-UBM/pad + 5 μm-Sn-solder around the peripherals on its top side, and 176 9 μm-UBM/pad + 2 μm-ENIG around the peripherals on its bottom side has been properly fabricated.

- The final assembly of the top chip, bottom chip, TSH interposer, package substrate, and PCB has also been properly fabricated. These are evidenced by the SEM and X-ray images.
- The structural integrity of the SiP has been demonstrated by drop test. Based on the JEDEC specification, after 10 drops there is no failure.
- The thermal cycling test conditions are $-55°C \leftrightarrow 125°C$, and one hour cycle (15 min ramp up and ramp down and 15 min at dwells). The characteristic life (at 63.2% failures) is found to be 1175 cycles. For the given test conditions, if the characteristic life is larger than 1000 cycles, then it is considered acceptable.
- The chips (especially the bottom chip) should be thinned down to less than 200 μm. In that case, the cavity in the package substrate is not needed.

8.6 References

[1] Lau, J. H., *Through-Silicon Via for 3D Integration*, McGraw-Hill Book Company, New York, NY, 2013.
[2] Lau, J. H., *Reliability of ROHS-Compliant 2D and 3D IC Integration*, McGraw-Hill Book Company, New York, NY, 2011.
[3] Lau, J. H., C. K. Lee, C. S. Premachandran, and A. Yu, *Advanced MEMS Packaging*, McGraw-Hill Book Company, New York, NY, 2010.
[4] Banijamali, B., S. Ramalingam, K. Nagarajan, and R. Chaware, "Advanced Reliability Study of TSV Interposers and Interconnects for the 28nm Technology FPGA," *Proceedings of IEEE/ECTC*, Orlando, FL, June 2011, pp. 285–290.
[5] Chaware, R., K. Nagarajan, and S. Ramalingam, "Assembly and Reliability Challenges in 3D Integration of 28nm FPGA Die on a Large High Density 65nm Passive Interposer," *Proceedings of IEEE/ECTC*, San Diego, CA, May 2012, pp. 279–283.
[6] Banijamali, B., S. Ramalingam, H. Liu and M. Kim, "Outstanding and Innovative Reliability Study of 3D TSV Interposer and Fine Pitch Solder Micro-bumps," *Proceedings of IEEE/ECTC*, May 2012, pp. 309–314.
[7] Xie, J., H. Shi, Y. Li, Z. Li, A. Rahman, K. Chandrasekar, D. Ratakonda, et al., "Enabling the 2.5D Integration," *Proceedings of IMAPS International Symposium on Microelectronics*, San Diego, CA, September 2012, pp. 254–267.
[8] Li, Z., H. Shi, J. Xie, and A. Rahman, "Development of an Optimized Power Delivery System for 3D IC Integration with TSV Silicon Interposer," *Proceedings of IEEE/ECTC*, San Diego, CA, May 2012, pp. 678–682.
[9] Banijamali, B., C. Chiu, C. Hsieh, T. Lin, C. Hu, S. Hou, S. Ramalingam, et al., "Reliability Evaluation of a CoWoS-enabled 3D IC Package," *Proceedings of IEEE/ECTC*, May 2013, pp. 35–40.
[10] Lin, L., T.-C. Yeh, J.-L. Wu, G. Lu, T.-F. Tsai, L. Chen, and A.-T. Xu, "Reliability Characterization of Chip-on-Wafer-on-Substrate (CoWoS) 3D IC Integration Technology," *Proceedings of IEEE/ECTC*, Las Vegas, NV, May 2013, pp. 366–371.
[11] Chuang, Y.-L., C.-S. Yuan, J.-J. Chen, C.-F. Chen, C.-S. Yang, W.-P. Changchien, C. C.C. Liu, et al., "Unified Methodology for Heterogeneous Integration with CoWoS Technology," *Proceedings of IEEE/ECTC*, San Diego, CA, May 2013, pp. 852–859.
[12] Kwon, W., M. Kim, J. Chang, S. Ramalingam, L. Madden, G. Tsai, S. Tseng, et al., "Enabling a Manufacturable 3D Technologies and Ecosystem using 28nm FPGA with Stack Silicon Interconnect Technology," *IMAPS Proceedings of International Symposium on Microelectronics*, October 2013, pp. 217–222.
[13] Lau, J. H., S. Lee, M. Yuen, J. Wu, C. Lo, H. Fan, and H. Chen, "Apparatus having thermal-enhanced and cost-effective 3D IC integration structure with through silicon via interposer". *US Patent No: 8,604,603*, Date of Patent: December 10, 2013, Filed on February 20, 2009.
[14] Lau, J. H., Y. S. Chan, S. W. R. Lee, "Thermal-Enhanced and Cost-Effective 3D IC Integration with TSV Interposers for High-Performance Applications," *ASME Paper no. IMECE2010-40975.*
[15] Lau, J. H., Y. S. Chan, and R. S. W. Lee, "3D IC Integration with TSV Interposers for High-Performance Applications," *Chip Scale Review*, Vol. 14, No. 5, September/October2010, pp. 26–29.

[16] Lau, J. H., P. Tzeng, C. Lee, C. Zhan, M. Li, J. Cline, K. Saito, et al., "Redistribution Layers (RDLs) for 2.5D/3D IC Integration," *IMAPS Transactions, Journal of Microelectronic Packaging*, Vol. 11, No. 1, First Quarter 2014, pp. 16–24.

[17] Lau, J. H., P. Tzeng, C. Lee, C. Zhan, M. Li, J. Cline, K. Saito, et al., "(Redistribution Layers (RDLs) for 2.5D/3D IC Integration," *Proceedings of the 46th IMAPS International Symposium on Microelectronics*, Orlando, FL, October 2013, pp. 434–441.

[18] Wu, S. T., H. Chien, J. H. Lau, M. Li, J. Cline, and M. Ji, "Thermal and Mechanical Design and Analysis of 3D IC Interposer with Double-Sided Active Chips," *IEEE/ECTC Proceedings*, Las Vegas, NA, May 2013, pp. 1471–1479.

[19] Tzeng, P. J., J. H. Lau, C. Zhan, Y. Hsin, P. Chang, Y. Chang, J. Chen, et al., "Process Integration of 3D Si Interposer with Double-Sided Active Chip Attachments," *IEEE/ECTC Proceedings*, Las Vegas, NA, May 2013, pp. 86–93.

[20] Ji, M., M. Li, J. Cline, D. Seeker, K. Cai, J. H. Lau, P. Tzeng, et al., "3D Si Interposer Design and Electrical Performance Study," *Proceedings of DesignCon*, Santa Clara, CA, January 2013, pp. 1–23.

[21] Li, L., P. Su, J. Xue, M. Brillhart, J. H. Lau, P. Tzeng, C. Lee, et al., "Addressing Bandwidth Challenges in Next Generation High Performance Network Systems with 3D IC Integration," *IEEE ECTC Proceedings*, San Diego, CA, May 2012, pp. 1040–1046.

[22] Lau, J. H., P. Tzeng, C. Zhan, C. Lee, M. Dai, J. Chen, Y. Hsin, et al., " Large Size Silicon Interposer and 3D IC Integration for System-in-Packaging (SiP)," *Proceedings of the 45th IMAPS International Symposium on Microelectronics,* September 2012, pp. 1209–1214.

[23] Wu, S. T., J. H. Lau, H. Chien, Y. Chao, R. Tain, L. Li, P. Su, et al., "Thermal Stress and Creep Strain Analyses of a 3D IC Integration SiP with Passive Interposer for Network System Application," *Proceedings of the 45th IMAPS International Symposium on Microelectronics,* September 2012, pp. 1038–1045.

[24] Chien, H., J. H. Lau, T. Chao, M. Dai, and R. Tain, "Thermal Management of Moore's Law Chips on Both sides of an Interposer for 3D IC integration SiP," *IEEE ICEP Proceedings*, Japan, April 2012, pp. 38–44.

[25] Chien, H., J. H. Lau, T. Chao, M. Dai, R. Tain, L. Li, P. Su, et al., "Thermal Evaluation and Analyses of 3D IC Integration SiP with TSVs for Network System Applications," *IEEE/ECTC Proceedings*, San Diego, CA, May 2012, pp. 1866–1873.

[26] Lau, J. H., P.-J. Tzeng, C.-K. Lee, C.-J. Zhan, M.-J. Dai, L. Li, C.-T. Ko, et al., "Wafer Bumping and Characterizations of Fine-Pitch Lead-Free Solder Microbumps on 12" (300mm) wafer for 3D IC Integration," *Proceedings of IMAPS International Conference*, Long Beach, CA, October 2011, pp. 650–656.

[27] Tsai, I., E. Wu, S. F. Yen, and T. H. Chuang, "Mechanical Properties of Intermetallic Compounds on Lead-Free Solder by Moire' Techniques," *Journal of Electronic Materials*, Vol. 35, 2006, pp.1566–1570.

[28] Kim, S., and S. Kang, "Sensitivity of Electroplating Conditions on Young's Modulus of Thin Film," *Journal of Applied Physics*, Vol. 47, No. 9, 2008, pp. 7314–7316.

[29] Zhang, X., J. H. Lau, C. S. Premachandran, S. Chong, L. Wai, V. Lee, T. C. Chai, et al., "Development of a Cu/Low-*k* Stack Die Fine Pitch Ball Grid Array (FBGA) Package for System in Package Applications," *IEEE Transactions on CPMT*, Vol. 1, no. 3, March 2011, pp. 299–309.

[30] Hsin, Y. C., C. Chen, J. H. Lau, P. Tzeng, S. Shen, Y. Hsu, S. Chen, et al., "Effects of Etch Rate on Scallop of Through-Silicon Vias (TSVs) in 200mm and 300mm Wafers," *IEEE ECTC Proceedings*, Orlando, Florida, June 2011, pp. 1130–1135.

[31] Wu, C., S. Chen, P. Tzeng, J. H. Lau, Y. Hsu, J. Chen, Y. Hsin, et al., "Oxide Liner, Barrier and Seed Layers, and Cu-Plating of Blind Through Silicon Vias (TSVs) on 300mm Wafers for 3D IC Integration," *IMAPS Transactions, Journal of Microelectronic Packaging*, Vol. 9, No. 1, First Quarter 2012, pp. 31–36.

[32] Chen, J. C., J. H. Lau, P. J. Tzeng, S. Chen, C. Wu, C. Chen, H. Yu, et al., "Effects of Slurry in Cu Chemical Mechanical Polishing (CMP) of TSVs for 3-D IC Integration," *IEEE Transactions on CPMT*, Vol. 2, No. 6, June 2012, pp. 956–963.

[33] Lau, J. H., C. Lee, C. Zhan, S. Wu, Y. Chao, M. Dai, R. Tain, et al., "Through-Silicon Hole Interposers for 3D IC Integration," *IEEE Transactions on CPMT*, Vol. 4, no. 9, September 2014, pp. 1407–1418.

[34] Wu, S., J. H. Lau, H. Chien, J. Hung, M. Dai, Y. Chao, R. Tain, et al., "Ultra Low-Cost Through-Silicon Holes (TSHs) Interposers for 3D IC Integration SiPs," *IEEE ECTC Proceedings*, San Diego, CA, May 2012, pp. 1618–1624.

[35] Wu, S. H., J. H. Lau, H.-C. Chien, R.-M. Tain, M.-J. Dai, and Y.-L. Chao, "Chip Stacking Structure and Fabricating Method of the Chip Stacking Structure," *US Patent No.: 8,519,524*, Date of Patent: August 27, 2013, Filed on August 16, 2012.

[36] JESD22-B111, *Board Level Drop Test Method of Components for Handheld Electronic Products*, JEDEC Standard, July 2003.

CHAPTER 9

Thermal Management of 2.5D/3D IC Integration

9.1 Introduction

One of the critical issues of 2.5D/3D IC integration is its thermal management. This is because (1) 3D circuits increase total power generated per unit surface area; (2) chips in the 3D stack may overheat if proper and adequate cooling is not provided; (3) the space between the 3D stack may be too small for cooling channels (i.e., no gap for air flow); and (4) thin chips may create extreme conditions for on chip hot spots. Thus, low-cost and effective thermal management solutions are desperately needed for widespread use of 2.5D/3D IC integration and are therefore also the focus of this chapter. Emphasis is placed on the design of a low-cost (with bare chips) and high thermal performance 2.5D/3D IC integration SiP. Some special cases of this new design will be presented. Furthermore, the thermal performance between a 2.5D integration and 3D IC integration will be discussed. Finally, an integrated liquid cooling system for 3D stacked TSV modules will be presented.

9.2 Design Philosophy

The objective is to design a generic thermal-enhanced, electrical-sound, mechanical-integrity, and cost-effective 2.5D/3D IC integration system for SiP applications. In other words, this design addresses the electronic packaging of 2.5D/3D IC integration with a TSV/RDL/IPD (through-silicon via/redistribution layer/integrated passive device) interposer for high-power, high-performance, high pin-count, ultra fine-pitch, and small real-estate applications. In order to achieve these goals, this design utilizes chip-to-chip interconnections through a TSV interposer in a 3D SiP format with excellent thermal management. Avoid to "dig" holes (TSVs) on the active-device chips, i.e., use whatever Moore's chips are made by the Foundries, and use a RDL/TSV interposer to perform the redistributions (to fan out the high pin-out and ultra fine-pitch circuitries of the chips) and most of the tasks, and then connect to the next level of interconnects (e.g., the package substrate) with much less pin-outs and coarse pitches.

9.3 The New Design

The proposed design is a 2.5D/3D IC integration system with high-power and low-power chips supported by a TSV/RDL/IPD interposer in a standard plastic ball grid array (PBGA) package for high-performance, low-cost, effective thermal management, and high-reliability applications, as shown in Fig. 9.1 [1–3]. The entire system consists of a silicon interposer with high-density TSVs, RDLs to connect various chips with pads that have different pitches, sizes, locations, and IPD to enhance electrical performance. A simple organic substrate with standard (in size and pitch) solder balls for printed circuit board (PCB) assembly supports the RDL/TSV interposer. All the high-power chips such as the MPU (microprocessor unit), GPU (graph-processor unit), ASIC (application specific IC), DSP (digital signal processor), MCU (microcontroller unit), RF (radio frequency), and high-power memory chips are on top of the RDL/TSV interposer in a flip chip format so that the back side of these chips can be attached to a heat spreader via a thermal interface material (TIM). In this case, most of the heat from all the high-power chips can be dissipated through the heat spreader (with a heat sink if it is necessary). All the low-power chips such as the MEMS (micro-electromechanical system), MOEMS (micro-optoelectromechanical system), CMOS image sensors, and memory chips are at the bottom side of the RDL/TSV interposer with either flip chip or wirebond formats.

A ring-stiffener connecting the organic substrate and the heat spreader is used to provide the adequate standoff for the 2.5D/3D IC integration with the RDL/TSV interposer and to support the heat spreader with or without the heat sink. Underfill encapsulants are, most likely, needed between the RDL/TSV interposer and the high- and low-power flip chips, and between the RDL/TSV interposer and the organic substrate. However, underfill is not needed between the 2.5D/3D IC integration module and the PCB. (This is a very important

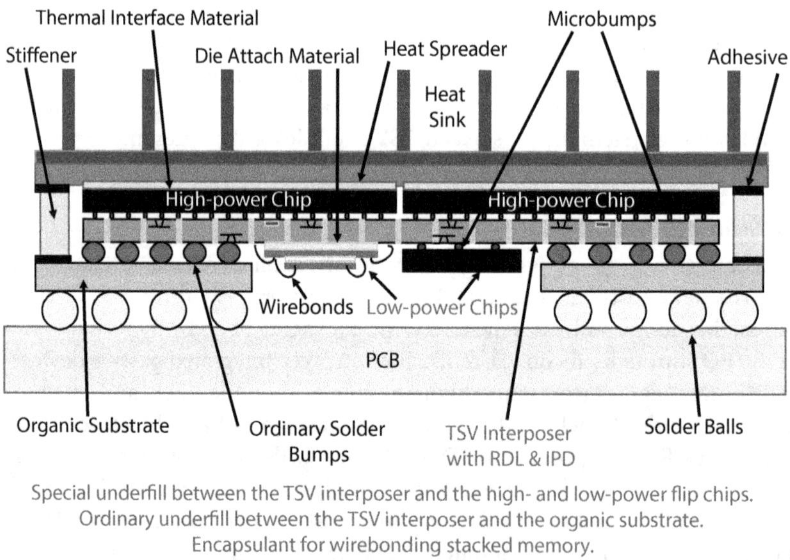

Figure 9.1 Apparatus having thermal-enhanced and cost-effective 3D IC integration structure with through-silicon via interposer. U.S. Patent No. 8,604,603; date of patent: December 10, 2013; filed on February 20, 2009.

feature for system houses who don't like to have underfills on their PCBs.) For wirebonding chips, encapsulants may be needed.

The appearance of the proposed 2.5D/3D IC integration SiP is very attractive to, e.g., IDM (integrated device manufacturers), OEM (original equipment manufacturing), and EMS (electronics manufacturing services). First of all, it is a standard PBGA package and has been used by the electronic industry for more than 20 years [4, 5]. This package is not only effective in thermal management [6–12] but its solder joints are very reliable [13–16]. Therefore, in conjunction with the proper design of the high-power and low-power chips above/below the RDL/TSV interposer inside the package, a SiP that is both cost-effective and displays high electrical and thermal performance can be achieved and manufactured.

9.4 Equivalent Model for Thermal Analysis

The equivalent Eqs. (2.19) and (2.20) in Chap. 2 can be used for the design and analysis of 2.5D/3D IC SiP by establishing an equivalent model. This equivalent model is used to replace the real detail TSV model for simplifying the thermal simulations. Figure 9.2 shows how a detail TSV model, Fig. 9.2(a), can be converted into an equivalent model, Fig. 9.2(b). The equivalent model regards the TSV group as an equivalent zone (block) which is represented with the calculated equivalent k_{xy} (2.19) and k_z (2.20). The solder balls (or bumps) and the traces are converted to other equivalent zones by using the common thermal resistance (series/parallel) calculations [17–21].

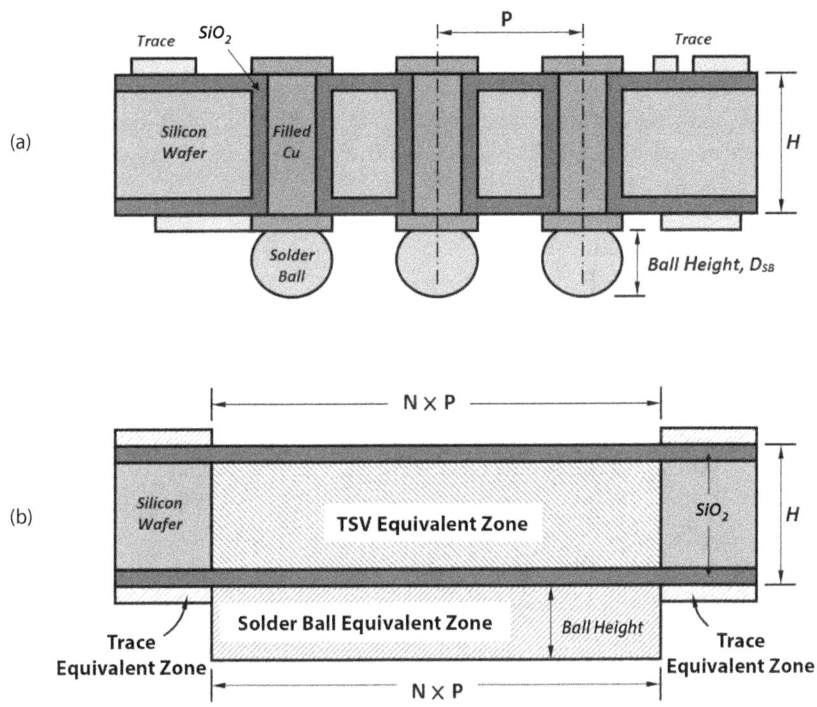

Figure 9.2 The conversion from (a) a detail model to (b) a simple equivalent model.

9.5 Interposer with Chip/Heat Spreader on Its Top Side and Chip on Its Bottom Side

9.5.1 The Structure

Figure 9.3 shows a TSV interposer with chips on both its sides. This is one of the degenerations of the new generic design (Fig. 9.1), i.e., there is no memory chip set and no cavity of the organic substrate. The dimension of the structure, material properties, and thermal boundary condition are shown, respectively, in Tables 9.1 and 9.2, and Fig. 9.4. It can be seen from Table 9.1 that the dimensions of the chips are 20 × 20 mm for chip #2 and 12 × 12 mm for chip #1. The dimensions of the TSV interposer are 30 mm × 30 mm × 100 μm. There is a heat spreader at the backside of chip #1.

9.5.2 Material Properties

For grease, silicon, and cooper, the thermal conductivity in cross-plane direction (k_z) and in-plane direction (k_{xy}) are equal due to their homogeneous properties (Table 9.2). For the bismaleimide triazene (BT) substrate and PCB, they are assumed as laminated materials in which there are several high conductive (Cu) layers inside the lamination, so their k_{xy} is obviously higher than k_z, as shown in Table 9.2. As for the interposer, microbump, ordinary bump, and solder ball, the anisotropic thermal conductivities could be calculated by using Eqs. (2.19) and (2.20) for interposer zone and using resistance series/parallel correlation for the zones of microbump, ordinary bump, and solder ball. The equivalent k_{xy} and k_z of microbump zone are nearly the same and equal to the value of underfill material ($k = 0.63$ W/m·K) because their pitch is too large and size is too small.

9.5.3 Boundary Conditions

The ambient temperature is assumed to be 20°C. The boundary condition on the top side and bottom side of the PCB is $h = 10$ W/m²K, which is to imitate a natural convection

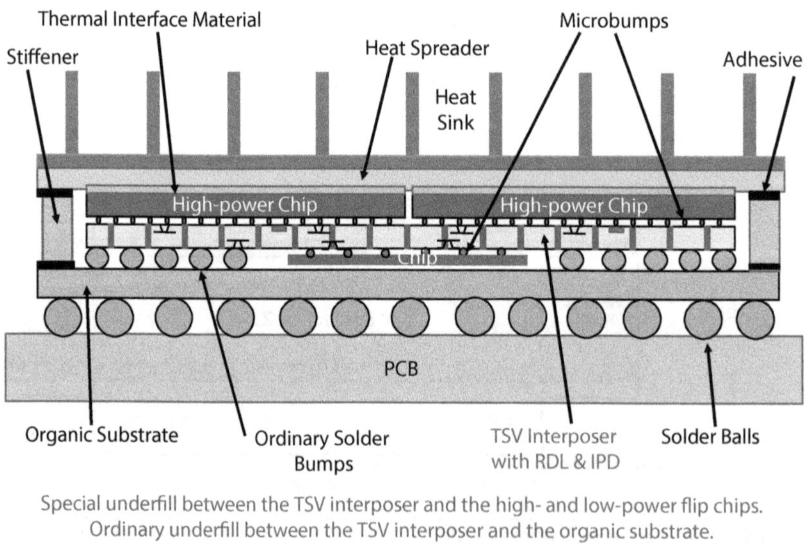

FIGURE 9.3 TSV interposer with chips on its top side and bottom side.

condition. For the heat spreader, we use the thermal resistance, R_{ca}, in place of cooling module for simplifying the simulations. The R_{ca} is assumed to be 0.3°C/W to imitate the heat spreader cooled by a pretty good heat sink with fan or a liquid cooling module. The power generation of chip #1 is 75 W and that of chip #2 is 15 W. The total power consumption of the SiP is 90 W. Icepak 12.1.6 is the simulation tool to solve the heat conduction problems using finite volume method.

Table 9.1 Dimensions of the Analyzed Structure

Chips and Substrates					
	Quantity	**Thickness (μm)**	**Length (mm)**	**Width (mm)**	**Power Consumption (W)**
Chip #1	1	150	12	12	75
Chip #2	1	150	20	20	15
Interposer	1	100	30	30	0
BT substrate	1	1000	40	40	0
PCB	1	2000	60	60	0

TSVs, Bumps, and Solder Balls				
	Diameter (μm)	**Pitch (μm)**	**Distribution**	**Notes**
TSV	20	100	Uniform	In interposer
Microbump	12	250	Uniform	Between chip #1 and interposer
Microbump	12	250	Uniform	Between chip #2 and interposer
Ordinary bump	200	750	Surround interposer edges	Between interposer and BT substrate
Solder ball	600	2000	Uniform	Between BT substrate and PCB

Table 9.2 Thermal Material Properties of the Analyzed Structure

	K_{xy} (W/m·K)	K_z	Thickness (μm)	Notes
Grease	1	1	50	Between chip #1 & heat spreader
Silicon	148	148	–	
Cooper	401	401	–	
Interposer	141.4	156.3	100	Calculated equi. properties
BT substrate	40.9	0.38	1000	
PCB	27.4	0.35	2000	
Microbump	0.63	0.63	12	Calculated equi. properties
Ordinary bump	0.63	2.79	200	Calculated equi. properties
Solder ball	0.05	3.53	600	Calculated equi. properties

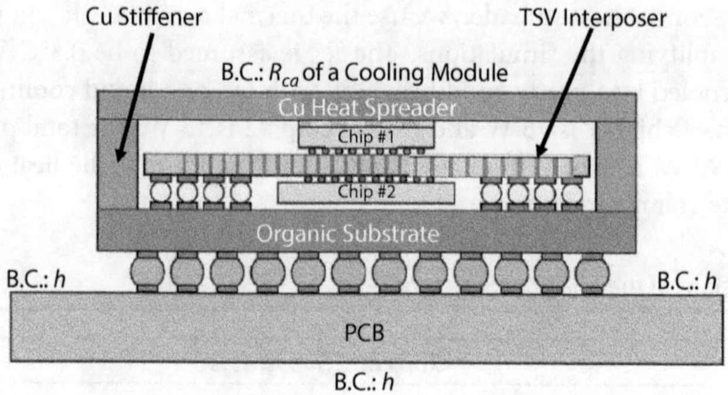

FIGURE 9.4 TSV interposer with chips on its top side and bottom side: boundary conditions.

9.5.4 Simulation Results

The SiP is under the boundary conditions of $R_{ca} = 0.3°C/W$ for heat spreader top side and $h = 10$ W/m·K for PCB both sides. The power consumption of each chip is shown in Table 9.1. Figure 9.5 shows the cross-section temperature distribution of the analyzed SiP. The average temperatures of chips #1 and #2 are 109°C and 121°C, respectively, under ambient temperature 20°C, as shown in Fig. 9.6. It is noted that even the chip #2 has a lower heat generation; however, its temperature is still higher than that of the chip #1. This is because chip #1 has a heat spreader at its back to conduct the heat out. Thus, a way to lower the temperature in chip #2 is necessary.

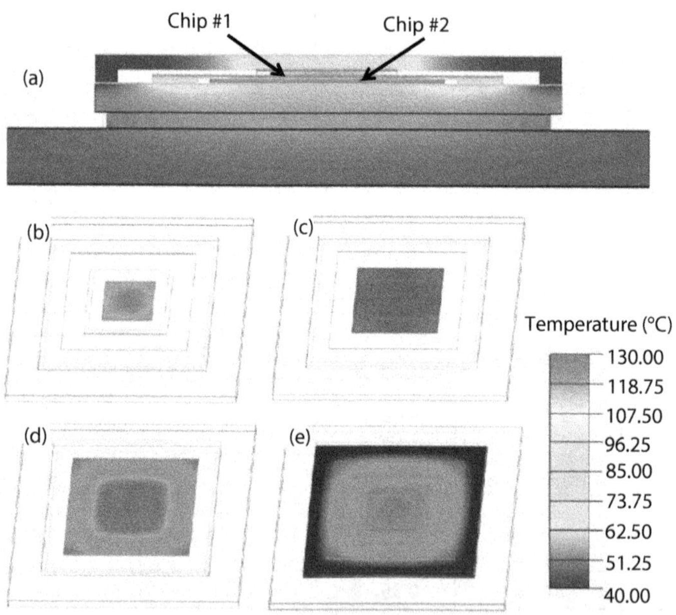

FIGURE 9.5 Temperature contour distribution of (a) cross-section of the analyzed structure, (b) chip #1, (c) chip #2, (d) interposer, and (e) the organic substrate.

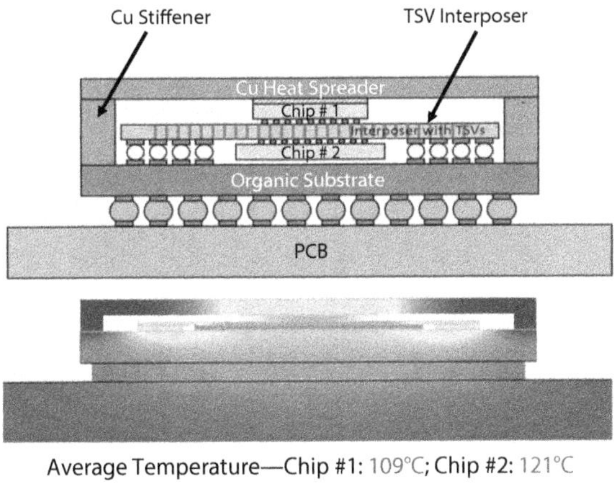

FIGURE 9.6 Average temperature of chip #1 (109°C) and chip #2 (121°C).

9.6 Interposer with Chip/Heat Spreader on Its Top Side and Chip/Heat Slug on Its Bottom Side

9.6.1 The Structure and Boundary Conditions

Figure 9.7 shows the bottom chip of Fig. 9.3 is attached to a Cu heat slug with TIM, while Fig. 9.8 shows the boundary condition for analysis. All the dimensions of the chips and interposer, and the boundary conditions are the same with those in Sec. 9.5 except there is a cavity of the PCB [20].

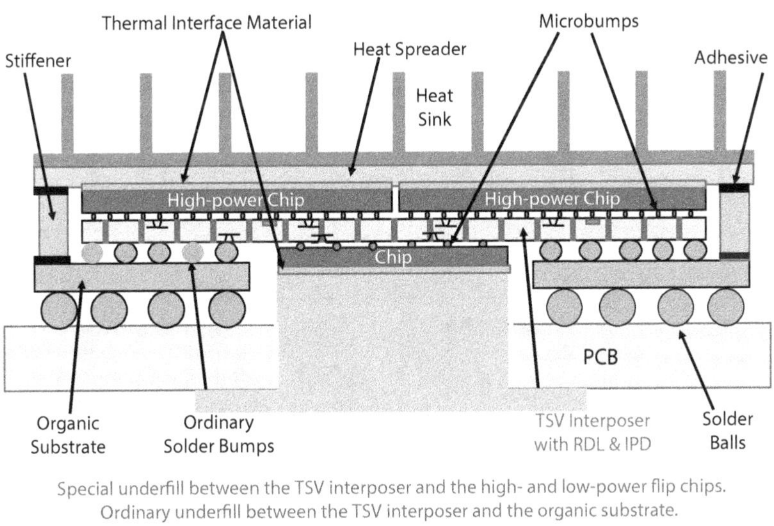

FIGURE 9.7 The bottom chip of Fig. 9.3 is attached to a heat slug.

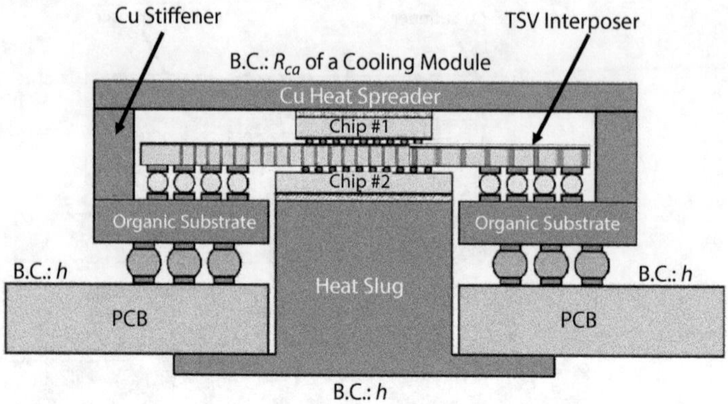

Figure 9.8 The bottom chip of Fig. 9.4 is attached to a heat slug: boundary conditions.

9.6.2 Simulation Results

Figure 9.9 shows the simulation results. It can be seen that chip #2 temperature is reduced from original (Sec. 9.5) 121°C to 98.0°C. Also, the temperature of chip #1 reduces from original 109°C to 102°C. This result shows that the chip at the bottom of an interposer can be cooled down effectively by using such a simple heat slug structure, as shown in Figs. 9.7 and 9.8.

To alter the boundary conditions applied on the heat slug, different temperatures of chip #2 are obtained: (1) 77°C for $h = 156.25$ W/m^{2}K, which imitates the slug is attached on a heat sink without fan and whose thermal resistance from slug bottom to ambient is 4.0°C/W; (2) 52.2°C for $h = 625$ W/m^{2}K, which imitates the slug is attached on a heat sink with working an whose thermal resistance is 1.0°C/W from slug bottom to ambient; and (3) 34°C for $h = 3125$ W/m^{2}K, which simulates the slug is cooled by a liquid cooling module whose thermal resistance is 0.2°C/W. Comparing these results with various boundary conditions to the

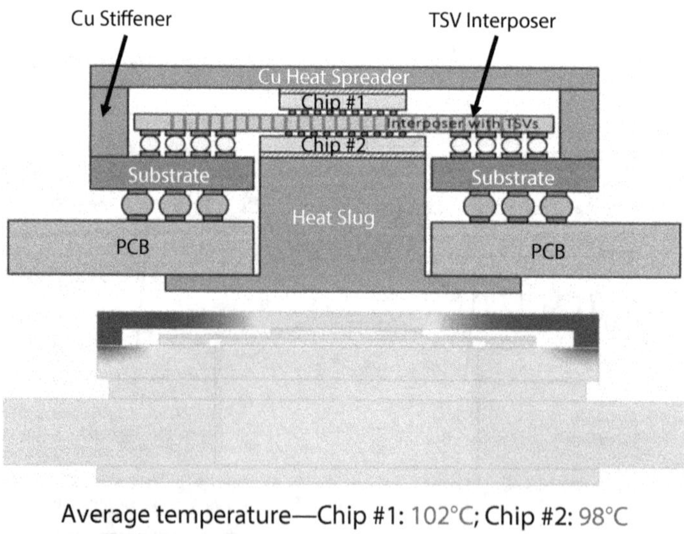

Figure 9.9 Average temperature of chip #1 (102°C) and chip #2 (98°C) (with heat slug at the bottom of chip #2).

heat slug under a natural convection condition, it is determined that by attaching any cooling module on the heat slug, the chip underneath the interposer can be chilled more effectively.

9.7 Interposer with Four Chips on Its Top Side with Heat Spreader

9.7.1 The Structure

Figure 9.10 is the simplest degeneration from Fig. 9.1, where there is no chip in the bottom-side of the interposer and there is no cavity of the organic substrate. The objective of this section is to show (via simulation) that for this system a heat spreader is necessary. Figure 9.11 shows the systems for analysis (a) with heat spreaders and (b) without heat spreader. There are two heat spreaders in Fig. 9.11(a), the smaller heat spreader is attached to the back side of chips with TIM and has the same horizontal dimensions as the TSV interposer. The larger heat spreader is attached to the smaller heat spreader and stiffener with adhesive.

9.7.2 Boundary Conditions

Table 9.3 lists the boundary and power conditions. There are four chips and the power consumption is 5 W for each chip. The boundary conditions on all the surfaces of PCB and organic package substrate are $h = 20\ \text{W/m}^2\text{K}$ (which means the substrate and PCB are cooled by a typical natural convection). As for the boundary condition applied to the top surfaces of the larger heat spreader, Fig. 9.11(a), and the chips, Fig. 9.11(b), the convection coefficient, h, is tunable for parametric study. The modulated coefficient h is depending on the thermal resistance R_{ca} that is changed from 0.2 to 4.0°C/W; the h can be calculated by the formula $h = 1/R_{ca} \cdot A$; here, A denotes the heat spreader area (or chip area). Typically, the cooling ability of liquid cooling is for $R_{ca} < 0.3$°C/W. For active heat sinks: $0.3 < R_{ca} < 1.0$, and for passive heat sinks: $1.0 < R_{ca}$.

Table 9.4 lists the dimensions of the systems for analyses. In performing the simulations, the detailed systems are replaced by equivalent models for simplification. The conversion from a detailed TSV system to an equivalent model has been described and demonstrated

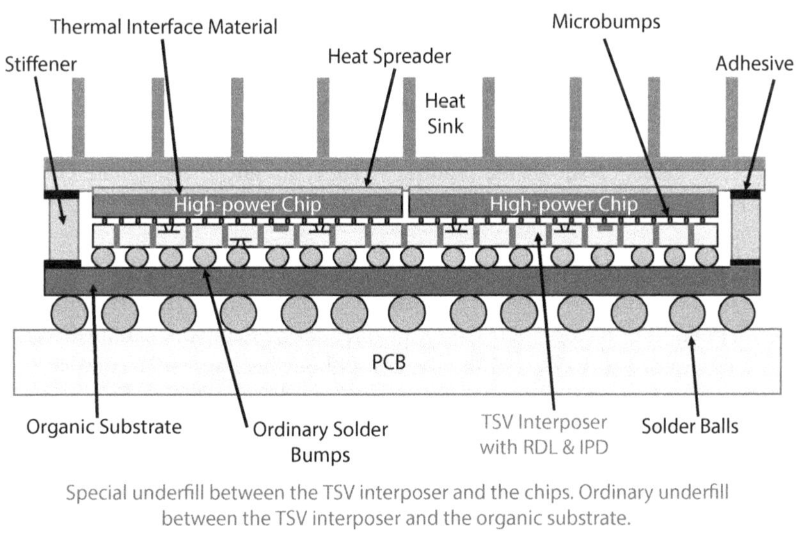

Figure 9.10 TSV interposer with chips on its top side only.

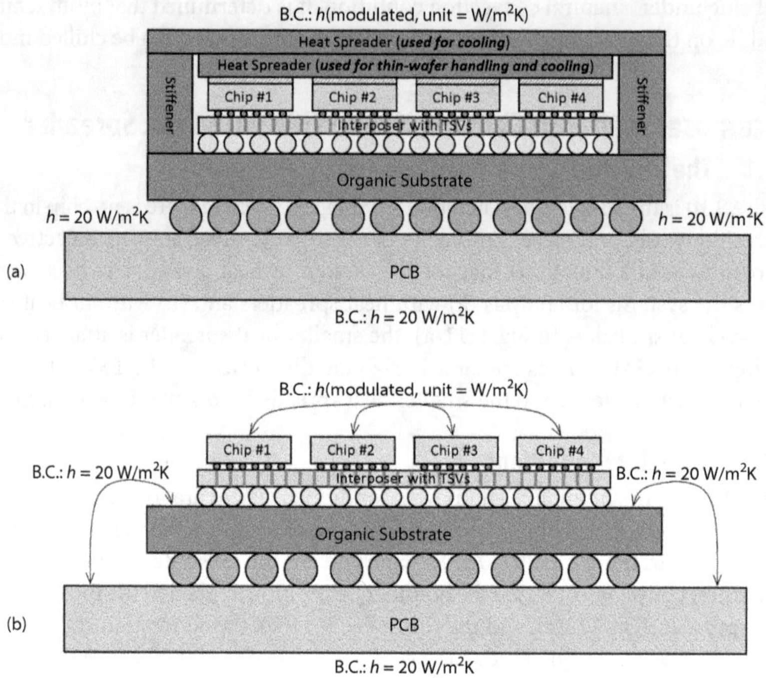

Figure 9.11 TSV interposer with chips on its top side only: boundary conditions.

Table 9.3 Power Loading and Boundary Conditions

Power load condition	5 W per chip (#1–4, total 20 W)
Boundary conditions	The thermal resistance (R_{CA}) of a heat sink: tunable for parameter study The convection coefficient $h = 20$ W/m^2·K on the both sides of PCB and organic substrate Other surfaces are regarded as adiabatic

in Sec. 9.5. Table 9.5 shows the materials properties and the calculated equivalent thermal properties of the TSV/solder bumps/solder ball structure.

9.7.3 Simulation Results

Figure 9.12 shows the temperature distributions of the two analyzed systems with the applied boundary condition ($R_{ca} = 2.0$°C/W) on spreader and chips. It can be seen that the average chip temperature (~105°C) of the system without heat spreader is much higher than that (~57°C) of the system with heat spreader. Thus, for this kind of structure and boundary condition, a heat spreader is usually needed.

Figure 9.13 shows the correlation between chip temperatures and the applied R_{ca}. Again, the average chip temperature of the system without the heat spreader is nearly 2.5 times higher than that in the system with heat spreaders. Besides, from Fig. 9.13, if the chip temperatures have to be cooled under 75°C (at ambient temperature of 20°C), the system without heat spreader

Table 9.4 Thermal Conductivity of the System

	K_{xy} (W/m·K)	k_z	Thickness (μm)	Notes
Grease	2	2	50	Between chips & heat spreader
Grease	2	2	50	Between the 2 heat spreaders
Interposer	147.27	141.46	–	Calculated equi. properties
Heat spreader	401	401	200	
BT substrate	40.9	0.38	800	
PCB	27.4	0.35	2000	
Microbump	0.63	4.36	15	Calculated equi. properties
Ordinary bump	0.63	27.3	150	Calculated equi. properties
Solder ball	0.05	14.14	600	Calculated equi. properties

Table 9.5 Dimensions of the System for Analyses

	Length	Width	Height	Diameter	Note
Chip #1–4	12 mm	7 mm	500 μm	—	The four chips are in the same sizes
Interposer	25 mm	31 mm	100 μm	—	
Organic substrate	35 mm	35 mm	800 μm	—	
PCB	50 mm	50 mm	2000 μm	—	
Heat spreader (for thin-wafer handling and cooling)	25 mm	31 mm	700 μm	—	
Heat spreader (for cooling)	35 mm	35 mm	300 μm	—	
TSV	Pitch = 250 μm			10 μm	
Microbump	Pitch = 45 μm			15 μm	Between chip and interposer
Ordinary bump	Pitch = 180 μm			150 μm	Between interposer and organic substrate
Solder ball	Pitch = 1000 μm			600 μm	Between organic substrate and PCB

should use an active heat sink (usually with fan), having a cooling ability of $R_{ca} < 1.2$°C/W. On the other hand, the system with heat spreaders should just use a passive heat sink (a small heat sink cooled by natural convection), having a cooling ability of $R_{ca} < 3.3$°C/W.

9.7.4 Summary and Recommendations

Some important results and recommendations are summarized as follows [17–21]:

- The structures shown in Figs. 9.10 and 9.11 are the simplest degeneration from Fig. 9.1, where there is no chip in the bottom side of the interposer and there is no cavity of the organic package substrate.

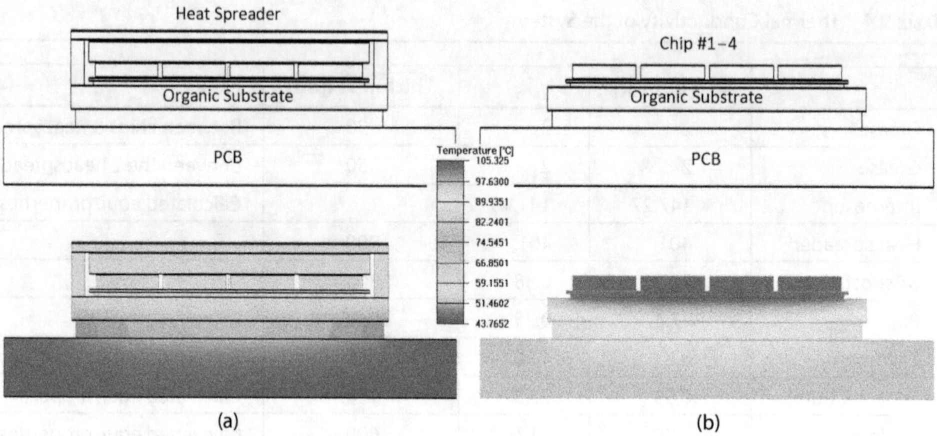

FIGURE 9.12 Temperature contour distributions of the TSV interposer with chips on its top side: (a) with heat spreader and (b) without heat spreader.

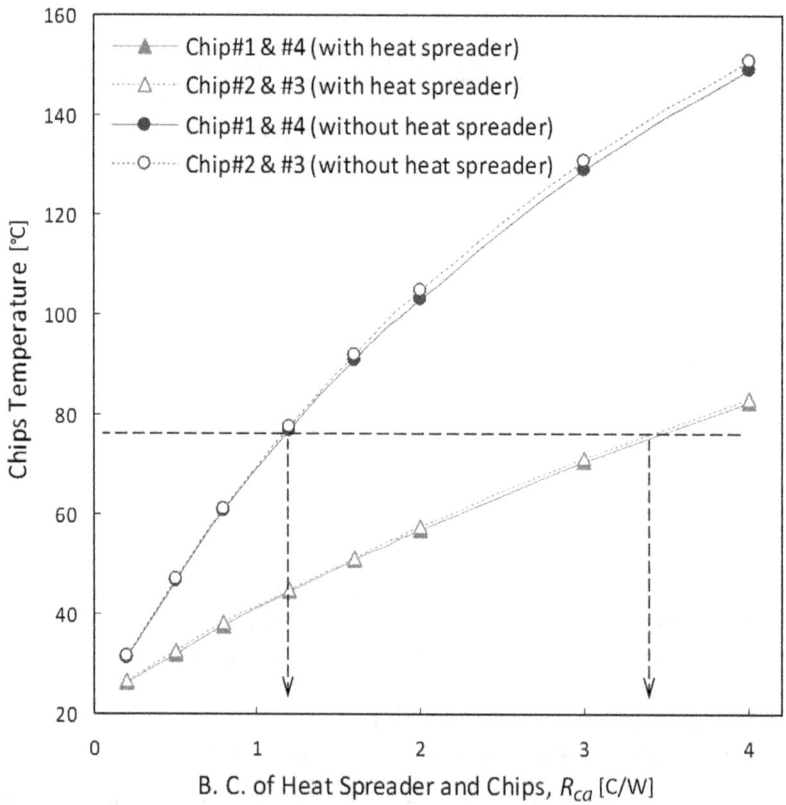

FIGURE 9.13 Average chip temperature vs. thermal resistance of heat spreader.

- The chips underneath an interposer have a major and severe thermal issue. They have quite a high and irreducible temperature because of temperature superposition effect and having a bad heat dissipation path from the chips to ambient via PCB. It has been found that only <10% of heat generated by the chips under an interposer could be dissipated via this path.
- A novel and simple design using a heat slug to resolve the major thermal issue has been provided. The slug having a high thermal conductivity penetrates the PCB through the BT substrate and then directly contacts the chips underneath an interposer. It is found that the slug has a pretty good effect for cooling the chips under the interposer as well as the chips above the interposer.
- Collaborate with an original cooling module attached on the heat spreader, the heat slug design with another powerful cooling module can chill all the chips inside the SiP package very effectively. In other words, the novel cooling design could help such a SiP package to have a much higher power consumption, which means to design a higher chip density or a more compact package or a higher system performance of a 3D IC SiP becomes possible.

9.8 Thermal Performance between 2.5D and 3D IC Integrations

The thermal performance between a 3D IC integration and a 2.5D IC integration (interposer) for low-power applications (overmolded in a package) and high-power applications (lid and heat sink) has been given in [22].

9.8.1 The Structures

The structures under consideration are shown in Figs. 9.14 and 9.15 [22]. It can be seen that the 3D stacking consists of the serdes chip (Die 2) which is on top of the logic chip (Die 1)

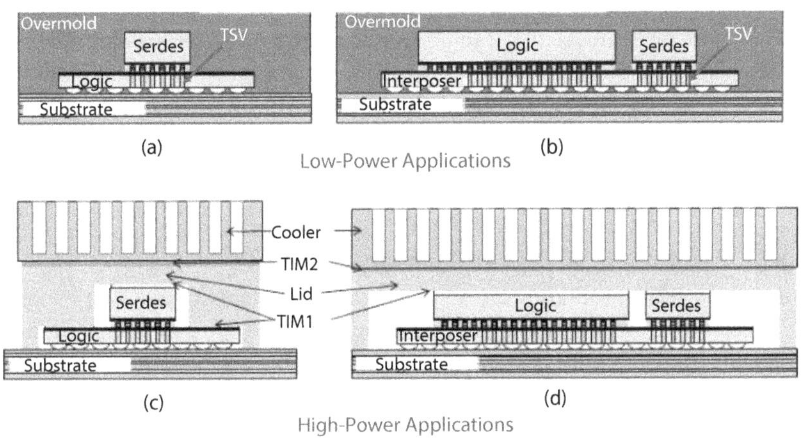

Figure 9.14 Structures for thermal performance comparison between 3D IC integration (stacking) (a) and (c) and 2.5D IC integration (Si interposer) (b) and (d). (a) and (b) are for low-power applications. (c) and (d) are for high-power applications.

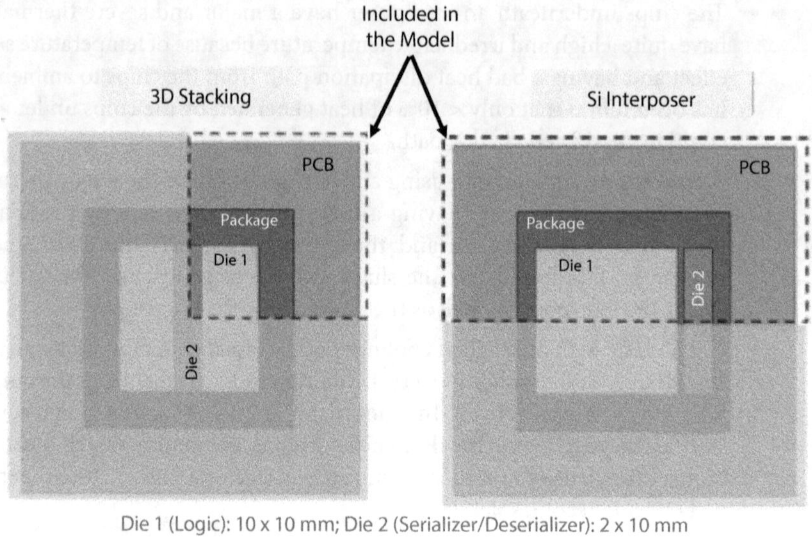

FIGURE 9.15 Double symmetry (1/4 of the structure) is used of the 3D stacking (left) and symmetry (1/2 of the structure) is used of Si interposer (right).

with TSVs. On the other hand, the Si interposer (2.5D) consists of the serdes chip and the logic chip which are TSV-less and side-by-side on a TSV interposer. Both structures are on a package substrate and on a PCB. The dimensions of the serdes chip and logic chip are, respectively, 2 × 10 mm and 10 × 10 mm as shown in Table 9.6, where other key elements' dimensions are also provided.

9.8.2 The Finite Element Models

Figures 9.16 and 9.17 show the finite element models for the thermal performance analyses [22]. Figure 9.16 shows the model for the Si interposer structure. Due to symmetry only half of the structure is modeled. It can be seen from Fig. 9.16 that the chips, laminate substrate, key elements from BEOL, solder bumps, PCB, microbumps, Cu-pillar, underfills, and solder balls are modeled. Figure 9.17 shows the model for the 3D stacking structure. Due to double symmetries, only one-fourth of the structure is modeled. It can be seen from Fig. 9.17 that the chip, underfills, Cu-pillar, package substrate, solder bumps, solder balls, PCB, silicon, and IMC are modeled.

9.8.3 Material Properties and Boundary Conditions

All the materials properties are assumed to be constants (Table 9.6) except the Si (chips and interposer), which is temperature dependent [23]. The thermal conductivity of Si (k) is equal to $k_o(T[K]/300K)^{-1.65}$, where $k_o = 148$ W/m·K is the thermal conductivity at room temperature.

The ambient temperature is considered 25°C. The uniform power dissipation is assumed as a heat flux in the active regions of the logic chip and serdes chip. In the low-power applications, a convective boundary condition is applied to the top and bottom side of the PCB and to the sides and top of the package using a heat transfer coefficient of 8 W/m^{2}K. A power

TABLE 9.6 Thermal Material Properties of the Analyzed Structures

Components	Parameters	3D stacking	Si interposer
Die 1 (logic die)	• Size • Thickness • Orientation	10 × 10 mm² 25 μm Face up	10 × 10 mm² 500 μm Face down
Die 2 (sercles die)	• Size • Thickness • Orientation	10 × 2 mm² 300 μm Face down	10 × 2 mm² 300 μm Face down
Package laminate	• Size • Thickness • k_{xy} / k_z	15 × 15 mm² 0.5 mm 15 / 2 W/mK	15 × 15 mm² 0.5 μm 15 / 2 W/mK
PCB	• Size • Thickness • k_{xy} / k_z	20 × 20 cm² 1.6 mm 16 / 0.67 W/mK	20 × 20 cm² 1.6 mm 16 / 0.67 W/mK
Cooling assumptions	• Convection PCB • Heat sink top 1. Low power 2. High power	*htc* 8 W/m²K *htc* 8 W/m²K R_{th} 0.25 K/W	*htc* 8 W / m²K *htc* 8 W/m²K R_{th} 0.25 K/W

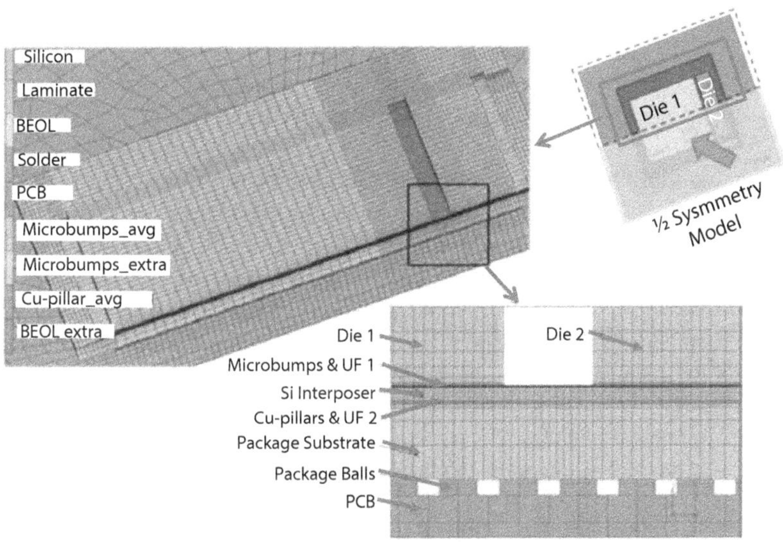

FIGURE 9.16 Finite element model of half of the Si interposer structure. The molding compound is not shown for visibility purposes.

dissipation of 2 W and 200 mW is assumed for the logic and serdes, respectively. In the case of the high-power applications the convective boundary condition applied to the lid represents the TIM-2 and the high performance heat sink. An equivalent convective heat transfer coefficient of 10,000 W/m²K is used for this boundary condition. This corresponds to a thermal resistance 0.4 K/W for the combination of the TIM-2 and the high performance heat

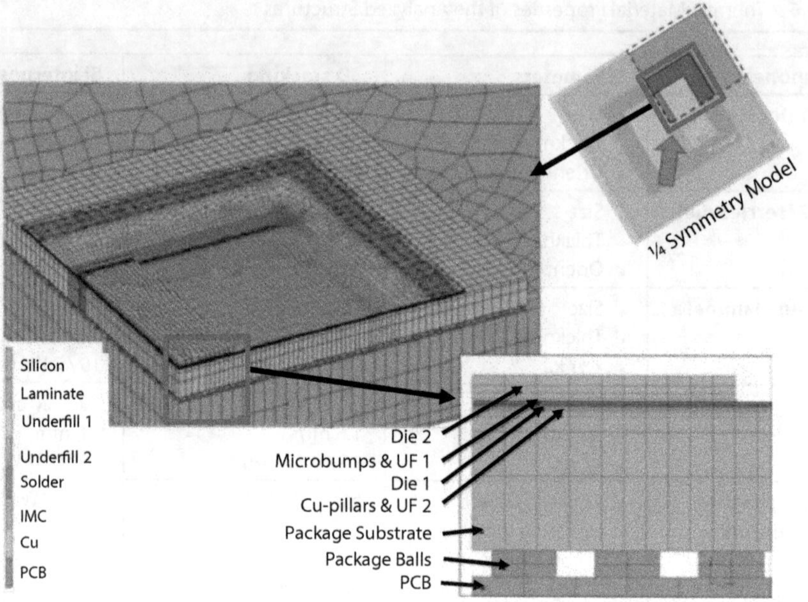

FIGURE 9.17 Finite element model of quarter of the 3D stacking structure. The molding compound is not shown for visibility purposes.

sink (0.25 K/W for the heat sink only) as shown in Table 9.6. A power dissipation of 50 W and 4 W is assumed for the logic and serdes, respectively.

9.8.4 Simulation Results—Low-Power Applications

Figure 9.18 shows the temperature contours in the 3D stacking structure (top) and Si interposer structure (bottom). These are due to the low-power dissipation of 2 W and 200 mW for the logic and serdes, respectively. The ambient temperature is assumed to be 25°C. It can be seen that (a) the maximum temperature in the logic (T_L) of the Si interposer structure (90°C) is smaller than that (109.6°C) in the 3D stacking structure, and (b) the maximum temperature in the serdes (T_S) of the Si interposer structure (88.9°C) is smaller than that (110°C) in the 3D stacking structure. This is because around 90% of the heat generated in the chips is removed by the PCB (i.e., the main heat path is downward from the chips to the PCB [23]) and the Si interposer structure has larger conduction area with the PCB.

9.8.5 Simulation Results—High-Power Applications

Figure 9.19 shows the temperature contours in the 3D stacking structure (top) and Si interposer structure (bottom). These are due to the high-power dissipation of 50 W and 4 W for the logic and serdes, respectively. The ambient temperature is assumed to be 25°C. It can be seen that (a) the maximum temperature in the logic (T_L) of the Si interposer structure (85.7°C) is smaller than that (106.1°C) in the 3D stacking structure, and (b) the maximum temperature in the serdes (T_S) of the Si interposer structure (81.1°C) is smaller than that (86.4°C) in the 3D stacking structure. This is because around 95% of the heat generated in the chips is removed by the lid, TIMs, and the heat sink (i.e., the main heat path is upward

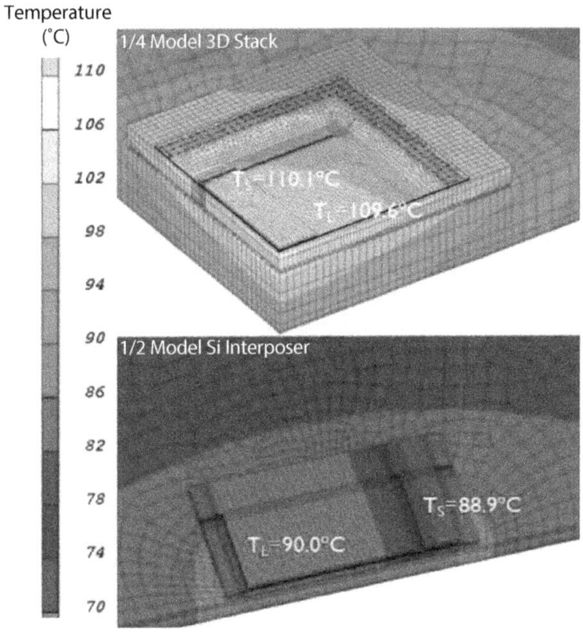

FIGURE 9.18 Temperature contour distributions (°C) in 3D stacking (top) and Si interposer (bottom) for the low-power case: 2 W for the logic chip and 200 mW for the serdes chip.

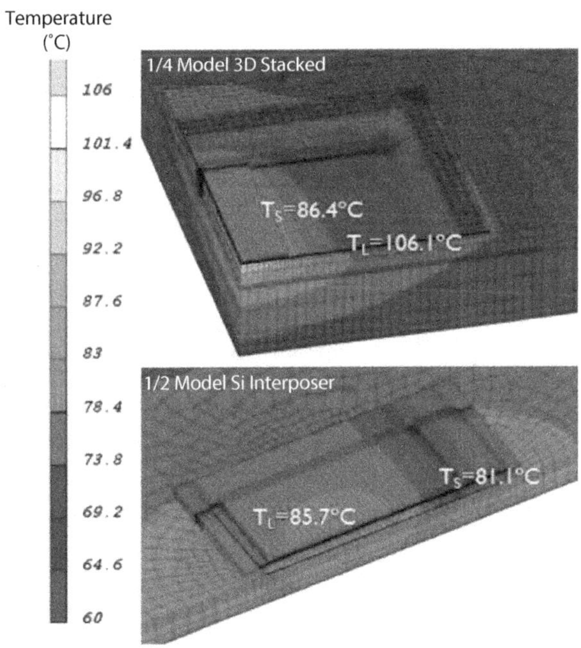

FIGURE 9.19 Temperature contour distributions (°C) in 3D stacking (top) and Si interposer (bottom) for the high-power case: 50 W for the logic chip and 4 W for the serdes chip.

from the chips to the lid and heat sink) and the Si interposer structure has larger conduction area with the lid and heat sink. For the coupling effect between the chips, please read [22] for more information.

9.8.6 Summary and Recommendations

Some important results and recommendations are summarized as follows:

- Since there is more conducting area of the Si interposer structure than the 3D stacking structure, thus the thermal performance of the Si interposer is better than that of the 3D stacking.
- For low-power application, the maximum temperature in the serdes chip of the Si interposer is 24% less than that of the 3D stacking.
- For high-power application, the maximum temperature in the logic chip of the Si interposer is 24% less than that of the 3D stacking.

9.9 Thermal Management System with TSV Interposers with Embedded Microchannels

A thermal management system is designed and fabricated in this section [24–33]. The pressure drops for different fluidic interconnects in this system are measured and compared with the analyzed results.

9.9.1 The Structure

Figure 9.20 schematically shows the structure under considered. It consists of two chips (chip 1 and chip 2), two TSV interposers (carrier 1 and carrier 2), Si spacer, adaptor, gasket, mini pump, heat exchanger and PCB. Both chips have the same size (10×10 mm) and dissipate 100 W of heat. Both carriers have the same size (15×15 mm) and have fluidic microchannels embedded and stacked vertically. The Si spacer is assembled between these two carriers and supported these carriers. The adaptor (50×50 mm) is used for connecting the microchannels in the carriers and mini sized ports of the pump, which splits the flow from the outlets of the pump to the inlets of the microchannel carriers. The external heat exchanger (50×50 mm) is used for heat removal to ambient/cold plate. The mini pump (50×50 mm) is commercial available and with a maximum operation pressure of 2 bar and a maximum flow rate of 500 mL/min for driving the coolant. How is the system work? The heated fluid is driven out from the stacked chip module to the heat exchanger by the pump, and the hot fluid is cooled down in the exchanger, then the cooled fluid is driven back to the module.

9.9.2 Adaptor

Figure 9.21 schematically shows the adaptor design for (a) bottom view, (b) top view, (c) channels for fluid rerouting at the outlet port, and (d) channels for fluid rerouting at the inlet port. At the bottom side of the adaptor, a cavity with depth of 1.5 mm is machined to house the chips and carriers, and to form the fluidic connection with the carriers, as shown in Fig. 9.21. On the top side of the adaptor, one inflow port is fabricated to deliver the fluid from the small pump to the carrier and another outflow port is drilled to convey the fluid

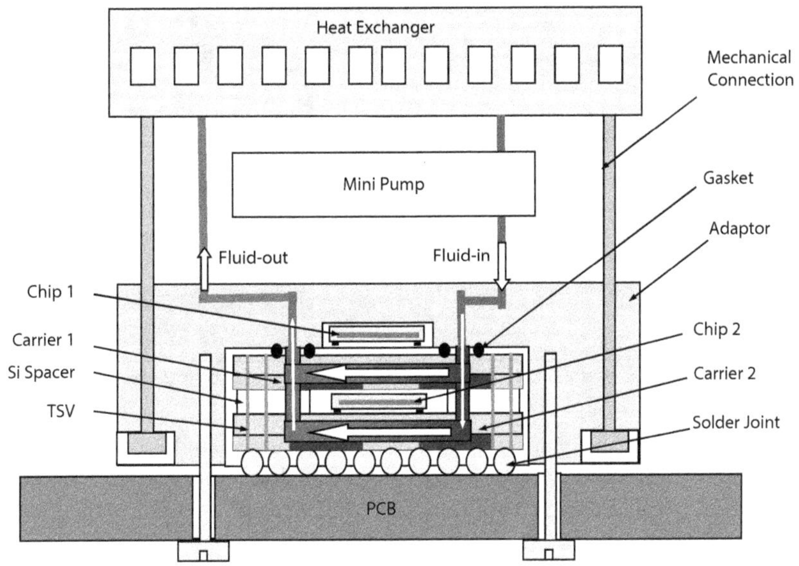

Figure 9.20 Schematic of the integrated liquid cooling system for 3D stacked module (not-to-scale).

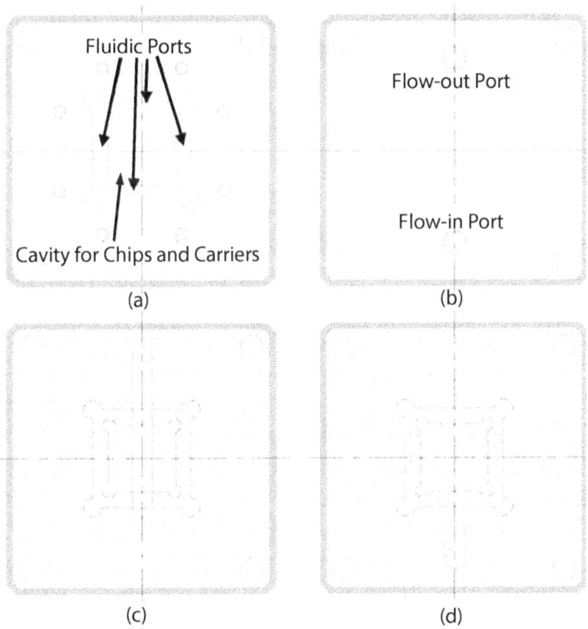

Figure 9.21 Adaptor design for fluid rerouting. (a) Bottom view. (b) Top view. (c) Channels for fluid rerouting at outlet. (d) Channels for fluid rerouting at inlet.

(a) (b)

Figure 9.22 Images of the fabricated adaptor. (a) Bottom view. (b) Top view.

from the carrier to the heat exchanger. These two ports are both fixed with one-fourth in tap for flexible tube connection and fluid sealing. The fluidic interface between the adaptor and the carrier is sealed using a rubber gasket. The image of the fabricated adaptor is shown in Fig. 9.22 for both (a) bottom-side view and (b) top-side view.

9.9.3 Heat Exchanger

In order to efficiently extract the heat from the package and dissipate the heat to the ambient, an optimized heat exchanger is imposed on the integrated system. The effective heat exchanger (50 × 50 mm) with parallel plate fin and single pass flow is shown in Fig. 9.23(a). The fin size is 1 mm in width, 10 mm in height, and 30 mm in length. The heat exchanger is machined by the wire cutting technique and is made of aluminum material due to its high thermal conductivity and light weight inlet and outlet ports are fixed with one-fourth in tap for flexible tube connection and fluidic sealing. The fabricated heat exchanger is shown in Fig. 9.23(b).

9.9.4 Carriers

The fluidic path design for the proposed carrier is illustrated in Fig. 9.24, which shows the schematic view of the silicon wafer for (a) the top-side view of the top wafer for both carriers; (b) the optimized microchannel patterns for both carriers; (c) the bottom view of the bottom wafer for carrier 1; and (d) the bottom view of the bottom wafer for carrier 2. From fluidic points of view, the carriers 1 and 2 are distinguishable only from their bottom wafers. For carrier 1, the bottom side has two inlet fluidic ports and two outlet fluidic ports [see Fig. 9.24(c)] while carrier 2 does not have these inlets and outlets but solder balls [see Fig. 9.24(d)]. The optimized microchannels cover an area of 10.5 × 10.5 mm with specific patterns to reduce the pressure drop and minimize the variation of chip temperature. The depth and the width of the channel are 350 μm and 100 μm, respectively. The detailed study for optimizing the microchannel design is described in [28].

In addition, TSVs are designed along the periphery of the carrier at a pitch of 500 μm for electrical interconnection. There are in total 144 TSVs with depth of 400 μm and diameter of 150 μm in the design. Sealing rings are deposited around both the fluidic path and the individual TSV to isolate and seal the fluid from the electrical interconnection. Moreover,

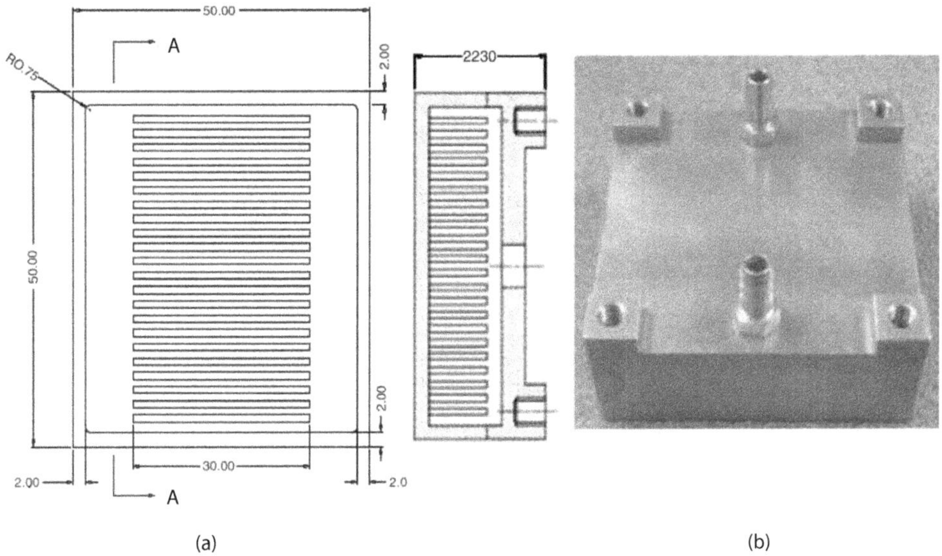

FIGURE 9.23 (a) Heat exchanger design. (b) Fabricated sample of the optimized heat exchanger.

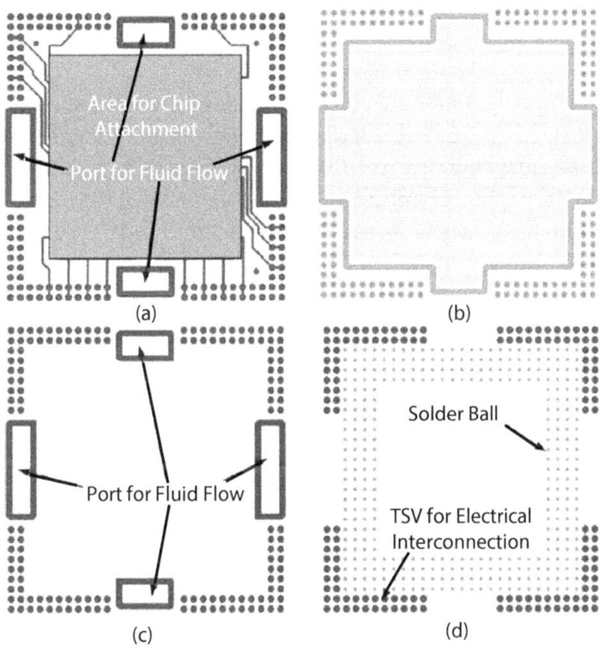

FIGURE 9.24 Carrier design for (a) top view of top wafer for both carriers, (b) optimized microchannel for both carriers, (c) bottom view of bottom wafer for carrier 1, and (d) bottom view of bottom wafer for carrier 2.

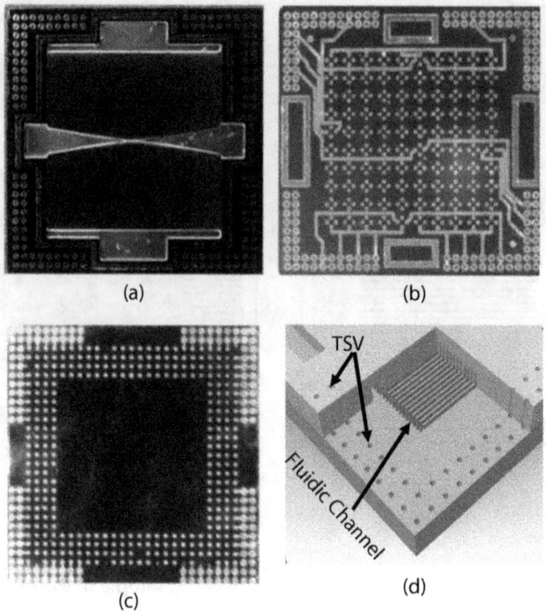

FIGURE 9.25 The fabricated carrier. (a) Optimized microchannels. (b) Top surface of carriers 1 and 2. (c) Bottom surface of carrier 2. (d) Cross-sectional view of the carrier.

the 3D silicon module is mounted on to a PCB for the electrical inputs. Spherical solder balls of size 250 μm are attached to the back side of the bottom wafer for carrier 2. The fabrication process and assembly are described in [26, 27]. Figure 9.25 shows the optical image of the fabricated carrier for (a) the optimized microchannels, (b) the top surface of carriers 1 and 2, and (c) the bottom surface of carrier 2. Figure 9.25(d) shows the SEM image of the TSV and microchannels. The patterns on both sides of the carriers provide electrical interconnection and rerouting.

9.9.5 System Integration

After the parts for the system are fabricated, the integration of the system is conducted. The overall foot print of the integrated system is 50 × 50 mm. Figure 9.26(a) shows the schematic process of integration for the cooling system. First, a thermal test die with a size of 10 × 10 mm is attached to the carrier. Second, two of such carriers are stacked vertically with a silicon spacer between them to house the die and support the carriers. Then, the assembled carriers are mounted on a PCB using spherical solder balls of size 250 μm [see Figs. 9.26(b) and 9.26(c)]. In parallel, the adaptor, mini pump, and heat exchanger are mechanically assembled together using mechanical supporters. The fluidic interconnects among the adaptor, the mini pump and the heat exchanger are formed using flexible tubes. At last, the module part and adaptor part are assembled together [see Fig. 9.26(d)] and fluidic sealing for the interface between the adaptor and module is achieved with a rubber gasket, which is machined by laser cutting.

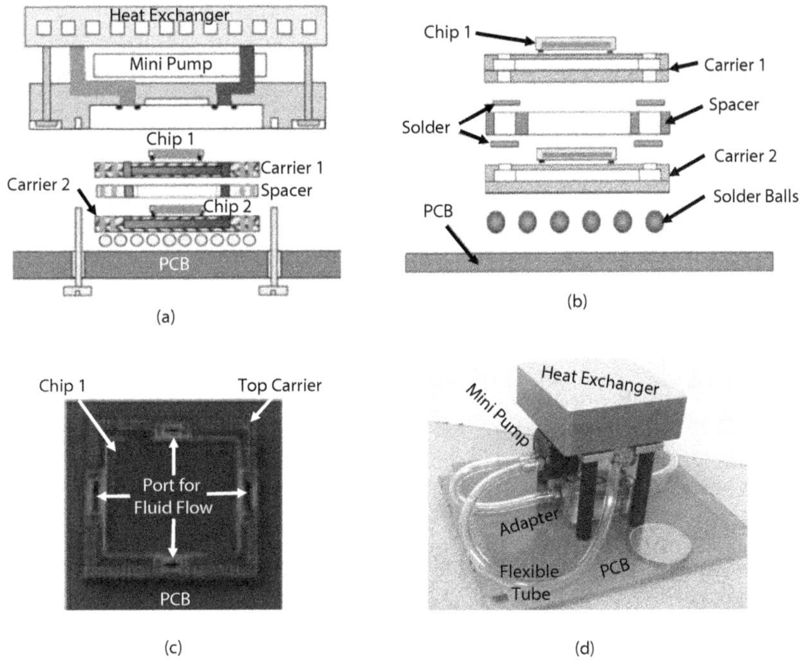

FIGURE 9.26 Process of the integration for the liquid cooling system. (a) Schematic drawing of the system. (b) Schematic drawing of the carriers and chips. (c) Image of the assembled module. (d) Image of the integrated cooling system.

9.9.6 Theoretical Analysis of the Pressure Drop

Neglecting the minor pressure drop of fitting and piping, the total pressure drop (ΔP_{total}), is a summation from the pressure drop of main parts in the cooling system, namely, the pressure drop in the adaptor (ΔP_a), the pressure drop in the microchannel carrier (ΔP_c), and the pressure drop in the external heat exchanger (ΔP_{he}), and can be expressed as

$$\Delta P_{\text{total}} = \Delta P_a + \Delta P_c + \Delta P_{he} \tag{9.1}$$

Generally, the pressure drop for each part (ΔP) mainly consists of different components, namely, the inlet contraction loss (ΔP_{in}), the friction loss in the channel (ΔP_{ch}), the outlet expansion loss (ΔP_{out}), and can be expressed as

$$\Delta P = \Delta P_{in} + \Delta P_{ch} + \Delta P_{out} \tag{9.2}$$

According to Cengel and Cimbala [35], the pressure drop in a straight channel can be estimated by

$$\Delta P_{ch} = f\frac{\rho u^2}{2}\left(\frac{L}{D_h}\right) \tag{9.3}$$

$$u = \frac{Q}{A} \tag{9.4}$$

where L, D_h, ρ, and u are the channel length in the direction of the flow, the channel hydraulic diameter, the liquid coolant mass density, and the flow average velocity in the channel, respectively. Q and A, respectively, represent the flow rate and cross-sectional area of the flow path. The friction factor (f), can be calculated by combining the friction factors for developing flow (f_{dev}) and fully developing flow (f_{fdev}), and can be expressed as

$$f = f_{\mathrm{dev}} + f_{\mathrm{fdev}} = \frac{K(\infty)}{4[L/(D_h R_e)]} + \frac{96/R_e}{\left[1+\frac{1}{\alpha}\right]^2 \left[1 - \frac{192}{\pi^5 \alpha} \sum_{n=1,3,5}^{\infty} \left[\frac{\tanh(\frac{n\pi\alpha}{2})}{n^5}\right]\right]} \tag{9.5}$$

where $K(\infty)$ is the constant value Hagenbach factor (effectively a loss coefficient), $R_e = \rho u D_h/\mu$ is the Reynolds number, and μ is the viscosity of the coolant. The constant value Hagenbach factor is a function of the aspect ratio ($0 < \alpha < 1$) of a channel given by

$$K(\infty) = 0.6796 + 1.2197\alpha + 3.3089\alpha^2 - 9.5921\alpha^3 + 8.9089\alpha^4 - 2.9959\alpha^5 \tag{9.6}$$

The inlet contraction and outlet expansion loss of the channels are estimated by the following equations:

$$\Delta P_{in} = K_{in} \frac{\rho u^2}{2} \tag{9.7}$$

$$\Delta P_{out} = K_{out} \frac{\rho u^2}{2} \tag{9.8}$$

where K_{in} is the loss coefficient for the abrupt entrance and K_{out} is the loss coefficient for the abrupt exit [34]. These loss coefficients are determined based on the experimental values widely available for circular tubes. Of the two loss coefficients, K_{out} is the most readily identified and is equal to 1. For any abrupt exit to a large reservoir (i.e., an exit from a microchannel to a large plenum), the geometry of the exit has no effect on the loss coefficient. The loss coefficient for the inlet, K_{in}, is highly dependent on the geometry of the inlet area. Typically, for a sharp-edged inlet (i.e., 90° corners), $K_{in} = 0.5$. For a well-rounded inlet (defined as $r/D_h > 0.2$), $K_{in} = 0.03$. For a slightly rounded inlet (defined as $r/D_h > 0.1$), $K_{in} = 0.12$ [34]. In this section, the actual value of r/D_h is unknown for the different parts, and the ratio certainly varies for each individual part. Therefore, a loss coefficient of $K_{in} = 0.5$ is used as an estimate for considering the worst cases.

To estimate the pressure drop of each part in the system, first, the friction factor of the flow passage for each part will be calculated using Eqs. (9.5) and (9.6). Then the pressure drop in the flow passage of each part can be obtained by Eqs. (9.3) and (9.4). Meanwhile, the pressure loss at the inlet and outlet of each part will also be calculated using Eqs. (9.7) and (9.8). As such, the pressure drop for each part and the whole system can be estimated by using Eqs. (9.1) and (9.2), respectively. The estimated pressure drop will be compared with the experimental measurement that is discussed in the following sections.

9.9.7 Experimental Process

An experimental rig is processed to conduct the hydraulic testing. The schematic diagram of the experimental setup for hydraulic characterization is shown in Fig. 9.27(a) for the entire

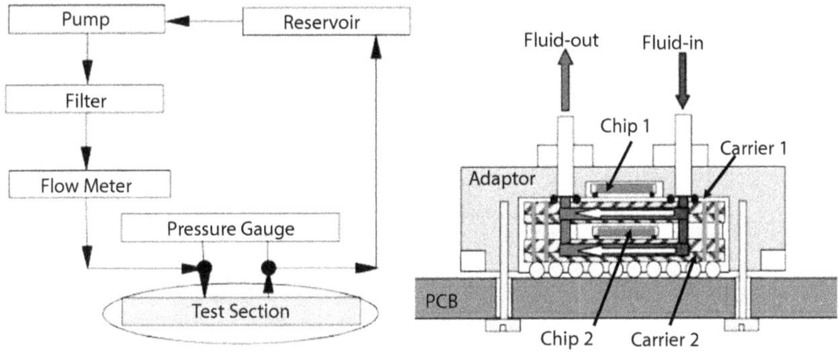

FIGURE 9.27 Schematic of the experimental setup. (Left) Experiment system. (Right) Test section.

system, and Fig. 9.27(b) for a close-up of test section. A variable-speed gear pump is used to provide the pressure head. A filter with a 10-μm mesh filter element is used (after the pump) to remove particles suspended in the liquid. Two variable area flowmeters are connected in parallel to measure the flow rates with a factory-supplied uncertainty within 2.5% in the flow range. The flowmeters are calibrated in-house to achieve a measurement error within 2%. An in-house calibration of the flowmeters is also conducted by collecting the liquid volume over a corresponding measured elapsed time. The pressure drop across the test section is measured with a piezo-resistive pressure transducer with a digital display. The uncertainty at the full range is estimated to be within 0.2%.

In the test section, the carrier with optimized microchannel geometry is placed into a test adaptor for connection to the piping. De-ionized water is used as the coolant. The experiments are conducted at different flow rates ranging from 0.08 to 0.4 L/min.

First, leakage testing is conducted by running the system at the maximum pressure condition (flow rate is 400 mL/min) for 10 min and the corresponding maximum pressure drop in the carrier is 9×10^4 Pa. No leakage is observed during this period, which indicates that the present sealing technique for fluidic path in this cooling system is working well. Then the total pressure drop due to the adaptor and carrier is measured. Consequently, the pressure drop, induced by the adaptor alone is measured with the carrier removed away from the adaptor, and the pressure drop of the carrier is then obtained by subtracting the adaptor pressure drop from the measured total pressure drop. In addition, the pressure drop of the heat exchanger is measured separately by replacing the adaptor and carrier with the heat exchanger.

9.9.8 Results and Discussions

The analytically predicted overall pressure drops in each part of the system vs. flow rate against experimentally measured values are compared in Figs. 9.28 through 9.30 in the adaptor, the carrier, and the heat exchanger, respectively. The predicted values plotted in these figures are obtained by Eqs. (9.2) through (9.8). It can be seen that in all of the main parts considered in this system, the experimentally measured pressure drop agrees reasonably well with the predicted value. The predicted results are slightly lower (5–25%) than the measured values. The lower pressure drop in the prediction can be attributed to the neglecting bends in the parts and rough surface in the flow paths, as well as to assuming the channels with a uniform length to replace the channel with a nonuniform length in the

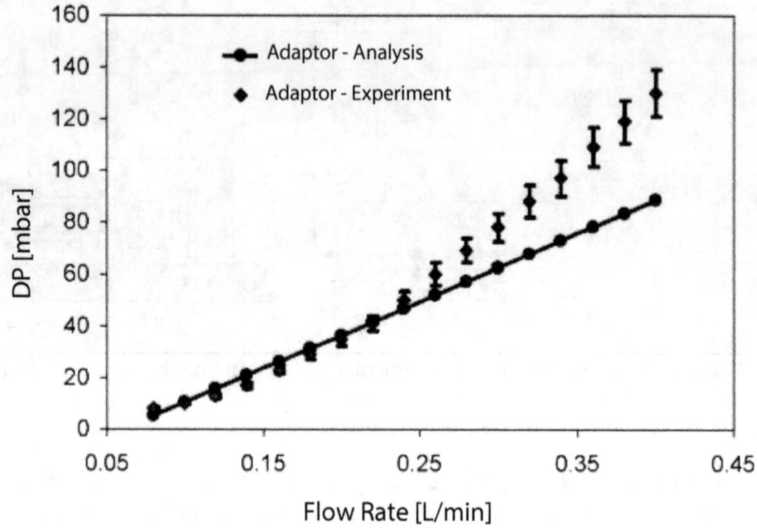

FIGURE 9.28 Pressure drop vs. flow rate in adaptor.

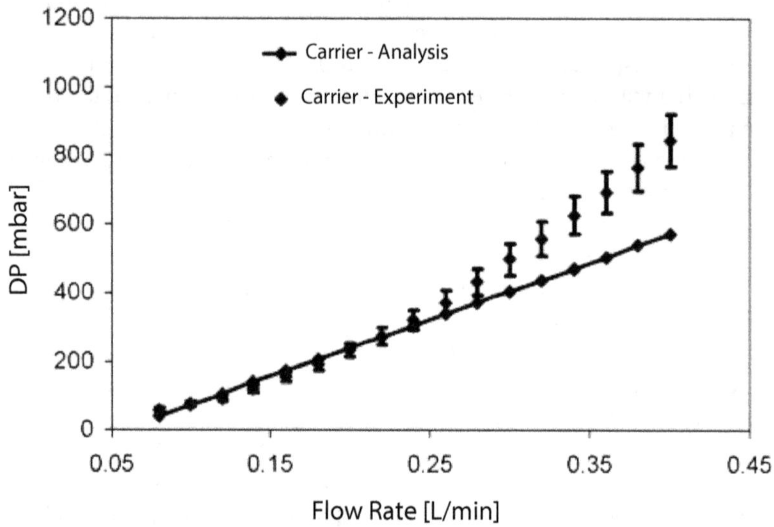

FIGURE 9.29 Pressure drop vs. flow rate in carrier.

carrier. In addition, the experimental results show a stronger parabolic behavior than those of the analysis, and it may be due to the effect of bending and contraction loss.

Taking a closer examination at Figs. 9.28 through 9.30, it can be seen that the carrier with embedded microchannels is the main contributor to the pressure drop of the system, accounting for around 87% of the system pressure drop. However, the adaptor and heat exchanger only account for around 12.5% and 0.5% of the system pressure drop, respectively. The proportions of system pressure drop in the carrier, adaptor, and external heat exchanger are shown in Fig. 9.31. Such a distribution of the system pressure drop indicates

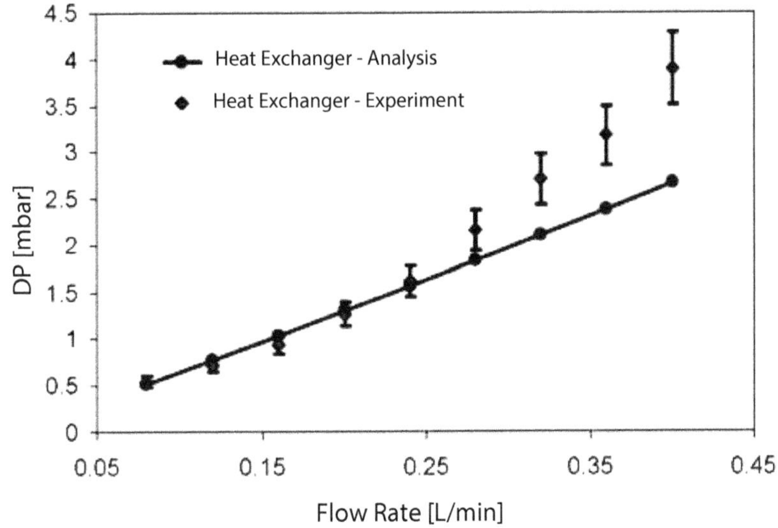

Figure 9.30 Pressure drop vs. flow rate in heat exchanger.

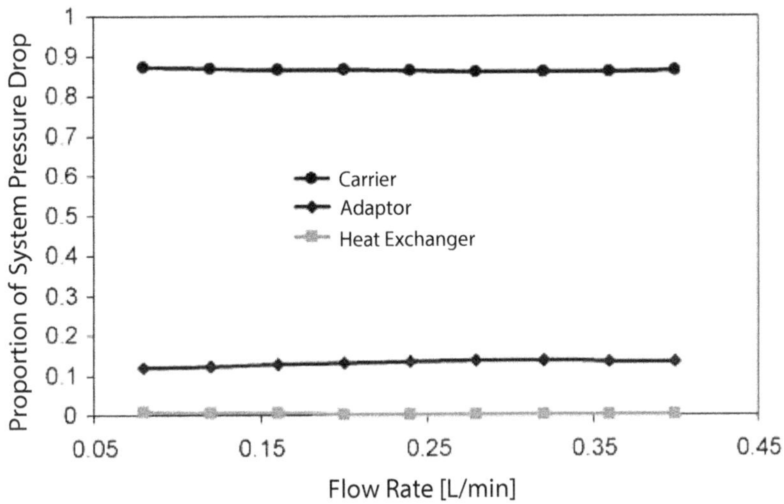

Figure 9.31 Proportion in system pressure drop vs. flow rate in adaptor, carrier, and heat exchanger.

that the hydraulic design of this system is efficient and the pressure drop in the adaptor and external heat exchanger has been reduced to a reasonable level.

In addition, the pressure drop in the conventional single inlet/outlet carrier is also measured to evaluate the improvement of hydraulic characteristics in the optimized two inlets/outlets carrier proposed in this chapter. In these experiments, the size of the optimized carrier and the conventional carrier are maintained the same. The comparison of the measured pressure drop in the optimized carrier with that of the conventional carrier is shown in Fig. 9.32. From this figure, it can be seen that the pressure drop in the optimized carrier

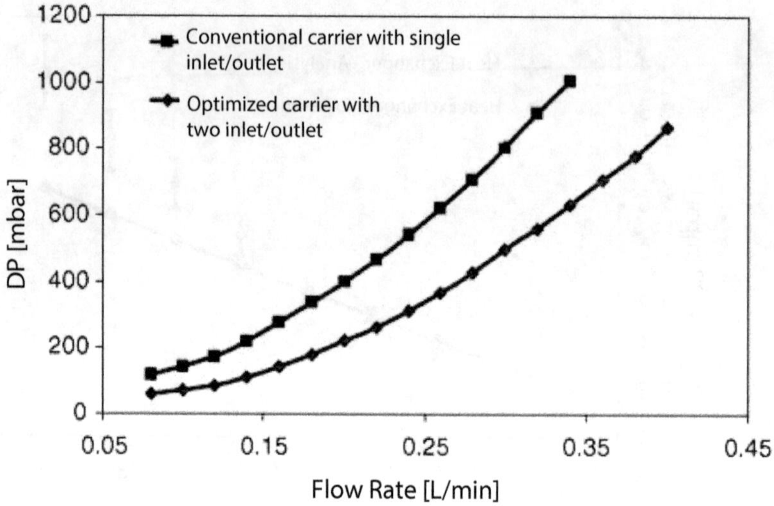

FIGURE 9.32 Comparison of the measured pressure drop vs. flow rate in the optimized carrier (two inlets/outlets) with that in the conventional carrier (single inlet/outlet).

decreases 30–50% as compared with that in the convectional carrier. Meanwhile, simulation for thermal performance of the cooling system with optimized carrier and conventional carrier are carried out and the results are compared. The simulation results show that the thermal resistance of the optimized carrier and that of the conventional carrier are consistent with each other. When the flow rate in each carrier is 230 mL/min and the power of each chip is 100 W, the thermal resistances for both carriers are about 0.18°C/W. However, the variation of the chip temperature for the optimized carrier is 8.1°C, which is much lower than that for the conventional carrier (14.1°C).

9.9.9 Summary and Recommendations

The fluidic interconnects in the integrated liquid cooling system for 3D stacked modules have been developed. The sealing technique for different parts is discussed, while the pressure drop for each part of the system is analyzed. The optimized fluidic interconnects which minimize the pressure drop have been designed and fabricated. In line with the fluidic interconnect design and analysis; an experiment process for characterizing the flow and heat transfer of the integrated cooling system has been developed. The pressure drop of different fluidic interconnects in this system has been measured. Some important results are summarized as follows [24–33]:

- An integrated cooling solution for a 3D silicon module developed has the following key features: (a) eliminating the bottlenecks of thermal interface material and spreading resistance; (b) forming fluidic interconnects connecting silicon microchannel; (c) eliminating external fluidic circuits; and (d) scalable to multiple number of 3D stacks.
- No leakage is observed at a high flow rate (400 mL/min) for a single carrier. The maximum pressure drop in the carrier is 9×104 Pa, which indicates that the present sealing technique for the fluidic path in this cooling system works well.

- The experimental results show that the pressure drop in the optimized carrier decreases 30–50% as compared with that in the convectional carrier with the same carrier size.
- The simulation results show that the thermal resistance of the optimized carrier and the thermal resistance of the conventional carrier are consistent with each other. The thermal resistances for both carriers are about 0.18°C/W. However, the variation of the junction temperature for the optimized carrier is 8.1°C, which is much lower than that for the conventional carrier, which is 14.1°C/W.
- The pressure drop in the adaptor and heat exchanger is only about 13% of the total system pressure drop, indicating that the adaptor and heat exchanger are properly designed.
- The total pressure drop in the integrated cooling system is around 1.2 bar at the maximum flow rate (400 mL/min), and this pressure drop is within the operation range (2 bar) of the selected mini pump.

9.10 References

[1] Lau, J. H., S. Lee, M. Yuen, J. Wu, C. Lo, H. Fan, and H. Chen, "Apparatus having thermal-enhanced and cost-effective 3D IC integration structure with through silicon via interposer," *US Patent No: 8,604,603*, Date of Patent: December 10, 2013, Filed on February 20, 2009.

[2] Lau, J. H., Y. S. Chan, S. W. R. Lee, "Thermal-Enhanced and Cost-Effective 3D IC Integration with TSV Interposers for High-Performance Applications," *ASME Paper no. IMECE2010-40975.*

[3] Lau, J. H., Y. S. Chan, and R. S. W. Lee, "3D IC Integration with TSV Interposers for High-Performance Applications," *Chip Scale Review*, Vol. 14, No. 5, September/October, 2010, pp. 26–29.

[4] Lau, J. H., *Ball Grid Array Technology*, McGraw-Hill, New York, NY, 1995.

[5] Lau, J. H., and R. Lee, *Microvias for Low Cost, High Density Interconnects*, McGraw-Hill, New York, NY, 2001.

[6] Lau, J. H., and K. L. Chen, "Thermal and Mechanical Evaluations of a Cost-Effective Plastic Ball Grid Array Package," *ASME Transactions, Journal of Electronic Packaging*, Vol. 119, September 1997, pp. 208–212.

[7] Lau, J. H., T. Tseng, and D. Cheng, "Heat Spreader with a Placement Recess and Bottom Saw-Teeth for Connection to Ground planes on a This Two-Side Single-Core BGA Substrate," *US Patent No. 6,057,601*, Date of Patent: May 2, 2000, Filed on November 27, 1998.

[8] Lau, J. H., and T. Chen, "Cooling Assessment and Distribution of Heat Dissipation of A Cavity Down Plastic Ball Grid Array Package—NuBGA," *IMAPS Transactions, International Journal of Microelectronics & Electronic Packaging*, Vol. 21, No. 1, 1998, pp. 20–28.

[9] Lau, J. H., and T. Chou, "Electrical Design of a Cost-Effective Thermal Enhanced Plastic Ball Grid Array Package—NuBGA," *IEEE Transactions on CPMT, Part B*, Vol. 21, No. 1, February 1998, pp. 35–42.

[10] Lau, J. H., T. Chen, and R. Lee, "Effect of Heat Spreader Sizes on the Thermal Performance of Large Cavity-Down Plastic Ball Grid Array Packages," *ASME Transactions, Journal of Electronic Packaging*, Vol. 121, No. 4, 1999, pp. 242–248.

[11] Lau, J. H., "Design, Manufacturing, and Testing of a Novel Plastic Ball Grid Array Package," *Journal of Electronics Manufacturing*, Vol. 9, No. 4, December 1999, pp. 283–291.

[12] Lau, J. H., and T. Chen, "Low-Cost Thermal and Electrical Enhanced Plastic Ball Grid Array Package—NuBGA," *Microelectronics International*, 1999.

[13] Lau, J. H., "Solder Joint Reliability of Flip Chip and Plastic Ball Grid Array Assemblies Under Thermal, Mechanical, and Vibration Conditions," *IEEE Transactions on CPMT, Part B*, Vol. 19, No. 4, November 1996, pp. 728–735.

[14] Lau, J. H., with R. Lee, "Design for Plastic Ball Grid Array Solder Joint Reliability," *Journal of the Institute of Interconnection Technology*, Vol. 23, No. 2, January 1997, pp. 11–13.

[15] Lau, J. H., with W. Jung, and Y. Pao, "Nonlinear Analysis of Full-Matrix and Perimeter Plastic Ball Grid Array Solder Joints," *ASME Transactions, Journal of Electronic Packaging*, Vol. 119, September 1997, pp. 163–170.

[16] Lau, J. H., and R. Lee, "Solder Joint Reliability of Cavity-Down Plastic Ball Grid Array Assemblies," *Journal of Soldering & Surface Mount Technology*, Vol. 10, No. 1, February 1998, pp. 26–31.

[17] Chien, H. C., J. H. Lau, Y. Chao, R. Tain, M. Dai, S. T. Wu, W. Lo, et al., "Thermal Performance of 3D IC Integration with Through-Silicon Via (TSV)," *IMAPS Transactions, Journal of Microelectronic Packaging*, Vol. 9, 2012, pp. 97–103.

[18] Chien, H. C., J. H. Lau, Y. Chao, R. Tain, M. Dai, S. T. Wu, W. Lo, et al., "Thermal Performance of 3D IC Integration with Through-Silicon Via (TSV)," *Proceedings of IMAPS International Conference*, Long Beach, CA, October 2011, pp. 25–32.

[19] Lau, J. H., H. C. Chien, S. T. Wu, Y. L. Chao, W. C. Lo, and M. J. Kao, "Thin-Wafer Handling with a Heat-Spreader Wafer for 2.5D/3D IC Integration," *Proceedings of the 46th IMAPS International Symposium on Microelectronics*, Orlando, FL, October 2013, pp. 389–396.

[20] Chien, H., J. H. Lau, T. Chao, M. Dai, and R. Tain, "Thermal Management of Moore's Law Chips on Both sides of an Interposer for 3D IC integration SiP," *IEEE ICEP Proceedings*, Japan, April 2012, pp. 38–44.

[21] Chien, H., J. H. Lau, Y. Chao, M. Dai, and R. Tain, "Thermal Evaluation and Analyses of 3D IC Integration SiP with TSVs for Network System Applications," *IEEE ECTC Proceedings*, San Diego, CA, May 2012, pp. 1866–1873.

[22] Oprins, H., B. Vandevelde, M. Badaroglu, M. Gonzalez, G. Van der Plas, and E. Beyne, "Numerical Comparison of the Thermal Performance of 3D Stacking and Si Interposer Base Packaging Concepts," *Proceedings of IEEE/ECTC*, Las Vegas, NV, May 2013, pp. 2183–2188.

[23] Palankovski, V., and S. Selberherr, "Thermal Models for Semiconductor Device Simulations", *The Third European Conference on High Temperature Electronics*, July 1999, pp. 25–28.

[24] Tang, G. Y., S. Tan, N. Khan, D. Pinjala, J. H. Lau, A. Yu, V. Kripesh, et al., "Integrated Liquid Cooling Systems for 3-D Stacked TSV Modules," *IEEE Transactions on CPMT*, Vol. 33, No. 1, March 2010, pp. 184–195.

[25] Tang, G., T. Pin, N. Khan, D. Pinjala, J. H. Lau, A. Yu, V. Kripesh, et al., "Fluidic Interconnects in Integrated Liquid Cooling Systems for 3-D Stacked TSV Modules", *IEEE/EPTC Proceedings*, Singapore, December 2008, pp. 552–558.

[26] Yu, A., N. Khan, G. Archit, D. Pinjala, K. Toh, V. Kripesh, J. H. Lau, et al., "Fabrication of Silicon Carriers With TSV Electrical Interconnections and Embedded Thermal Solutions for High Power 3-D Packages," *IEEE Transactions on CPMT*, Vol. 32, No. 3, September 2009, pp. 566–571.

[27] Yu, A., N. Khan, G. Archit, D. Pinjalal, K. Toh, V. Kripesh, J. H. Lau, et al., "Fabrication of Silicon Carriers with TSV Electrical Interconnection and Embedded Thermal Solutions for High Power 3-D Package," *IEEE/ECTC Proceeding*, Orlando, FL, May 2008, pp. 24–28.

[28] Tan, S. P., K. C. Toh, N. Khan, G. Y. Tang, D. Pinjala, V. Kripesh, and J. H. Lau, "Thermal and Hydraulic Design and Characterization of a Liquid-Cooled 3D Silicon Module," *IEEE/EPTC Proceedings*, Singapore, December 2008, pp. 350–354.

[29] Khan, N., L. Yu, P. Tan, S. Ho, N. Su, H. Wai, J. H. Lau, et al., "3D Packaging with Through Silicon Via (TSV) for Electrical and Fluidic Interconnections," *IEEE Proceedings of ECTC*, San Diego, CA, May 2009, pp. 1153–1158.

[30] Khan, N., H. Li, S. Tan, S. Ho, V. Kripesh, D. Pinjala, J. H. Lau, et al., "3-D Packaging With Through-Silicon Via (TSV) for Electrical and Fluidic Interconnections," *IEEE Transactions on CPMT*, Vol. 3, No. 2, February 2013, pp. 221–228.

[31] Lau, J. H., and G. Y. Tang, "Effects of TSVs (through-silicon vias) on thermal performances of 3D IC integration system-in-package (SiP)," *Journal of Microelectronics Reliability*, Vo. 52, No. 11, November 2012, pp. 2660–2669.

[32] Lau, J. H., and Tang, G., "Thermal Management of 3D IC Integration with TSV (Through Silicon Via)," *IEEE Proceedings of ECTC*, San Diego, May 2009, pp. 635–640.

[33] Hoe, Y., Y. Tang, D. Pinjala, T. Chai, J. H. Lau, X. Zhang, and V. Kripesh, "Effect of TSV Interposer on the Thermal Performance of FCBGA Package," *IEEE/EPTC Proceedings*, Singapore, December 2009, pp. 778–786.

[34] Cengel, Y. A., and J. M. Cimbala, *Fluid Mechanics—Fundamentals and Applications*. McGraw-Hill, New York, NY, 2006.

CHAPTER 10
Embedded 3D Hybrid Integration

10.1 Introduction

A low-cost (with bare chips) and high-performance (optical, electrical, thermal, and mechanical) optoelectronic system embedded into a PCB (printed circuit board) or an organic laminated substrate is designed and described in this chapter. This system consists of a rigid PCB (or a substrate) with an embedded optical polymer waveguide, an embedded VCSEL (vertical cavity surface emitted laser), an embedded driver chip, an embedded serializer, an embedded photodiode detector, an embedded TIA (transimpedance amplifier), an embedded deserializer, embedded heat slugs, and a heat spreader.

The bare VCSEL, driver chip, and serializer chip are 3D stacked and then attached to one end of the embedded optical polymer waveguide in the PCB. Similarly, the bare photodiode detector, TIA chip, and deserializer chip are 3D stacked and then attached to the other end. The back side of the driver or serializer and the TIA or deserializer chips is attached to a heat slug with or without a spreader. This novel structural design offers potential solutions for low-cost and high-performance semiconductor circuits with optical devices to realize wide-bandwidth and low-profile optoelectronic packaging for chip-to-chip optical interconnect applications. Optical, thermal management and mechanical performances are demonstrated by simulations based on optic theory, heat-transfer theory, and continuum mechanics.

A couple of old designs (2D system) are discussed first. These are (a) high-frequency data link on FR4 PCB using optical waveguides and (b) embedded board-level optical interconnects. Finally, a design of semi-embedded through-silicon via (TSV) interposer with stress relief gap is presented.

10.2 Trends of Optoelectronic Products

The major trend in the optoelectronic industry today is to make products more personal by making them smarter, lighter, smaller, thinner, shorter, and faster, while at the same time making them more friendly, functional, powerful, reliable, robust, innovative, creative, and economical [1, 2]. As the trend toward miniature and compact products continues, the introduction of innovative products that are more user-friendly and contain a wider variety of functions will provide growth in the market [1, 2]. One of the key technologies that helps make these product design goals possible is to have a bare chip on optical PCB [3, 4]. It is even better to stack up a few chips (in three dimensions) on the board to enhance performance and save more cost and real-estate [5]. In order to make super-thin and super-light products, e.g., mobile phones, the waveguide [6–8] and chips should be embedded in the optical PCB, which is the focus of this chapter.

As integrated circuit (IC) technology advances, the speed and complexity of interconnects increase tremendously. As the number of devices per chip, the number of chips per board, the modulation speed, and the degrees of integration continue to increase, electrical interconnects face their fundamental bottlenecks, e.g., speed, data rate, and power dissipation.

Optical interconnect is an attractive solution to the bandwidth limitation of the electrical transmission line for board-level interconnections. However, the implementation of optical waveguide in a commercial PCB imposes severe restrictions on the materials and processes such as low propagation loss, high thermal stability, and cost-effective fabrication. Currently, polymer-based waveguide is the most popular choice for optical interconnect due to its low cost, suitability for mass production, high thermal stability, and low optical loss [3–32]. The typical loss for the waveguide used in optical PCBs, depending on the fabrication process, materials, and its cross-sectional area, ranges from 0.6 to 0.05 dB/cm. Depending on the fabrication process and materials, polymer properties can have a thermal stability of more than 260°C, which is adequate for lead-free assembly process [33].

VCSEL is a specialized laser diode that is gaining popularity as a light source for optical interconnect applications. Unlike older edge emitting diodes that emit IR radiation in the plane of the semiconductor, the VCSEL produces a nearly circular beam of symmetrical laser light with low threshold current and small output beam divergence angle that is perpendicular to the plane of the semiconductor. These devices may be constructed on the surface of a fabricated wafer and tested for their optical and electrical properties on the wafer, and have the advantage of combining large 2D emitter arrays and active devices such as CMOS drivers with conventional technology. However, when the VCSEL is integrated with the optical PCB, it requires an out-of-plane 45° reflector to couple the optical signal to the polymer waveguide due to its surface emitting characteristic. This problem is also encountered by the surface illuminated photodetector. The optical polymer waveguide is usually designed in multimode and has a larger aperture for reasons of low cost and compatibility with PCB process.

The speed, data rate, power, and size of optoelectronic devices generally scale with the wavelength of light used. The complexity of the present-day 2D photonic modules is limited by their substrate sizes and the difficulty in connecting a large number of electrical connections. By progressing to multilayer interconnects, the IC density can be increased greatly. Since fewer connections are required between bare chips, a higher performance will result. Some devices can actually be made smaller when integrated vertically rather than laterally. Vertical stacking also allows more chips per wafer. Complete optoelectronic integration of devices of different materials onto a single substrate (like the optical PCB) will achieve a dramatic cost and labor reduction as well. Novel combinations of materials/geometries may lead to the production of devices far superior than their in-plane counterparts. Additional flexibility can be afforded in package design, interconnect routing, and package placement through 3D architectures. These factors have the potential to revolutionize the way 3D hybrid modules are produced and used.

10.3 The Old Design—High-Frequency Data Link on PCB Using Optical Waveguides

Figure 10.1 shows a low-cost, simple OECB (optical/electrical circuit board) using multi-mode polymer waveguide on FR4 PCB [3, 4]. Similar to [31, 34], the design of this OECB uses only a 45°-ended waveguide to couple and decouple the optical signal directly between the optical devices and the waveguide. The 45° mirror is formed using an excimer laser process on a multimode waveguide with temperature stability at reflow temperature. The optical waveguide is attached to a diced channel in the FR4 PCB using adhesive to form a completely planar circuit. This allows the laser diode and the photodiode to be assembled directly above the input and output of the waveguide using precision flip chip technology, which provides good alignment accuracy. This helps increase the mechanical reliability of the circuit and minimize assembly requirements.

The high-frequency electronic circuit is designed [3] using 2.5 Gbps components from Maxim Integrated Products with −3 dB bandwidth of 1.7 GHz. The VCSEL and PIN are flip chip devices procured from U.L.M Photonics. The VCSEL has a 3 dB modulation bandwidth of 6 GHz, typical relative intensity noise of −130 dB/Hz, and output power of 1 mW. The photodiode has a 3 dB modulation bandwidth of 5 GHz, responsivity of 0.5 A/W, capacitance of 0.4 pF, and an active area diameter of 70 μm. The preliminary testing result shows that this simple OECB designed on FR4, without using any submodule or micro-lens can achieve a transmission speed of 1.25 Gbps at a distance of 82 mm.

10.3.1 Polymer Optical Waveguide

The cross-sectional area of the waveguide is an important parameter that determines the total coupling efficiency and insertion loss [35, 36]. It has to be optimized so as to balance the tradeoff between input and output coupling efficiency. A large cross-sectional area helps increase the input misalignment tolerance, hence improving the input coupling efficiency. However, it decreases the output misalignment tolerance, thereby deteriorating the output coupling efficiency. To overcome this tradeoff, a tapered waveguide with its input cross-sectional area decreasing gradually toward the output has been used [37]. Such waveguides have been measured to have propagation losses of 0.1 dB/cm, showing that the tapering profile does not cause significant loss. Currently, most of the optical waveguides

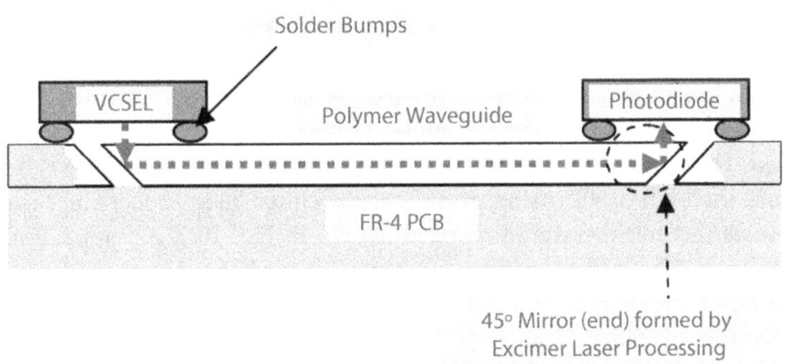

Figure 10.1 Cutoff side view of the integrated planar optical waveguide PCB.

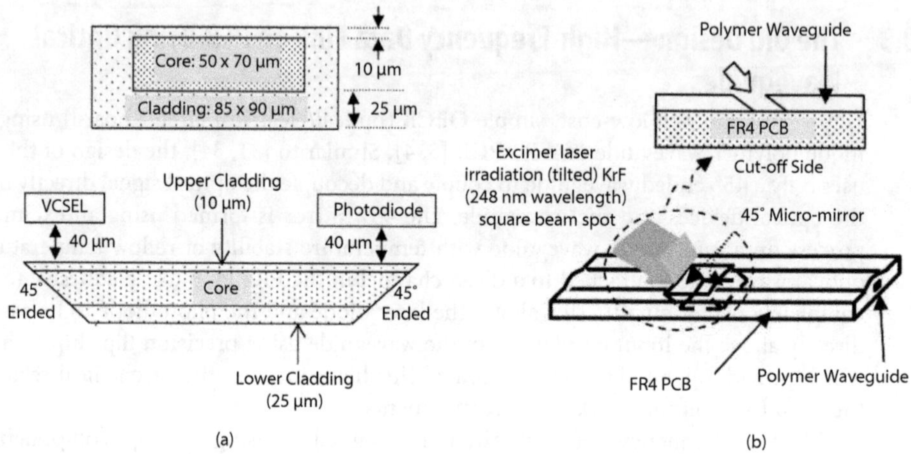

FIGURE 10.2 (a) Waveguide cross-sectional and side views. (b) 45° micro-mirror fabricated by excimer laser.

are designed to operate in the multimode for better input coupling efficiency and ease of fabrication. Intermodal dispersion is not critical as the transmission distance is very short, typically less than 10 cm for interchip on module or less than 50 cm for intermodule on board [38].

The optical waveguide for the present OECB is designed to operate in the multimode, with a constant core height of 50 μm and a width of 70 μm, as shown in Fig. 10.2. The cross-sectional area is optimized for the VCSEL beam profile with a divergence angle of 18° and the photodiode detecting area diameter of 70 μm. The waveguide profile is not tapered as this will cause the design of the waveguide to be modified and build for different lengths of waveguide. The refractive indices for the core and cladding are 1.5622 and 1.5544, respectively, and the materials are proprietary to Mitsui Chemicals Inc. The waveguide is designed to couple the optical signal from the top. The upper cladding layer is 10 μm thick while the lower cladding is 25 μm thick. The two edges of the 45°-ended waveguide are exposed so that the external material is air, which creates total internal reflection when the optical signal is directed 45° on it at both ends. In this way, the required out-of-plane reflection is realized at both ends without any additional optical components.

The assembling of the polymer waveguide consists mainly of the following steps. A U-type groove of 100 μm is first made on the PCB within which the polymer waveguide fabricated separately and diced earlier can be fitted exactly. The dicing channel location is identified from the VCSEL and photodiode pads on the FR4 PCB and no special markers are used. The polymer waveguide is attached to the PCB using adhesive. Both the waveguide ends are then cut at 45° using proprietary excimer laser processing technology of Mitsui Chemicals Inc. to form the mirrors, as shown in Fig. 10.2. No optical mirror coatings are incorporated. The excimer laser used has a square beam spot area. The square beam is to ensure that the surface of the mirror is flat, which is required for good total internal reflection. During this fabrication process, it is very crucial to dice accurately at the desired position with reference to the bonding pads of the optoelectronic devices, as shown in Fig. 10.3. The location of the 45° mirror is determined by the VCSEL and photodiode electrical pads.

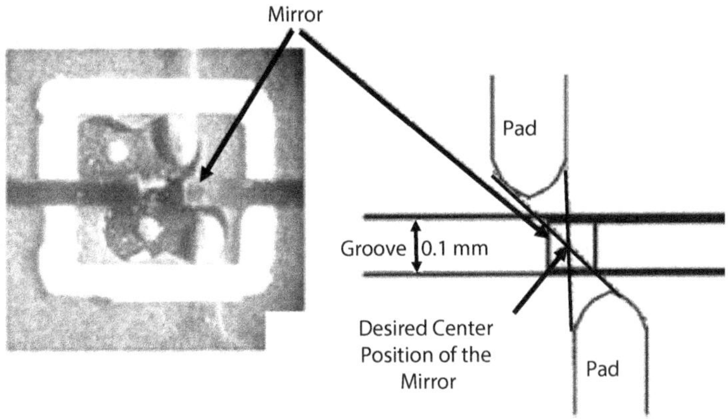

FIGURE 10.3 Top view of the fabricated 45° micro-mirror on the polymer waveguide with its location reference to the bonding pads of the optoelectronic device

With this fabrication process, the mirror center achieves a typical deviation of ~3.5 μm from the center of the two bonding pads. This can be considered a very good alignment, since the waveguide has a width of 70 μm.

The embedded waveguide with its out-of-plane 45° reflectors forms a complete planar IC. This allows the required VCSEL and the photodiode to be flip chip bonded into the PCB, as shown in Fig. 10.1. The fabricated 45°-ended optical waveguide is first characterized before assembling the various components. The optical source with a wavelength of 850 nm is coupled to the waveguide through a single-mode fiber, and the output is measured using a wide area photodiode. The measured waveguide losses using the cutback method range from 0.1 to 0.3 dB/cm.

The surface of the out-of-plane 45° reflector or mirror fabricated using the excimer laser process has a roughness of ~60 nm. This gives an average measured optical reflection loss of 0.6 dB. The measured total optical loss for the 82-mm long 45°-ended optical waveguide on both sides is less than 4 dB.

10.3.2 Simulations—Optical Coupling Models

Both the photodiode and the VCSEL used for this OECB are from ULM Photonics, which have an operating speed of 5 Gbps. The photodiode has a sensitive area radius of 35 μm. The VCSEL has a full-width beam divergence of 18°. With these optoelectronic components and the above-mentioned waveguide parameters, the optical link between the VCSEL and the photodiode is simulated using commercial optical design software, ASAP (Advanced Systems Analysis Program) from Breault Research Organization using ray tracing method. In the simulation, the VCSEL is considered an incoherent source of 14 μm and 18° divergence [28]. The loss of the waveguide of 0.3 dB/cm, which is determined from the earlier measurement, is incorporated into the simulation through the waveguide medium absorption coefficient. Similarly, the loss of the mirror of 0.6 dB is also incorporated into the simulator. The heights of the VCSEL and the photodiode are set to 40 μm, which is determined by the diameter of the solder ball specified by the supplier datasheet. During the simulation, only the position of the component under test is varied while the rest are set at their ideal condition.

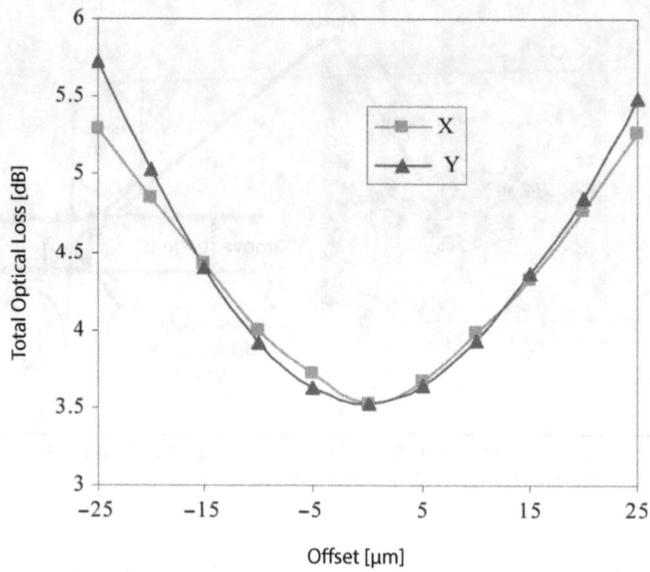

Figure 10.4 Simulated total optical loss as a function of the photodiode lateral misalignment.

Figure 10.4 shows the total optical loss as a function of the photodiode misalignment in the x and y directions. The simulated results show that the total optical loss is about 3.6 dB at ideal condition. The loss increases symmetrically for x-axis for both positive and negative directions. On the other hand, for the y-axis the loss increases at a slightly higher rate at higher degree of misalignment for the negative direction as compared to the positive direction. This is due to the waveguide termination in the negative direction. The total optical loss of both x and y-axes is sensitive to the photodiode misalignment. This is because the diameter of the photodiode active area, 70 μm, is comparable to the rectangular spot size of the waveguide output signal of approximately 60 × 76 μm, as shown in Fig. 10.5.

For the photodiode vertical misalignment, the total optical loss increases as the misalignment increases positively, as shown in Fig. 10.6. This is due to the dispersion characteristic of the optical signal when it emerges from the waveguide, which increases the spot size as the propagating distance increases. This leads to less optical signal being captured by the photodiode.

For the VCSEL misalignment in its lateral axis, as shown in Fig. 10.7, there is a window of ±15 μm misalignment tolerance for the y-axis and ±25 μm for the x-axis where the optical loss does not increase. This is mainly due to the large input aperture of the waveguide as compared to the VCSEL beamwidth, with a spot width of approximately 24 μm, as shown in Fig. 10.8.

Similarly, for the VCSEL vertical misalignment, the optical loss does not increase when the misalignment increases positively up to 100 μm, as shown in Fig. 10.9. Again this is due to the large input aperture of the waveguide. The above-simulated optical loss shows that the alignment of the photodiode is much more critical.

After the flip chip assembly process, the solder bumps may not give a constant height, and this will result in the VCSEL and photodiode being not planar with respect to the

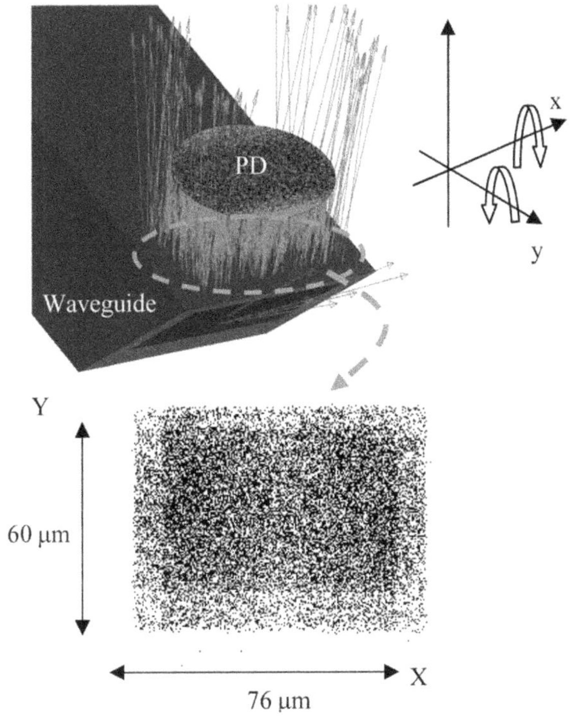

Figure 10.5 Simulated spot diagram of optical waveguide output signal with rectangular shape.

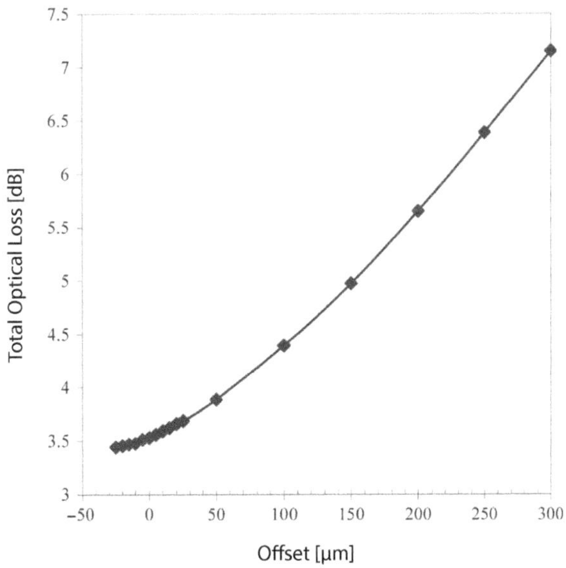

Figure 10.6 Simulated total optical loss as a function of photodiode vertical (*z*-axis) misalignment.

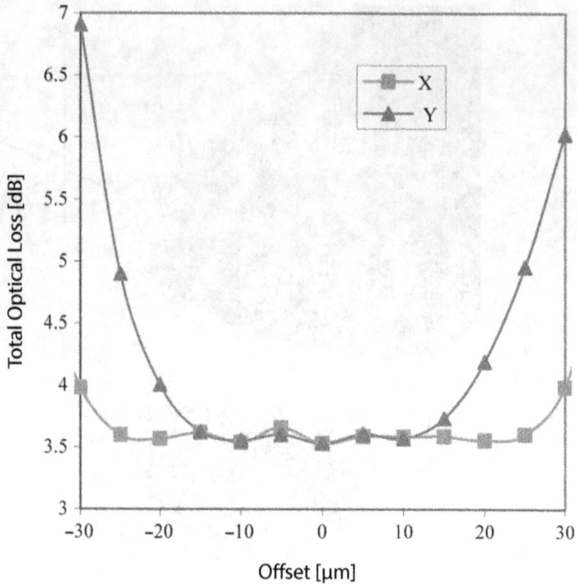

Figure 10.7 Simulated total optical loss as a function of VCSEL lateral misalignment.

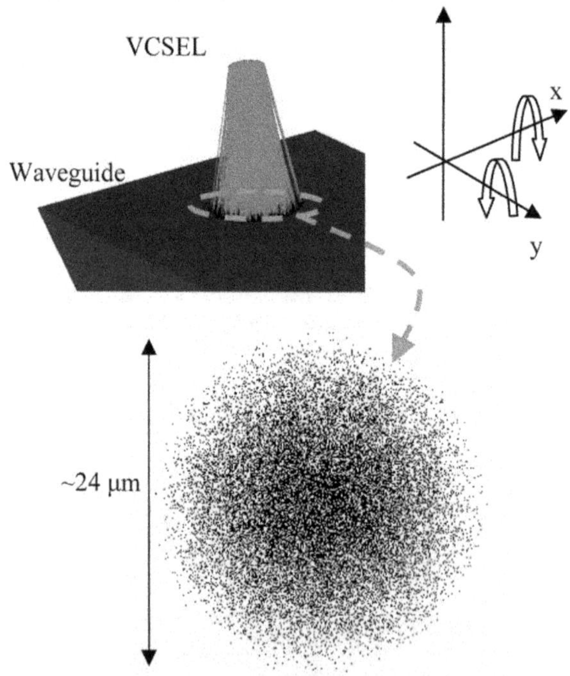

Figure 10.8 Simulated spot size of VCSEL at the surface of the waveguide.

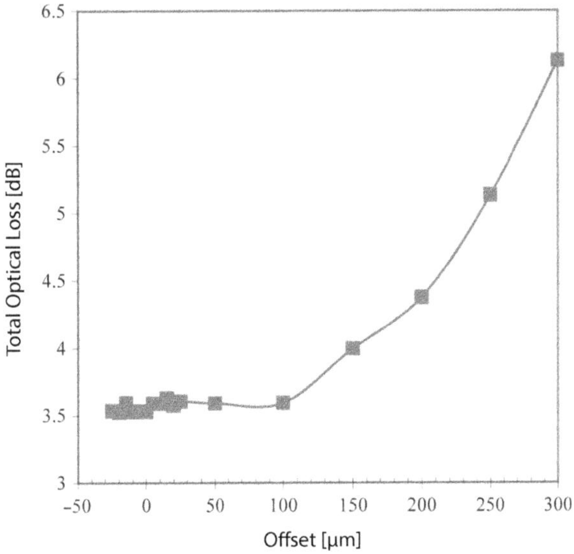

Figure 10.9 Simulated total optical loss as a function of VCSEL vertical (*z*-axis) misalignment.

waveguide top surface. The effects of this nonplanarity are simulated. During the simulation, the photodiode is tilted around the *x*- and *y*-axes at its center position, while the VCSEL is positioned at its ideal location. The axis of titling is shown in Fig. 10.5. The simulated results of the total coupling loss as a function of various tilting angles and axes are shown in Fig. 10.10. The results show that the tilting of the photodiode around the *x*- and *y*-axes has insignificant effect on the total coupling efficiency. This is because the maximum change of the active area of the photodiode with respect to the optical signal due to the titling is only less than 2%.

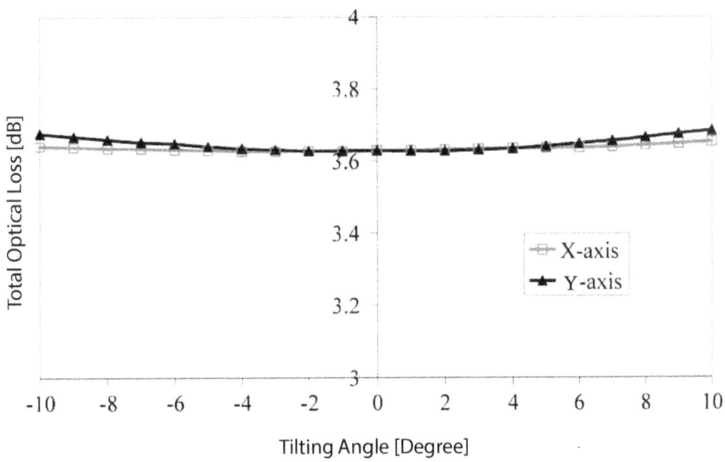

Figure 10.10 Simulated total coupling loss with respect to the photodiode rotation error in the *x*- and *y*-axes.

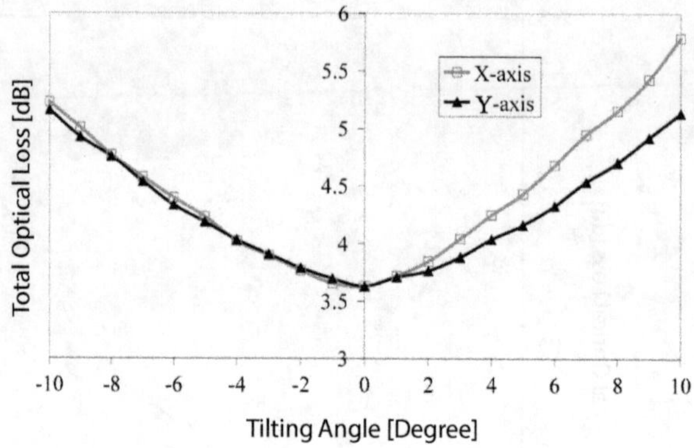

Figure 10.11 Simulated total coupling loss with respect to various VCSEL titling angle in the *x*- and *y*-axes.

Similarly, the VCSEL is then tilted around the *x*- and *y*-axes at its center position, while the photodiode is positioned at its ideal location. The axis of rotation is shown in Fig. 10.8. The simulated results in Fig. 10.11 show that the tilting of the VCSEL increases the total coupling loss—the total coupling loss increases by about 0.6 dB for a tilting angle error of $\pm 4°$. The tilting error will generally have a greater impact on the coupling loss if the distance between the optoelectronic component and the waveguide is large. The slight nonsymmetrical response for the *y*-axis tilting is due to the optical signal reaching the edge of the waveguide reflector when it is tilted with a positive angle error.

Another optical design concern is the additional coupling loss due to the tilting angular error of the 45° reflector during the fabrication process. The total coupling losses are simulated as a function of the titling angular error for both the 45° reflector at the VCSEL end (TX) and photodiode end (RX) separately. The total optical loss vs. the titling error is shown in Fig. 10.12. It shows that the total coupling loss has an additional loss of 2.8 dB when the 45° reflector at the VCSEL end has a tilting error of $\pm 4°$, while the total coupling loss has an additional loss of 1.5 dB when the 45° reflector at the photodiode end has a tilting error of $\pm 4°$.

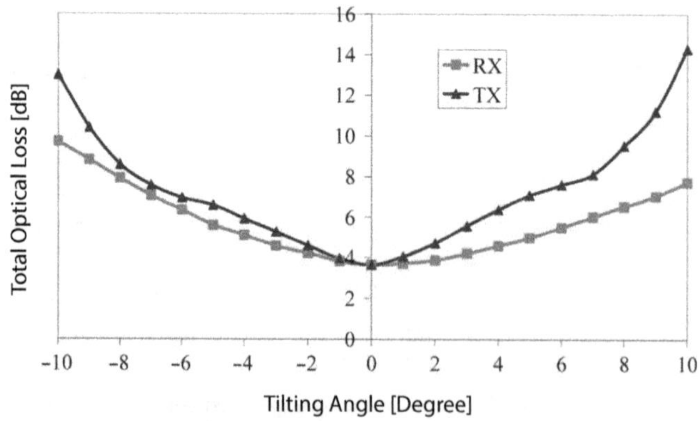

Figure 10.12 Simulated total coupling loss as a function of titling error of the waveguide 45° reflector at the VCSEL end (TX) and the photodiode end (RX).

10.3.3 Simulations—System Link Design

The OECB is designed to operate at 2.5 Gbps with a BER (bit error rate) of 10^{-12} using commercial available components and ICs. The TIA MAX3271 and LA (limiting amplifier) MAX3272 are both from MAXIM. The TIA has a typical input-referred RMS noise current (i_n) of 0.495 μA and a transimpedance gain (Z_{TIA}) of 2.8 kΩ. The input sensitivity of LA is 15 mVp-p. For BER of 10^{-12}, an SNR (input signal to ratio) of 14.1 is required [39]. The typical responsivity (ρ) of the photodiode is 0.5 A/W.

$$\mathrm{OMA} = \frac{i_n \mathrm{SNR} + \left(\dfrac{V_{\mathrm{TH}}}{Z_{\mathrm{TIA}}}\right)}{\rho}$$

$$P_{\mathrm{AVG}} = \frac{\mathrm{OMA}(r_e + 1)}{2(r_e - 1)}$$

Using the above equations [39], the minimum OMA (optical modulation amplitude) calculated is 24.6 μW. Assuming the extinction ratio (r_e) of 8, the average optical power calculated is then 15.8 μW or −18 dBm. With a typical output power of 0 dBm from the VCSEL, the maximum optical loss for the whole design is 18 dB.

Assume the assembly processes have typical alignment errors of ±20 μm in the lateral x- and y-axes, and also a ±20 μm in the vertical z-axis. In addition, assume the VCSEL and the photodiode have titling error of ±4° both in the y- and x-axes. The estimated losses from various processes are given in Table 10.1. The computed root mean square error due to these assembly misalignments is about 4.5 dB. This result shows that the optical loss due to the typical assembly error is much lower than the maximum allowable coupling loss of 18 dB for 2.5 Gbps bit rate with 10^{-12} BER OECB design.

Table 10.1 Simulated Losses due to Various Misalignments and Tilting Errors

	Misalignment/Tilting	Error	Loss (dB)
1	PD lateral X	± 20 μm	4.8
2	PD lateral Y	± 20 μm	5
3	PD vertical Z	± 20 μm	4.8
4	PD titling Y	± 4°	3.7
5	PD titling X	± 4°	3.7
6	VCSEL lateral X	± 20 μm	3.6
7	VCSEL lateral Y	± 20 μm	4.2
8	VCSEL vertical Z	± 20 μm	3.6
9	VCSEL titling Y	± 4°	4.2
10	VCSEL titling X	± 4°	4
11	PD 45° reflector titling	± 4°	5.1
12	VCSEL 45° reflector titling	± 4°	6.4

10.3.4 Assembly of the OECB

The waveguide is first manually assembled into the grove of the OECB to form a complete planar circuit. The 45° reflectors are then formed at both ends of the waveguide using excimer laser process as described in Sec. 10.3.1. In order to ensure that all the components can be assembled using the standard commercial assembly process, the bare chip TIA is first packaged into a ceramic SOIC (small outline IC)-16 that has superior high-frequency performance compared to the plastic packages. The packaged TIA and all the other electrical components are then assembled into the OECB.

The VCSEL and the photodiode are assembled on the FR4 PCB using flip chip bonding. Both the VCSEL and the photodiodes are supplied with predeposited 60-μm-diameter AuSn solder bump on their respective pads. After assembly, the distance between the optoelectronic active area and the waveguide top surface is about 40 μm. In order to achieve maximum mechanical stability, four solder bumps on each optoelectronic component are used for the assembly, as shown in Figs. 10.13 and 10.14. Both the standard optoelectronic components have an array of 1 × 2; however, only one element is used for this single-channel communication. During assembly, it was found that there was some inherent misalignment due to the PCB fabrication inaccuracy. The design pad size for both the VCSEL and photodiode is 100 μm; however, the fabricated PCB pad size is reduced to approximately 52 to 67 μm. This resulted in some assembly difficulties and alignment errors.

For the VCSEL, even though the pad size in the PCB is much smaller, the assembly is not greatly affected and good assembly accuracy can be obtained, as seen in Fig. 10.14(a). On the other hand, the photodiode that has a bigger pad pitch cannot be assembled with good accuracy, as seen in Fig. 10.13(a). To overcome this mishap, the photodiode has to be assembled with a slight tilting angle laterally.

As shown in Figs. 10.13(a) and 10.14(a), it is not possible to measure the amount of misalignment as the reference out-of-plane 45° reflector is not visible from the X-ray pictures. In addition, the photodiode active area and the pads on the PCB are not visible. However,

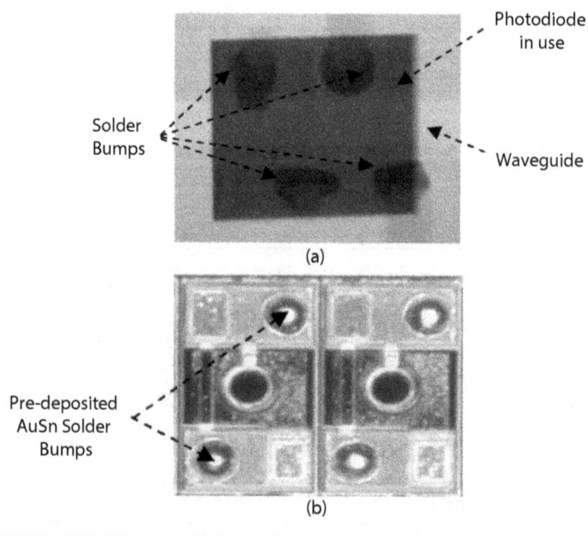

FIGURE 10.13 (a) X-ray image of assembled photodiode. (b) Top view of the photodiode.

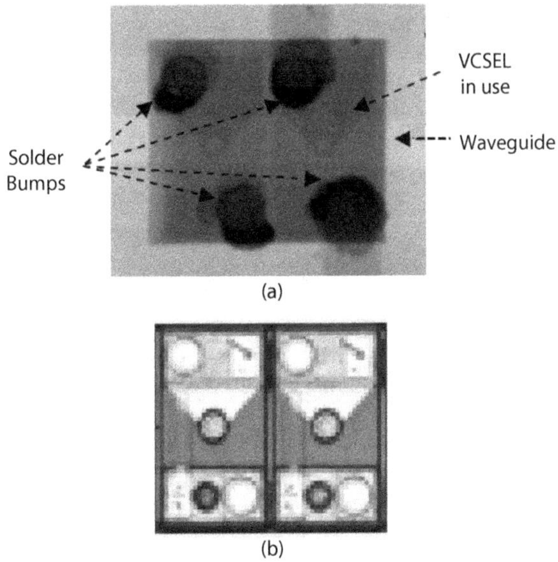

FIGURE 10.14 (a) X-ray image of assembled VCSEL. (b) Top view of the VCSEL.

using the waveguide and the VCSEL emitting area, as shown in Fig. 10.14(a), the assembly accuracy in the x-axis is estimated to be about ± 5 μm.

10.3.5 Measurement Results of the OECB

The assembled OECB as shown in Fig. 10.15 is measured using Anritsu Data Analyzer to generate the PRBS NRZ and Tektronix Digital Sampling Oscilloscope to capture the output signal. The measured result at 1.25 Gbps shows a large eye pattern, as shown in Fig. 10.16(a). To further investigate the performance of the OECB, the transmission rate is increased to

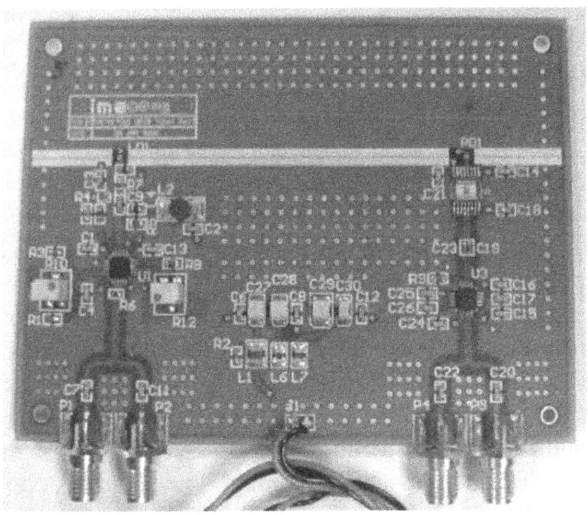

FIGURE 10.15 Assembled OCEB using SMT.

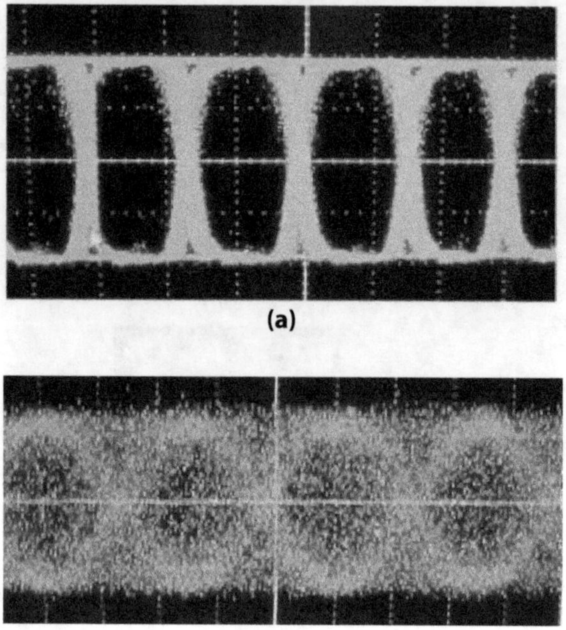

(a)

(b)

Figure 10.16 Measured eye-diagrams of the OECB. (a) 1.25 Gb/s. (b) 2.5 Gb/s.

2.5 Gbps. However, the measured eye pattern at 2.5 Gbps, as shown in Fig. 10.16(b), is very noisy and not capable of transmitting any data.

The measured optical insertion loss for the 82-mm-long waveguide with the 45° reflectors at both ends is approximately 4 dB; hence, the probable cause for the low bit rate is the high coupling loss between the optoelectronic components and the waveguide, which is caused by the inherent misalignment described in Sec. 10.3.4. The total coupling loss of the circuit is highly sensitive to the alignment accuracy of the photodiode as shown in the simulation result in Fig. 10.4. On the other hand, the VCSEL has a larger misalignment tolerance, as shown in Fig. 10.7; moreover, its X-ray picture has shown that the assembly has good accuracy, as shown in Fig. 10.14. Lower optical loss can be achieved with proper PCB fabrication control to give the desired optoelectronic pads dimension, which determines the optical coupling efficiency. In addition, the titling angular error of the 45° reflector, as shown in the simulated results in Fig. 10.12, can also contribute to significant coupling loss. The degree of tilting error is not measured and the additional coupling loss due to these tilting angular errors is not included in the measured waveguide loss of 4 dB.

The other probable cause is some electrical design flaws in the SOIC package of the TIA. The design of the TIA in the SOIC has led to additional parasitic inductance and capacitance due to the bonding wire length and pads, respectively. These parasitic components reduce the bandwidth and lead to poor measurement results. Finally, poor performances may also be caused by the thermal issues, as there is no thermal control design in the electrical and optical circuitries.

10.3.6 Summary and Recommendations

A simple and low-cost OECB with a transmission rate of 1.25 Gbps and a length of 82 mm is demonstrated on FR4 PCB. Some important results and recommendations are summarized as follows [3]:

- Its simple design allows the OECB to be fabricated and assembled with current industrial technology.
- It does not have any focusing lens and it utilizes only the precision of the flip chip bonder to align both the VCSEL and the photodiode with the optical waveguide.
- All the electronic components, i.e., LDD, TIA and the limiting amplifier, are assembled directly on the FR4 PCB.
- The optical simulation results and its electrical design show that the direct couple design is capable of achieving 2.5 Gbps with a BER of 10^{-12}. However, due to the fabrication error and some designs oversight, the measured response is only up to 1.25 Gbps.

10.4 The Old Design—Embedded Board-Level Optical Interconnects

Figure 10.17 shows a single-channel OECB with embedded waveguide using traditional PCB manufacturing process [6–8]. The OECB is made up of four electrical layers and one optical layer embedded below a 60-μm-thick BT substrate. Two optical vias with 100 μm diameter are formed to guide the optical beam from VCSEL to the 45° mirror coupler. Likewise, optical beam exiting the waveguide is diverted from the 45° mirror coupler through the optical via and received by the photodetector.

A 10-cm-long embedded polymer waveguide consists of a 70 × 70 μm core and a 15-μm-thick top and bottom cladding. Two 45° mirror couplers are formed at both corners of the waveguide by using 90° diamond dicing blade. These mirrors convert optical beam emitted from VCSEL in the vertical path to the planar direction and into the waveguide.

10.4.1 Fabrication of Polymer Waveguide

Various fabrication processes are used to fabricate the flexible polymer waveguide, such as photolithography [30, 31], RIE (reactive ion etching), hot embossing, and soft molding [40]. In this section the soft-molding method is discussed, as it is cost effective and convenient to

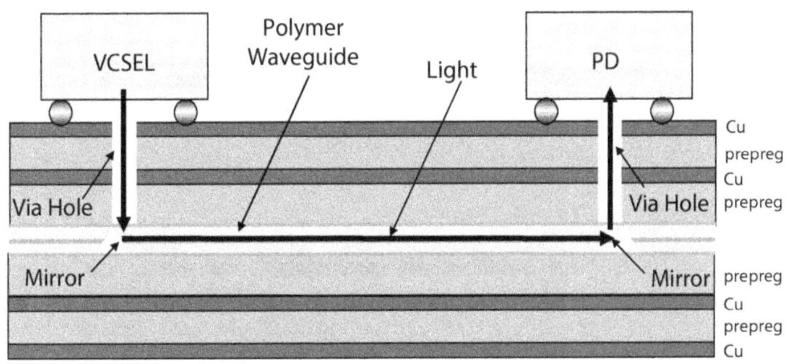

Figure 10.17 Schematic of fully embedded board-level optical interconnect system.

fabricate micro and nanostructures [30–31, 40–42]. Soft molding is suitable for replicating 3D structures in single-step and for mass production or small quantity in prototyping.

The fabrication process of the waveguide using the UV-curable fluorinated acrylate material is divided into four steps, as shown in Fig. 10.18. First, a layer of bottom cladding material (WIR30-RI) is spin coated on top of the copper polyimide film with a supporting substrate. Second, the PDMS (poly dimethylsiloxane) mold is placed on top of the bottom cladding material. It is then filled up with the core material and is UV cured inside a nitrogen environment. The polymer waveguide is 10 cm long with a 70 × 70 μm core cross-section. Third, the soft mold is removed from the polymer substrate. Finally, it is spin coated with another layer of cladding material as top cladding and then it is UV cured.

10.4.2 Fabrication of the 45° Micro-Mirror

To construct a 3D optical interconnection between waveguides and active optical devices, integrated micro-mirrors are often introduced to assist in deflecting the optical light into/out of the waveguide. Hence, the micro-mirror is an essential basic component for 3D optical interconnections [29]. Till now, the fabrication processes of the micro-mirror have been done by laser ablation [3, 43], grayscale lithography [44], dicing [45], X-ray lithography [46], reactive ion etching [47], and molding [48]. The dicing method is discussed in this section.

After the waveguide is fabricated, the next process fabricates the 45° micro-mirror by mechanical dicing using 90° diamond dicing blade. A few key design issues including the

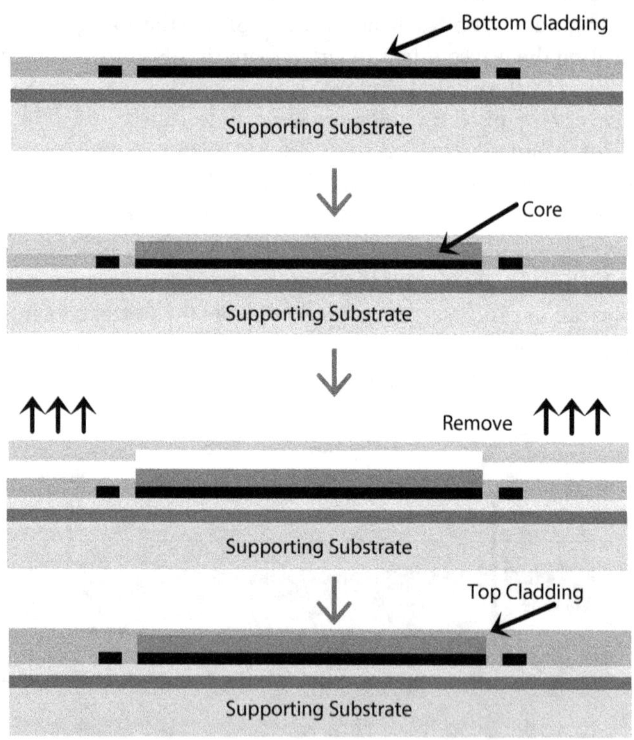

FIGURE 10.18 Steps in fabricating the polymer waveguide.

Table 10.2 Dicing Parameters for Dicing Mirror

Samples	Spindle speed (RPM)	Feed rate (mm/s)
X1	35,000	1
X2	35,000	5
X3	45,000	1
X4	45,000	5

positioning of the mirror, the tilting of the mirror, the angle of the mirror, and the mirror surface roughness should be noted when fabricating the micro-mirror because these issues affect the coupling efficiency. Therefore, a set of different dicing parameters are selected for the fabrication of the 45° micro-mirror (Table 10.2). This table provides the spindle speed and the feed rate, which affect the roughness and the angle of the micro-mirror. The result will determine which set of dicing parameter is able to achieve the expected requirement.

The optical waveguide supported by the substrate is mounted on the dicing machine. It is then aligned using the alignment marking, as seen in Fig. 10.19. The micro-mirror is diced at the dicing line position and the 45° micro-mirror is formed. Figure 10.20 shows the side view of the optical waveguide with the micro-mirror.

After the 45° micro-mirror is formed, the next step is to ensure that the angle of the micro-mirror is within the range. To obtain the angle of the micro-mirror, the optical waveguide is cross-sectioned and is determined using SEM (scanning electron microscopy). When the image is captured using the SEM, 3D graphical software is used to measure the angle of the micro-mirror, as seen in Fig. 10.21. The following steps explain how the angle is measured:

1. A baseline is drawn parallel to the plane.
2. Mirror line is drawn along the edge of the mirror. Two mirrors are drawn, which intersect at a point.

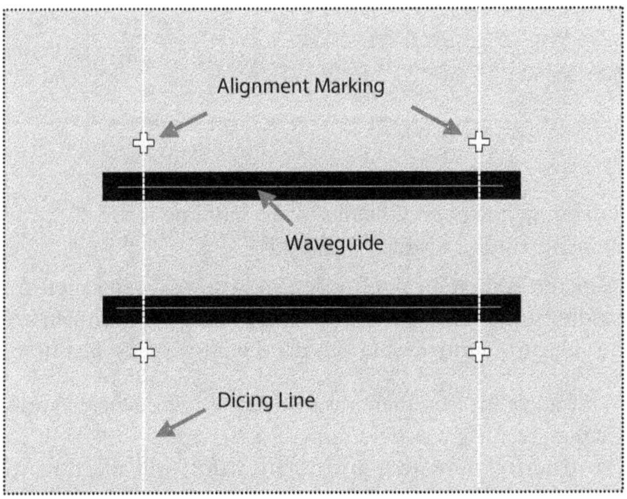

Figure 10.19 Top view of the optical waveguides and alignment marks.

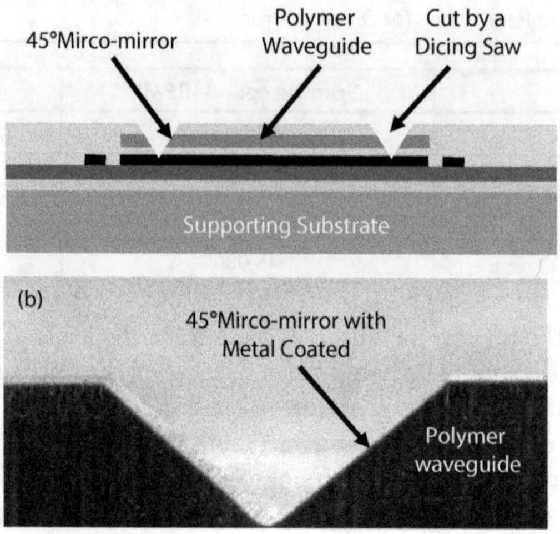

FIGURE 10.20 (Top) Schematic of the 45° micro-mirror. (Bottom) Side view of the 45° micro-mirror.

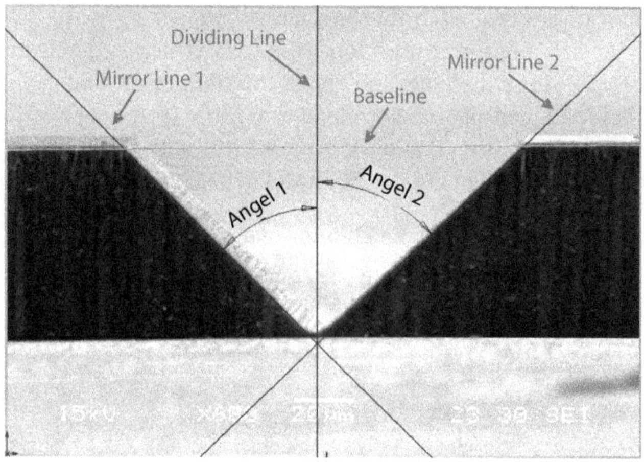

FIGURE 10.21 Measurement of the 45° micro-mirror.

3. A dividing line perpendicular to the baseline is drawn through the intersect point of the mirror lines, which forms 90°.
4. Using the 3D graphical software measuring tool, the angle from mirror line 1 to the dividing line is measured, which creates Angle 1, as shown in Fig. 10.21. For Angle 2, the angle from mirror line 2 to the dividing line is measured.

Finally, the angle measurement results (Angle 1 and Angle 2) are tabulated according to the set of dicing parameters as determined earlier on.

There is no perfect 45° micro-mirror. The defect will affect the coupling efficiency. Due to the defect form, the mirror area becomes smaller and the coupling light source is unable to deflect accurately into the waveguide core and coupling out to the receiver, which increases

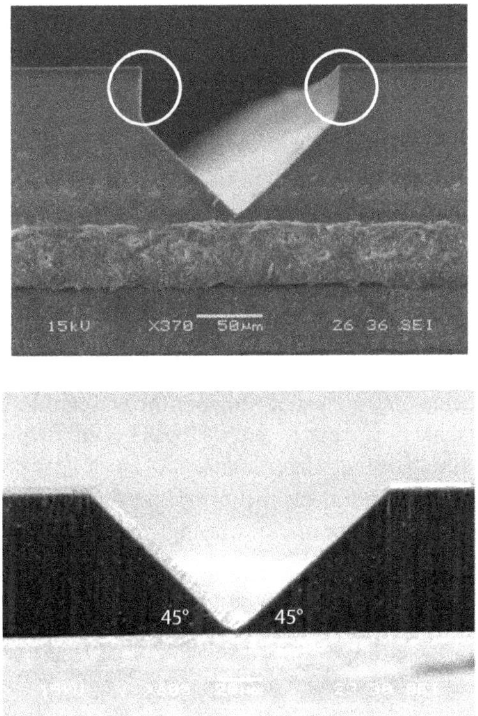

Figure 10.22 (Top) Side view of the 45° micro-mirror (half cure). (Bottom) Side view of the 45° micro-mirror (fully cure).

the optical loss. Due to this, the analysis of the waveguide is carried out. To ensure that the curing condition of the waveguide causes the defect, a comparison test is carried out. Both waveguides are fabricated with different curing conditions, one is fully cured and the other is partially (half) cured. After this the waveguides are diced to form the micro-mirrors. The partially cured waveguide has the defect form, as shown in the top of Fig. 10.22, and the fully cured waveguide has a perfect 45° micro-mirror shape without defect, as shown in the bottom of Fig. 10.22. Therefore, the waveguide must be fully cured, and the optimizing curing condition for the waveguide is UV intensity of 15 to 18 mW/cm^2 for 1.5 h in a nitrogen environment.

The reason that causes the defect is that the aspect ratio of the spin rate and feeding speed provides sufficient removal rate to remove the waveguide materials at the tip of the V-groove. However, due to the increase in the removal volume at the top surface, the aspect ratio is no longer sufficient for the removal rate. Therefore, during the dicing process, the dicing blade pushes aside the waveguide material instead of removing the waveguide materials. Then, it recovers after the blade loading is removed, which results in the defects as shown in the top of Fig. 10.22. This shows that the waveguide materials experience the elastic deformation during the dicing process. This does not happen for cured waveguide materials, as shown in the bottom of Fig. 10.22. Therefore, it is believed to be due to the uncured waveguide material that is still in soft and elastic state.

Under the SEM, another issue is found. From the top view of the waveguide, as seen in the top of Fig. 10.23, a slight tilting of the mirror is observed. The positioning and the tilting of the angle of the mirror affect the optical coupling efficiency, as a result there is an increase in optical loss. In order to correct the displacement error in placing of the core of the waveguide, precision alignment bonder is used to minimize the waveguide tilting; the bottom of Fig. 10.23 shows how the measurement was carried out.

A set of alignment marks, as seen in Fig. 10.19 (top view of the waveguide substrate), is used to align the PDMS soft mold, which is used to fabricate the core of the waveguide. On the PDMS mold, there is a set of alignment marks similar to the marking on the waveguide substrate. The two sets of marking are placed on top of each other; this is the perfect scenario for a perfect alignment without errors. But error always occurs and alignment system is implemented to minimize the misalignment. For normal placement of PDMS mold, there will be misalignments that will cause tilting of the waveguide. Table 10.3 shows the different sample misalignments of the placement of PDMS mold by manual alignment; the angle of tilting causes the tilting of the mirror angle.

As the dicing machine moves in the straight line, normal to the waveguide materials with reference to the alignment marks, the angle is formed by the placement of the core that

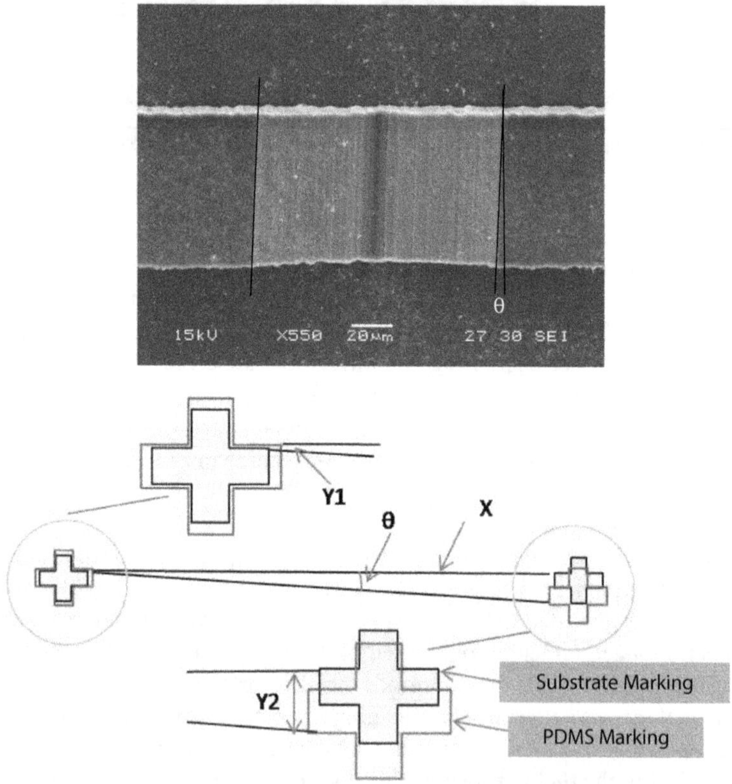

FIGURE 10.23 (Top) Mirror titling (without alignment). (Bottom) Measurement of the position of waveguide.

Table 10.3 Results of the Alignment Difference in *y*-Axis (Alignment System)

Samples	Y1 (μm)	Y2 (μm)	θ (°)
1	0	3	0.008
2	1	3	0.006
3	1	4	0.008
4	0	1	0.003
Average			0.006

Table 10.4 Results of the Alignment Difference in *y*-Axis (Manual Placement)

Samples	Y1 (μm)	Y2 (μm)	θ (°)
1	10	123	0.32
2	15	44	0.30
3	8	54	0.45
Average			0.36

causes the tilting of the mirror. That is why placement of the core is crucial, and therefore, high accuracy alignment machine is used to place the core so that the slanting is minimized. Because of the tilted angle of the mirror, the coupling light will not completely couple to the core. Small amount of the coupling light will be lost due to the angle error.

Using the alignment system provides a huge improvement in alignment issue. Table 10.3 shows a few sets of measurement by alignment system. According to the result in Table 10.3, the average angle of the tilting for alignment system is 0.006. This minimizes the tilting angle and improves the mirror angle, reducing the coupling loss; the average angle of the tilting for manual alignment is 0.36 (Table 10.4).

Different spindle speed and feed rate affect the angle of the mirror and the roughness of the mirror. After dicing using a set of dicing parameters given in Table 10.2, an optimized dicing parameter is chosen. Table 10.5 lists the results of the angles of micro-mirrors of different sets of dicing parameters, which are measured using SEM and 3D graphical software as mentioned earlier. From Table 10.5, the angle that meets the requirement is obtained for the dicing parameter: spindle speed of 35,000 rpm and feed rate of 1 mm/s. For this micro-mirror Angle 1 is 45.23° and Angle 2 is 45.85°, which are within the range of $45° \pm 1°$.

Table 10.5 Result of the Angle of the Micro-Mirror

Samples	Spindle speed (RPM)	Feed rate (mm/s)	Angle 1 (°)	Angle 2 (°)
X1	35,000	1	45.23	45.85
X2	35,000	5	46.79	45.73
X3	45,000	1	39.93	46.58
X4	45,000	5	43.98	42.39

TABLE 10.6 Result of the Roughness of the Micro-Mirror

Samples	Spindle speed (RPM)	Feed rate (mm/s)	Roughness (nm)
X1	35,000	1	316
X2	35,000	5	355
X3	45,000	1	340
X4	45,000	5	389

TABLE 10.7 Roughness Result for Spindle Rate (35,000 rpm) and Feed Rate (1 mm/s)

	Roughness (nm)
1	0.313
2	0.307
3	0.312
4	0.309
5	0.312
6	0.315
7	0.317
8	0.322
9	0.325
10	0.324
Average	0.316
StdDev	0.006

For dicing parameter of spindle speed of 45,000 rpm and feed rate of 1 and 5 mm/s, the angle of the micro-mirror is out of the range, and affects the coupling efficiency. The angle is either smaller or bigger than 45°, which affects the optical loss as the coupling light is unable to accurately deflect into the core of the waveguide.

Next, the roughness of the mirror surface is measured, taking 10 different areas of the surface of the mirror for measurement and getting the mean of the roughness of the surface. Table 10.6 provides the result of the roughness in these 10 different areas of the mirror surface. The average of these values is 316 nm with standard deviation of 0.006. Table 10.7 shows that the best roughness result is for spindle speed of 35,000 rpm and feed rate of 1 mm/s as compared to the rest. The remaining dicing parameters, even though their roughness is close to 316 nm, are not able to achieve the angle requirement. Therefore, the spindle speed of 35,000 rpm and feed rate of 1 mm/s is the preferred dicing parameter.

10.4.3 Assembly Process of the OECB

Figure 10.24 shows the optical images of the micro-mirror and polymer waveguide on a polyimide film. Two critical challenges are needed to be resolved in assembling the OECB: (a) adhesion of lamination between waveguide and prepreg material, and (b) precision

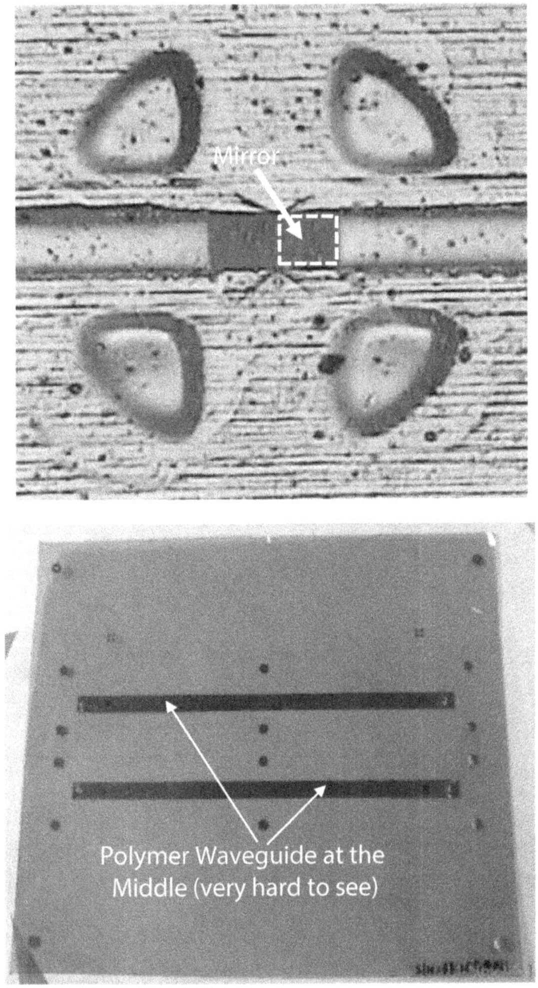

Figure 10.24 Optical images of the micro-mirror and polymer waveguide on a polyimide film.

of layer-to-layer alignment between waveguide mirror, PCB vertical-optical channel, and optical devices packaging. In consideration of these two points, the fabrication process is designed and developed [9] and shown in Figs. 10.25 and 10.26.

Before laminating, surface pretreatment, such as plasma, micro-etching, or pumice, is conducted on the surface of waveguide layer. This pretreatment can increase the surface roughness; therefore, adhesion between waveguide layer and prepreg can be increased as well. Then, stack up with waveguide layer located in the center between two pieces of prepreg and copper foils, perform the laminating process, and form the sandwich structure of copper–prepreg–waveguide–prepreg–copper, which is called L2/L3, as shown in Fig. 10.25. The laminating condition is 200°C for 100 min. After laminating, X-ray is used to detect alignment marks and to drill out alignment holes used for L2/L3 patterning, and the same method is used for L1/L4 laminating. Also, use X-ray to drill out alignment holes, and

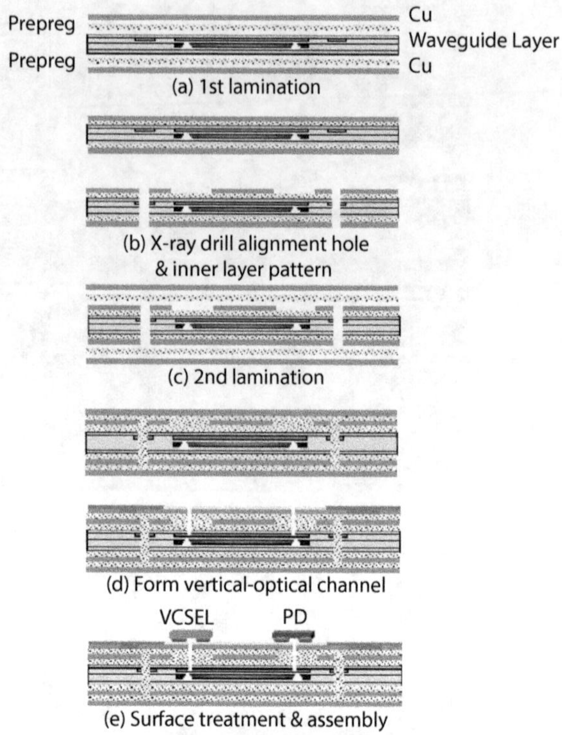

FIGURE 10.25 Fabrication process of the OPCB.

proceed drilling through the holes and copper plating. Then form vertical-optical channel to make the mirror exposed. Finally, outer layer patterning, surface finishing, and a 4 + 1 layer OECB are completed.

10.4.4 Fabrication Process of Vertical-Optical Channel

Fabrication of vertical-optical channel is the most important step of the whole process. A special process is designed and discussed as follows. Three key points are discussed in this section. The first is how to align with mirror in the inner layer of PCB from the top layer; the second is how to control the roughness of the channel wall; the third is how to control the shape of the channel. The special fabrication process is shown in Fig. 10.26.

First, use UV laser to open a copper conformal mask on the desired location of vertical-optical channel, and press the dry film on the surface. The UV laser can precisely find the location and drill directly. Use CO_2 laser to ablate prepreg and form a blind via; the roughness of the wall can be optimized by adjusting laser parameter. Then, remove bottom copper by flash etching, strip the dry film, and form the vertical-optical channel. Finally, use the conventional HDI (high density interconnect) process and finish the OECB.

10.4.5 Final Assembly

After OECB is completed, SMT (surface mount technology) is used to do the final assembly. Solder paste SAC305 is used, and the maximum temperature is 240°C. The solder pad size of

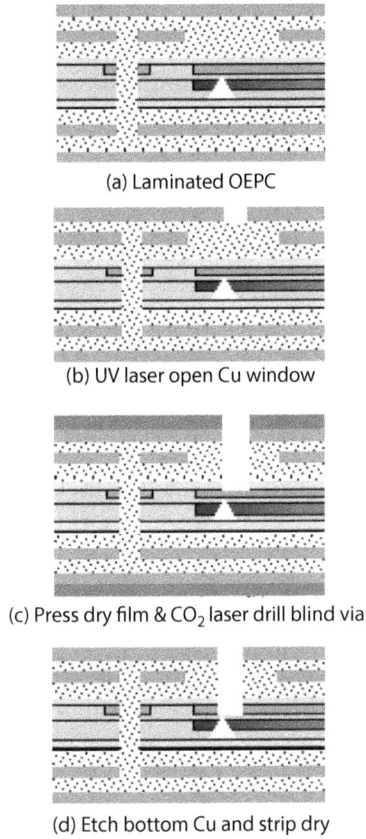

FIGURE 10.26 Fabrication process of the vertical-optical channel.

VCSEL and PD is only 80 μm, so micro-lens is used for soldering. The selected VCSEL can emit laser beam with 850 nm wave length, frequency can reach 5 Gbps, diameter of emitter opening is 13 μm, and light dissipation angle is about 17°. The assembled sample is shown in Fig. 10.27 and the size is 12 by 5 cm. The test setup is shown in Fig. 10.28 and some of the measurement results can be found in [6, 9, 32].

10.4.6 Summary and Recommendations

Some important results and recommendations are summarized as follows [6–8]:

- The mirror surface roughness is achieved (316 nm) with standard deviation of 0.006 by optimizing the spindle speed of 35,000 rpm and feed rate of 1 mm/s.
- The angle of the mirror of about $45° \pm 1°$ is achieved by optimizing the spindle speed of 35,000 rpm and feed rate of 1 mm/s.
- The positional misalignment of the micro-mirror in the waveguide (~3 μm) is achieved by employing the alignment marking at four corners and vision alignment system (Fig. 10.19).

- The angle of the tilting mirror of average 0.006° is achieved by using the alignment system.
- Perfect V shape of the micro-mirror is obtained by optimizing the curing process parameter (UV intensity of 15 to 18 mW/cm^2 for 1.5 h in a nitrogen environment).
- The fabrication of waveguide layer, measurements of optical characterization, fabrication of 4 + 1 layer OECB (including vertical-optical channel), assembly of surface devices, and final optical characterization on OECB are successfully completed.
- The CTE (coefficient of thermal expansion) of the polymer waveguide is more than two times of all other materials. Thus, the thermal expansion mismatch among all the materials during lamination causes defects such as delaminating between the waveguide and PI or copper and the waveguide being twisted and even broken [9].
- The effect of swollen material and shrinkage of the polymer waveguide induced by the lamination is to increase the scale measurements (100.30%) and deform from rectangle to rhombus or trapezoid shapes [9].

FIGURE 10.27 Top view of the assembled OECB.

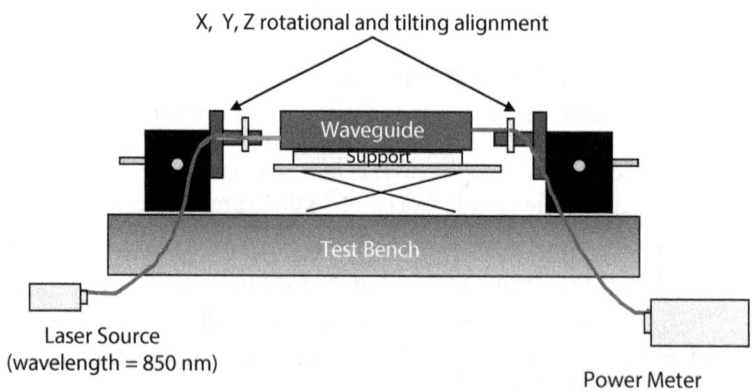

FIGURE 10.28 Measurement setup for propagation loss and insertion loss.

10.5 The New Designs

In general, this new design relates to the optoelectronic packaging of 3D IC integration embedded into an optical polymer waveguide PCB (or substrate) for high-performance, low-profile, light-weight, compact, and low-cost applications. In particular, this design relates to embedded chip-to-chip optical interconnect in a 3D hybrid IC integration SiP (system-in-package) format.

A 3D hybrid optoelectronic system with electrical and photonics ICs for super-thin, super-light, high-performance, and low-cost applications, as shown in Fig. 10.29, is proposed [49, 50]. The entire system consists of a rigid PCB made of FR4 material (or a substrate made of, e.g., BT materials) with an embedded optical polymer waveguide, an embedded VCSEL, an embedded LD (laser driver), an embedded serializer, an embedded photodiode detector, an embedded TIA, an embedded deserializer, and heat slugs with or without heat spreader. The bare VCSEL, LD, and serializer chips are stacked and then attached on one end of the embedded optical polymer waveguide in the PCB. It is noted that since the VCSEL is so close to the mirror of the optical polymer waveguide, an optical lens is optional. Similarly, the bare photodiode detector, TIA, and deserializer chips are stacked and then attached on the other end of the embedded optical polymer waveguide in PCB. Both the 3D stacked chip sets are encapsulated by special underfills, e.g., transparent polymer.

As shown in Fig. 10.29, the VCSEL (with light emitting from the front side of the chip) with TSV is flip chip bumped with any materials on the VCSEL driver chip with TSV, which is solder bumped flip chip on the serializer chip. Larger bumps with any materials are mounted on the serializer chip while it is in a wafer format.

After dicing the 3D hybrid IC chip set, it is placed on the PCB (or substrate) on top of the polymer waveguide. Special encapsulants such as the transparent polymer may be needed

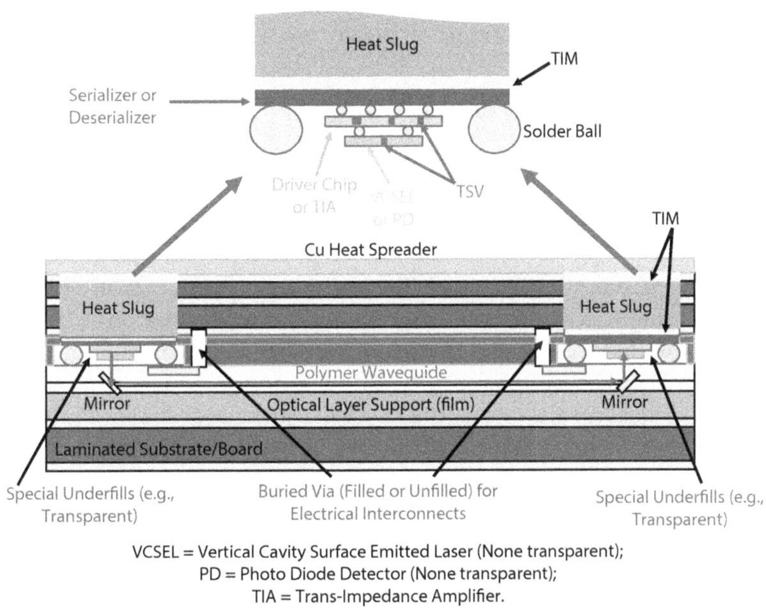

FIGURE 10.29 Embedded hybrid 3D integration for opto-electronic interconnects.

to protect the chip set. If it is needed, a heat slug or any conductive material can be attached to the back side of the serializer chip by a TIM (thermal interface material). Again, if it is needed, a heat spreader can be attached to the top side of the heat slug with a TIM. If the light emits from the back side of the VCSEL, then there is no need to have TSV in the VCSEL for electrical feed through.

Similarly, the photodetector chip is solder bumped flip chip on the TIA chip, which is then solder bumped flip chip on the deserializer chip. The thermal management techniques are the same as those of the VCSEL chip set. In this case, underfills may be needed between the VCSEL chip and VCSEL driver chip; the VCSEL driver chip and serializer chip; the PD chip and TIA chip; and the TIA chip and deserializer chip.

In many applications the serializer and deserializer chips may not be needed or have to be placed outside the PCB. Also, when heat dissipation of the driver and TIA chips is not very large, then the heat spreader may not be necessary. Furthermore, if the heat dissipation of the driver and TIA chips is very small, then the heat slug or conductive materials are not even needed.

10.6 An Embedded 3D Hybrid Integration Design Example

A low-cost embedded 3D hybrid optoelectronic system with electrical and photonics ICs for board-level applications, as shown in Fig. 10.29, is designed and analyzed [49, 50]. The entire system consists of a rigid PCB made of FR4 material with an embedded optical polymer waveguide, a 10 Gbps VCSEL (0.31 × 0.4 × 0.2mm, 2.2 V, 33 mW, 16 W/cm^2), a 10.7 Gbps LD (2 × 2 × 0.3 mm, 3.3 V, 0.35 W, 6.35 W/cm^2), a 10 Gbps 16:1 serializer or multiplexer (4.5 × 4.5 × 0.8 mm, 3.3 V, 2.5 W, 2.5–6.7 W/cm^2), a 10 Gbps photodiode detector (0.31 × 0.4 × 0.23 mm), a 10.7 Gbps transimpedance amplifier or TIA (2 × 2 × 0.3 mm), a 10 Gbps 1:16 deserializer or demultiplexer (4.5 × 4.5 × 0.8 mm, 3.3 V, 2.5 W, 2.5–6.7 W/cm^2), and heat spreaders. The optical, thermal, and mechanical design philosophy and their metrics of the proposed 3D optoelectronic system are given and the analysis results are discussed.

10.6.1 Optical Design, Analysis, and Results

(A) Optical Design Philosophy

Besides the heat slug/spreader, Fig. 10.29 shows that the top most layer of the 3D system contains a parallel to serial electrical converter called serializer. The serializer converts the 16-bit 622 Mbps parallel signals into a 10 Gbps serial signal which is fed to the next layer (the VCSEL LD) in the 3D system. The signal inputs to the serializer and VCSEL LD are re-routed using wafer level RDL (redistribution layer) deposition techniques. The VCSEL itself is solder bumped on the LD, which converts the incoming signal into an electrical driving signal for the VCSEL.

Again, Fig. 10.29 shows that the receiver schematic is similar to that of the transmitter except that the internal ICs are the receiving elements. The 3D stacked transmitter and receiver can be used for board-level optical interconnects with the transmitter output coupling to one end of the polymer waveguide, and the output on the other end of the polymer waveguide coupling to the receiver elements, as shown in Fig. 10.29.

For cost reduction, direct coupling (without lens) is used for optical coupling design in this study. The optical signal is coupled from the VCSEL directly into the waveguide through its mirror on one end, and from the waveguide directly to the photodiode through the mirror on its other end. For direct coupling, due to the divergence nature of the optical signal, it is critical to maintain the optical propagating distance as short as possible in order to achieve the desired coupling efficiency.

Based on a given structure, this minimum distance is determined by the assembly design of the optoelectronic components. For wirebonding component, the minimum distance or height is determined by the bonding wire loop, which is typically approximately 200 μm. On the other hand, for flip chip back illuminated optoelectronics components, the minimum height can be as close as possible (~100 μm for conventional solder bumps and ~25 μm for microbumps [29, 30]) without damaging the active area of the optoelectronic components.

(B) Optical Simulation

The design of the optical waveguide is shown in Fig. 10.30. The waveguide is embedded in the FR4 PCB to give a complete planar design. The cross-sectional design of the waveguide is shown in Fig. 10.30 and the refractive indices for the core and cladding are, respectively, 1.5622 and 1.5544. At both ends of the waveguide, there is an out-of-plane reflector or mirror, which reflects the optical signal at 90°.

(C) Results

The optical simulation is done using the commercial software, ASAP, and Fig. 10.30 shows the layout used in the simulation. The VCSEL used in the simulation has a beam divergence of 30°, and the photodiode has an active area of 40 μm. The mirror on both sides of the waveguide is simulated with a loss of 0.6 dB, and the waveguide loss is not simulated. Using the above parameters, the coupling losses are 8.5 dB and 5 dB, respectively, for 250 μm and 50 μm heights (distances of the VCSEL and photodiode from the waveguide). Figure 10.31 shows the coupling losses for other heights. As expected, the greater the height, the larger the coupling loss.

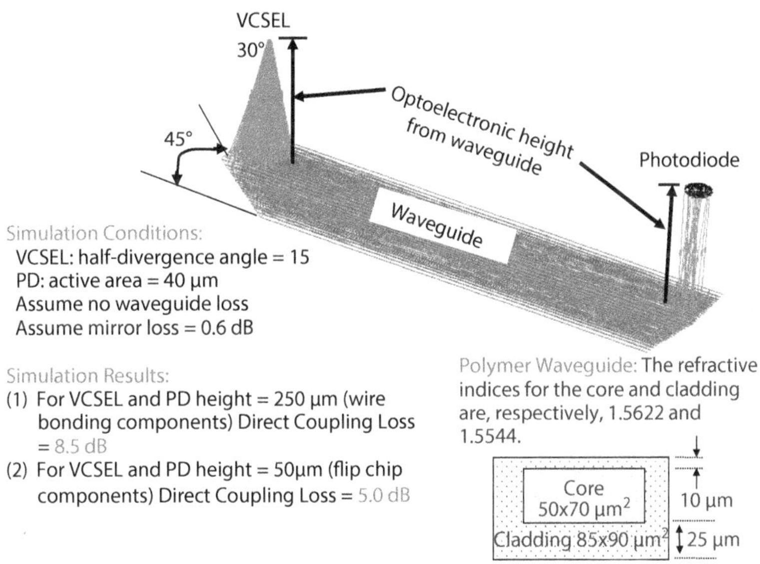

FIGURE 10.30 Simulated model showing the direct coupling of the optical signal.

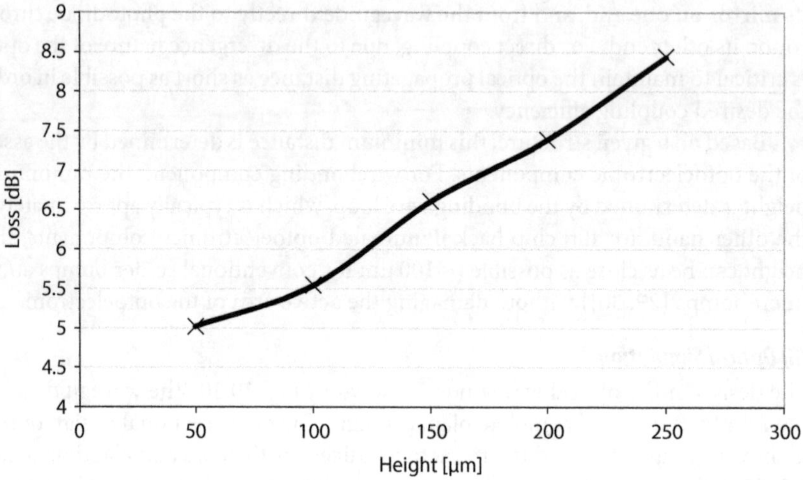

Figure 10.31 Coupling loss vs. the height of VCSEL.

The typical optical power from the VCSEL is 3 dBm and the responsitivity for the GaAs photodiode at 850 μm is 0.6 A/W. Assuming the waveguide loss is 0.1 dB/cm, the optical loss of a 10-cm long waveguide will be 1 dB. Also assuming that the minimum input optical power required for a 10 Gbps receiver is −13 dBm, then the maximum optical acceptable coupling loss is approximately 15 dB. Therefore, the direct coupling schemes are feasible for the 10 Gbps OECB design.

10.6.2 Thermal Design, Analysis, and Results

(A) Design Philosophy and Metrics

The choice of a photonic package configuration requires recognition not only of critical cost issues and optical performances but also those of thermal performance and mechanical reliability. Since low-cost bare (VCSEL, PIN, LD, TIA, serializer, and deserializer) chips are used to stack up the optoelectronic structure, as shown in Fig. 10.29, thermal management is a critical issue. This is compounded with the high power dissipation of the 10 Gbps laser (2.2 V, 33 mW, and 16 W/cm^2), the 10.7 Gbps LD and TIA chips (3.3 V, 0.35 W, and 6–35 W/cm^2), and the 10 Gbps 1:16 serializer and the 10 Gpbs 16:1 deserializer (3.3 V, 2.5 W, and 2.5–6.7 W/cm^2); thus, how to take the heat out from these chips before they burn themselves is a challenge. In particular, a stable thermal environment is required for the VCSEL. Excursions from the desired operating temperature may alter the wavelength of the emitted light, consequently degrading the VCSEL's performance. In this study, the target temperature at the laser of the 10 Gpbs VCSEL is 85°C.

(B) Thermal Simulations

Figure 10.32 shows a new design for detailed analyses of a 3D hybrid IC integration SiP for opto-electronic interconnects in organic substrates. It consists of the VCSEL, LD, and serializer on one side and the PIN, TIA, and deserializer on the other side (their dimensions are mentioned earlier). Since the transmitter and the receiver are similar, only half of

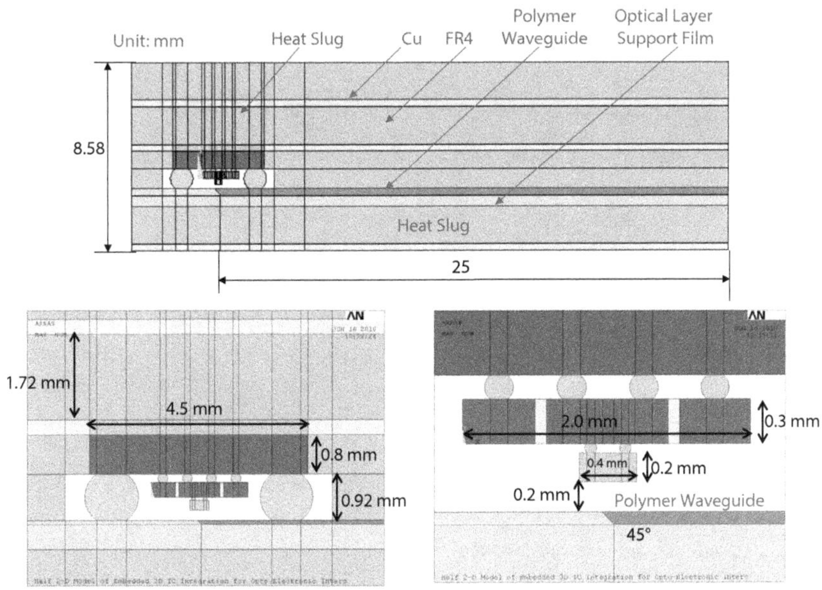

FIGURE 10.32 Finite element modeling of the embedded 3D SiP for opto-electronic interconnects in organic substrate.

the structure is modeled. It can be seen that there are three copper layers, four FR4 epoxy layers, one polymer waveguide, one optical layer support film, and a copper heat slug. The thickness of copper layers is 0.3 mm; of FR4 is 1.72 mm; of polymer waveguide is 0.085 mm; and of optical layer support film is 0.5 mm. The length of the polymer waveguide is 50 mm and the total thickness of the OECB is 8.37 mm. The material properties of the structure are given in Table 10.8.

The thermal boundary condition is that 1 W of heat is applied to the VCSEL chip and the environmental temperature is 25°C, as shown in Fig. 10.33. In service, most VCSEL will

TABLE 10.8 Material Properties for Modeling of the 3D Hybrid Integration

Component	PCB	Heat spreader	Multi-plexer	Laser driver	VCSEL	Copper layer	Solder bump	Polymer wave-guide
Material	FR4	Cu	Si	Si	GaAs	Cu	63Sn-37Pb	Polymer Wave-guide
Thermal conductivity (W/m-K)	0.8	390	150	150	68	390	50.9	1.5
TCE (ppm/K)	15	17	2.7	2.7	5.6	17	24.5	251
Modulus (MPa)	21000	110000	131000	131000	85000	110000	25000	50
Poisson ratio	0.18	0.34	0.28	0.28	0.3	0.34	0.35	0.3

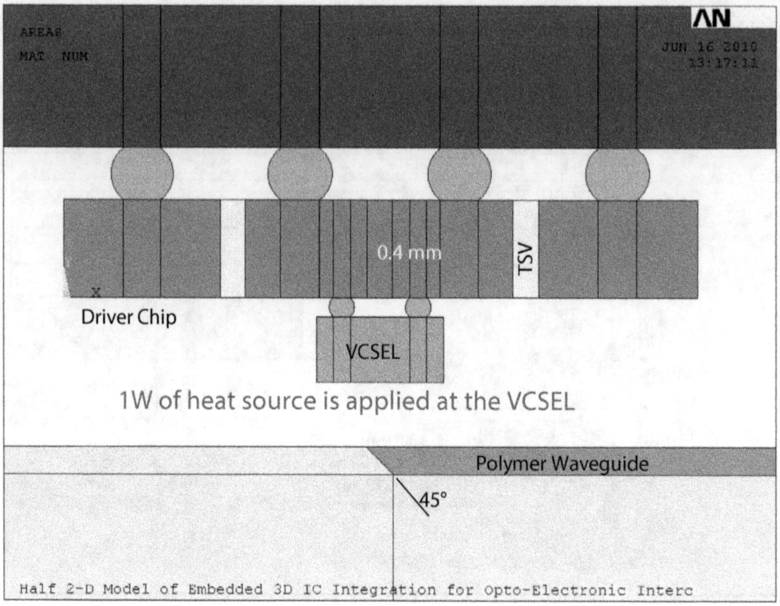

FIGURE 10.33 Thermal boundary condition.

be switched on and off at a frequency of millions (or more) of cycles per second. Since the thermal time constant of the VCSEL structure is significantly greater than the period of the power-on/power-off cycling, the structure will equilibrate to a near-uniform operating temperature. We conservatively presume that the VCSEL has continuous power generation of 1 W that is uniformly distributed within the VCSEL. Also, since it is a linear heat-transfer analysis, one unit (1 W) is applied. In real application, most VCSEL is much less than 1 W. There is heat transfer between the package and the air. The heat convection conditions are applied on all the external package surfaces, where the environmental temperature is 25°C and convection coefficient is 0.3 W/m^{2}K. For the embedded portion, there is no heat transfer between VCSEL chip and the air, since it is embedded in the PCB.

(C) Results

Figure 10.34 shows the temperature distribution in half of the OECB and at the critical area of the transmitter. It can be seen that the maximum temperature location is at the VCSEL and the magnitude is 64°C, which is below the allowable specification (85°C). Thus, the VCSEL should perform as expected.

10.6.3 Mechanical Design, Analysis, and Results

(A) Design Philosophy and Metric

As mentioned earlier, the choice of a photonic package configuration requires recognition not only of critical cost issues and optical and thermal performances but also those of mechanical reliability. Since low-cost bare chips with different material properties (e.g., CTE) are used to stack up the optoelectronic structure, as shown in Fig. 10.29, the mechanical deflections of the VCSEL with respect to the mirror of the polymer waveguide,

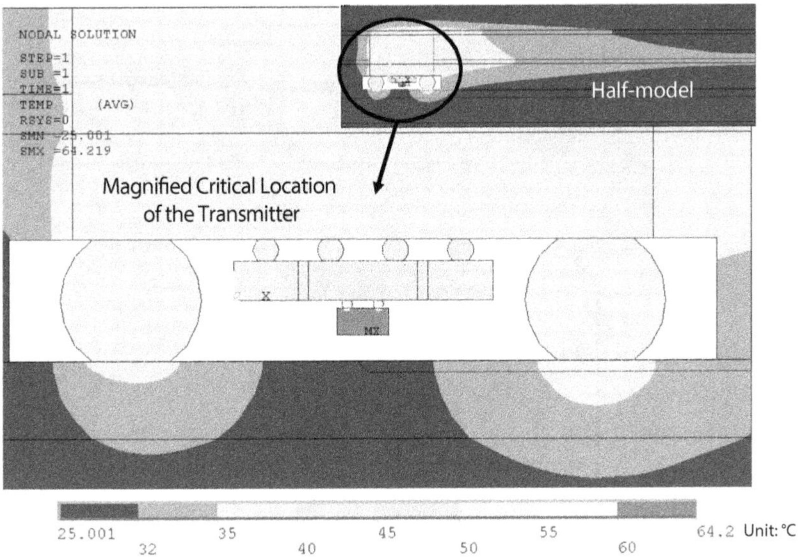

FIGURE 10.34 Temperature distribution at the critical location of the 3D hybrid IC SiP.

and stresses and strains of the solder joints of the OECB, which are caused by the global and local thermal expansion mismatches of the structure under the environmental stress condition (e.g., temperature cycling with −25°C to +140°C, one hour per cycle), are critical issues. The main concern is the mismatch between the VCSEL and the mirror on the polymer waveguide. In this study, the allowable mismatch is 2 μm.

One of the functions of solder ball is to increase the stand-off height of the 3D stacked system and thus to relax the stress and strain acting on the solder joints. However, based on the optical performance of the laser of the VCSEL without lens, the stand-off height should be as small as possible. Thus, it is necessary to design a height of the solder ball such that the structure has enough compliance (for the solder-joint reliability) and still remain within the optical performance target. In this study, the target for the strain and stress per temperature cycle is, respectively, 2% and 45 MPa.

(B) Mechanical Simulations

The simulation model is exactly the same as that for thermal simulation, except the symmetric boundary conditions that are applied at the center line. The thermal cycling boundary condition is shown in Fig. 10.35. It can be seen that the whole OECB is subjected to −25°C for 15 min and ramp up to 140°C for 15 min, then ramp down to −25°C. The ramp up and ramp down times are 15 min. The reference temperature is 25°C, i.e., the stress-free condition. The residual stress is not considered, because the thermal expansion effect is the focus of this section.

(C) Results

Figure 10.36 shows the undeformed shape (left) and deformed shape (right) of the critical area of the transmitter of the OECB due to temperature cycling and the definition of mismatch between the VCSEL and the mirror on the polymer waveguide. Figure 10.37 shows

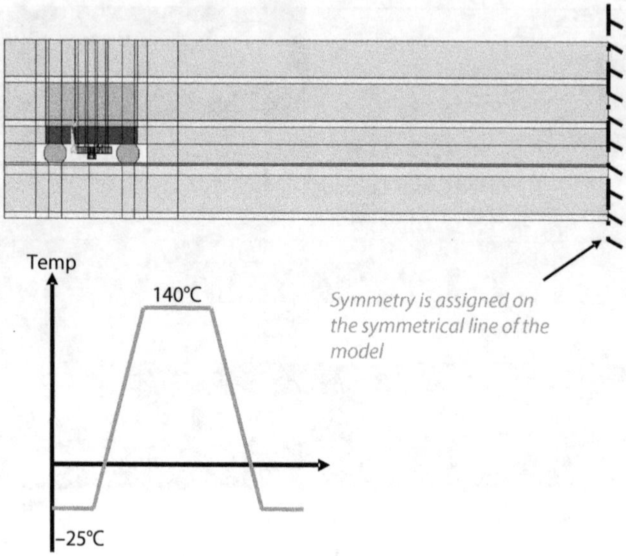

Figure 10.35 Geometry and temperature loading boundary conditions for mechanical analyses.

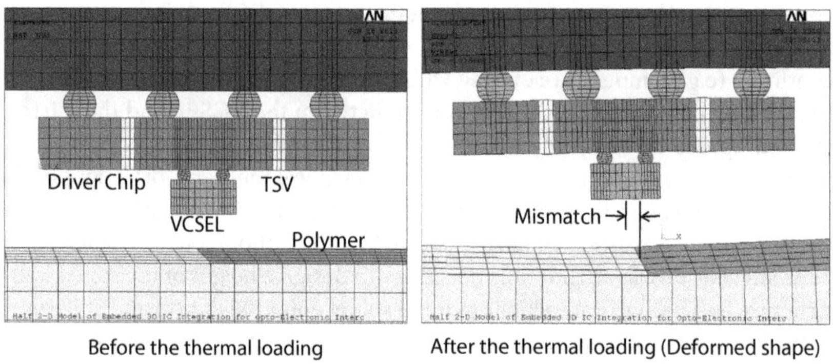

Figure 10.36 Undeformed (left) and deformed (right) shapes of the transmitter (100×).

the simulation results of the mismatch for various polymer waveguide lengths (25 μm, 50 μm, and 100 μm). It can be seen that the longer the polymer waveguide the larger the mismatch. However, they all are within the allowable specifications (2 μm).

Figure 10.38 shows the distribution of von Mises strain (upper) and stress (lower) acting on the solder joint. Again, their maximum values (0.14% and 35 MPa) are within the allowable specifications, and should be reliable under most of the operating conditions.

10.6.4 Summary and Recommendations

A generic lost-cost and high (optical, electrical, thermal, and mechanical) performance 3D hybrid IC optoelectronic system embedded into a PCB or an organic laminated substrate with an optical polymer waveguide is proposed. In addition, a special design is provided

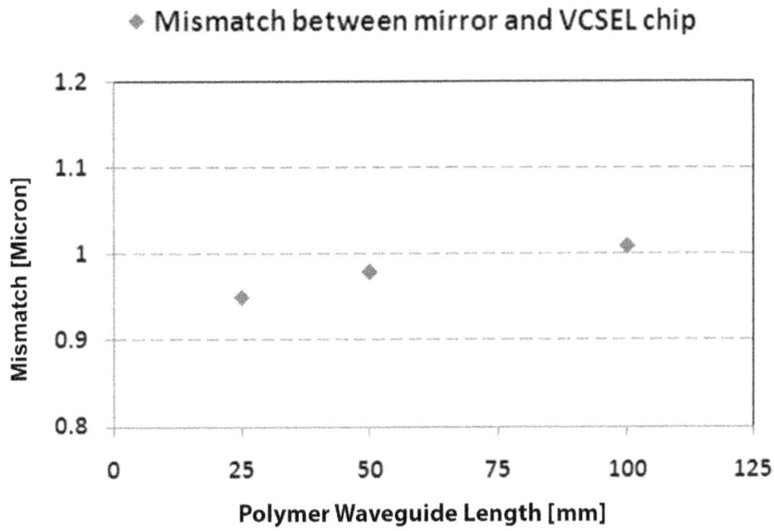

Figure 10.37 Mismatch between the VCSEL and the mirror on the polymer waveguide as a function of waveguide length.

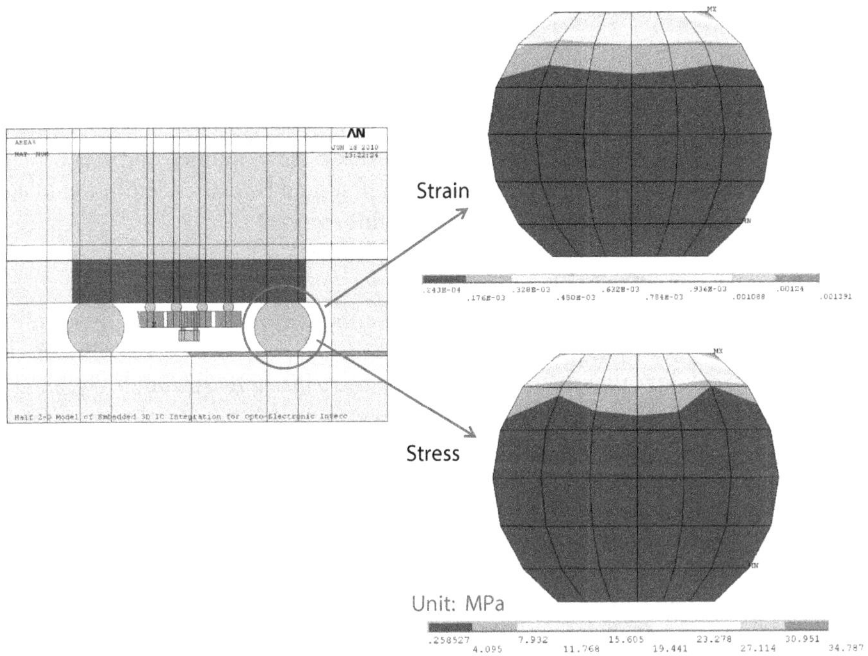

Figure 10.38 von Mises strain and stress at the critical solder joint of the transmitter.

for demonstration through simulations. Some important results and recommendations are summarized as follows [49, 50]:

- Optical analyses show that the current design of the cross-section of the optical polymer waveguide meets the optical acceptable coupling loss (15 dB). Therefore, the direct coupling schemes (without lens) are feasible for the 10 Gbps optical board-level interconnect.
- Thermal analyses show that when 1 W of heat is applied to the 10 Gbps VCSEL, with the current design, the maximum temperature at the VCSEL is only 65°C, which is smaller than the specification (85°C).
- Mechanical analyses show that when the whole embedded 3D hybrid IC SiP is subjected to an environmental temperature cycling loading (−25°C to 140°C), with the current design, the maximum mismatch (1 μm) between the VCSEL and the mirror of the polymer waveguide with different lengths is within the allowable specification (2 μm). The longer the length of the waveguide the larger the mismatch.
- Mechanical analyses show that when the whole embedded 3D hybrid IC SiP is subjected to an environmental temperature cycling loading (−25°C to 140°C), with the current design, the maximum effective strain (0.14%) and stress (35 MPa) at the critical solder joint are less than the allowable specifications (2% and 45 MPa).
- With TSV for vertical electrical feed through interconnection, every chip can have two surfaces with circuits such as the driver chip, as shown in Fig. 10.29. This concept opens the doors for many innovative applications.

10.7 Semi-Embedded TSV Interposer with Stress Relief Gap

One of the trends in mobile products is to make them thinner and thinner. In order to make super-thin products, the TSV interposer should be embedded in the PCB/substrate, as shown in Fig. 10.39, which is the focus of this section.

10.7.1 Design Philosophy

Embedding devices in the substrate/PCB are the best way to make thin and low-profile products. The basic requirements of embedded structures are (a) reworkable; (b) thermal management; (c) minimum disruption (since it is a disruptive technology); and (d) reliable.

Figure 10.39 shows a semi-embedded interposer (with a stress relief gap) supporting chips on its top surface. The advantages of this design are (1) low profile; (2) free to use of any Moore's law chips without TSVs; (3) short design cycle; (4) low manufacturing cost; (5) RDLs allow chips to talk to each other at short distance; (6) many TSVs can be used for powers, grounds, and some signals; (7) reworkable (test the chip-interposer module on substrate/PCB before underfilling); (8) the heat can get out from chips' back side through a heat spreader/sink and/or a heat slug with a spreader (not shown) conducting the heat from the solder bumps to the bottom side of the substrate/PCB; (9) reliable (because the stress relief gap reduces the global thermal expansion mismatch between the embedded interposer ($6 - 8 \times 10^{-6}$/°C) and the organic substrate/PCB ($15 - 18.5 \times 10^{-6}$/°C); and (10) potentially low system cost [51].

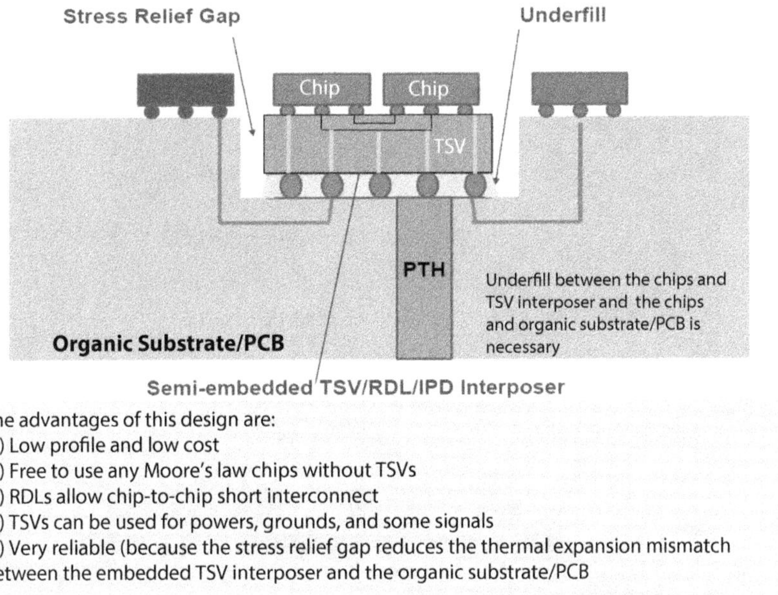

Figure 10.39 Chips on a TSV interposer semi-embedded on a substrate/PCB with stress relief gap.

10.7.2 Problem Definition

Figure 10.40 schematically shows a TSV interposer (24 × 12 mm) that is supporting two chips with the same size (10 × 10 mm). The interposer is semi-embedded on top of an organic substrate or PCB. Underfill is used between the chips and the TSV interposer and the interposer and the substrate/PCB. There is a horizontal gap (~1 mm) between the interposer and the substrate/PCB.

There are two boundary-value problems. The first one is to determine the temperature distribution and the maximum stress and strain of this structure when it is subjected to an operating condition; the power dissipation of the logic chip is 10 W and the memory chip is 2 W, as shown in Fig. 10.41. The ambient temperature is 298 K. The second one is to determine the creep strain energy density per cycle of the microsolder joint in the structure when the whole structure is subjected to an environmental condition; temperature cycling which is −25°C ⇆ 125°C on a 40-min cycle (stress-free at 25°C) [51].

10.7.3 Semi-Embedded TSV Interposer Subjected to Operating Condition

(A) Semi-Embedded TSV Interposer without Heat Spreader

First of all, there is no way to model all TSVs in great details. Thus, an equivalent model is used to model the TSV interposer as shown in Fig. 10.42. The equivalent equations for the TSV in the in-plane and out-of-plane directions have been developed in chapters 2 and 9 and [52], where the microsolder bumps have also been modeled, as shown in Fig. 10.42.

The material properties of the structure are given in Table 10.9, including the mechanical material properties such as the Young's modulus, Poisson's ratio and coefficient of thermal

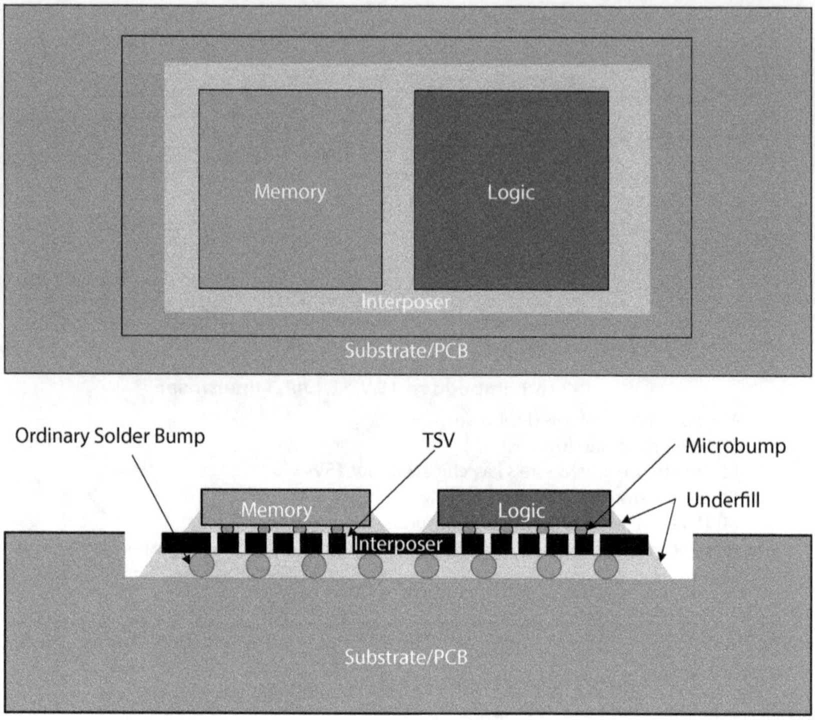

Figure 10.40 Schematic of the structure under consideration.

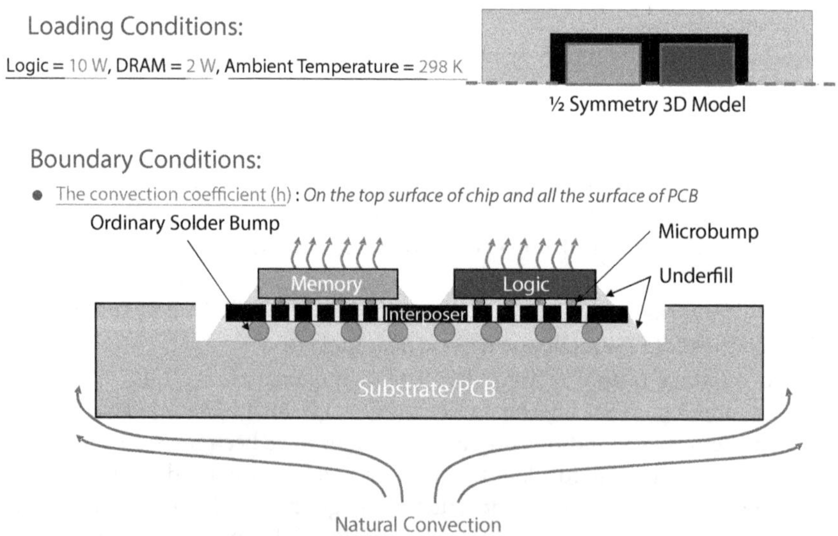

Figure 10.41 The structure under operating boundary conditions.

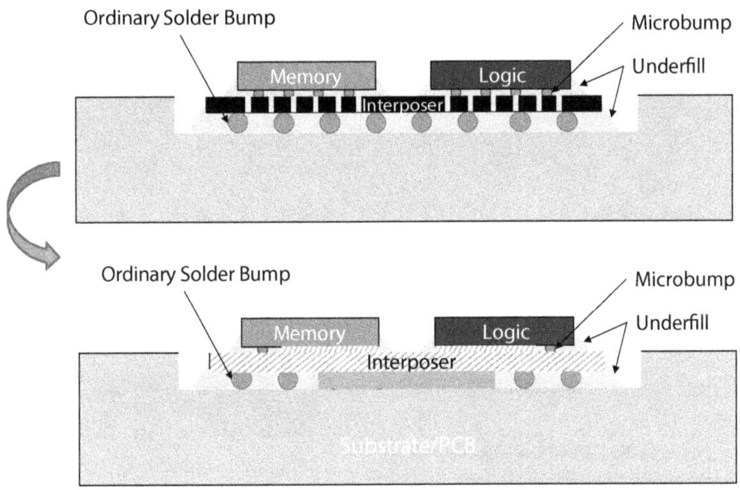

FIGURE 10.42 Equivalent thermal conductivity models for TSVs and solder joints for operating boundary condition.

TABLE 10.9 Material Properties for the Structure

	Young's modulus (GPa)	Poisson's ratio	CTE (ppm/K)	Thermal conductivity (W/m·K)
Silicon	130	0.28	2.8	148
FR4	22	0.28	18	X, Y : 56.9 Z : 0.36
Underfill	9.07	0.3	40.75	0.23
SAC 305	Emperature dependent	0.35	Temperature dependent	X, Y : 0.63 Z : 14.1
Electroplated Cu	70	0.34	18	400
Al pad	72	0.33	23	237
Cu	110	0.34	17.5	400

expansion Since lead-free solders are temperature and time dependent, the following nonlinear constitutive equations are used:

$$E(\text{GPa}) = 49 - 0.07 \times T(^\circ\text{C})$$

$$\alpha(\text{ppm}/^\circ\text{C}) = 21.301 + 0.017 \times T(^\circ\text{C})$$

$$\frac{d\varepsilon}{dt} = C_1[\sinh(C_2\sigma)]^{C_3} \exp\left(-\frac{C_4}{T}\right)$$

C1 = 50,000 (s^{-1})
C2 = 0.01 (MPa^{-1})
C3 = 5
C4 = 0.5 (K)

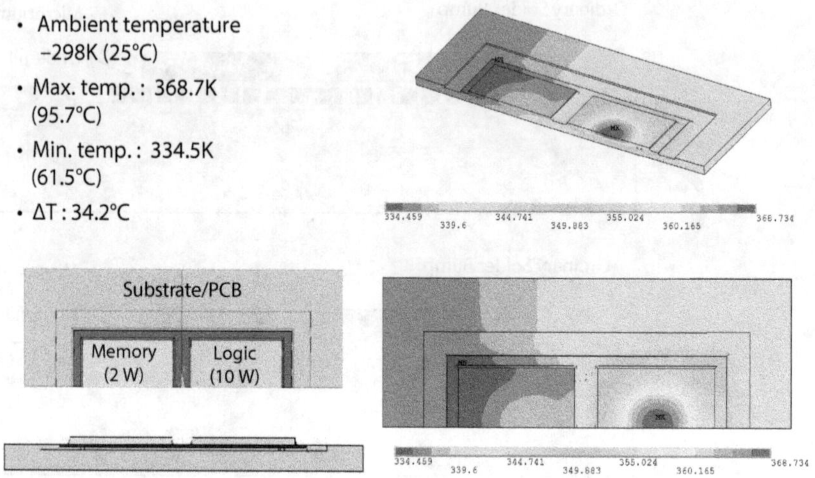

Figure 10.43 Temperature contours at the whole structure (without a heat spreader).

In these equations, E is the Young's modulus, α is the coefficient of thermal expansion, ε is the creep strain and t is time.

The finite element code used for this analysis is ANSYS Multiphysics, ver. 13.1. Due to symmetry only half of the structure is modeled and analyzed. The power dissipation of the logic chip is 10 W and the memory chip is 2 W. The temperature distribution of the structure without a heat spreader is shown in Fig. 10.43. It can be seen that the maximum temperature (95.7°C) occurs at the logic chip. The minimum temperature in the memory chip is 61.5°C.

Figure 10.44 shows the von Mises stress contours and warpage contours of the structure at 125°C. It can be seen that the maximum von Mises stress (199 MPa) occurs at the corner

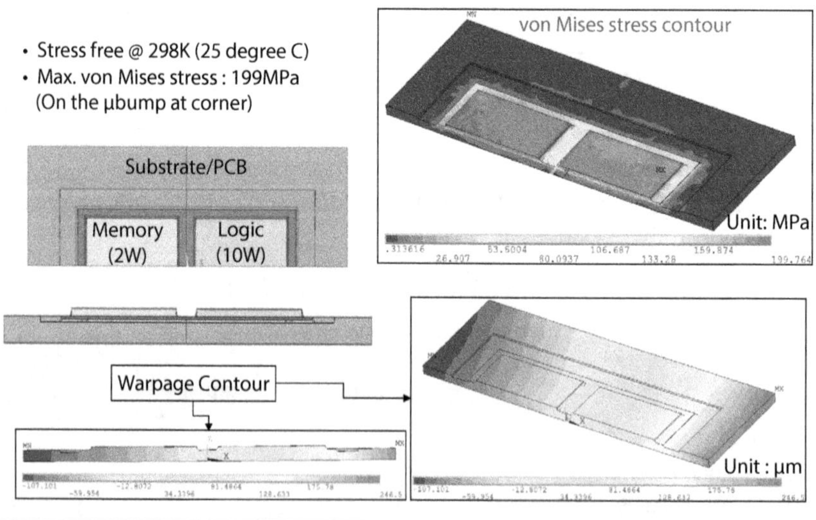

Figure 10.44 Von Mises stress contours at the structure at 125°C.

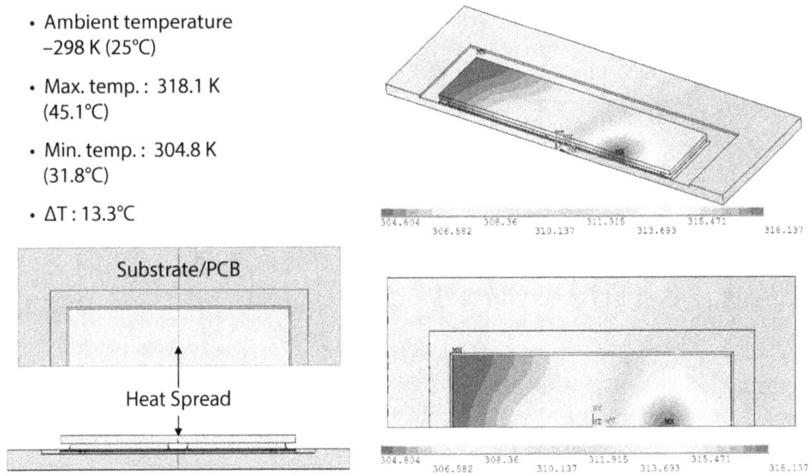

Figure 10.45 Temperature contours at the whole structure with a heat spreader.

microsolder joint of the logic chip. The maximum deflection of the structure (at the tip of the substrate/PCB) is 246 μm.

(B) Semi-Embedded TSV Interposer with Heat Spreader

Figure 10.45 shows the temperature contours acting on the whole structure with a heat spreader. The power dissipation of the logic chip is 10 W and the memory chip is 2 W, and the ambient temperature is 298 K. It can be seen that the maximum temperature (45.1°C) occurs at the logic chip and its value is less than half of that without a heat spreader. The minimum temperature in the memory chip is 31.8°C.

Figure 10.46 shows the von Mises stress contours and warpage contours of the structure at 125°C with a heat spreader. It can be seen that the maximum von Mises stress

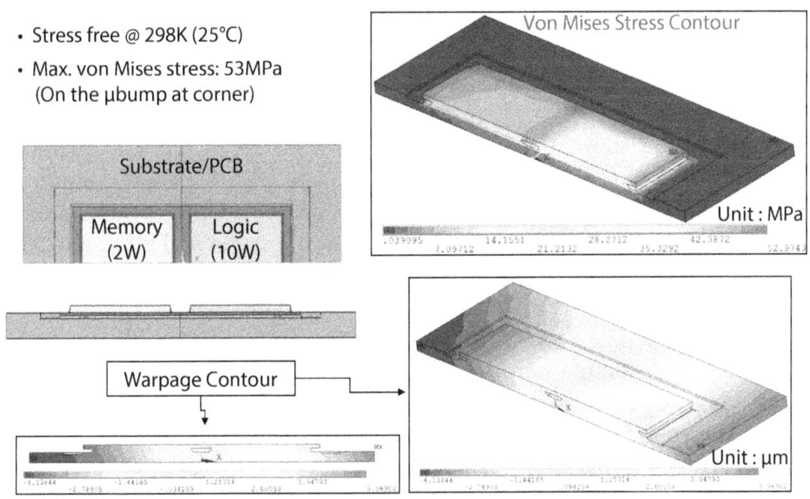

Figure 10.46 Von Mises stress contours at the structure at 125°C (with a heat spreader).

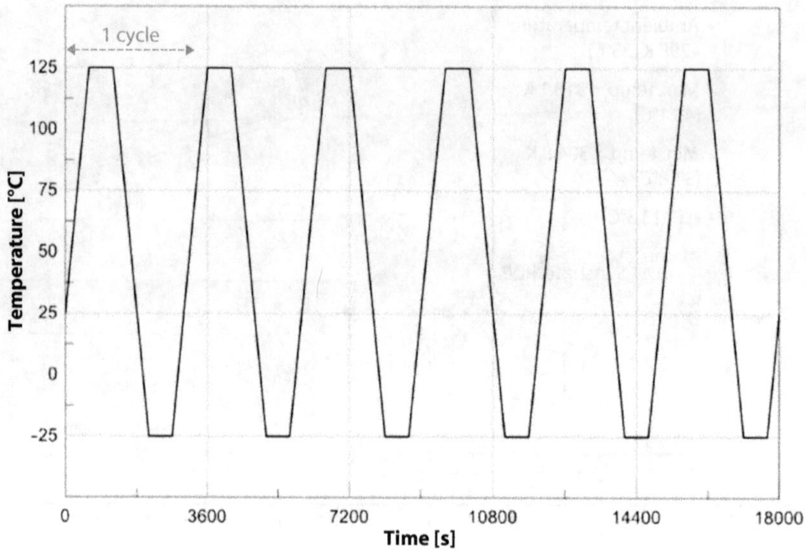

FIGURE 10.47 Temperature loading condition.

(53 MPa) occurs at the logic's corner microsolder joint. The maximum deflection of the whole structure is only 6 μm. These small values show the effect of the heat spreader, which reduce the values of temperature distribution, stress distribution, and deflection distribution.

10.7.4 Semi-Embedded TSV Interposer Subjected to an Environmental Condition

The second boundary-value problem is to determine the creep strain energy density per cycle of the corner microsolder joint of the structure (without the heat spreader) subjected to a thermal cycling loading (environmental) condition, as shown in Fig. 10.47. Due to double symmetry only a quarter of the whole structure, as shown in Fig. 10.48, is modeled.

Figure 10.49 shows the von Mises stress contours of the whole structure at −25°C during the first cycle (stress-free at 25°C). It can be seen that the maximum value occurs at the interface between the Al/Cu-low-k pad and the corner microsolder joint and the value is 134 MPa. The maximum warpage occurs at the chip center and is equal to 26 μm.

Figure 10.50 shows the von Mises stress contours of the whole structure at 125°C during the first cycle (stress-free at 25°C). It can be seen that the maximum value occurs at the interface between the Al/Cu-low-k pad and the corner microsolder joint and the value is 210 MPa. The maximum warpage occurs at the substrate/PCB and is equal to 172 μm.

The maximum creep strain energy density time-history acting at the corner microsolder joint is shown in Fig. 10.51. It can be seen that the creep strain energy density per cycle is 0.03 MPa which is much less than the specification (0.1 MPa) and thus it should be reliable for most of the environmental conditions. This result shows that underfill is a must for the reliability of microsolder joints.

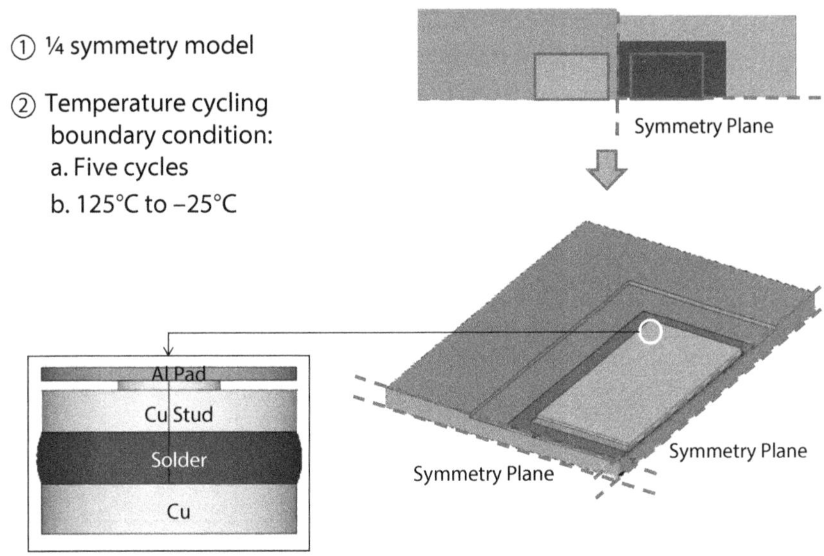

FIGURE 10.48 Quarter model for analysis.

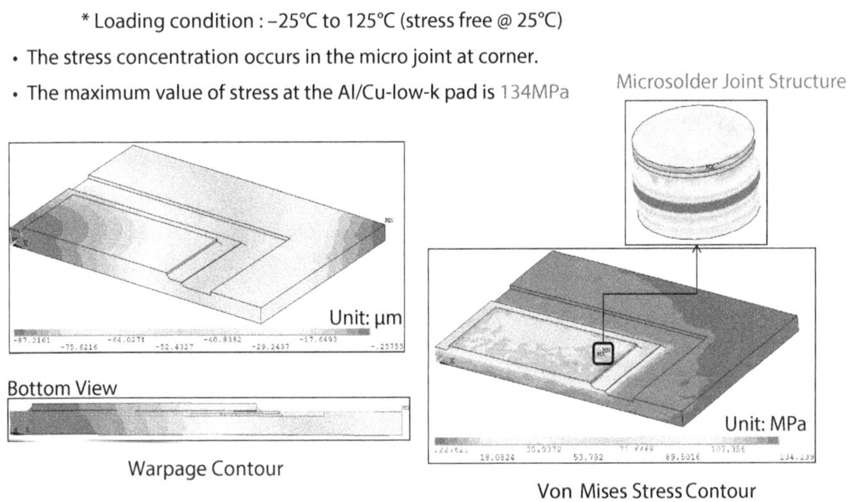

FIGURE 10.49 Von Mises stress contours and warpage contours of the structure at −25°C during the first cycle (stress-free at 25°C).

10.7.5 Summary and Recommendations

The 3DIC integration with a semi-embedded TSV interposer with stress relief gap under thermal operating and environmental conditions is designed and investigated. Some important results and recommendations are summarized as follows [51]:

- A super thin and low-profile 3D IC integration with semi-embedded TSV interposer supporting Moore's law chips is proposed.

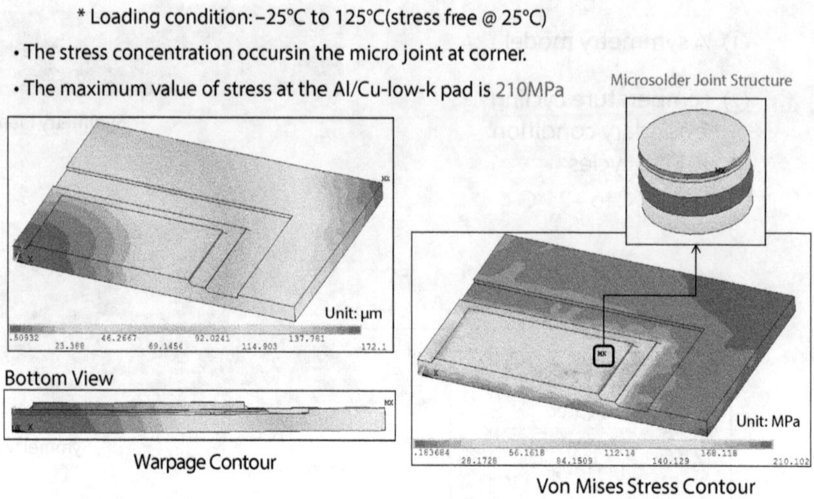

Figure 10.50 Von Mises stress contours and warpage contours of the structure at 125°C during the first cycle (stress-free at 25°C).

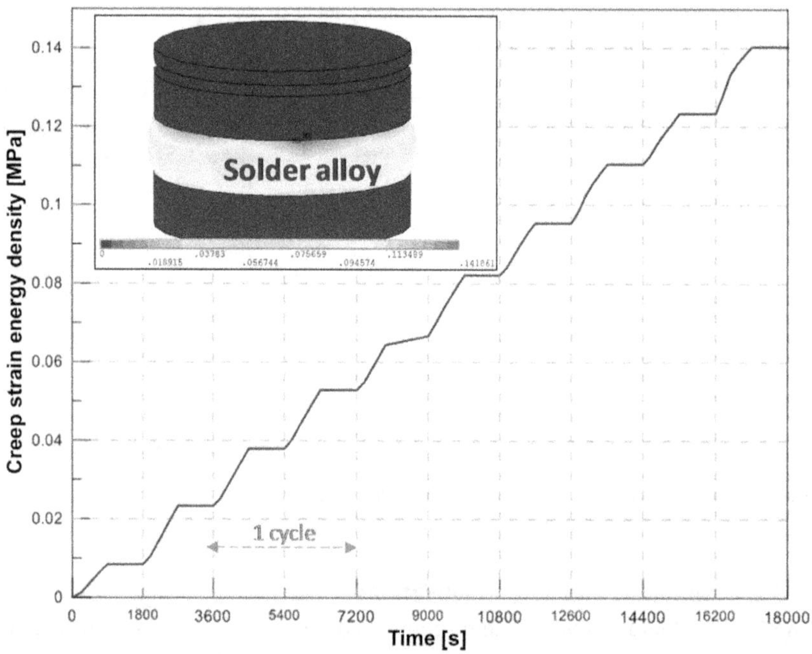

Figure 10.51 Creep strain energy density time-history at the corner microsolder joint.

- A stress relief gap is proposed to reduce the thermal expansion mismatch between the TSV interposer and the surrounding organic substrate or PCB.
- Under the operating condition, with a heat spreader, the values of the temperature distribution and the stress distribution are within the specifications.
- Under the environmental condition, with the underfill, the values of the creep strain energy density per cycling and the wargpage are too small to create any reliability concerns.

10.8 References

[1] Lau, J. H., *Flip Chip Technology*, McGraw-Hill, New York, NY, 1995.

[2] Lau, J. H., *Low-Cost Flip Chip Technology*, McGraw-Hill, New York, NY, 2000.

[3] Lim, T. G., B. Lee, T. Shioda, H. Kuruveettil, J. Li, K. Suzuki, J. H. Lau, et al., "Demonstration of high frequency data link on FR4 PCB using optical waveguides," *IEEE Transactions of Advanced Packaging*, Vol. 32, May 2009, pp. 509–516.

[4] Ramana, P., H. Kuruveettil, B. Lee, K. Suzuki, T. Shioda, C. Tan, J. H. Lau, et al., "Bi-directional Optical Communication at 10 Gb/s on FR4 PCB using Reflow Solderable SMT Transceiver," *IEEE/ECTC Proceedings*, May 2008, pp. 244–249.

[5] Lau, J. H., Y. Lim, T. Lim, G. Tang, K. Houe, X. Zhang, P. Ramana, et al., "Design and Analysis of 3D Stacked Optoelectronics on Optical Printed Circuit Boards," *Proceedings* of *SPIE, Photonics Packaging, Integration, and Interconnects VIII*, Vol. 6899, San Jose, CA, January 19–24, 2008, pp. 07.1–07.20.

[6] Chai, J., G. Yap, T. Lim, C. Tan, Y. Khoo, C. Teo, J. H. Lau, et al., "Electrical Interconnect Design Optimization for Fully Embedded Board-level Optical Interconnects," *IEEE/EPTC Proceedings*, December 2008, pp. 1126–1130.

[7] Lim, L, C. Teo, H. Yee, C. Tan, O. Chai, Y. Jie, J. H. Lau, et al., "Optimization and Characterization of Flexible Polymeric Optical Waveguide Fabrication Process for Fully Embedded Board-level Optical Interconnects," *IEEE/EPTC Proceedings*, December 2008, pp. 1114–1120.

[8] Teo, C., W. Liang, H. Yee, L. Lim, C. Tan, J. Chai, J. H. Lau, et al., "Fabrication and Optimization of the 45° Micro-mirrors for 3-D Optical Interconnections" *IEEE/EPTC Proceedings*, December 2009, pp. 1121–1125.

[9] Chang, C., J. Chang, J. H. Lau, A. Chang, T. Tang, S. Chiang, M. Lee, et al., "Fabrication of Fully Embedded Board-level Optical Interconnects and Optoelectronic Printed Circuit Boards," *IEEE/EPTC Proceedings*, December 2009, pp. 973–976.

[10] Yuzo, I., S. Koike, Y. Arai, and Y. Ando, "SMT-Compatible Large-Tolerance "OptoBump" Interface for Interchip Optical Interconnections," *IEEE Transactions on Advanced Packaging*, Vol. 26, No. 2, May 2003, pp. 122–127.

[11] Johnson, E., and D. Caletka, "Passive Alignment of VCSELs to Waveguides in Opto-Electronics Cards and Printed Circuit Boards," *US Patent No. 2005/0089264A1*, filing date: October 24, 2003, publication date: April 28, 2005.

[12] Laurent Schares, "Terabus: Terabit/Second-Class Card-Level Optical Interconnect Technologies," *IEEE Journal of Selected Topic in Quantum Electronics*," Vol. 12, No. 5, September/October 2006, pp. 1032–1044.

[13] Doany, F., C. Schow, C. Tsang, N. Ruiz, R. Horton, D. Kuchta, C. Patel, et al., "300-Gbps 24-Channel Bidirectional Si Carrier Transceiver Optochip for Board-Level Interconnects," *IEEE/ECTC Proceedings*, 2008, pp. 238–243.

[14] Lu, D., J. Tang, J. He, and E. Zarbock, "Chip-to-Chip Optical Interconnect," *US Patent No. 7373033*. May 13, 2008.

[15] Chen, R. T., "Packaging Enhanced Board Level Opto-Electronic Interconnects," *US Patent No. 6243509B1*, June 5, 2001.

[16] Chen, Y. M., C., L. Yang, L. C. Yao, H. H. Chen, Y. C. Chen, Y. Chu, and T. E. Hsieh, "10Gbps Multi-Mode Waveguide for Optical Interconnect," *IEEE Proceedings of Electronic Components and Technology Conference*, 2005, pp. 1739–1743.

[17] Shioda T., "Recent Progress and Potential Markets for Optical Circuit Boards," *IEEE Proceedings of Polytronic Conference*, 2007, pp. 1–3.

[18] Hiramatsu, S., K. Miura, and K. Hirao, "Optical Backplane Connectors Using Three-Dimensional Waveguide Arrays," *IEEE Journal of Lightwave Technology*, Vol. 25, No. 9, September 2007, pp. 2776–2782.

[19] Chang, G. K., "High-speed, High-density Optical Interconnects for Next Generation Computing and Communications Systems," *IME Seminar*, April 13, 2006.

[20] Glebov, A. L., C. J. Uchibori, and M. G. Lee, "Direct Attach of Photonic Components on Substrates with Optical Interconnects," *IEEE Photonics Technology Letters*, Vol. 19, No. 8, April 2007, pp. 547–549.

[21] Holden, H. T., "The Developing Technologies of Integrated Optical Waveguides in Printed Circuits," *Circuit World*, Vol. 29, No. 4, 2003, pp. 42–50.

[22] Pugliano, N., N. Chiarotto, J. Fisher, N. Heiks, T. Ho, G. Khanarian, M. Moynihan, et al., "Progress Toward the Development of Manufacturable Integrated Optical Data Buses," *SPIE Proceedings of Photonics Packaging and Integration IV*, Vol. 5358, 2004, pp. 71–79.

[23] Wang, L., J. Choi, X. L. Wang, R. T. Chen, D. Hass, and J. Magera, "Thin Film Optical Wavegide and Optoelectronic Device Integration for Fully Embedded Board Level Optical Interconnects," *SPIE Proceedings of Photonics Packaging and Integration IV*, Vol. 5556, 2004, pp. 1–13.

[24] Lee, B., R. Pamidigantham, and C. S. Premachandran, "Prototype Development for Chip-Chip Interconnection by Multimode Waveguide," *IEEE Proceedings of Electronic Packaging and Technology Conference*, 2005, pp. 488–491.

[25] Schow, C. L., Doany, F. E., Liboiron-Ladouceur, O., Baks, C., Kuchta, D. M., Schares, L., John, R., et al., "160-Gb/s, 16-Channel Full-Duplex, Single-Chip CMOS Optical Transceiver," *Proceedings of OThG4*, 2007, pp. 1–3.

[26] Dellmann, L., C. Berger, R. Beyeler, R. Dangel, M. Gmur, R. Hamelin, F. Horst, et al., "120 Gb/s Optical Card-to-Card Interconnect Link Demonstrator with Embedded Waveguides," *IEEE Proceedings of Electronic Components and Technology Conference*, 2007, pp. 1288–1293.

[27] Mikawa, T., M. Kinoshita, K. Hiruma, T. Ishitsuka, M. Okabe, S. Hiramatsu, H. Furuyama, et al., "Implementation of Active Interposer for High-Speed and Low-Cost Chip Level Optical Interconnects," *IEEE Journal of Selected Topics in Quantum Electronics*, Vol. 9, No. 2, 2003, pp. 452–459.

[28] Toffano, Z., A. Gholami, M. Fez, A. Destrez, and M. Marec, "VCSEL Short Reach Communications: Behavioural Modelling of High Speed Optoelectronic Modules," *Proc. 4th Int. Conf. Numerical Simulation Optoelectronic Devices*, Aug. 2004, pp. 49–50.

[29] Park, J., E.-D. Sim, and Y.-S. Beak, "Improvement of Fabrication Yield and Loss Uniformity of Waveguide Mirror," *IEEE Photonics Technology Letters*, Vol. 17, No. 4, April 2005, pp. 807–809.

[30] Matsubara, T., K. Oda, K. Watanabe, K. Tanaka, M. Maetani, Y. Nishimura, S. Tanahashi, "Three Dimensional Optical Interconnect on Organic Circuit Board," *IEEE Electronic Components and Technology Conference*, May 2006, pp. 789–794.

[31] Immonen, M., M. Karppinen, J. Kivilahti, "Fabrication and Characterization of Polymer Optical Waveguuides with Integrated Micromirrors for 3-D Board Level Optical Interconnects," *IEEE Transactions on Electronics Packaging Manufacturing*, Vol. 28, No. 4, 2005, pp. 304–311.

[32] Chandrappan, J., H. Kuruveettil, T. Wei, C. Liang, P. Ramana, K. Suzuki, T. Shioda, et al., "Performance Characterization Methods for Optoelectronic Circuit Boards," *IEEE Transactions on CPMT*, Vol. 1, No. 3, March 3, 2011, pp. 318–326.

[33] Lau J. H., C. P. Wong, N. C. Lee, R. Lee, *Electronics Manufacturing with Lead-Free, Halogen-Free, and Adhesive Materials*, McGraw-Hill, New York, NY, 2003.

[34] Wang, L., X. Wang, W. Jiang, J. Choi, H. Bi, and R. Chen, "45 Polymer-Based Total Internal Reflection Coupling Mirrors for Fully Embedded Intraboard Guided Wave Optical Interconnects," *Applied Physics Letters*, Vol. 87, September 2005, pp. 141110–141110-3.

[35] Glebov, A. L., D. Bhusari, P. Kohl, M. S. Bakir, J. D. Meindl, and M. G. Lee, "Flexible Pillars for Displacement Compensation in Optical Chip Assembly," *IEEE Photonics Technology Letters*, Vol. 18, No. 8, April 2006, pp. 974–976.

[36] Cho, H. S., K.-M. Chu, S. Kang, S. H. Hwang, B. S. Rho, W. H. Kim, J.-S. Kim, et al., "Compact Packaging of Optical and Electronic Components for On-Board Optical Interconnects," *IEEE Transactions on Advanced Packaging*, Vol. 28, No. 1, February 2005, pp. 114–120.

[37] Yoon, K. B., I.-K. Cho, S. H. Ahn, M. Y. Jeong, D. J. Lee, Y. U. Heo, B. S. Rho, et al., "Optical Backplane System using Waveguide-Embedded PCBs and Optical Slots," *Journals Lightwave Technology*, Vol. 22, September 2004, pp. 2119–2127.

[38] Chang, G.-K., D. Guidotti, F. Liu, Y.-J. Chang, Z. Huang, V. Sundaram, D. Balaraman, et al., "Chip-to-Chip Optoelectronics SOP on Organic Boards or Packages," *IEEE Transactions on Advanced Packaging*, Vol. 27, No. 2, May 2004, pp. 386–397.

[39] Maxim Application Notes, "Accurately Estimating Optical Receiver Sensitivity," HFAN-3.0.0, Rev 0.

[40] Choi, C., L. Lin, Y. Liu, J. Choi, L. Wang, D. Hass, J. Magera, et al., "Flexible Optical Waveguide Film Fabrications and Optoelectronic Devices Integration for Fully Embedded Board-Level Optical Interconnects," *IEEE Journal of Lightwave Technology*, Vol. 22, No. 9, 2004, pp. 2168–2176.

[41] Wang, L., X. L. Wang, J. Choi, D. Haas, J. Magera, and R. T. Chen, "Low-loss Thermally Stable Waveguide with 45° Micromirrors Fabricated by Soft Molding for Fully Embedded Board-level Optical Interconnects," *Proceedings of SPIE*, Vol. 5731, 2005, pp. 87–93.

[42] Xia, Y., and G. M. Whitesides, " Soft-lithography," *Annual Review Of Materials Science*, Vol. 28, 1998, pp. 84–153.

[43] Kim, J. T., B. C. Kim, M. Jeong, and M. Lee, "Fabrication of a Micro-Optical Coupling Structure by using Laser Ablation," *Journal of Materials Processing Technology*, Vol. 146, 2004, pp. 163–166.

[44] Garner, S., S.-S. Lee, V. Chuyanov, A. Chen, A. Yacoubian, W. Steier, and L. Dalton, "Three-Dimensional Integrated Optics using Polymers," *IEEE Journals Of Quantum Electronics*, Vol. 35, No. 8, August 1999, pp. 1146–1155.

[45] Kim, J.-S., and J.-J. Kim, "Stacked Polymeric Multimode Waveguide Arrays for Two-Dimensional Optical Interconnects," *IEEE Journal of Lightwave. Technology*, Vol. 22, No. 3, March 2004, pp. 840–844.

[46] Kim, J.-S., and J.-J. Kim, "Fabrication of Multimode Polymeric Waveguides and Micromirrors using Deep X-ray Lithography," *IEEE Photonics Technology Letters*, Vol. 16, No. 3, March 2004, pp. 798–800.

[47] Kim, J.-H., and R. T. Chen, "A Collimation Mirror in Polymeric Planar Waveguide Formed by Reactive Ion Etching," *IEEE Photonics Technology Letters*, Vol. 15, No. 3, March 2003, pp. 422–424.

[48] Lehmacher, S., and A. Neyer, "Integration of Polymer Optical Waveguides into Printed Circuit Boards," *Electronics Letters*, Vol. 36, No. 12, June 2000, pp. 1052–1053.

[49] Lau, J. H., M. S. Zhang, and S. W. R. Lee, "Embedded 3D Hybrid IC Integration System-in-Package (SiP) for Opto-Electronic Interconnects in Organic Substrates," *ASME Transactions, Journal of Electronic Packaging*, Vol. 133, September 2011, pp. 1–7.

[50] Lau, J. H., S. Lee, M. Yuen, J. Wu, C. Lo, H. Fan, and H. Chen, "Apparatus Having an Embedded 3D Hybrid Integration for Optoelectronic Interconnects in Organic Substrate". US 20100215314A1, Notice of Allowance on 02/09/2015, Filed on February 20, 2009.

[51] Lau, J. H., S. T. Wu, and H. C. Chien, "Nonlinear Analyses of Semi-Embedded Through-Silicon Via (TSV) Interposer with Stress Relief Gap Under Thermal Operating and Environmental Conditions," *IEEE EuroSime Proceedings, Chapter 11: Thermo-Mechanical Issues in Microelectronics*, Lisbon, Portugal, April 2012, pp. 1/6–6/6.

[52] Chien, H. C., J. H. Lau, Y. Chao, R. Tain, M. Dai, S. T. Wu, W. Lo, et al., "Thermal Performance of 3D IC Integration with Through-Silicon Via (TSV)," *IMAPS Transactions, Journal of Microelectronic Packaging*, Vol. 9, 2012, pp. 97–103.

CHAPTER 11

3D LED and IC Integration

11.1 Introduction

3D LED (light-emitting diode) and IC integration packages are discussed in this chapter. These packages consist of the single- or multi-LEDs and either active IC (integrated circuit) chip such as the ASIC (application-specific IC), LED driver, processor, memory, RF (radio frequency) sensor, and power controller in a 3D manner or passive interposer in a 2.5D manner. The assembly processes of these packages are also presented and discussed. Specifically, 2.5D IC and LED integrations, such as (a) LED packaging using silicon substrate with cavities for phosphor printing and copper-filled TSVs for interconnection; (b) silicon substrate with cavity and TSVs for LED packaging; and (c) LED on silicon substrate with cavity and TSVs, are discussed. Finally, a thermal management system for 3D IC and LEDs integration packages is proposed. Status and outlook of Haitz's law are briefly mentioned first.

11.2 Status and Outlook of Haitz's Law

Dr. Roland Haitz, a now-retired scientist at HP/Agilent, known for Haitz's law, in 1999 predicted [1, 2]: "Every decade, the cost per lumen (unit of useful light emitted) falls by a factor of 10 and the amount of light generated per LED package increases by a factor of 20." Due to the rapid progress of the white LED and the pressure of environmental initiatives, the amount of light generated per LED package has increased by a factor of ~30 in the past 10 years (Figs. 11.1 [1–11] and 11.2). Consequently, Haitz (still very active in the LED arena) and Tsao commented [12] on the progress of SSL (solid-state lighting) through the past 10 years and made a set of new predictions for the coming 10 to 20 years. Their conclusions and recommendation are summarized as follows:

1. The expression "solid-state lighting" (SSL) is now commonly used in the United States and Japan to describe this area. It is analogous to the expression "solid-state electronics," which became common in the 1950s and 1960s to describe the semiconductor-transistor-based technologies that were then displacing vacuum tubes for electronics. Other, perhaps more descriptive, expressions are often used in other languages, e.g., "LED lighting" in Korean, Russian, and German, and

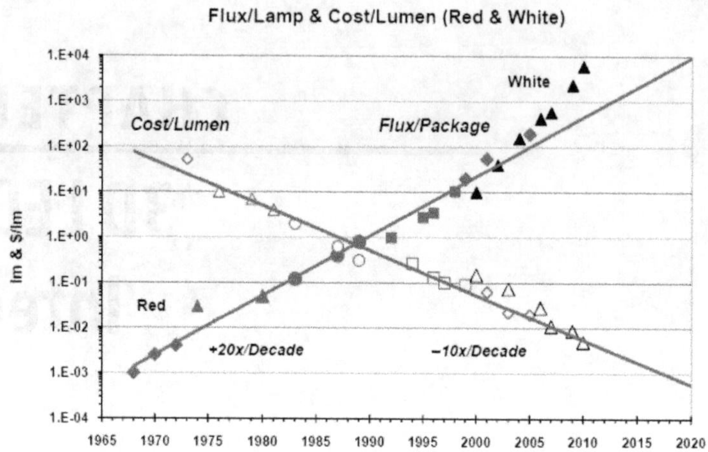

Outlook of Haitz's Law

Haitz's Law is an observation/prediction about the steady improvement over the years of LEDs. It states that every decade, the cost per lumen (unit of useful light emitted) falls by a factor of 10, the amount of light generated per LED package increases by a factor of 20-30, for a given wavelength (color) of light. It is considered the LED counterpart to Moore's Law.

What drives $/lm down over time?

- **Production cost of LED**
- **From growth to fabrication to package to test**
- **Performance breakthrough**
- **Yield**
- **Competition**
- **Volume and automation**

Figure 11.1 Haitz's law.

"semiconductor lighting" in Chinese. In these comments, we will interchangeably use these expressions.

2. Note that, unlike Haitz's law, Moore's law does not specify cost, but it is well recognized that Moore's law has been the driver behind the enormous reduction in the cost of computing, in the range of 70% per year over the past three decades. With the continued insatiable demand for computing power at steadily reduced cost, Moore's law does not face a foreseeable application-based limit.
3. One might argue that the trend toward greater flux/lamp may ultimately reverse, if finer control over the temporal and spatial placement of light is desired because such control is facilitated by larger numbers of spatially distributed but lower-flux lamps.
4. The color-mixing approach preferred in the white paper could have used a broad production distribution of LEDs of any of the selected four to six colors. Ambient temperature variations and their corresponding color shift, flux stability over time, differential aging, etc., could all have been corrected, with drive current changes

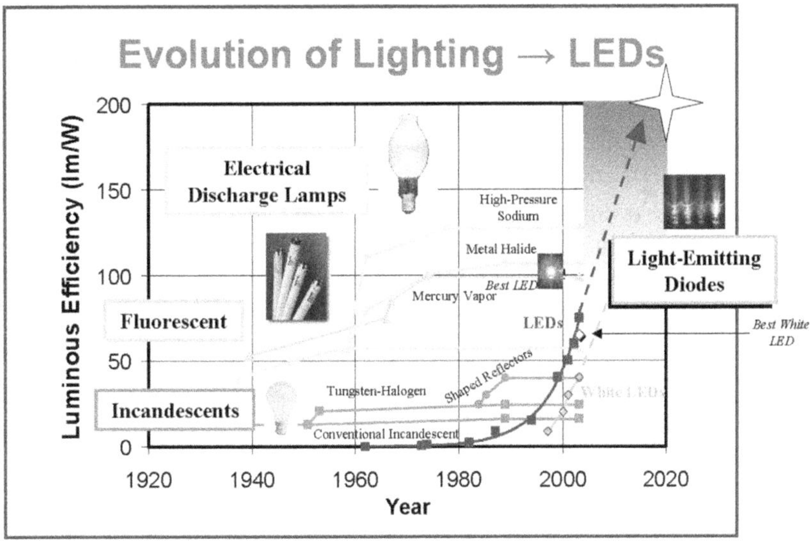

Lumileds

Figure 11.2 Evolution of lighting—LEDs (light-emitting diodes).

controlled by an active feedback loop. If digital color tuning becomes a market need, then this approach to color control may have to be implemented.

5. Color mixing will play a role in niche sections of the lighting market where limited or extensive color variations are needed. Its market share will depend on progress in efficacy improvements in red, yellow, and green and on a satisfactory design of the feedback loops. Currently, the cost of the power converter/driver is comparable to the cost of the basic high-volume LED lamp. Even if the power converter can be shared for all colors, the cost of four to six independent drivers will present a significant cost hurdle.
6. Market forecasts during a rapid technology-driven conversion have to be taken with a grain of salt. Economic hiccups in consumer markets are notorious and faster-than-expected price erosions resulting from overinvestments during an industry boom are classic problems in the semiconductor industry.
7. Please note that these market numbers refer to simple packaged LED lamps only, without secondary optics, drivers, sockets, etc. This phenomenal growth has to be seen in the context of the fairly static \$15B/year market for all conventional lightbulbs during the past 15 years. This conventional lightbulb market will start shrinking as SSL takes over and may eventually disappear nearly completely!
8. We note that these prices are consistent with US DOE forecasts. Recently, Jim Brodrick presented DOE's expectations on SSL relative costs for the 2009–2015 time frames [10]. DOE expects a cost reduction for the LED element of 8 xs for the six-year window. The red trend line cost for 2009 in Fig. 11.1 is \$8/klm. A reduction by 8 xs in 2015 gets to the same range that we predict in Fig. 11.1.

9. In 2003, Haitz [6] predicted that by 2015 LED lamps would start to compete for any electricity-based light source application on earth, even sports stadiums with TV coverage. In a recent press release (http://www.philipslumileds.com/newsandevents/releases/PR128.pdf), Philips announced solar/battery-based lighting systems for small soccer training fields in remote areas without access to an electricity grid.
10. Note that this is not to imply that thermal issues will disappear. It will always be advantageous to drive LEDs as hard as possible so as to amortize LED cost over greater lumens produced. In the absence of 100% efficiency, thermal issues will generally be the factor that limits how hard LEDs can be driven.

11.3 LED Has Come a Long Way!

The global market for LEDs (including standard, AlInGaP and InGaN) in 2011 reached over \$10 billion, and has been forecasted to increase to \$45 billion LED sales in 2015. Exterior applications, driven by maintenance/labor, safety/security, and "Green-Tech" issues will continue to dominate the use of LEDs in solid-state general lighting. Today, the top LED suppliers (by revenue) are Nichia, Samsung LED, Osram, LG Innotek, Seoul Semiconductor, Philips/Lumileds, and Cree. LED has come a long way!

One hundred years ago, the discovery of electroluminescence by Henry Joseph Round in 1907 [13] and the invention of the first LED by Oleg Vladimirovich Losev in the 1920s [14, 15] *foreshadowed* the development of generations of LED to come. The invention of the first practical visible-spectrum red LED by Nick Holonyak, Jr. in 1962 [16], the first yellow LED by M. George Craford in 1972 [17], and the first high brightness (HB) blue LED by Shuji Nakamura in 1993 [18–33] *excited* the development of LEDs in a broader range of technologies. The use of a transparent contact made of indium-tin-oxide (ITO, e.g., Sn2O3:In2O3 = 1:9) on (e.g., AlGaInP/GaAs) LED to enhance the efficiency of HB LEDs was demonstrated by Alberto Barbieri in 1995. The existence of the HB blue LEDs and high-efficiency HB LEDs quickly led to the development of the first white LED by Nakamura, who employed a Y3Al5O12:Ce, or YAG:Ce yellow phosphor coating on blue LEDs to produce light that appears white [34–36].

Nakamura's invention has been the most important and powerful driver for the development of the LED industry in the past 10+ years (as shown in Fig. 11.2 with the sharp turn) and opened the doors for LEDs in many applications, such as (1) indicators and signs, e.g., traffic lights and vehicle lighting; (2) lighting, e.g., street lamps and backlighting for LCD (liquid crystal display) televisions; (3) smart lighting, e.g., to transmit broadband data and wireless routers for data transport; (4) environmentally sustainable lighting, e.g., nontoxic; (5) economically sustainable lighting, e.g., to reduce cost; (6) nonvisual applications, e.g., sensors and light matter interaction and communication; and (7) light sources for machine vision systems, e.g., barcode scanners. Figures 11.3 and 11.4 show these applications. All these earned Nakamura (with Isamu Akasaki and Hiroshi Amano) the 2014 Nobel Prize in Physics.

A few years later, the demonstration of high-power (HP) AlGaInN flip chip (FC) LEDs (HP FCLED) by Jonathan Wierer and his colleagues at Lumileds [37] pushed the doors even wider for LEDs in more new applications such as those shown in Fig. 11.4. 3D LED and IC

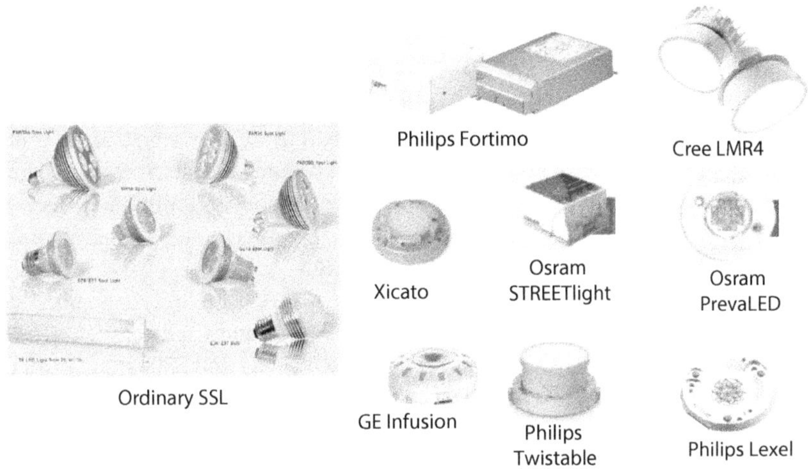

Figure 11.3 Examples of solid-state lighting. (Left) Ordinary ones. (Right) More options and higher levels of integration.

Figure 11.4 Beautiful examples of LEDs. (Upper left) The Gateway to Sohar, Oman; (Lower left) Guangzhou TV Tower, China; (Right) Mega Bridge, Bangkok, Thailand.

integration with TSV (through-silicon via) technology will make LED products with better performance, lower cost, less footprint, lighter weight, and smaller form factor [38]. LED is a clean (green) product because LED lighting is less harmful to human beings. Also, LED lighting can last for a very long time. Thus, unlike the current lighting system, much less LEDs will end up in the land-field to harm the environment. It is an exciting time for LEDs.

11.4 Four Key Segments of LED Products

Figure 11.5 shows the four key segments in making an LED product, namely, substrate and epitaxial deposition, LED device fabrication, packaging assembly and test, and final product assembly. LED products are more expensive than those they are going to replace; thus, the only way to lower the cost and be competitive is to integrate all these four segments together and form a vertically integrated company such as Lumileds/Philips and Cree.

11.4.1 Substrates for LED Epitaxial Deposition

The most used substrate for LED is sapphire (Al_2O_3), even SiC, GaN, and GaAs have also been used. For example, Lumileds/Philips uses Al_2O_3 and Cree uses SiC. The thermal conductivity (W/m°C) of Al_2O_3 and SiC are, respectively, 24 and 250. The thermal coefficient of expansion of Al_2O_3 and SiC are, respectively, 6×10^{-6}/°C and 3×10^{-6}/°C. The elastic modulus of Al_2O_3 and SiC are, respectively, 310 and 500 GPa. The electrical resistivity of Al_2O_3 and SiC are, respectively, $>10^{14}$ Ωcm and 0.02 Ωcm.

The epi depositions of LED on the substrate are grown in a reactor (e.g., Fig. 11.6 shows the VECO by Veeco Instruments) by MOCVD (metal-organic chemical vapor deposition). The process temperature ranges from 750°C to 1100°C. The well-known equipment vendors of MOCVD for LED applications are Aixtron (Germany), Veeco (USA), Nippon Santo and Nissin Electric (Japan). China has also started their production of MOCVD. Currently, China has more than 1000 units of MOCVD. After the LED epi deposition, the wafer is patterned and etched.

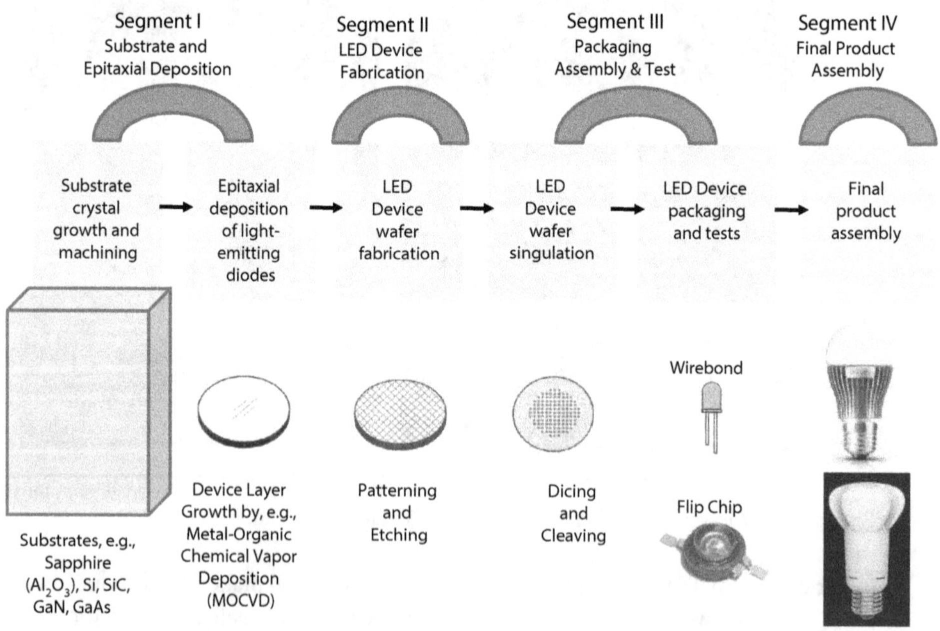

Figure 11.5 Four key segments of LED semiconductor: (a) substrate and epitaxial deposition; (b) LED wafer fabrication; (c) packaging, e.g., dicing, cleaving, wire bonding, solder bumped flip chip, assembly and test; and (d) final product assembly.

FIGURE 11.6 Metal-organic chemical vapor deposition reactor.

11.4.2 LED Device Fabrication

The key tasks of LED device fabrication are patterning and etching, which are just like those used by IC semiconductor fabrication. The multiple quantum wells are basically achieved by photolithographic patterning and plasma etching techniques.

11.4.3 Packaging Assembly and Test of LED

Packaging of the LED devices is a big task. In addition to the routine dicing and cleaving (singulation) of the LED device wafer, the individual LED device is housed into a package, which is supposed to do four major tasks: (1) to provide a path for the electrical current that powers the LED device, (2) to distribute the signals onto and off the LED device, (3) to remove the heat generated by the LED, and (4) to support and protect the LED device from hostile environments. Today, there are two ways to connect the LED device to its outside world (next level of interconnect): one is by wirebonding (the so-called "bullet-head") as shown in Fig. 11.5 and the other is by solder bumped flip chip, as shown in Figs. 11.5 and 11.7. The LED packaging trends are brighter, smaller, smarter, and cheaper as shown in Fig. 11.8.

11.4.4 LED Final Product Assembly

Is LED equal to a lightbulb or lamp in SSL application? No, LED is not just a lightbulb or lamp for SSL application, but much more! The final assembly of LED bulb or lamp consists of at least a socket, heat sink, driver (AC to DC), LED, and optic. This is one of the reasons why the lightbulb is cheaper than the LED bulb. The factors affecting the LED bulb efficiency are the drivers, optics, and thermal. Due to the very small size of the LED devices (i.e., very small conducting surface), it is very difficult to conduct the heat out and usually (out-of-proportional) heat sink is needed to keep the LED temperature below 85°C so as to maintain the LED lumen performance. Figures 11.9 through 11.11 show the temperature distribution of LED package and assembled final products. It can be seen that LEDs are hot (red color)!

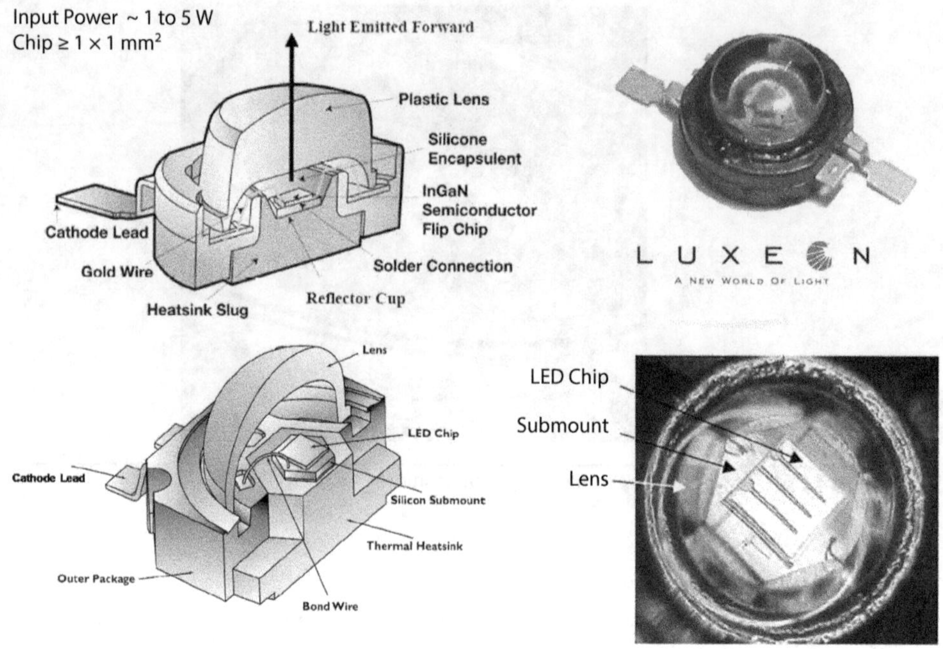

FIGURE 11.7 Philips' HPFC LEDs.

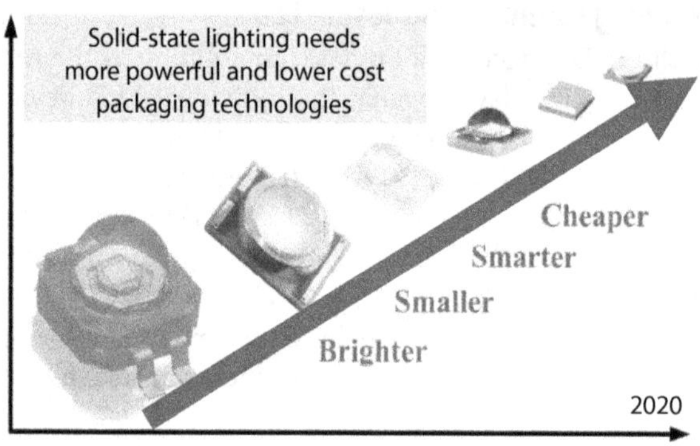

FIGURE 11.8 LED SSL packaging types and trends.

11.4.5 Outlook of LED Products

With the momentum of LED and the environmental initiatives for green products from different countries, there should not be any problem to follow (or do better than) the Haitz's law. However, there are two big concerns from the LED industry. One is the poor-quality LED products and the other is unfair pricing.

As mentioned earlier, LED products will be more expensive than those they are going to replace; thus, people are expecting high-quality LED products that would be brighter and would last longer. Some established companies are afraid that the poor LED products made

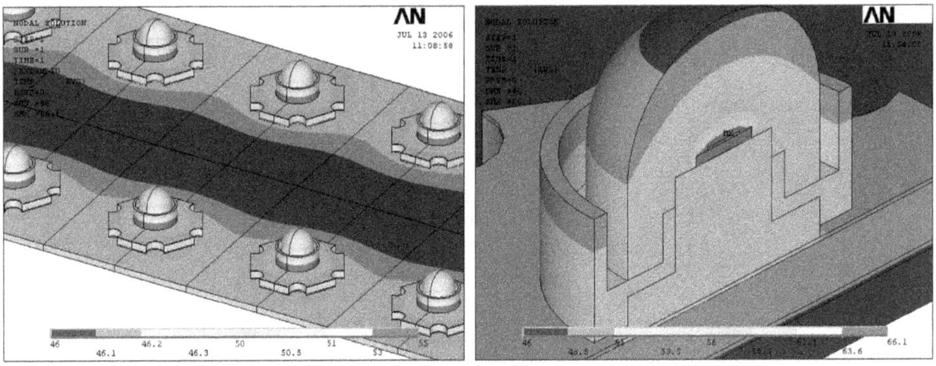

FIGURE 11.9 Temperature distributions in LED packaging.

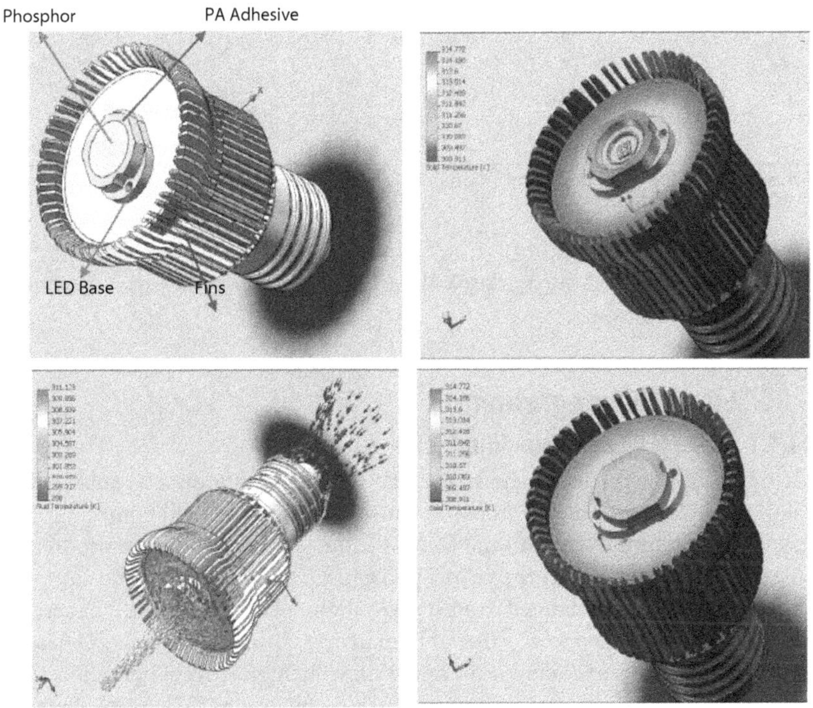

FIGURE 11.10 System-level thermal modeling and analysis.

by other companies will give the users around the world bad experience leading them to even stop using the LED products. Table 11.1 shows the qualification tests and two different sets of detailed test conditions. They are good examples and should be performed on all LED products.

Unfair pricing is another concern of some established companies since they have to report the earning, quarter-by-quarter, to their shareholders. If they lose money because

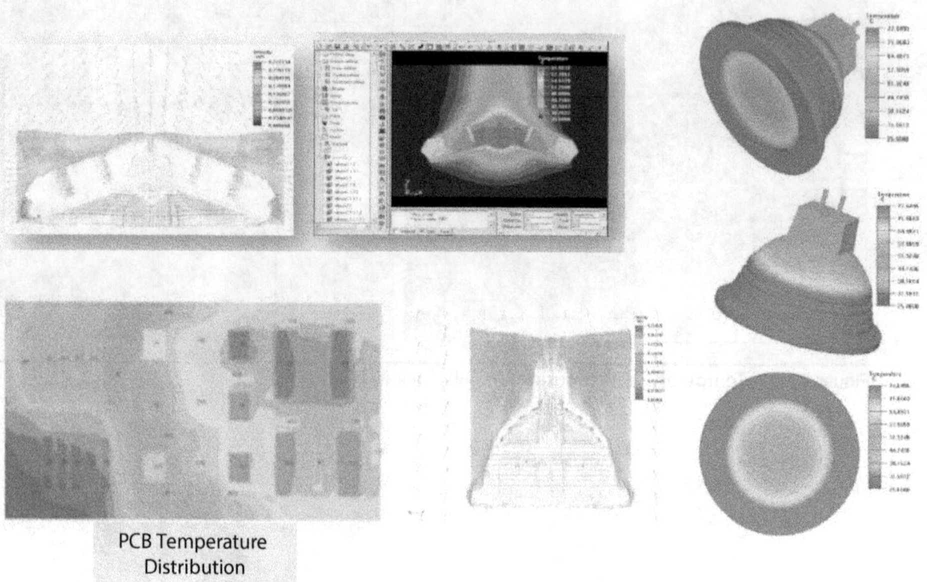

FIGURE 11.11 Temperature distribution in LED lighting products.

some companies flood the market with unfair pricing, then they may have to lose and eventually stop their business.

11.5 3D LED and IC Integration

11.5.1 HP FCLED and Thin-Film FCLED

Figure 11.12 schematically shows the HP FCLED sitting on a Si submount proposed by Lumileds [37] (now called Philips). It shows that the blue light comes out from the MQW (multiple quantum wells) and the GaN is extracted through the transparent sapphire substrate which is coated with the yellow phosphor. This yellow phosphor absorbs a portion of the blue light and is stimulated to emit yellow light, which mixes with the rest of the blue light to generate light that appears white to human eyes. Also, the HP FCLED has a large emitting area and an optimized contacting scheme allowing high current operation with low forward voltages, and therefore high-power conversion efficiencies [37]. Since there is no wirebond and it has solder bumps, this package will have better heat extraction and optical efficiency. This technology has been in mass production and some of the products are shown in Fig. 11.7. Recently, Philips removed the sapphire and roughened the back side of the n-GaN, as shown in Fig. 11.13, which is called thin-film FCLED. In this case, they can achieve higher optical extraction efficiency.

Usually, the FCLED with the Si submount is packaged in a reflector cup with metal coating (but some are without the cup as shown in the lower left-hand corner of Fig. 11.7), attached to a interconnect, e.g., housing and/or PCB (printed circuit board), filled with a transparent silicone encapsulant, and then covered by a plastic lens (either curve or flat).

Table 11.1 High-power LEDs' Packaging Qualification/Reliability Tests

Test method	Stress conditions: SPEC A	Stress conditions: SPEC B
RTOL	Ambient Temperature: 45°C Forward current: Maximum in data sheet Test period: 1008 hours	Ambient temperature: 55°C Forward current: Maximum in datasheet Test period:1000 hours
HTOL	Ambient temperature: 85°C Forward current: Maximum in data sheet Test period: 1008 hours	Ambient temperature: 85°C Forward current: Maximum in datasheet Test period: 1000 hours
LTOL	40°C Forward current: Maximum in data sheet Test period: 1008 hours	55°C Forward current: Maximum in datasheet Test period: 1000 hours
WHTOL	Ambient temperature: 85°C Forward current: Maximum in datasheet Humidity: 85% relative humidity (RH) Time period: 1008 hours (cycles)	Precondition at 60°C and 60% RH prior to reflow soldering, 120 hours Autoclave: 121°C, 100% RH, 15psig, 96 hours
Thermal shock	40°C to 125°C Dwell time: 15 minutes Transfer time: <20 seconds Cycles: 200 cycles	40°C to110°C Dwell time: 15 minutes Transfer time: <20 seconds Cycles: 1000 cycles
Mechanical shock	Shock: 1500 G Pulse width: 0.5 ms Direction: 5 each, 6 axis (30 total)	Shock: 1500 G Pulse width: 0.5 ms Direction: 5 each, 6 axis
Salt atmosphere	Ambient temperature: 35°C Salt deposit: 30 g/m^2/day Test period: 48 hours	Ambient temperature: 35°C Test period: 48 hours

The input power ranges from 1 to 5 W and the LED chip size is $\geq 1 \times 1$ mm^2. It should be pointed out that in order to improve the light output and quality, Philips/Lumileds removed the sapphire substrate, as shown in Fig. 11.13.

Most of the system houses design and assembly the LED module (component) on one location of the PCB and the ASIC, LED driver, sensor, RF, power controller, processor, or memory, etc. on another location of the PCB, as shown in the top of Fig. 11.14. (For simplicity the cover lens, encapsulant, thermal management system, etc. of the LED module are not shown.)

11.5.2 3D LED and IC Integration Packages

The major trend in the electronic industry today is to make products more personal by making them smarter, lighter, smaller, thinner, shorter, and faster, while simultaneously making them more friendly, functional, powerful, reliable, robust, innovative, and economical. As the trend toward miniature and faster products continues, the introduction of innovative

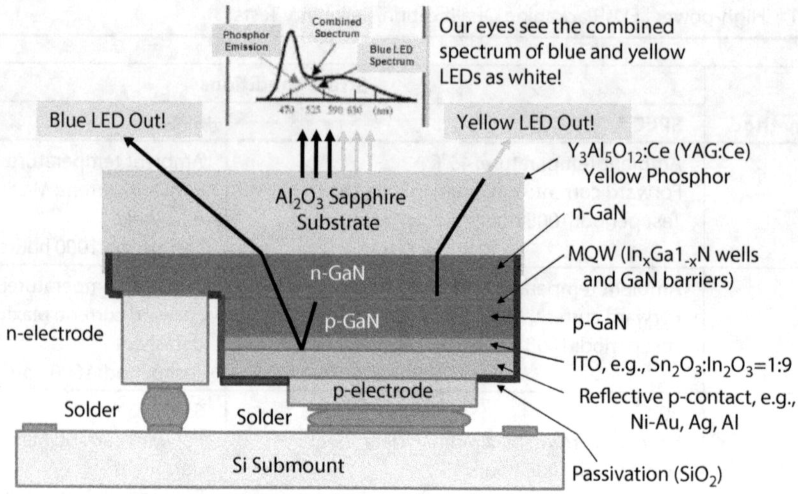

FIGURE 11.12 High-power AlGaInN flip chip LED.

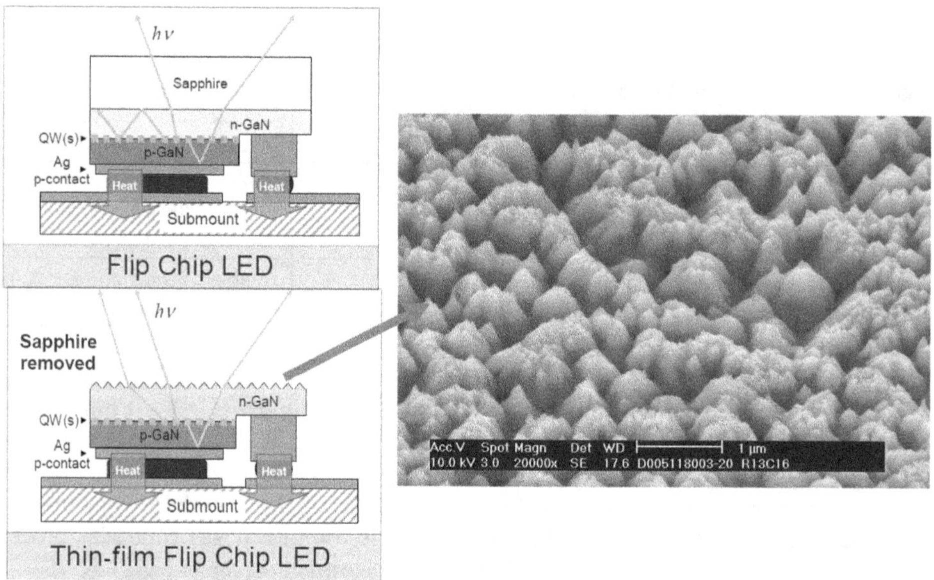

FIGURE 11.13 Thin-film flip chip LED (with sapphire removed).

products that are more user-friendly and contain a wider variety of functions will be a driving factor in providing growth in the market. One of the key technologies that will help to make these product design goals possible is 3D IC integration; see for examples [38–55].

The bottom schematic of Fig. 11.14, for example, shows the passive Si submount replaced by an IC chip such as the ASIC, LED driver, processor, power controller, sensor, RF, etc., i.e.,

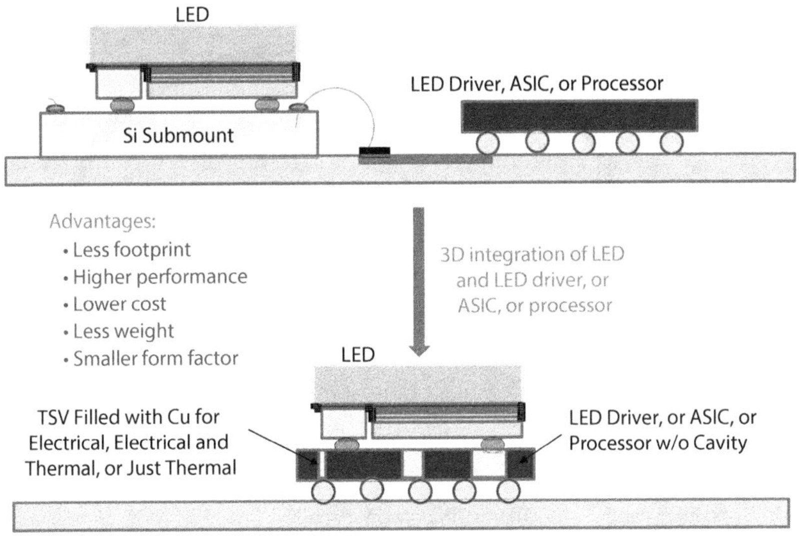

FIGURE 11.14 3D LED and IC (e.g., LED driver, ASIC, memory, processor, sensor, power controller, and RF) integration.

integrating the LEDs and the IC chip together in a 3D manner [38]. Their electrical feed-through and some of the thermal paths can be effectively achieved by the through-silicon vias (TSV) filled with copper on the IC chip. The advantages of this 3D LED and IC integration package are (1) better performance, (2) lower cost, (3) less footprint, (4) lighter weight, and (5) smaller form factor; and it is the focus of this chapter.

Figures 11.15 and 11.16 show two 3D LED and IC integration packages. Figure 11.15 shows a package without a cup but a curve lens. Figure 11.16 shows a package with a reflector cup and a flat plastic lens. Both packages are able to support yellow-phosphor multi-LEDs and active chips. It can be seen that the passive Si submount (shown in Figs. 11.12 through 11.14) has been replaced by an active IC chip, which could be, e.g., the ASIC, LED driver, power controller, sensor, RF, processor, and memory. This active IC chip can be used to support the yellow-phosphor multi-LEDs for many functions, e.g., dimming and lighting control, step-up and step-down topologies, power conversion, electrical feed-through, and thermal management. This active IC chip has TSVs filled with copper (and most likely with RDL) for pure electrical feed-through, electrical feed-through as well as thermal path, and pure thermal path (i.e., large dummy vias filled with copper). The thermal conductivity of Si is 150 W/m-K and of Cu is 390 W/m-K.

11.5.3 Manufacturing Process of 3D LED and IC Integration

The manufacturing processes of these two 3D LED and IC integration packages are basically the same, except one has the cavity (Fig. 11.16), which is detailed in this book and shown in Fig. 11.17. The fabrication of the reflector cup (cavity) of the active silicon IC wafer is by wet anisotropic etched with the etchants such as KOH (potassium hydroxide) solution, EDP (ethylenediamine pyrocatechol) solution, and TMAH (tetramethylammonium hydroxide). Depending on the dimensions of the mask opening, a V-groove or trapezoidal basin can

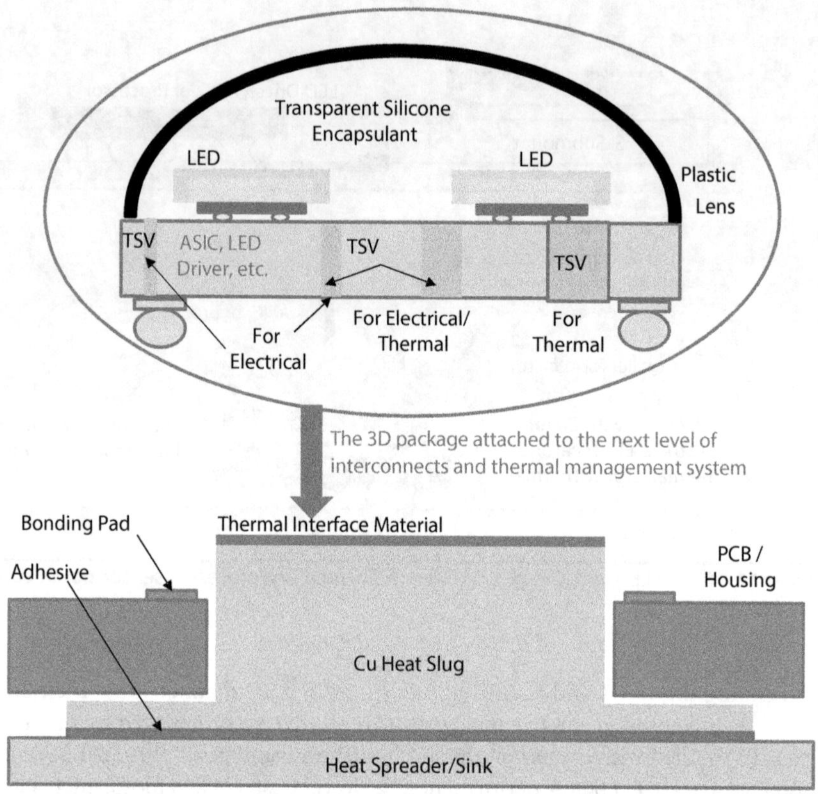

FIGURE 11.15 3D LED and IC integration package (without a cavity) on an ordinary thermal management system.

be formed in the active IC wafer. KOH is the most commonly used etchant. It is much less dangerous (toxic) than others, easy to handle, readily available, and etches fast. The greatest disadvantages are that KOH is IC incompatible and that the selectivity to PECVD (plasma-enhanced chemical vapor deposition) oxide is rather poor. TMAH is nontoxic and IC compatible, but fewer studies exist on this system. EDP is not easy to handle. It is toxic, and the solution degrades if it comes in contact with oxygen. Thus, EDP is mainly used in research laboratories and is not used in mainstream semiconductor fabrications.

The most frequently used mask material for KOH etchant is LPCVD (low-pressure chemical vapor deposition) silicon nitride/thermal oxide, since it has large etching selectivity with respect to silicon in the KOH etchant. With the KOH anisotropic wet etchant to the most commonly used (100) orientation silicon wafer, the sidewalls ({111} planes) of the cavity are at an angle of 54.74 degrees to the wafer surface and at 90 degrees for the less frequently used (110) silicon wafers. To facilitate concept development in silicon crystal orientation, the four types of commercially available silicon wafers are illustrated in Fig. 11.18(a), whereas various planes in a {100} n-type silicon wafer are sketched in Fig. 11.18(b). Depending on the dimensions of the mask opening, a V-groove or trapezoidal basin is formed in the [100] wafer, as shown in Fig. 11.18(c).

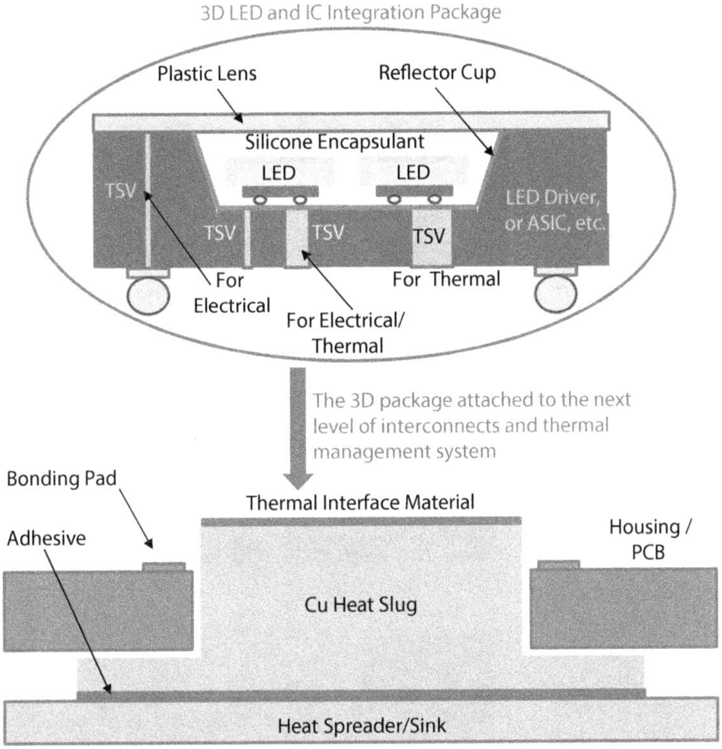

Figure 11.16 3D LED and IC integration package on an ordinary thermal management system.

With an addition of IPA (isopropyl alcohol) to the KOH etchant (KOH/IPA), the sidewalls ({110} planes) of the cavity of a (100) orientation silicon wafer are at an angle of 45 degrees to the wafer surface, as shown in Fig. 11.19 [56]. It should be pointed out that EDP and KOH/IPA solutions behave alike but smoother surfaces are obtained with EDP [56]. Also, by utilizing the knowledge of the etch rates for different crystallographic planes of the starting wafers, a variety of micromachined structures can be formed by properly choosing the masks and etchants [56–59]. After the cavity of the IC wafer is formed by the wet anisotropic etching with the KOH/IPA etchant, the surface of the cavity is coated with a metal, such as Ag, Al, or NiAu, and become a reflector.

TSV and RDL are the most important key enabling technologies for 3D IC integration [38–55]. There are six major steps (via formation, dielectric isolation layer deposition, barrier/adhesion and seed metal layer, via filling with copper, copper polishing, and TSV Cu revealing) to make just the TSV and more steps to make the RDL, UBM (under bump metallurgy), solder bumps, and metallization as shown in Chaps. 2 through 8 in this book.

There are at least two ways to bond the yellow-phosphor LEDs (chips) to the IC wafer (C2W) with TSV as shown by the methods in Chap. 7. One is to use the P&P (pick & place) machine to place all the LEDs (w/o flux) on the IC wafer (w/o flux) and then reflow all the LEDs at once. The other is to use a flip chip bonder to P&P and reflow the LEDs (w/o flux), one at a time, on the IC wafer (w/o flux).

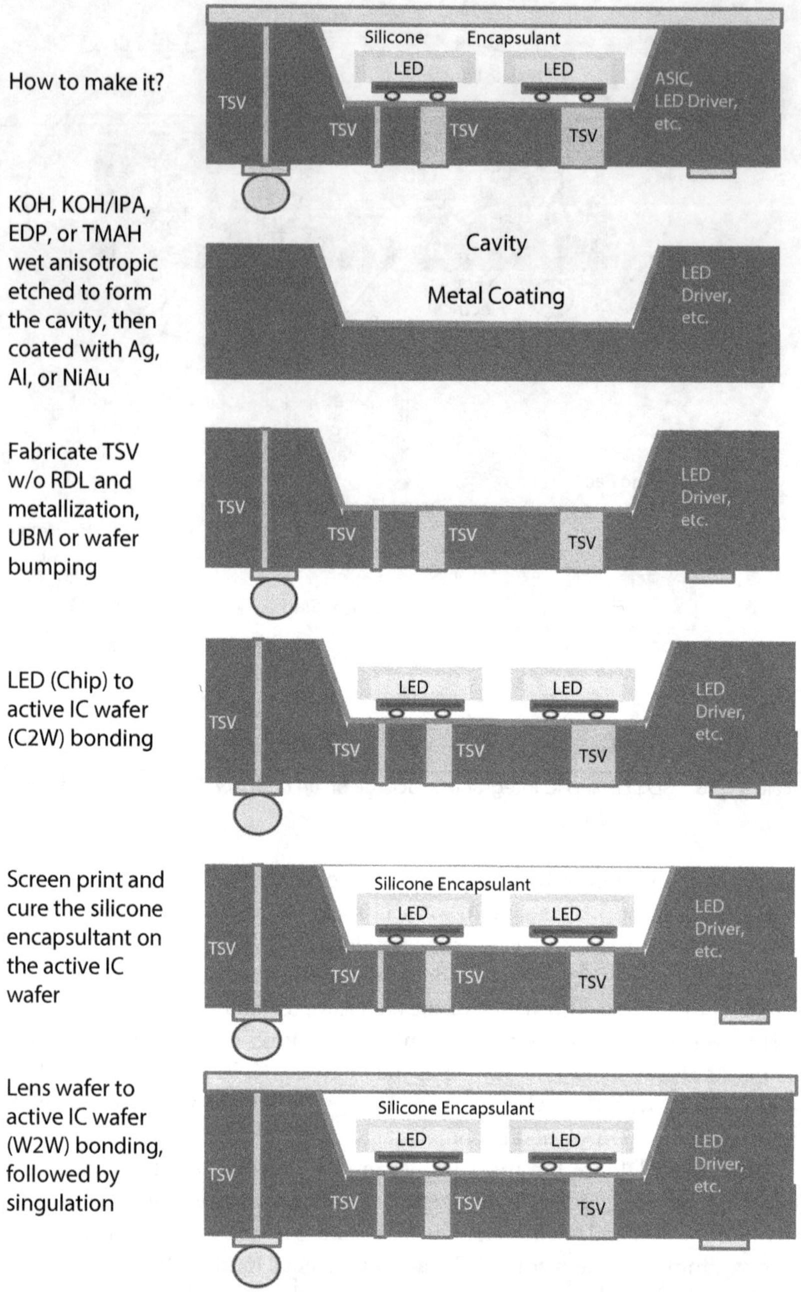

Figure 11.17 Manufacturing process of 3D LED and IC integration package.

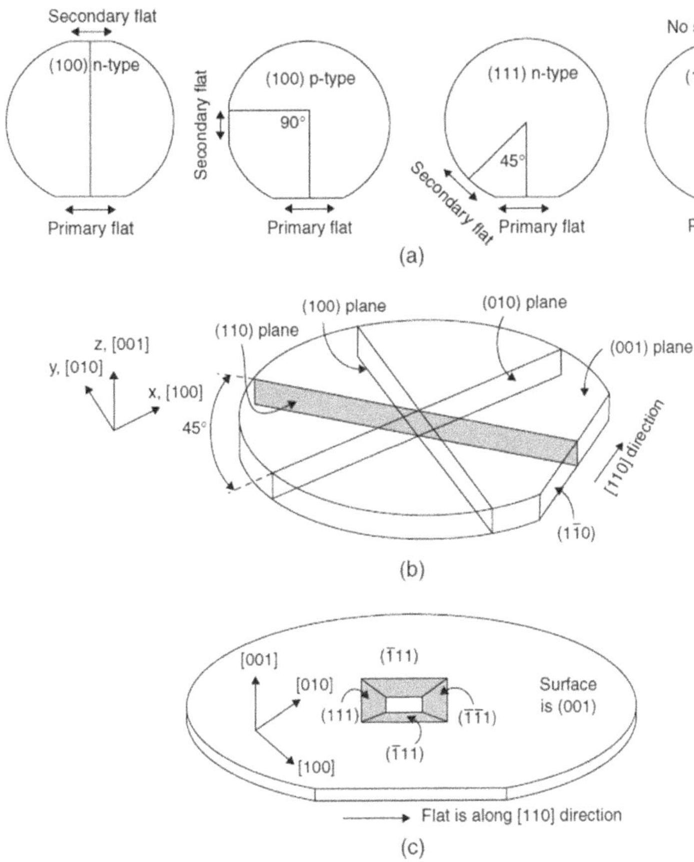

FIGURE 11.18 (a) Drawings showing the primary and secondary flats of {100} and {111} wafers for both n- and p-type doping. (b) Various planes in a wafer of {100} orientation (the wafer thickness is exaggerated). (c) Perspective view of a {100} wafer and a KOH-etched pit bounded by {111} planes. (Maluf and Williams.).

After the yellow-phosphor chip (LEDs) to IC wafer (C2W) bonding, it is followed by silicone encapsulation. Figure 11.20 shows the simple process steps. It can be seen that a stainless steel stencil with openings that are slightly larger than the top opening of the cavity on the IC wafer is used to screen print the silicone encapsulant into the cavity. Then, cure the encapsulant.

After the cavity of the IC wafer is filled with encapsulant, it is ready for the lens wafer to IC wafer (W2W) bonding as shown by the methods in Chap. 6. This step is very similar to the W2W bonding of the cavity-cap wafer and the ASIC wafer (with the MEMS devices already attached on its top surface), which is detailed in Chap. 12 of this book and [40]. Then, singulate the bonded wafers by either mechanical dicing saw or lasers.

If the plastic lens is not in a wafer form, i.e., individual lens, then there are two ways to seal the package. One is by chip (lens) to IC wafer bonding, then singulation. The other is to singulate the IC wafer (into individual IC chips) first as shown in the bottom of

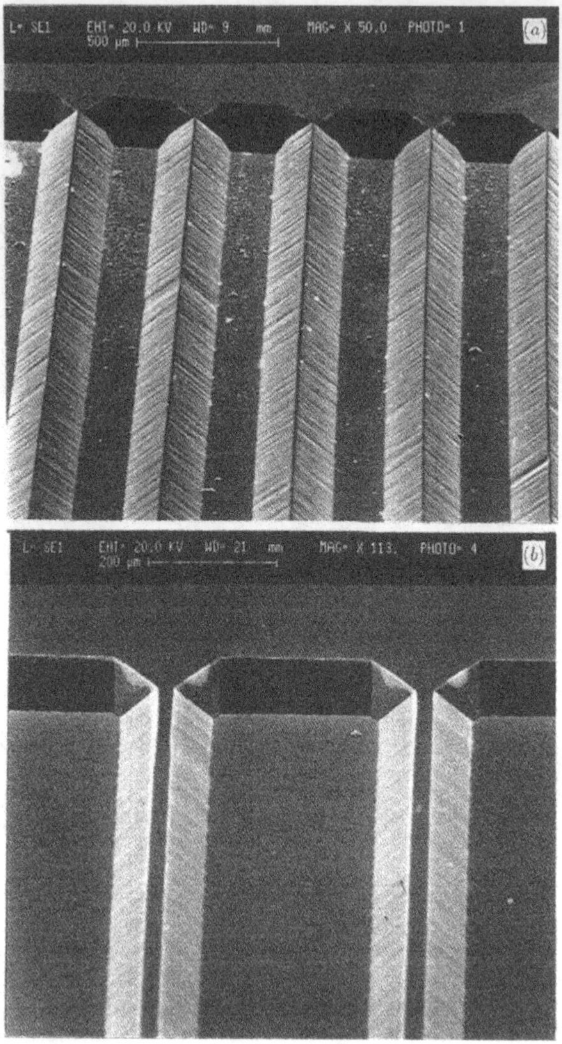

Figure 11.19 SEM micrographs of cavities by KOH/IPA and EDP etchants. The sidewalls are {110} planes with an inclination of 45° to the (110) surface.

Fig. 11.20 and perform chip (plastic lens) to IC chip (C2C) bonding as shown by the methods in Chap. 8. Now the 3D packages are ready for the next level of interconnects and thermal management.

11.5.4 Summary and Recommendations

Some important results and recommendations are summarized as follows [38]:

- By replacing the Si submount of the HP FCLEDs with an active IC chip such as the ASIC, LED driver, processor, memory, RF, sensor, and power controller, i.e., integrating the LEDs and the active chip together into a 3D LED and IC integration

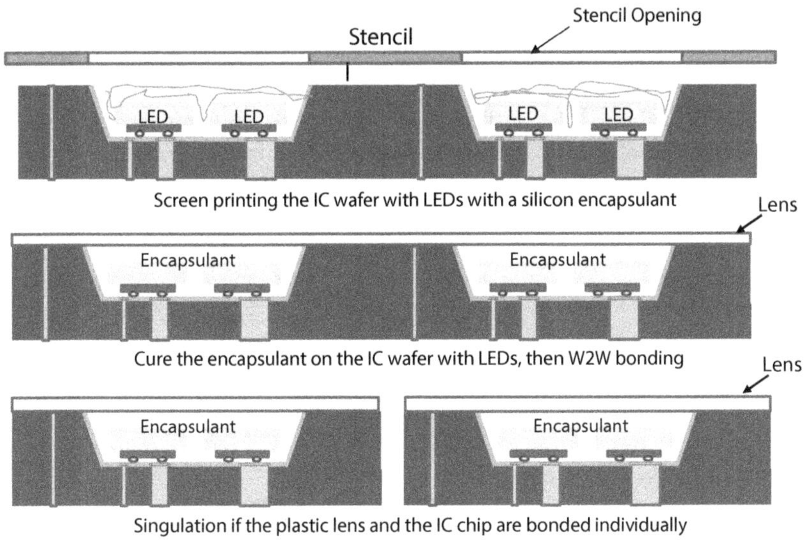

Figure 11.20 Screen printing of silicone encapsulant on IC wafer with LEDs.

package. The advantages of these 3D integration packages are (1) better performance, (2) lower cost, (3) less footprint, (4) lighter weight, and (5) smaller form factor.

- By utilizing the knowledge of the etch rates for different crystallographic planes of the starting wafers, a variety of micromachined structures (reflector cups) can be formed by properly selecting the masking materials, such as the Si_3N_4, SiO_2, Au, Pt, Cr, Cu, Ag, and Ta, and etchants, such as the EDP, KOH/IPA, or TMAH.
- The copper-filled TSV in the active IC chip can be used for not only electrical feed-through, but also as a thermal path. Since the thermal conductivity of copper is more than two times of silicon, some large dummy TSVs filled with copper are recommended.
- It is recommended that the LEDs be picked and placed on the IC wafer (C2W) with flux and reflowed all at once to increase throughputs and reduce costs.
- The silicone encapsulant can be screen printed with a stainless steel stencil with openings slightly larger than the opening of the cavity of the IC wafer.
- It is recommended that the plastic lens be in wafer form, which allows IC wafer to plastic lens wafer (W2W) bonding to increase throughputs and save costs.

11.6 2.5D IC and LED Integration

As of today, there is no research paper that demonstrates the feasibility of 3D IC and LED integration. Recently, however, a few 2.5D IC and LED integrations have been demonstrated [60–65].

11.6.1 LED Packaging Using Si-Substrate with Cavities and Cu-Filled TSVs

Figure 11.21 shows the schematic of the cross-section of a Si-substrate with cavity for phosphor printing and Cu-filled TSVs for interconnection [60, 61]. It can be seen that the Si-substrate is about 400 μm thick with 3-μm-thick low-temperature oxide on both sides. The dimensions of the cavity are 1.3 × 1.3 × 0.22 mm. The diameter of the TSV is 100 μm and is filled with Cu. The exposed tip of the Cu TSV is 30 μm and plated with solder.

The key assembly process flow of the LED package is shown in Fig. 11.22. First, deep reactive ion etch (DRIE) is applied to make the blind vias from the back side and the cavity

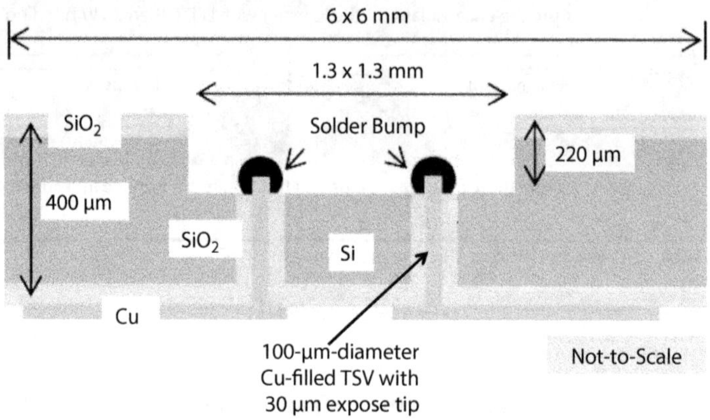

FIGURE 11.21 Si-substrate with a cavity and TSVs.

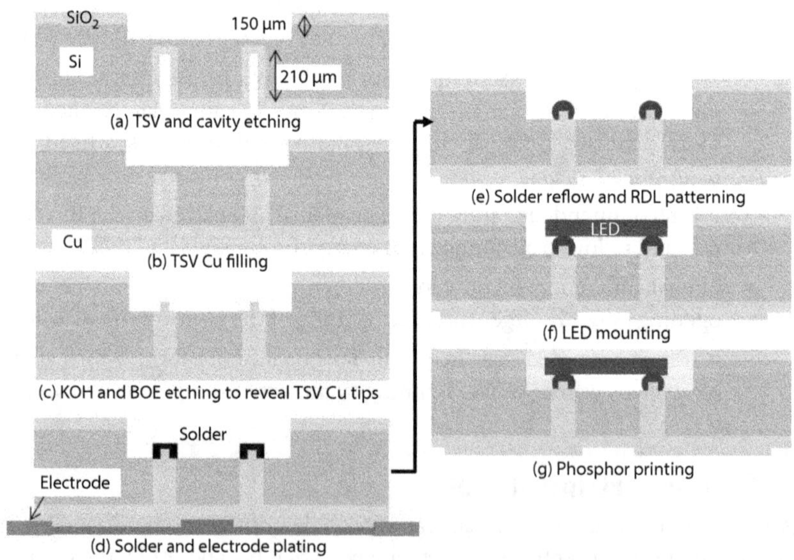

FIGURE 11.22 Fabrication process flow of LED package with Si-substrate.

on the front side, Fig. 11.22(a). It is followed by dielectric layer, barrier layer, and seed layer deposition and Cu plating to fill the vias, Fig. 11.22(b). Then apply the KOH and BOE (buffered oxide etch) to further etch the cavity on the front side in order to expose the tip of the Cu-filled TSV, Fig. 11.22(c). Solder plating on the tip of the TSV and then electroplating the electrode on the back side, Fig. 11.22(d). Solder reflow and then RDL patterning and plating on the back side, Fig. 11.22(e). It is followed by pick and place the flip chip ($1 \times 1 \times 0.07$ mm) blue LED device on the solder and reflow to form the solder joints, Fig. 11.22(f). Finally, phosphor is printed to fill the cavity as shown in Fig. 11.22(g).

Figure 11.23(a) shows the image of the tips of Cu-filled TSVs. Figure 11.23(b) shows the image of the plated solder on the tip of the Cu-filled TSVs (before solder reflow) and Fig. 11.23(c) shows the image of the plated solder on the tip of the Cu-filled TSV (after solder reflow). It can be seen that they are properly done.

Figures 11.24(a) and 11.24(b) show the images, respectively, of the top view and cross-sectional view of the LED package. The Si-substrate, the RDL, the cavity, and the TSV tips with solder bumps (before LED mounting) can be clearly seen.

Figure 11.25(a) schematically shows the squeegee printing of the dry YAG:Ce yellow phosphor powders to encapsulate the LED and fill the cavity. First, a small amount (1 μL) of UV-curable epoxy is dispensed into the cavity and then pre-cured for 10 s by a UV lamp with a power density of 75 mW/cm^2. The phosphor powder is spread on top of the Si-substrate and printed to fill the cavity by a squeegee blade. Finally, the epoxy is fully cured for 6 minutes to bind the phosphor and to encapsulate the LED device. Figures 11.25(b) and 11.25(c), respectively, show the images of the cavity before phosphor printing and after.

Figures 11.26(a) and 11.26(b) show, respectively, the top view and cross-section view of the final LED package. It can be seen that the LED device is embedded in the cavity and

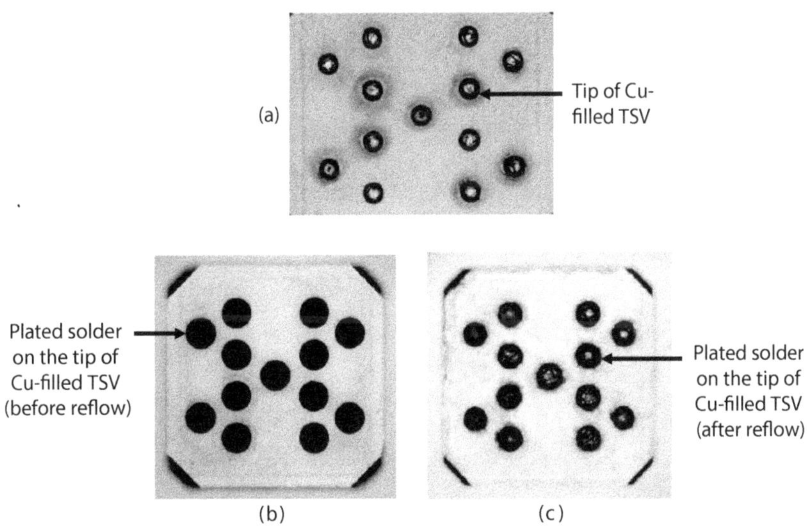

FIGURE 11.23 (a) Tips of Cu-filled TSVs. (b) Plated solder on the tip of Cu-filled TSV (before reflow). (c) Plated solder on the tip of Cu-filled TSV (after reflow).

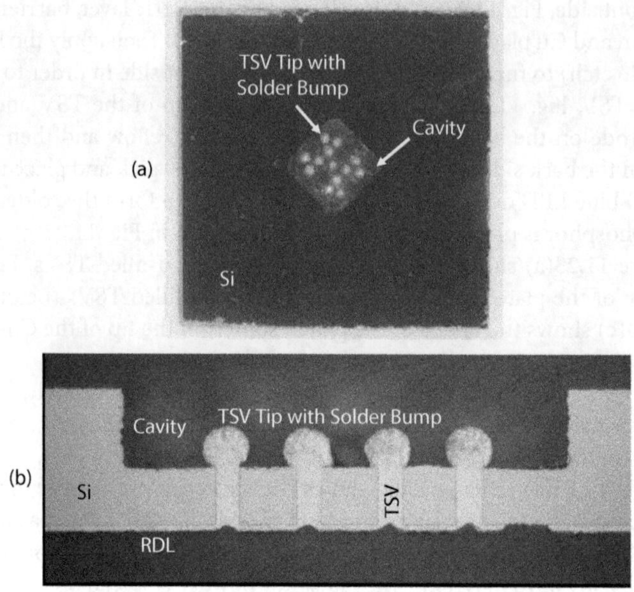

FIGURE 11.24 Top view of the Si-substrate with cavity and TSV tips with solder bumps. (b) Cross-sectional view of the Si-substrate with cavity and TSV tips with solder bumps.

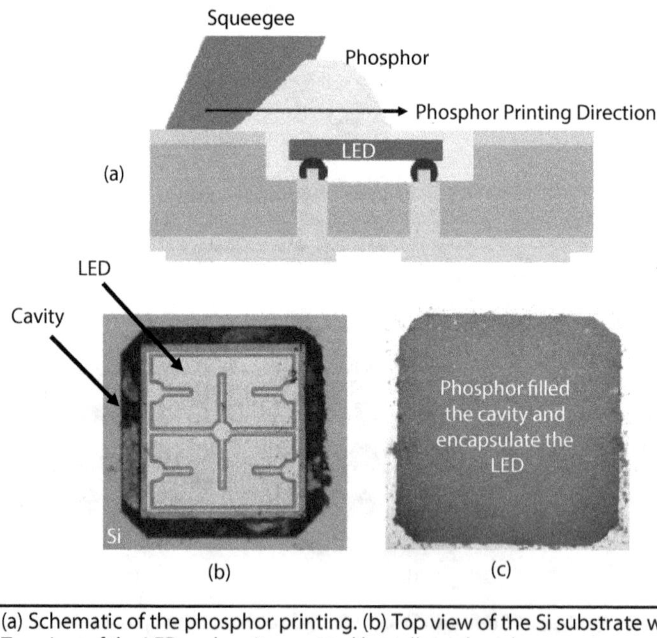

FIGURE 11.25 (a) Schematic of the phosphor printing. (b) Top view of the Si substrate with cavity and LED device. (c) Top view of the LED and cavity covered by yellow phosphor.

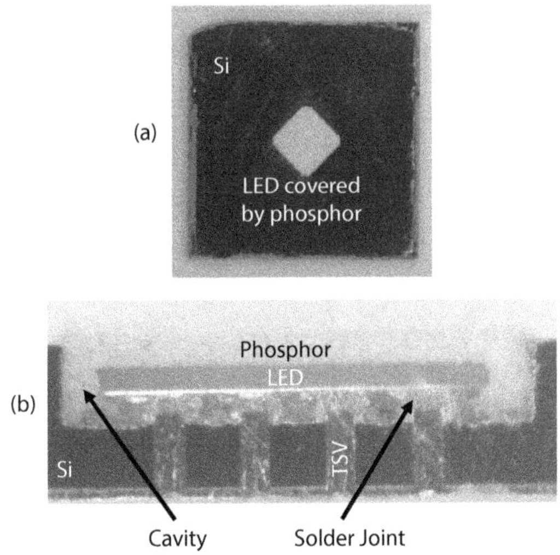

Figure 11.26 (a) Top view of the Si-substrate with cavity and LED covered by phosphor. (b) Cross-sectional view of the Si-substrate with cavity and LED covered by phosphor.

encapsulated by the yellow phosphor. The encapsulation is formed within the cavity and achieves a flat profile.

Figure 11.27(a) and 11.27(b) show, respectively, the lit-up blue LED without the phosphor and with the yellow phosphor. It can be seen that the yellow phosphor does convert the blue LED into white LED.

11.6.2 Si-Substrate with Cavity and TSVs for LED Packaging

Figure 11.28 shows a conventional LED packaging. It can be seen that the LED device is mounted on a Cu slug and wirebonded to a lead frame. The LED is encapsulated by a yellow phosphor and surrounded by a silicone in a lens. A new LED package shown in Fig. 11.29 is proposed [62, 63] to replace the conventional one shown in Fig. 11.28. It can be seen that (a) the Si-substrate is about 370 μm; 190 μm for the cavity and 180 μm for the TSVs; (b) the cavity has a cup-shape and the TSV a cone-shape; (c) the TSVs are filled with Cu; and (d) the cavity houses the LED and is filled with phosphor and epoxy mixture.

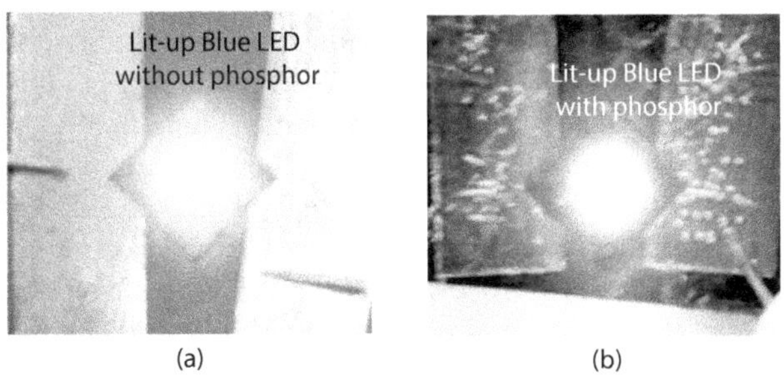

Figure 11.27 (a) Lit-up blue LED without phosphor. (b) Lit-up blue LED with phosphor (white light).

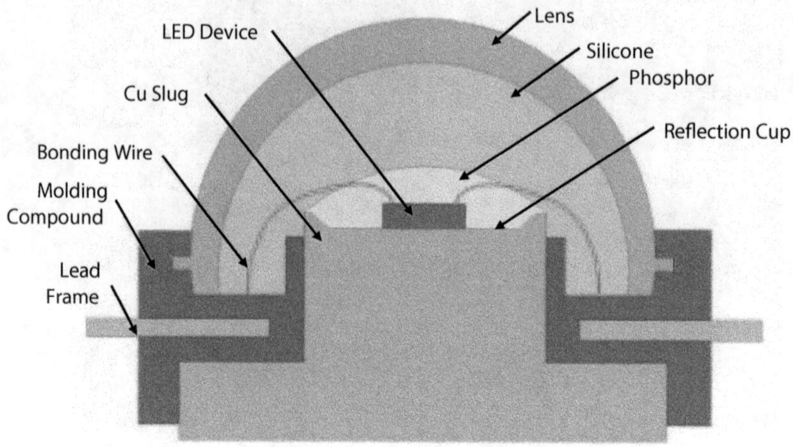

FIGURE 11.28 Schematic cross-sectional view of conventional leadframe based LED package and thermal management system.

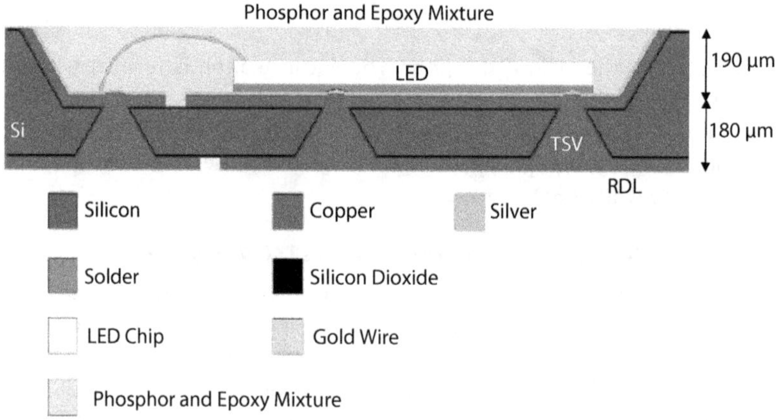

FIGURE 11.29 Schematic cross-sectional view of proposed Si-substrate with cavity and TSVs for LED packaging.

The assembly process flow of the new LED package is shown in Fig. 11.30. First, grow (by thermal oxidation) a 0.8-μm-thick silicon dioxide on the Si-substrate and then open the silicon dioxide on both sides of the Si-substrate by RIE, Fig. 11.30(a). The wafer is placed in a bath with 40 wt% KOH and 40 vol% IPA solutions for 4 h, Fig. 11.30(b). The depth of the cavity is about 190 μm and that of the TSV is about 180 μm. Grow another silicon dioxide and it is followed by Cu sputtering for the seed layer on both sides of the wafer, Fig. 11.30(c). Then the Cu layer is thickened and the vias are filled by electroplating, Fig. 11.30(d). After electroplating, a lithography process is used to form the lift-off mask of the silver layer on the cavity side of the wafer. Another etching mask is formed on the other side of the wafer to allow the creation of the Cu RDL with given pattern. The anode and cathode of the circuit are separated after the Cu etching, Fig. 11.30(e). Figures 11.31(a) and 11.31(b) show,

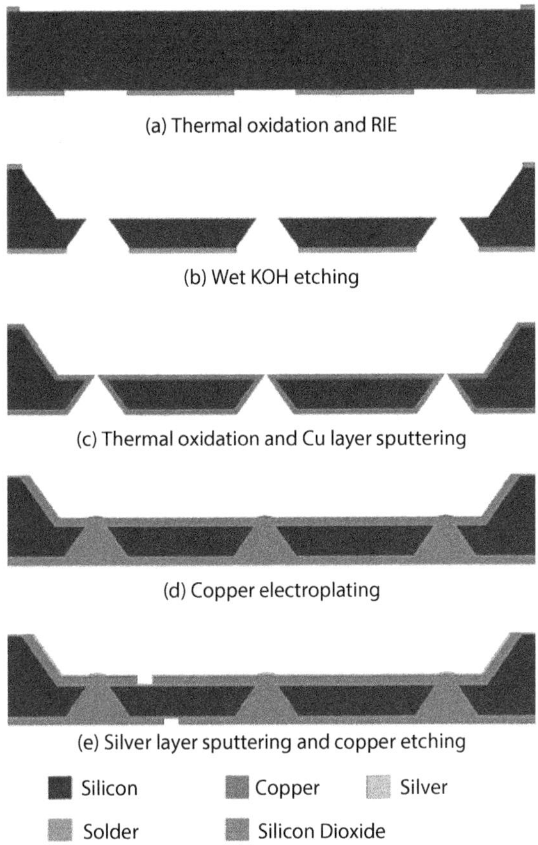

FIGURE 11.30 Fabrication process flow of the Si-substrate with cavity and TSVs.

respectively, the top view and cross-sectional view of the Si-substrate with cavity and TSVs. It can be seen that the cone-shaped TSV is about 180 μm.

Current density plays a very important role. If the current density is too high, the wide end of the cone-shaped TSV seals very rapidly and there is not enough time to achieve TSV filling without seams or voids as shown in Fig. 11.32(a). If the current density is too low, there is no seam and void, Fig. 11.32(b), but it takes a very long time to fill the via (12 h with a current density of 1 to 2 ASD [Amperes per square decimeter]).

Figure 11.33(a) shows schematically the printing of the yellow phosphor and epoxy mixture on the LED and to fill the cavity; Fig. 11.33(b) shows the cross-section image of the Si-substrate with cavity to house the LED and TSVs for interconnection. Figure 11.33(c) shows the lit-up LED package with a small current.

The performance of the LED package is shown in Fig. 11.34 (Mentor Graphics' T3Ster). The 1 W LED is from SemiLEDs and its luminous power is ~90 lm with 350 mA drive current when color temperature is 4200 K. It can be seen that the thermal resistance of the substrate (only) is only 1.3 K/W, and for the whole LED package, including the LED device, solder layer, substrate, thermal interface materials layer, and heat sink, it is about 8 K/W.

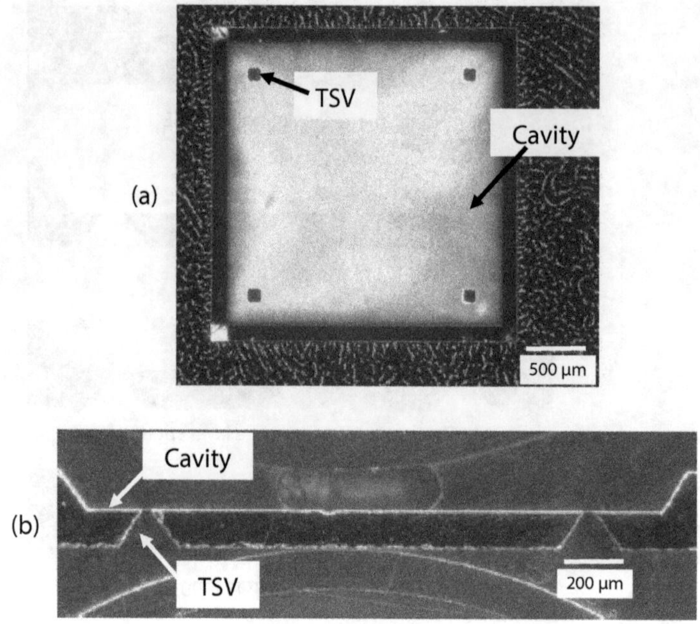

FIGURE 11.31 (a) Top view of the Si-substrate with cavity and TSVs. (b) Cross-sectional view of the Si-substrate with cavity and TSVs.

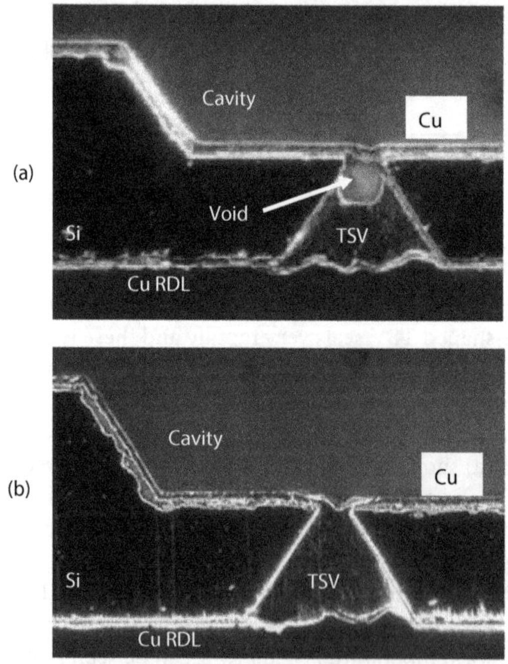

FIGURE 11.32 (a) Void in the Cu-filled TSV due to very high current density and short plating time. (b) Voidless TSV due to low current density and very long plating time.

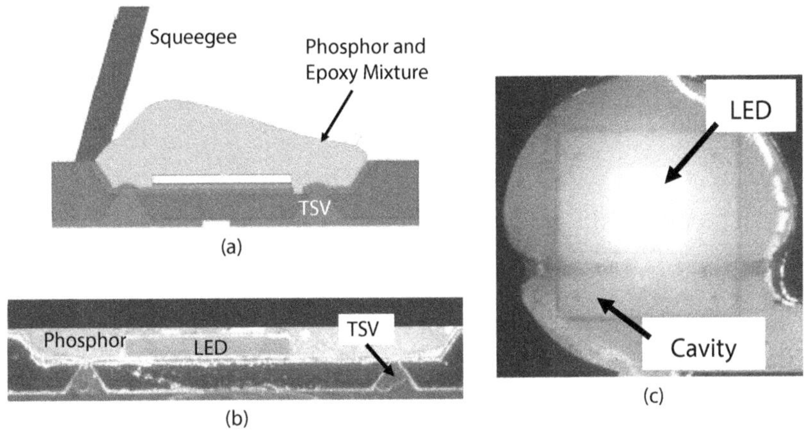

Figure 11.33 (a) Schematic of phosphor and epoxy mixture printing. (b) Cross-sectional view of the Si-substrate with a cavity to house the LED and TSVs. (c) Lit-up the LED.

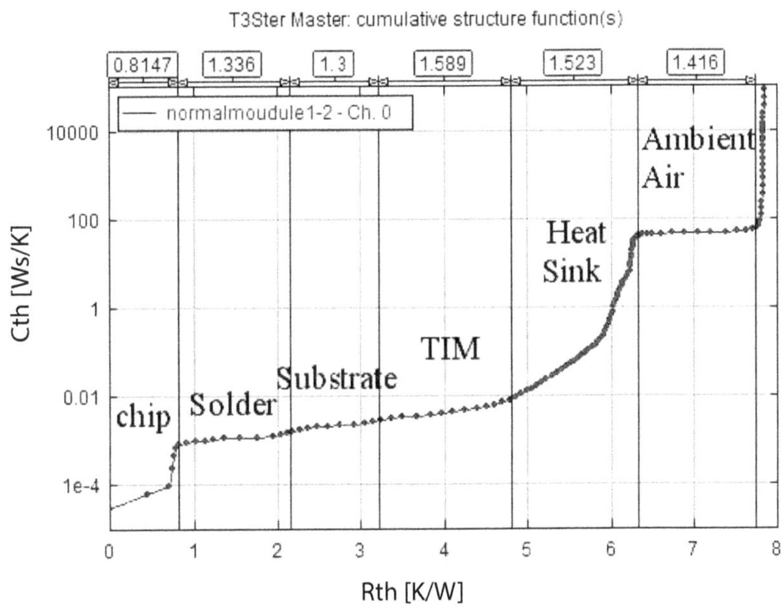

Figure 11.34 Thermal measurement results of the LED package.

11.6.3 LED Wafer-Level Packaging

Figure 11.35 schematically shows a wafer level packaging with LED devices in the cavities of a Si substrate [64, 65]. It can be seen that the Si-substrate has cavities on its top side to house the LED devices and TSVs to connect the LEDs to the RDLs on its bottom side. The Si-substrate is covered by a glass with yellow phosphor. For the test vehicle used in [64], the 1 W LED chip is shown in Fig. 11.36. It can be seen that the size of the LED is 885 × 885 μm and the size of the P-pad and N-pad is, respectively, 100 × 200 μm and 100 × 100 μm.

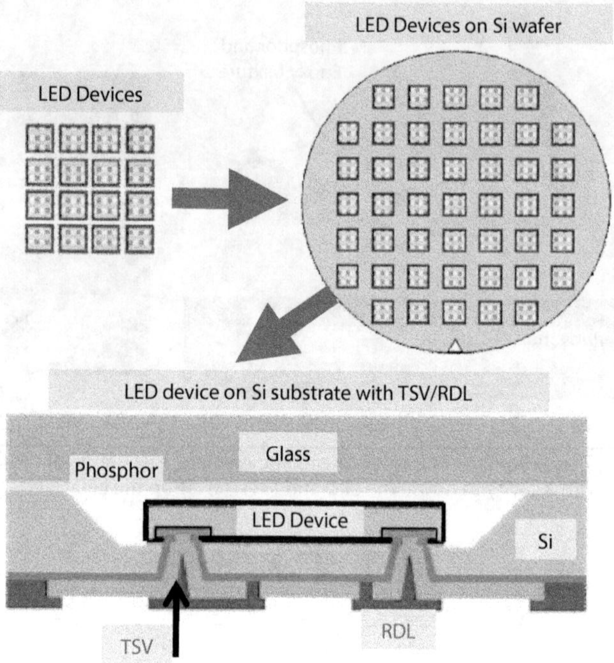

Figure 11.35 Si-substrate with cavity and TSVs for LED packaging.

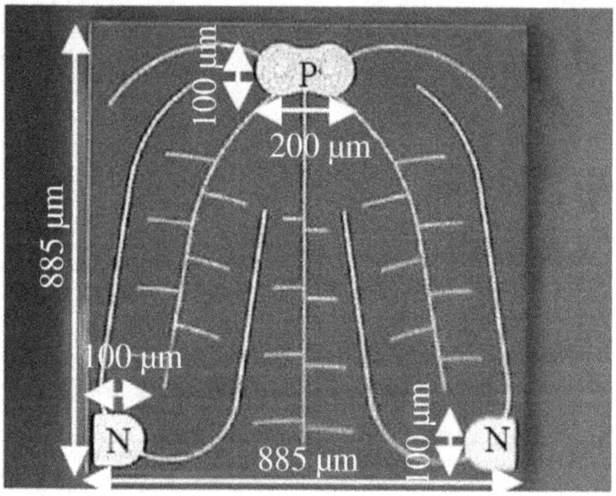

Figure 11.36 LED device for the test vehicle.

The assembly process flow of the proposed LED package is shown in Fig. 11.37. First, DRIE is applied to form the cavity of the top side of the Si-substrate, Fig. 11.37(A). It is followed by sputtering the reflective metal, Fig. 11.37(B). Then, attach the LED to the bottom of the Si-substrate cavity with polymer adhesive, Fig. 11.37(C). Optical glass with phosphor wafer is bonded to Si-substrate with LED wafer (W2W), Fig. 11.37(D). Then, backgrind the

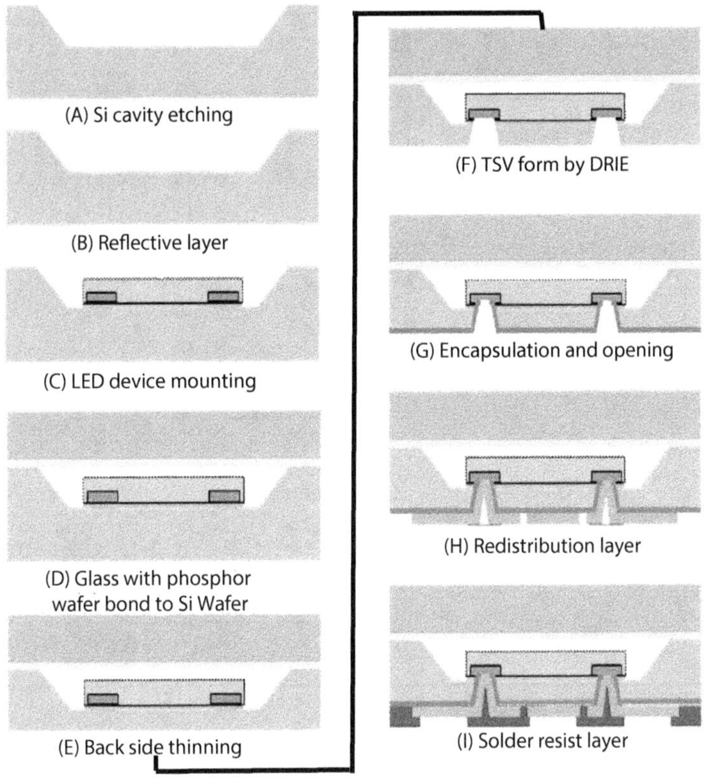

FIGURE 11.37 Fabrication process flow for the Si-substrate with cavity and TSVs for LED packaging.

bottom side of the Si-substrate, Fig. 11.37(E). It is followed by photoresist, mask, lithographing, patterning, and DRIE to form the TSVs, Fig. 11.37(F). Encapsulation layer and opening are formed by spin coating and then laser drilling, Fig. 11.37(G). It is followed by sputtering the seed layer, lithographing, and then plating the RDL, Fig. 11.37(H). Finally, the solder resist layer is formed by photo lithography, Fig. 11.37(I). The photo images are shown in Fig. 11.38.

The packaged wafer is singulated into individual LED package (2000 × 2000 μm). Figure 11.39 shows the SEM images of the cross-section of the LED package without phosphor. It can be seen that (a) the LED device is attached on the bottom of the cavity of the Si-substrate, (b) the glass is bonded on the top of the Si-substrate, (c) the LED pads are connected to the RDL through the TSV with contactor size of about 20 μm, and (d) the encapsulation separates the Si-substrate from the RDL, which prevents electrical shortage between the cathode and the anode.

Figure 11.40(a) shows the LED package with phosphor on an MCPCB (metal-core printed circuit board). The back side of the MCPCB is attached to a heat spreader with a TIM (thermal interface material), Figs. 11.40(b) and 11.40(c). Figure 11.40(d) shows the lit-up LED with phosphor. It can be seen that the light is of mixed white and blue color, which means the phosphor volume/process needs to be optimized.

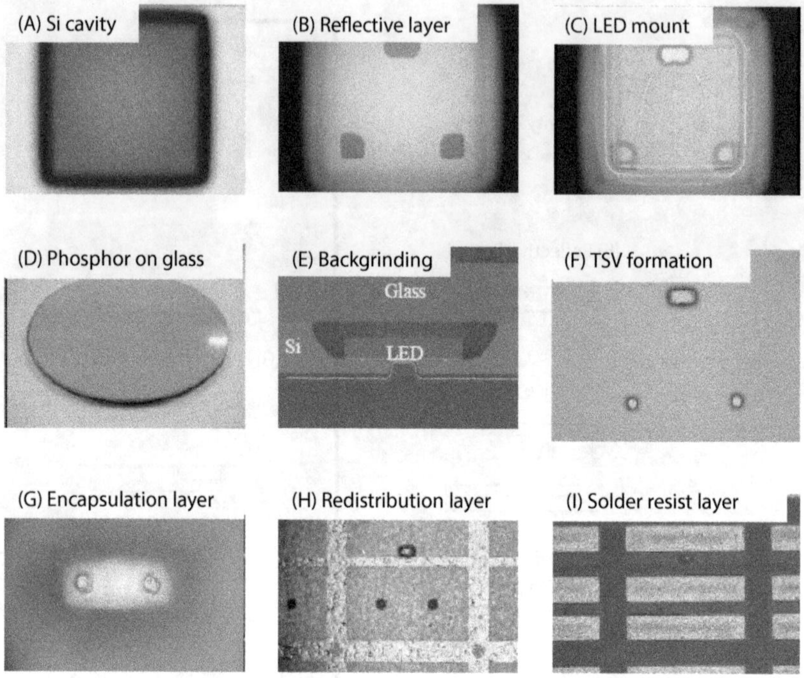

Figure 11.38 Photo images of the fabrication processes of the LED package.

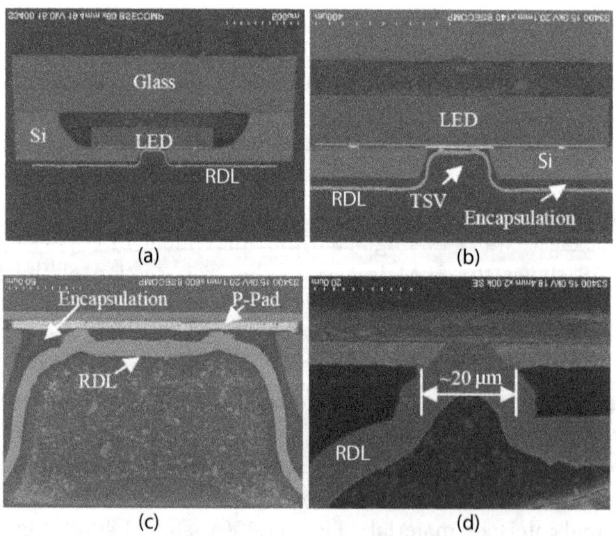

Figure 11.39 SEM images of the cross-sectional view of the LED package. (a) The Si-substrate with a cavity to house the LED and TSV to connect to the RDL. (b) Zoom-in view of (a). (c) Zoom-in view of (b). (d) Zoom-in view of (c).

Figure 11.40 (a) Attach the LED package with phosphor on MCPCB. (b) Coat the TIM at the back side of the MCPCB. (c) Attach the heat spreader to the back side of the MCPCB. (d) Lit-up the LED.

Figure 11.41 shows the Labsphere integrating sphere test results with different electricity currents: 50 mA, 200 mA, and 350 mA of the LED package with phosphor. It can be seen that (a) the light power increases with the input electrical current, (b) the light power is 189.98 mW at 350 mA, which is about 85% of the raw LED device (this means valid electrical connection has been formed through wafer-level packaging), and (c) the light power efficacy decreases with the input electrical current (this means the junction temperature of the LED device increases with the input current). Thus, the thermal management of the present LED packaging system needs to be optimized.

11.6.4 Summary and Recommendation

Some important results and recommendations are summarized as follows [60–65]:

- A few 2.5D IC and LED integration examples have been provided and discussed. Basically, they consist of a Si-substrate (-interposer) with cavity to house the LED and TSVs to connect the LED to the next level of interconnections. A lot of work still remains to be done; however, the feasibility of using a Si-interposer to support the LED has been demonstrated.
- The next step is to replace the Si-substrate (-interposer) by active device chip with TSVs; a true 3D IC and LED integration.

11.7 Thermal Management of 3D LED and IC Integration

One of the key challenges of FCLED packaging is to take the heat out from a very tiny LED chip "area" before it burns itself or affects its light performance. Figures 11.15 and 11.16 show the most conventional thermal management system, which uses a copper heat slug

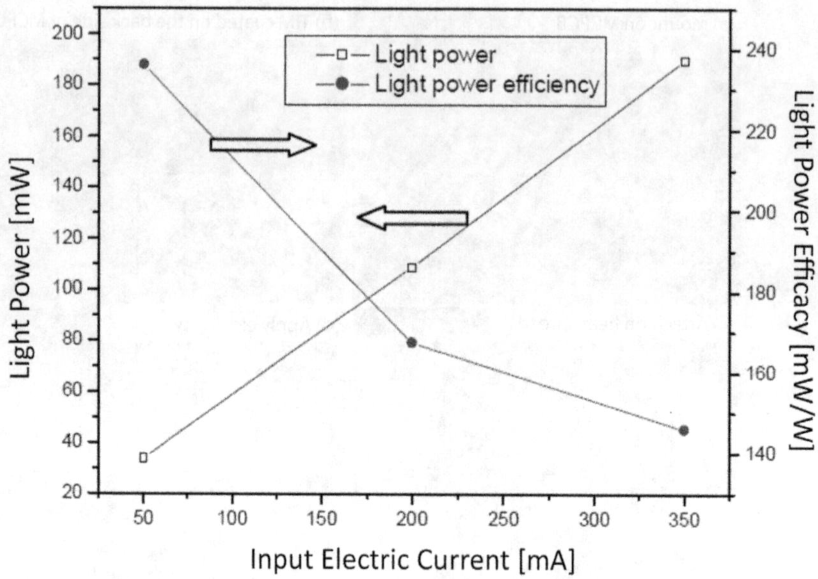

Figure 11.41 Light power and light power efficacy vs. input electrical current of the test vehicle.

to attach to the bottom of the Si submount through a TIM. If that is not enough to take the heat out, then add a copper heat spreader, and if that is still not enough, then add a heat sink.

A thermal management system that consists of a microchannel carrier, a mini-pump, a heat exchanger, and an adapter for flow distribution (Fig. 11.42) for FCLEDs is proposed [66]. The silicon microchannel carrier consists of two silicon chips and an optimized liquid cooling channel structure is embedded in between, as shown in Figs. 11.43 and 11.44. Silicon is chosen as carrier material because it is a suitable material for integration of

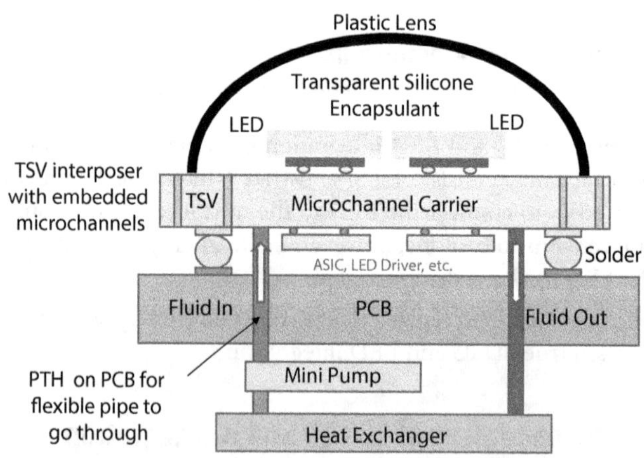

Figure 11.42 TSV interposer with embedded microchannels to support multiple LEDs on top and high-performance logics at the bottom.

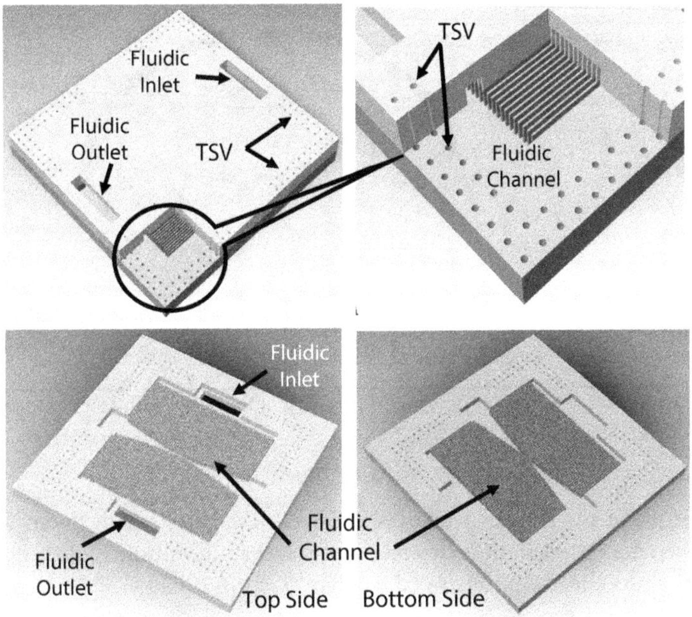

Figure 11.43 Interposer (carrier) with TSV for electrical feed-through and fluidic microchannels for thermal management.

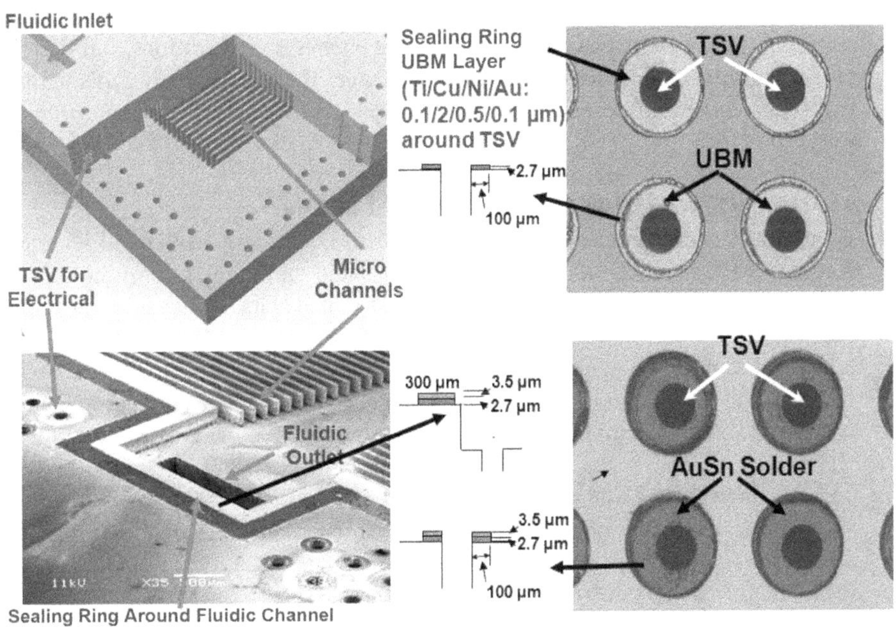

Figure 11.44 Bonding materials and geometry of TSV wafers (carrier) with embedded microchannels.

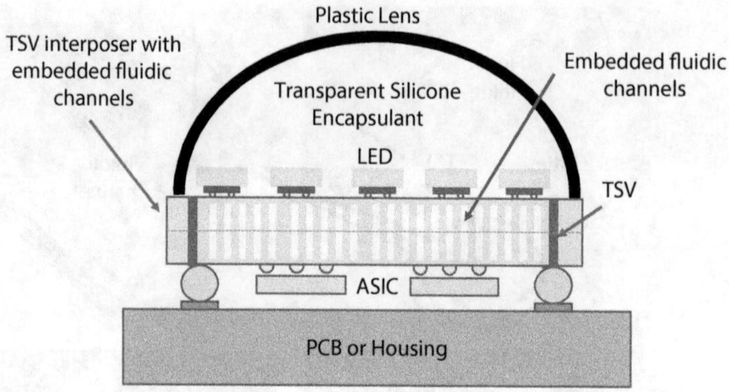

FIGURE 11.45 A low-cost high-power 3D LED and IC integration with TSV interposer with embedded fluidic channels.

both electrical and fluidic structures in the same substrate with micro-fabrication process. The difference between these two pieces of Si is that the bottom Si does not have outlets. This technology has been demonstrated in Chap. 9 of this book.

11.7.1 The New Designs

The proposed design is a 3D IC integration with all the LEDs. MEMS, COMS image sensors, and IC chips supported by a TSV/RDL/IPD interposer with embedded fluidic microchannels for high-performance, low-cost, effective thermal management, and high-reliability applications, as shown in Fig. 11.45. The entire system consists of a silicon interposer with TSVs and RDLs to connect various LEDs and chips (on its both sides just like those discussed in Chap. 8) with pads that have different pitches, sizes, locations, and IPD (integrated passive devices) to enhance electrical performance. There are not any TSVs in the active IC chips and let the interposer be the workhorse and most cost-effective integrator for 3D IC integration [66]!

11.7.2 3D IC and LED Integration: A Design Example

Figure 11.46 shows an example for detailed analysis of the 3D LED and IC integration. It consists of 100 identical LEDs uniformly distributed on the top side of the TSV interposer and 4 identical logics (e.g., ASIC) uniformly distributed at its bottom side. The dimensions of the LEDs are 1 mm × 1 mm × 300 μm, the IC chips 6 mm × 6 mm × 300 μm, and the TSV interposer 25 × 25 × 1.4 mm. Two geometries of microchannels are considered: (1) the channel height = 700 μm, and (2) the channel height = 350 μm. For both cases, the fin width is 0.1 mm and on a 1.25 mm pitch. The fluidic inlet/outlet openings are the same (20 × 1.5 mm).

11.7.3 Boundary-Value Problem

The material properties of the 3D IC and LED integration are as follows: (1) for water; $\rho =$ 989 kg/m^3; $Cp = 4177$ J/kg-K; $k = 0.6367$ W/m-K; and viscosity $= 5.77 \times 10^{-4}$ kg/m-s; and (2) for Si; $\rho = 2330$ kg/m^3; $Cp = 660$ J/kg-K; and $k = 148$ W/m-K.

The boundary conditions are as follows: (1) all exposed surfaces are adiabatic (the worst case for thermal responses); (2) the powers of each LED are 1 W, 1.5 W, and 2 W; (3) the

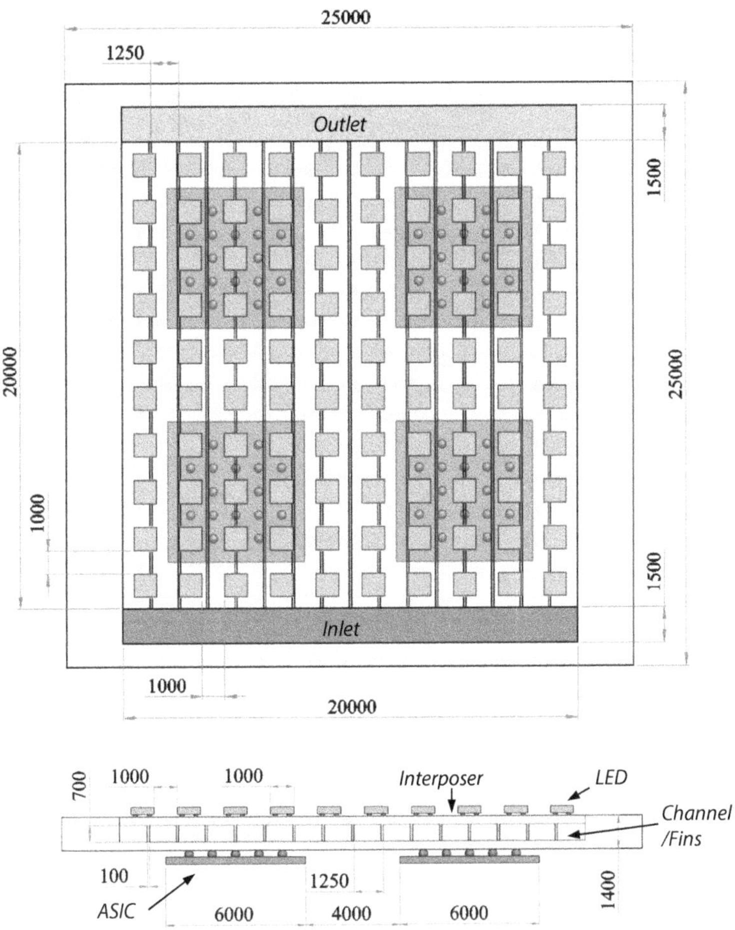

FIGURE 11.46 3D LED and IC integration (unit: μm). 100 LEDs on top and 4 ASICs at the bottom of the TSV interposer with embedded fluidic microchannels.

powers of each ASIC are 5 W and 10 W; and (4) the flow rates are 0.18, 0.36, 0.54, 0.72, 0.9, 1.08, and 1.26 L/min. The finite-volume model is used for the analyses, and the computer code is ANSYS ICEPAK 12.1.6.

11.7.4 Simulation Results (Channel Height = 700 μm)

Figures 11.47 and 11.48 show the simulation results for channel height = 700 μm. It can be seen that (1) the temperature of the LEDs and chips near the inlet is the smallest and that near the outlet is the largest (the water carries the heat out); (2) the temperature of the LEDs and chips near the interposer center is larger than that near both sides (parallel to the fluid direction) of the interposer; and (3) the temperature in the LEDs is, in general, larger than that in the ASIC.

Table 11.2 summarizes the temperature distributions in LEDs, ASICs, inlet, and outlet with various LEDs and chips powers for a flow rate = 0.54 L/min and pressure drop = 8.62 mbar.

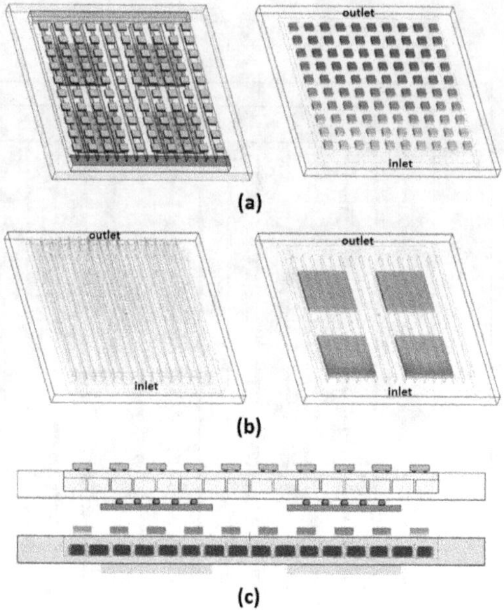

Figure 11.47 Typical results: (a) left (CAD model), right (LED temperature distribution); (b) left (water flow path), right (ASIC temperature); (c) upper (CAD model), lower (temperature distribution).

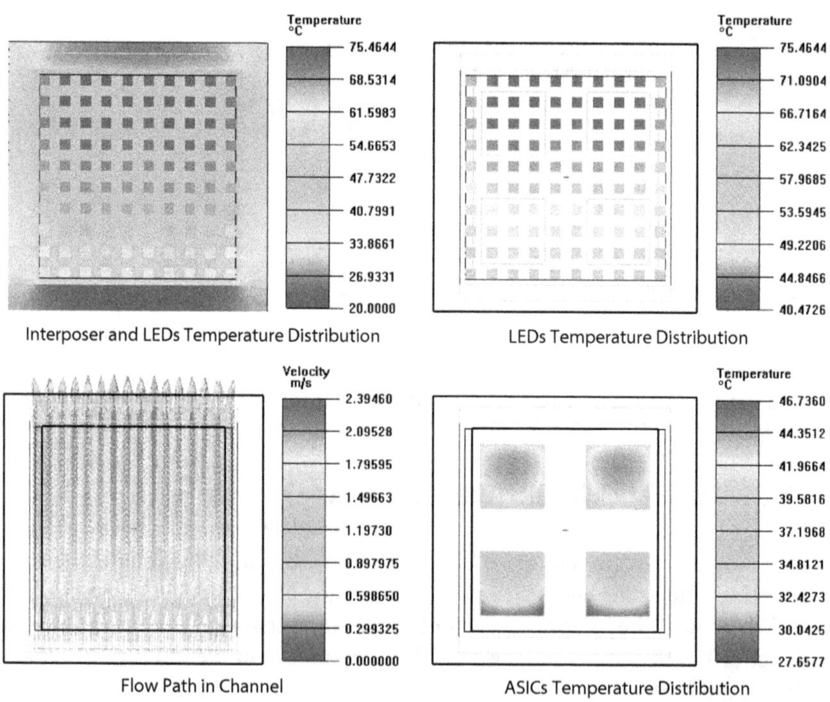

Figure 11.48 For channel height = 700 μm, LED = 2 W, ASIC = 10 W, flow rate = 1.26 L/min: top left (interposer and LEDs temperature distribution); top right (LEDs temperature distribution); bottom left (flow path in channel); bottom right (ASICs temperature distribution).

TABLE 11.2 Temperature Distributions in LEDs, ASICs, Inlet and Outlet with Various LEDs and Chips Powers

Power (W)			LED Temperature (C)			ASIC Temperature (C)			Temperature (C)			Flow rate (L/min)	Pressure Drop (mbar)
LED	ASIC	Total Power	Tmax	Tmin	Tavg	Tmax	Tmin	Tavg	Inlet	Outlet	Average		
1	5	120	54.3	32	46.5	39.6	25.9	34.8	20	24.9	22.45	0.54	8.62
1	10	140	57.4	32.6	48.3	47.1	27.8	40.5	20	25.6	22.8		
1.5	5	170	70	37.7	58.8	45.6	27.9	39.4	20	27	23.5		
1.5	10	190	73	38.3	60.7	53.1	29.8	45.1	20	27.7	23.85		
2	5	220	85.7	43.3	71.2	51.7	29.9	44	20	29	24.5		
2	10	240	88.7	44	73	59.2	31.8	49.7	20	29.8	24.9		

Also, Fig. 11.49 shows the relationship between total (LEDs and ASICs) chips power and its components' average temperature. It can be seen and expected that (1) the larger the LED power, the higher the LED temperature; (2) the larger the ASIC power, the higher the ASIC temperature; (3) the larger the total chips power, the higher the LED, ASIC, and water outlet temperature; and (4) the rate of change of average temperature of LEDs is larger than that of ASIC and water outlet. Also, it should be pointed out that with only 0.54 L/min flow rate and inlet water temperature = 20°C, it can take away 240 W of power consumption from the interposer; the maximum temperature in the LEDs is 88.7°C and in the ASICs is 59.2°C. This shows the current design has very high cooling ability.

The effect of flow rate (for LED = 2 W and ASIC = 10 W, a total of 240 W) on the LED, ASIC, and water outlet temperature is shown in Table 11.3. The average temperature of LEDs, ASICs, and water outlet for various flow rates is also shown in Fig. 11.50. It can be

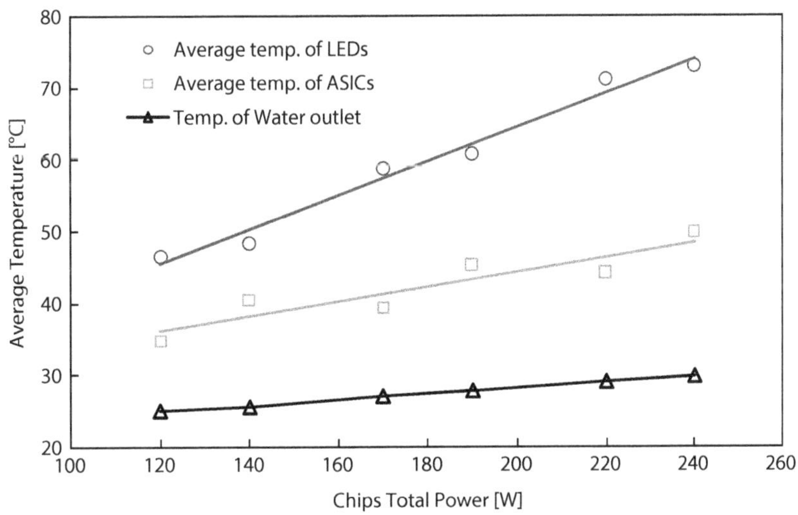

FIGURE 11.49 3D IC and LED integration total chips power vs. its components' average temperature.

Table 11.3 Effect of Flow Rates (LEDs = 2 W and ASICs = 10 W, a Total of 240 W Chips Power) on the Temperature in LEDs, ASICs, and Water Outlet as well as Pressure Drop

Flow Rate (L/min)	Power (W)			LED Temperature (C)			ASIC Temperature (C)			Temperature (C)			Pressure Drop (mbar)
	LED	ASIC	Total power	Tmax	Tmin	Tavg	Tmax	Tmin	Tavg	Inlet	Outlet	Average	
0.18				115.6	54.1	93.9	84.8	43.6	70.4	20	44	32	1.609
0.36				97.2	46.8	79.4	67.2	35.1	56.1	20	33.5	26.75	4.52
0.54				88.7	44	73	59.2	31.8	49.7	20	29.8	24.9	8.62
0.72				83.5	42.5	69.3	54.4	30	45.9	20	29	24.5	13.82
0.9				80	41.6	66.8	51.1	28.9	43.3	20	26.7	23.35	20.21
1.08				77.4	40.9	65	48.7	28.2	41.4	20	25.9	22.95	27.57
1.26	2	10	240	75.4	40.5	63.7	46.7	27.7	39.9	20	25.3	22.65	36.01

seen that (1) the average temperature of LEDs, ASICs, and water outlet is higher for smaller flow rates; (2) the average temperature of LEDs, ASICs, and water outlet is more sensitive to smaller flow rates (<0.8 L/min); and (3) for all flow rates, the average temperature of LEDs is larger than that of ASICs.

The pressure drop is a nonlinear function of the flow rate and is shown in Fig. 11.51. It can be seen that the larger the flow rate, the larger the pressure drop. Thus, there is always a trade-off between the flow rate, pressure drop, and temperature. The larger the flow rates, the lower the 3D IC LED integration temperatures, which is good. However, the larger the

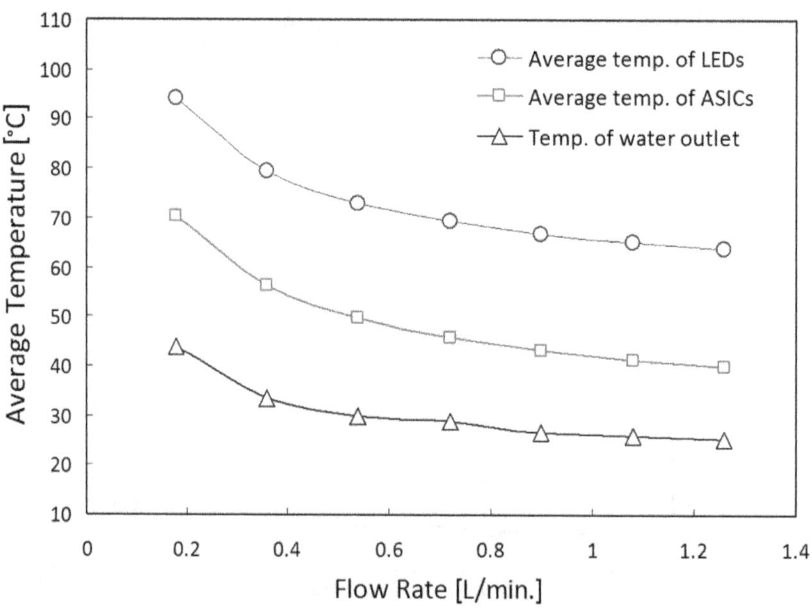

Figure 11.50 Flow rate vs. average temperature in LEDs, ASICs, and water outlet (LED = 2 W and ASIC = 10 W).

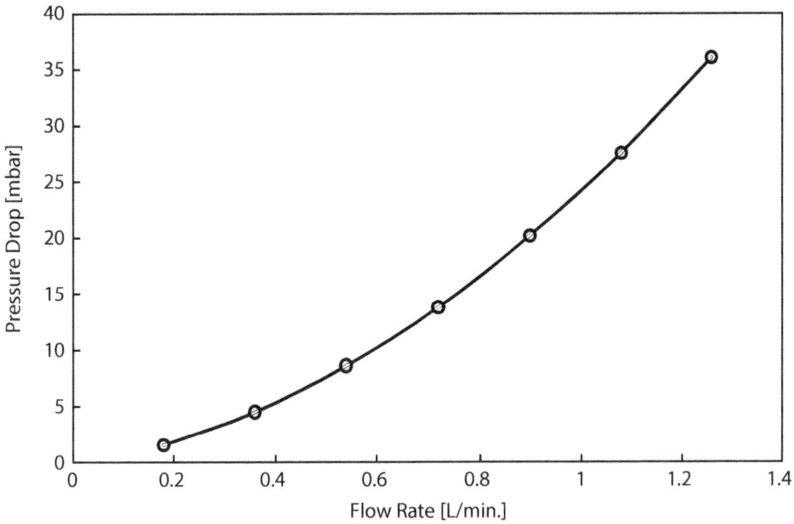

FIGURE 11.51 Flow rate vs. pressure drop (LED = 2 W and ASIC = 10 W).

flow rates, the larger the pressure drops, which means a larger (higher power, or higher cost) mini pump is needed.

11.7.5 Simulation Results (Channel Height = 350 μm)

The temperature distribution of the LEDs, ASICs, and water outlet for the case of channel height = 350 μm is shown in Fig. 11.52 and Table 11.4. It can be seen that (1) the temperature behavior of the LEDs, ASICs, and water outlet is almost same between the 350 μm and 700 μm channel heights; (2) the maximum LED temperature reduces by 10% when the flow rate is 0.54 L/min and 4% when the flow rate is 1.26 L/min; (3) the maximum ASIC temperature reduces by 10% when the flow rate is 0.54 L/min and only 2% when the flow rate is 1.26 L/min; (4) for both flow rates, the maximum water outlet temperature does not change; and (5) for both flow rates, the pressure drop with channel height = 350 μm is 3.5 to 4 times of that with channel height = 700 μm.

Based on these results, there is another trade-off in designing the TSV interposer with embedded fluidic microchannels. The advantages of smaller channel height are (1) a stronger and stiffer interposer, (2) less fabrication time, and (3) save manufacturing cost. The disadvantages of smaller channel height are (1) higher pressure drop, (2) larger power and size pump, and (3) higher-cost pump.

11.7.6 Summary and Recommendations

A microchannel cooling method for the thermal management of 3D LED and IC integration is proposed. An example is also provided. Some important results and recommendations are summarized as follows [66]:

- The proposed microchannel carrier not only cools the IC chip with LEDs but also allows electrical feed-through. It is a real 3D LED and IC integration.
- The thermal performance of the microchannel carrier is much better than that of the conventional thermal management method such as the heat slug/sink.

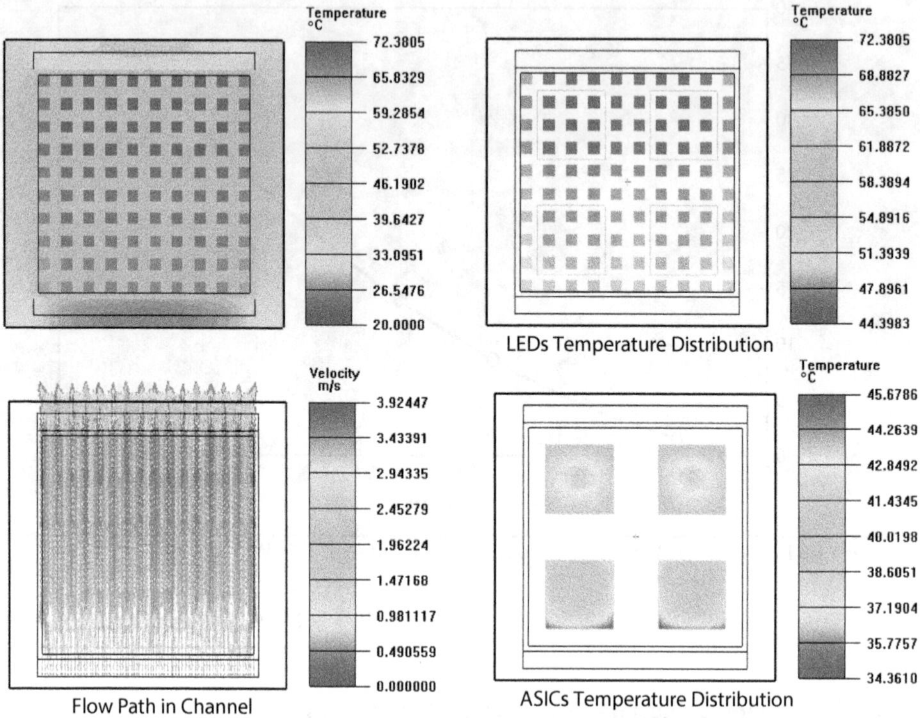

FIGURE 11.52 For channel height = 350 μm, LED = 2 W, ASIC = 10 W, flow rate = 1.26 L/min: top left (interposer and LEDs temperature distribution); top right (LEDs temperature distribution); bottom left (flow path in channel); bottom right (ASICs temperature distribution).

TABLE 11.4 Temperature Distribution of LEDs, ASICs, and Water Outlet for Two Different Channel Heights (350 μm and 700 μm) and Two Different Flow Rates (0.54 L/min and 1.26 L/min)

Channel Height (μm)	Flow Rate (L/min)	Power (W)			LED Temperature (C)			ASIC Temperature (C)			Temperature (C)			Pressure Drop (mbar)
		LED	ASIC	Total Power	Tmax	Tmin	Tavg	Tmax	Tmin	Tavg	Inlet	Outlet	Average	
700	0.54	2	10	240	88.7	44	73	59.2	31.8	49.7	20	29.8	24.9	8.62
	1.26				75.5	40.5	63.7	46.7	27.7	39.9	20	25.3	22.65	36.01
350	0.54				79.4	46	68.7	53	36	47.2	20	26.7	23.35	33.83
	1.26				72.4	44.4	64	45.7	34.4	42.1	20	25.3	22.65	124

- The TSV interposer with embedded fluidic microchannels, which supports any active chips (ASICs) and optical devices (LEDs), is the gut and workhorse of the current design.
- The LED and ASIC temperatures are smaller near the water inlet and larger near the outlet.
- For all the cases considered, the LED temperature is higher than the ASIC temperature.

- As expected, the larger the LED powers the higher the LED temperatures. This is also true for the ASIC temperature.
- For all the chips total power, the rate of change of average temperature of LEDs is larger than that of ASIC.
- The current 3D LED and IC integration design has very high cooling ability; with only 0.54 L/min flow rate and inlet water temperature at 20°C, it can take away 240 W of power consumption from the interposer and the maximum temperature in the LEDs = 88.7°C and in ASICs = 59.2°C.
- The average temperature of LEDs and ASICs is more sensitive to smaller flow rates (<0.8 L/min).
- For all the flow rates, the average temperature of LEDs is larger than that of ASICs.
- The pressure drop is a function of the flow rates, and the larger the flow rates the larger the pressure drops.
- The effect of (reducing) channel heights is increase in the pressure drop.

11.8 References

[1] Haitz, R., F. Kish, J. Tsao, and J. Nelson, "The Case for a National Research Program on Semiconductor Lighting," presented at the *Annual Forum of the Optoelectronics Industry Development Association*, Washington, DC, 1999.

[2] Haitz, R., F. Kish, J. Tsao, and J. Nelson, "The Case for a National Research Program on Semiconductor Lighting," presented at *Strategies in Light*, Burlingame, CA, 2000.

[3] Krames, M. R., O. B. Shchekin, R. Mueller-Mach, G. O. Mueller, L. Zhou, G. Harbers, and M. G. Craford, "Status and Future of High-Power Light-Emitting Diodes for Solid-State Lighting," *IEEE Journal of Display Technology*, Vol. 3, 2007, pp. 160–175.

[4] Tsao, J. Y., M. E. Coltrin, M. H. Crawford, and J. A. Simmons, "Solid State Lighting: An Integrated Human Factors, Technology, and Economic Perspective," *Proceedings of IEEE*, Vol. 98, No. 7, 2010, pp. 1162–1179.

[5] Workshop, Electric Power Research Institute (EPRI), Rochester, NY, 1995, *Proceedings of ALITE* '95 Workshop, EPRI-TR-106022.

[6] Haitz, R., "Another Semiconductor Revolution: This Time It's Lighting!," in *Advances in Solid State Physics,* edited by B. Kramer, Vol. 43, Springer, Berlin, 2003, pp. 35–50.

[7] Christensen, C. M., *The Innovator's Dilemma: When New Technologies Cause Great Firms to Fail*, Harvard Business School Press, Boston, MA, 1997.

[8] EERE, "*Solid-State Lighting Multi-Year Program Plan FY'10-FY'16*," U.S. Department of Energy, Office of Energy Efficiency and Renewable Energy Building Technologies Program, 2010.

[9] Steele, R., "High-Brightness LED Market Review and Forecast," in *Strategies in Light*, Strategies Unlimited, Santa Clara, CA, 2010.

[10] Brodrick, J., "DOE SSL Roadmap," in *Strategies in Light*, Strategies Unlimited, Santa Clara, CA, 2010.

[11] Schubert, E. F., and J. K. Kim, "Solid-State Light Sources Getting Smart," *Science*, Vol. 308, 3005, 2005, pp. 1274–1278.

[12] Haitz, R. and J. Y., Tsao, "Solid-state Lighting: 'The Case' 10 Years after and Future Prospects," *Physica Status Solidi (a)*, Vol. 208, 2011, pp. 17–29.

[13] Round, H. J., "A Note on Carborundum," *Electrical World*, Vol. 19, 1907, p. 309.

[14] Losev, O. V., "CII. Luminous Carborundum Detector and Detection Effect and Oscillations with Crystals," *Philosophical Magzine*, Vol. 6, 1928, pp. 1024–1044.

[15] Soviet patent No. 12191 granted in 1929.

[16] Holonyak, Jr. H., and S. F. Bevaqua, "Coherent (Visible) Light Emission from Ga(As1-xPx) Junctions," *Applied Physics Letters*, Vol. 1, No. 4, 1962, pp. 82–83.

[17] Craford, M. G., R. W. Shaw, W. O. Groves, and A. H. Herzog, "Radiative Recombination Mechanisms in GaAsP Diodes With and Without Nitrogen Doping," *Journal of Applied Physics*, Vol. 43, October 1972, pp. 4075–4083.

[18] Nakamura, S., "GaN Growth Using GaN Buffer Layer," *Japanese Journal of Applied Physics,* Vol. 30, 1991, p. L1705.
[19] Nakamura, S., T. Mukai, and M. Senoh, "High-Power GaN p-n Junction Blue-Light-Emitting Diodes," *Japanese Journal of Applied Physics,* Vol. 30, 1991, p. L1998.
[20] Nakamura, S., Y. Harada, and M. Senoh, "Novel Metalorganic Chemical Vapor Deposition System for GaN Growth," *Applied Physics Letters,* Vol. 58, 1991, p. 2021.
[21] Nakamura, S., M. Senoh, and T. Mukai, "Highly P-Type Mg-Doped GaN Films Grown with GaN Buffer Layers," *Japanese Journal of Applied Physics,* Vol. 30, 1991, p. L1708.
[22] Nakamura, S., T. Mukai, M. Senoh, and N. Iwasa, "Thermal Annealing Effects on P-Type Mg-Doped GaN Films," *Japanese Journal of Applied Physics,* Vol. 31, 1992, p. L139.
[23] Nakamura, S. and T. Mukai, "High-Quality InGaN Films Grown on GaN Films," *Japanese Journal of Applied Physics,* Vol. 31, 1992, p. L1457.
[24] Nakamura, S., T. Mukai, and M. Senoh, "Si- and Ge-Doped GaN Films Grown with GaN Buffer Layers," *Japanese Journal of Applied Physics,* Vol. 31, 1992, p. 2883.
[25] Nakamura, S., N. Iwasa, M. Senoh, and T. Mukai, "Hole Compensation Mechanism of P-Type GaN Films," *Japanese Journal of Applied Physics,* Vol. 31, 1992, p. 1258.
[26] Nakamura, S., M. Senoh, and T. Mukai, "P-GaN/N-InGaN/N-GaN Double-Heterostructure Blue-Light-Emitting Diodes," *Japanese Journal of Applied Physics,* Vol. 32, 1993, p. L8.
[27] Nakamura, S., T. Mukai, and M. Senoh, "Candela-Class High-Brightness InGaN/AlGaN Double-Heterostructure Blue-Light-Emitting-Diodes," *Applied Physics Letters,* Vol. 64, 1994, p. 1687.
[28] Nakamura, S., "Growth of InXGa1-XN Compound Semiconductors and High-Power InGaN/AlGaN Double Heterostructure Violet-Light-Emitting Diodes," *Microelectronics,* Vol. 25, 1994, p. 651.
[29] Nakamura, S., M. Senoh, N. Iwasa, and S. Nagahama, "High-Power InGaN Single-Quantum Well-Structure Blue and Violet Light-Emitting Diodes," *Applied Physics Letters,* Vol. 67, 1995, p. 1868.
[30] Nakamura, S., M. Senoh, N. Iwasa, S. Nagahama, T. Yamada, and T. Mukai, "Superbright Green InGaN Single-Quantum-Well-Structure Light-Emitting Diodes," *Japanese Journal of Applied Physics,* Vol. 34, 1995, p. L1332.
[31] Nakamura, S., M. Senoh, N. Iwasa, and S. Nagahama, "High-Brightness InGaN Blue, Green and Yellow Light-Emitting Diodes with Quantum Well Structures," *Japanese Journal of Applied Physics,* Vol. 34, 1995, p. L797.
[32] Nakamura, S., M. Senoh, S. Nagahama, N. Iwasa, T. Yamada, T. Matsushita, Y. Sugimoto, et al., "Room-Temperature Continuous-Wave Operation of InGaN Multi-Quantum-Well Structure Laser Diodes," *Applied Physics Letters,* Vol. 69, 1996, p. 4056.
[33] Nakamura, S., "Characteristics of Room Temperature-cw Operated InGaN Multi-Quantum-Well-Structure Laser Diodes," *MRS Internet Journal of Nitride Semiconductor Research,* Vol. 2, No. 5, 1997, p. 449.
[34] Japan Patent Application No. 10-194156 (194156, 1998) filed on July 9, 1998.
[35] Japan Patent Application No. 10-316169 (316169, 1998) filed on November 6, 1998.
[36] Japan Patent Application No. 10-321605 (321605, 1998) filed on November 12, 1998.
[37] Wierer, J., D. A. Steigerwald, M. R. Krames, J. J. O'Shea, M. J. Ludowise, G. Christenson, Y.-C. Shen, et al., "High-power AlGaInN Flip-chip Light-emitting Diodes," *Applied physics Letters,* Vol. 78, No. 22, May 28, p. 2001.
[38] Lau, J. H., R. Lee, M. Yuen, and P. Chan, "3D LED and IC Wafer Level Packaging," *Journal of Microelectronics International,* Vol. 27, No. 2, 2010, pp. 98–105.
[39] Lau, J. H., *Reliability of RoHS Compliant 2D & 3D Interconnects*, McGraw-Hill, 2011.
[40] Lau, J. H., C. Lee, C. Premachandran, and A. Yu, *Advanced MEMS Packaging,* McGraw-Hill, New York, NY, 2010.
[41] Zhang, X., T. Chai, J. H. Lau, C. Selvanayagam, K. Biswas, S. Liu, D. Pinjala, et al., "Development of Through Silicon Via (TSV) Interposer Technology for Large Die (21 × 21 mm) Fine-pitch Cu/low-*k* FCBGA Package," *IEEE Proceedings of Electronic Components and Technology Conference,* San Diego, CA, May 2009, pp. 305–312.
[42] Chai, T. C., X. Zhang, J. H. Lau, C. S. Selvanayagam, D. Pinjala, Y. Hoe, Y. Ong, et al., "Development of Large Die Fine-Pitch Cu/low-*k* FCBGA Package with through Silicon via (TSV) Interposer," *IEEE Transactions on CPMT,* Vol. 1, No. 5, May 2011, pp. 660–672.

[43] Yu, A., J. H. Lau, S. Ho, A. Kumar, W. Hnin, D. Yu, M. Jong, et al., "Study of 15 μm Pitch Solder Microbumps for 3D IC Integration," *IEEE Proceedings of Electronic Components and Technology Conference*, San Diego, CA, May 2009, pp. 6–10.
[44] Yu, A., J. H. Lau, S. Ho, A. Kumar, H. Yin, J. Ching, V. Kripesh, et al., "Three dimensional interconnects with high aspect ratio TSVs and fine pitch solder microbumps," *IEEE Proceedings of Electronic Components and Technology Conference*, San Diego, CA, May 2009, pp. 350–354.
[45] Yu, A., J. H. Lau, S. Ho, A. Kumar, W. Hnin, W. Lee, M. Jong, et al., "Fabrication of High Aspect Ratio TSV and Assembly with Fine-Pitch Low-Cost Solder Microbump for Si Interposer Technology with High-Density Interconnects," *IEEE Transactions on CPMT*, Vol. 1, No. 9, September 2011, pp. 1336–1344.
[46] Ho, S., S. Yoon, Q. Zhou, K. Pasad, V. Kripesh, and J. H. Lau, "High RF performance TSV for silicon carrier for high frequency application," *IEEE Proceedings of Electronic Components and Technology Conference*, Orlando, FL, May 27–30, 2008, pp. 1946–1952.
[47] Khan, N., V. Rao, S. Lim, S. Ho, V. Lee, X. Zhang, J. H. Lau, et al., "Development of 3D Silicon Module with TSV for System in Packaging," *IEEE Proceedings of Electronic Components and Technology Conference*, Orlando, FL, May 27–30, 2008, pp. 550–555.
[48] Khan, N., H. Li, S. Tan, S. Ho, V. Kripesh, D. Pinjala, J. H. Lau, et al., "3-D Packaging With Through-Silicon Via (TSV) for Electrical and Fluidic Interconnections," *IEEE Transactions on CPMT*, Vol. 3, No. 2, February 2013, pp. 221–228.
[49] Khan, N., V. Rao, S. Lim, H. We, V. Lee, X. Zhang, J. H. Lau, et al., "Development of 3-D Silicon Module With TSV for System in Packaging," *IEEE Transactions on CPMT*, Vol. 33, No. 1, March 2010, pp. 3–9.
[50] Selvanayagam, C., J. H. Lau, X. Zhang, S. Seah, K. Vaidyanathan, and T. Chai, "Nonlinear Thermal Stress/Strain Analysis of Copper Filled TSV (Through Silicon Via) and their Flip-chip Microbumps," *IEEE Transactions on Advanced Packaging,* Vol. 32, No. 4, 2009, pp. 720–728.
[51] Khan, N., H. Li, S. Tan, S. Ho, N. Su, W. Hnin, J. H. Lau, et al., "3D Packaging with Through Silicon Via (TSV) for Electrical and Fluidic Interconnections," *IEEE Proceedings of Electronic Components and Technology Conference*, San Diego, CA, May 2009, p. 1153–1158.
[52] Tang, G., S. Tan, N. Khan, D. Pinjala, J. H. Lau, A. Yu, V. Kripesh, et al., "Fluidic Interconnects in Integrated Liquid Cooling Systems for 3-D Stacked TSV Module," *IEEE Proceedings of Electronic Packaging and Technology Conference*, December 2008, pp. 552–558.
[53] Tang, G., O. Navas, D. Pinjala, J. H. Lau, A. Yu, and V. Kripesh, "Integrated Liquid Cooling Systems for 3-D Stacked TSV Modules," *IEEE Transactions on Components and Packaging Technologies*, Vol. 33, No. 1, 2010, pp. 184–195.
[54] Lau, J. H., and G. Tang, "Thermal management of 3D IC integration with TSV (through silicon via)," *IEEE Proceedings of Electronic Components and Technology Conference*, San Diego, CA, May 2009, pp. 635–640.
[55] Yu, A., N. Khan, G. Archit, D. Pinjalal, K. Toh, V. Kripesh, J. H. Lau, et al., "Development of Silicon Carriers with Embedded Thermal Solutions for High Power 3-D Package," *IEEE Transactions on Components and Packaging Technology*, Vol. 32, No. 3, September 2009, pp. 566–571.
[56] Y. Backlund, and L. Rosengren, "New Shapes in (100) Si using KOH and EDP Etches," *Journal of Micromechanics and Microengineering*, Vol. 2, 1992, pp. 75–79.
[57] Tiggelaar, R., T. Veenstra, R. Sanders, J. Gardeniers, M. Elwenspoek, and A. Berg, "A Light Absorption Cell for Micro-TAS with KOH/IPA Etched 45 Degrees Mirrors in Silicon," *Journal of Micromechanics and Microengineering*, Vol. 17, 2007 pp. 137–142.
[58] Pal, P., K. Sato, and S. Chandra, "Fabrication Techniques of Convex Corners in a (100)-Silicon Wafer using Bulk Micromachining: A Review," *Journal of Micromechanics and Microengineering*, Vol. 17, 2007, pp. R111–R133.
[59] An, S. Lee, B. O, H. Kim, S. Park, and E. Lee, "The Effect of KOH and KOH/IPA Etching on the Surface Roughness of the Silicon Mold to be used for Polymer Waveguide Imprinting," *Proc. of the SPIE*, V 6897, 2008, pp. 689717–689717-7.
[60] Zhang, R., R. Lee, D. Xiao, and H. Chen, "LED Packaging using Silicon Substrate with Cavities for Phosphor Printing and Copper-filled TSVs for 3D Interconnection," *Proceeding of IEEE/ECTC*, Orlando, FL, May 2011, pp. 1616–1621.
[61] Zhang, R., and R. Lee, "Moldless Encapsulation for LED Wafer Level Packaging using Integrated DRIE Trenches," *Journal of Microelectronics Reliability*, Vol. 52, 2012, pp. 922–932.

[62] Lv, Z., X. Liu, L. Yang, J. Yuan, X. Wang, and S. Liu, "Silicon Substrate with TSV for Light Emitting Diode Packaging," *IEEE Proceedings of EMAP*, November 2012, pp. 1–4.
[63] Lv, Z., X. Wang, L. Yang, J. Yuan, J. Fang, B. Cao, and S. Liu, "Study on Packaging Method Using Silicon Substrate With Cavity and TSV for Light Emitting Diodes," *IEEE Transactions on CPMT*, Vol. 3, No. 7, July 2013, pp. 1123–1129.
[64] Chen, D., L. Zhang, Y. Xie, K. Tan, and C. Lai, "A Study of Novel Wafer Level LED Package Based on TSV Technology," *IEEE Proceedings on ICEPT*, August 2012, pp. 52–55.
[65] Xie, Y., D. Chen, L. Zhand, K. Tan, and C. Lai, "A Novel Wafer Level Packaging for White Light LED," *IEEE Proceedings on ICEPT*, August 2013, pp. 1170–1174.
[66] Lau, J. H., H. C. Chien, and R. Tain, "TSV Interposers with Embedded Microchannels for 3D IC and LED Integration," *ASME Paper: InterPACK2011-52204*, pp. 1–8.

CHAPTER 12
3D MEMS and IC Integration

12.1 Introduction

MEMS stands for *m*icro*e*lectro*m*echanical *s*ystems, which is the integration of mechanical elements such as sensors and actuators and electronics on a common silicon substrate through the use of microfabrication technology [1–4]. The same basic technology and process steps, such as etching, patterning, doping, and connecting, employed for making conventional electronic ICs (integrated circuits), such as CMOS (complementary metal-oxidesemiconductors), can be used to fabricate (micromachining) micromechanical elements by selectively etching away parts of a silicon wafer or adding new structural layers to form MEMS devices. However, there is a fundamental difference between MEMS and IC devices—namely, MEMS devices must remain physically movable. MEMS devices use electronics to move their mechanical elements (parts)!

Most people consider the next big thing after smartphones is the IoTs (Internet of Things). IoTs are sometimes called the M2M (machine-to-machine) revolution, and MEMS is one of the important groups of machines which will play an essential role in expansion of the boom of the IoTs. For example, MEMS sensors allow equipment to gather and digitize real-world data that then can be shared on the Internet. The IoTs promote a major new growth opportunity for the MEMS market.

12.2 MEMS Packaging

Unlike electronics IC packaging [21–30], MEMS packaging [5–20] is not trivial, is expensive (60% to 80% of the product total cost), and is custom-built. Electronics IC packaging is a well-developed technology (e.g., wirebonding, tape automated bonding, and the flip chip). However, MEMS packaging is a specially designed packaging process that is difficult owing to moving structural elements, and it is the most expensive process in micromachining.

Unlike electronics IC packaging [21–30], MEMS packaging [5–20] needs a cap. In order for the moving elements of a MEMS device to move effectively (e.g., with minimum damping and stiction) in a well-controlled atmosphere, hermetic sealing of the MEMS device in a cap is necessary. For some MEMS devices (e.g., resonators, infrared bolometers, and gyroscopes), vacuum packaging is even required. Usually, the cap is processed at the wafer level, and therefore MEMS packaging is truly a wafer-level packaging. The top MEMS suppliers are BOSCH, STMicroelectronics, Texas Instruments, Hewlett-Packard, Robert Lexmark, Seiko Epson, Freescale Semiconductor, Canon, Analog Devices, and Systron Donner.

Just like electronics IC devices [21–30], MEMS devices [5–20] are not isolated islands. They must communicate with other IC chips in a circuit through an I/O (input/output) system of interconnects. Furthermore, the MEMS device needs to be powered up, and its embedded circuitry and moving elements are delicate, requiring the package to both carry and protect the device. Consequently, the major functions of MEMS packaging are (1) to provide a path for the electrical current that powers the circuit and moves the moving elements, (2) to distribute the signals onto and off the MEMS device, (3) to remove the heat generated by the circuit, and (4) to support and protect the MEMS device from hostile environments.

Most of the system houses design and assembly the MEMS modules (component) on one location of the PCB (printed circuit board) and the active chips such as the ASIC (application specific IC) on another location of the PCB as shown in the top of Fig. 12.1. However, it would be nice if the ASIC chip can be placed directly underneath the MEMS device (bottom of Fig. 12.1). In this case (3D MEMS and IC integration), the advantages are (1) better performance, (2) lower cost, (3) less footprint, (4) lighter weight, and (5) smaller form factor; and it is the focus of this chapter.

In this chapter, the design of 10 different configurations of 3D MEMS packaging that meet these requirements are presented [31, 32]. Also, these packages are supposed to yield low-cost, high-performance, and small footprints. Furthermore, the packaging assembly processes of these 3D packages are provided and discussed. Emphasis is put on the low-temperature bonding, MEMS chip to ASIC wafer (C2W) bonding, and ASIC wafer with MEMS chip to cap wafer (W2W) bonding. A 3D IC MEMS integration with low-temperature bonding example is provided. Finally, recent advances in MEMS packaging, such as (a) TSVs for wafer-level packaging of RF MEMS devices, (b) zero-level packaging for RF-MEMS

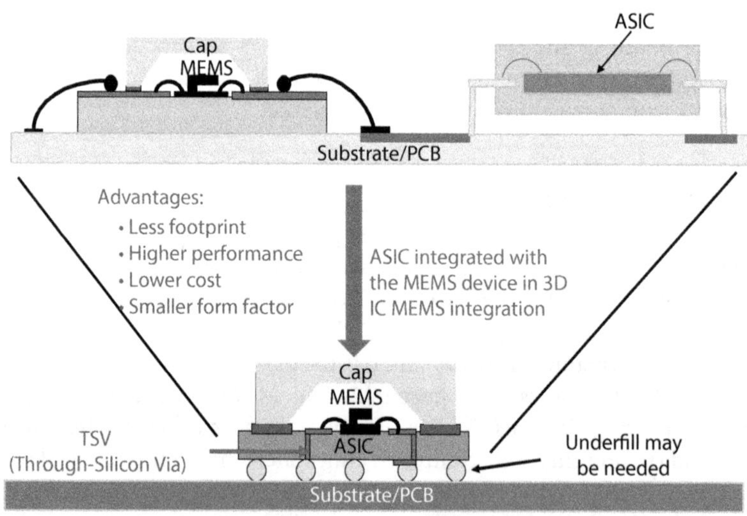

FIGURE 12.1 3D MEMS and IC integration.

implementing TSVs and metal bonding, (c) MEMS package based on Si-interposer wafer with Cu-filled TSVs, and (d) wafer-scale packaging for FBAR-based oscillators, are briefly presented.

12.3 Design of 3D MEMS and IC Integration

Advanced 3D IC and MEMS integration usually involves three wafers, namely, the MEMS device wafer, the ASIC wafer, and the cavity cap wafer [31, 32]. Figure 12.2 shows nine different combinations (designs) of 3D MEMS packaging from the MEMS wafer (with wirebonding pads, solder-bumped TSV substrate, or solder-bumped flip chip without TSV), the ASIC wafer (with or without TSV), and the cavity-cap wafer (with or without TSV).

12.3.1 3D MEMS and IC Integration with Lateral Electrical Feed-Through

Case 1: The MEMS device is die attached and then wirebonded on the ASIC chip or wafer. The cavity-cap chip/wafer is attached to the ASIC chip/wafer with a sealing ring, as shown in Fig. 12.3. The signal lines go through beneath the sealing ring to the wirebonding pad on the ASIC chip periphery. Another set of wirebonding pads connects the MEMS/ASIC 3D stack to a substrate in a package or on a PCB.

Case 2: The microsolder-bumped MEMS device with a TSV substrate is first attached to the ASIC, as shown in Fig. 12.4. The rest is the same as in case 1.

Case 3: The solder-bumped flip chip MEMS device is first attached on the ASIC, as shown in Fig. 12.5. The rest is the same as in case 1.

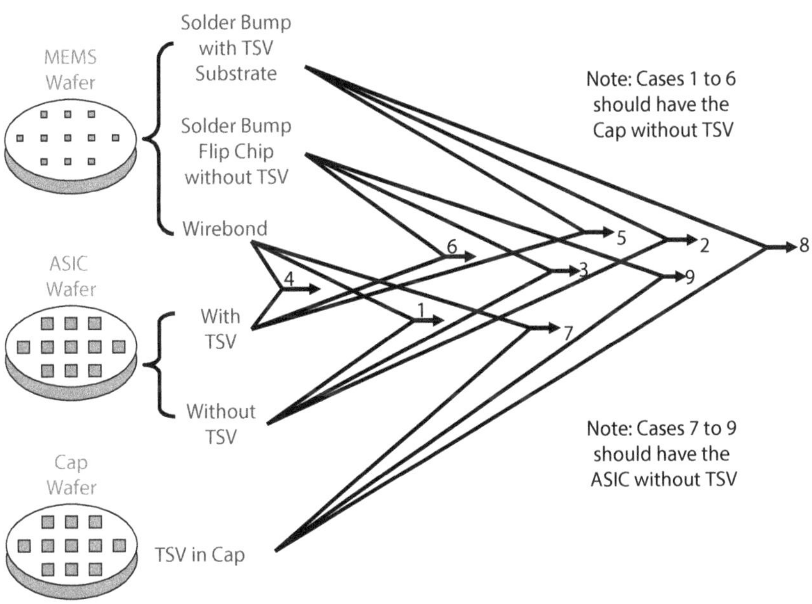

FIGURE 12.2 Nine different designs of 3D MEMS packaging from the MEMS, ASIC, and cavity-cap wafers.

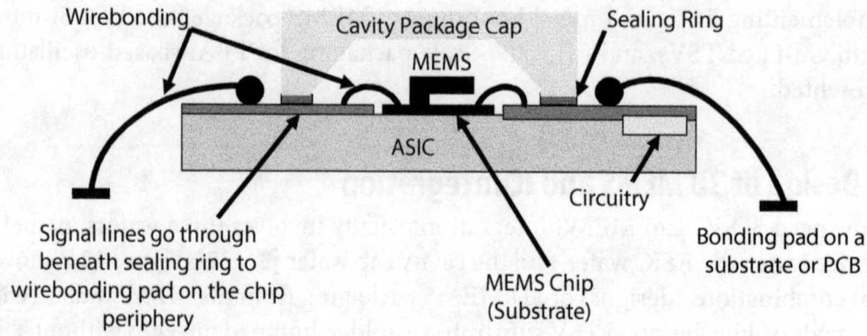

FIGURE 12.3 Wire-bonding 3D MEMS packaging with lateral electrical feed-through.

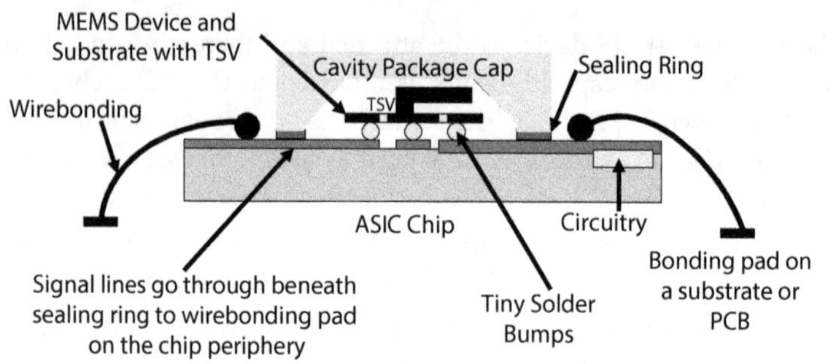

FIGURE 12.4 Solder-bumped MEMS device with TSV substrate on an ASIC chip with lateral electrical feed-through.

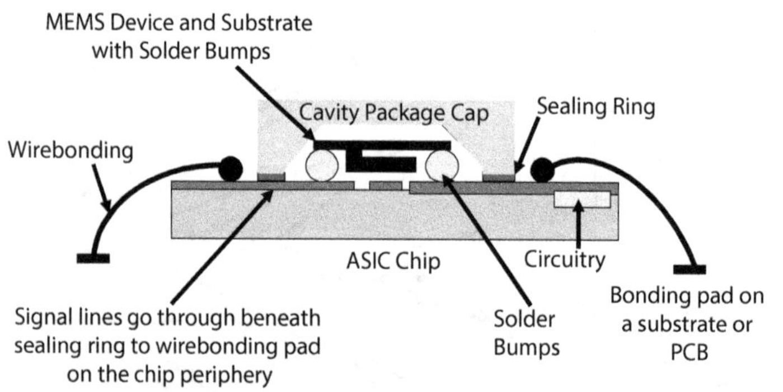

FIGURE 12.5 Solder-bumped flip-chip MEMS device on an ASIC with lateral electrical feed-through.

12.3.2 3D MEMS and IC Integration with Vertical Electrical Feed-Through in ASIC

Case 4: The MEMS device is die attached and wirebonded on the ASIC chip with TSV and ordinary solder bumps, as shown in Fig. 12.6. A cavity cap is attached on the ASIC with a sealing ring. The MEMS/ASIC 3D stack is then attached (solder reflowed) to a substrate in a package or on a PCB.

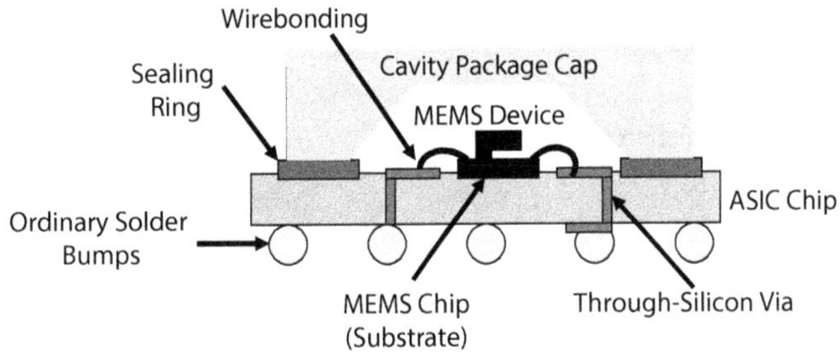

FIGURE 12.6 MEMS device wire bonded on an ASIC with vertical electrical feed-through TSV.

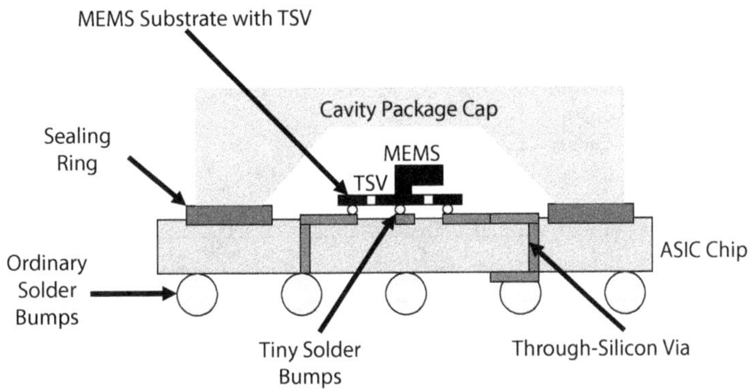

FIGURE 12.7 TSV MEMS device solder bonded on an ASIC with vertical electrical feed-through TSV.

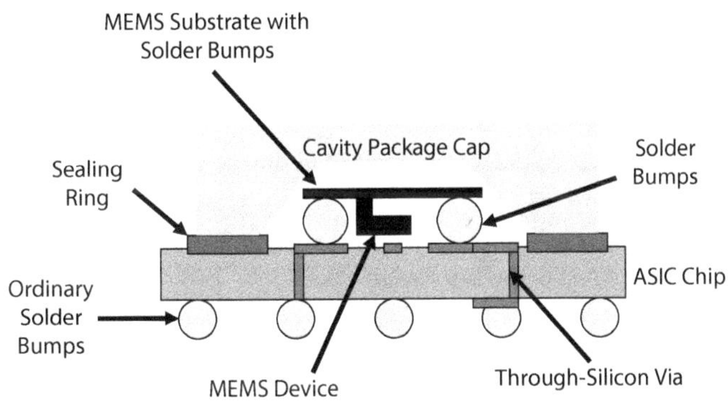

FIGURE 12.8 Solder-bumped MEMS device flip chip on an ASIC with vertical electrical feed-through TSV.

Case 5: The microsolder-bumped MEMS device with a TSV substrate is first attached to the ASIC chip/wafer with TSV and ordinary solder bumps, as shown in Fig. 12.7. The rest is the same as in case 4.

Case 6: The solder-bumped MEMS device is first attached to the TSV ASIC with ordinary solder bumps, as shown in Fig. 12.8. The rest is the same as in case 5.

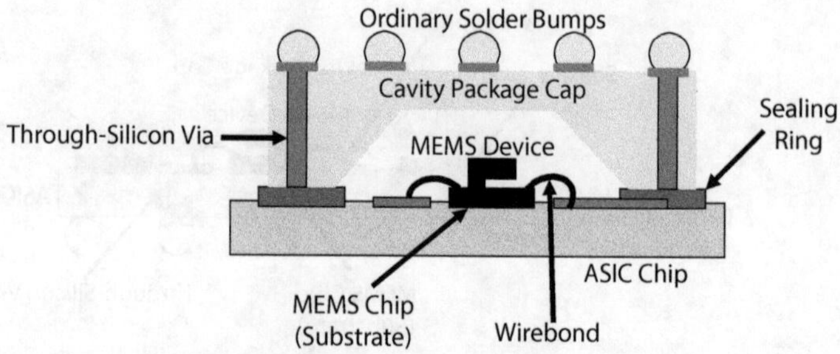

FIGURE 12.9 MEMS device wire bonded on an ASIC (electrical feed-through TSV is in the package cap).

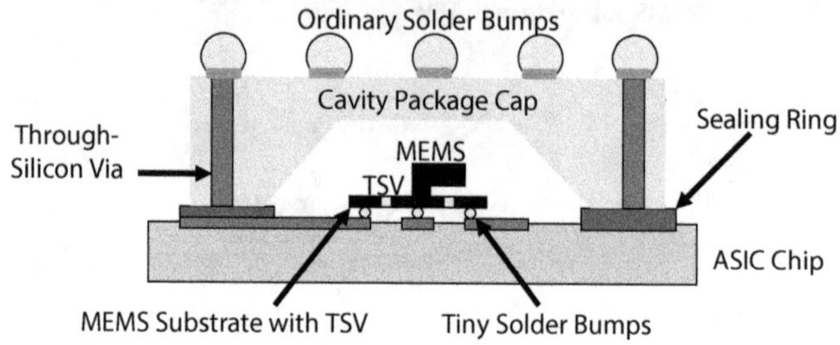

FIGURE 12.10 TSV MEMS device solder bonded on an ASIC (electrical feed-through TSV is in the cap).

12.3.3 3D MEMS and IC Integration with Vertical Electrical Feed-Through in the Package Cap

Case 7: The MEMS device is die attached and then wirebonded on the ASIC chip/wafer. The cavity cap with TSV and ordinary solder bumps is attached to the ASIC with a sealing ring, as shown in Fig. 12.9. Then the whole 3D carrier is attached (solder reflowed) to a substrate in a package or on a PCB.

Case 8: The microsolder-bumped MEMS device with TSV substrate is first attached to the ASIC as shown in Fig. 12.10. The rest is the same as in case 7.

Case 9: The solder-bumped MEMS device is first attached to the ASIC, as shown in Fig. 12.11. The rest is the same as in case 7.

12.3.4 3D MEMS and IC Integration with MEMS on ASIC with TSVs

Case 10: The MEMS device is attached to the solder-bumped ASIC with TSVs and a sealing ring, as shown in Fig. 12.12. Then the whole 3D stack is attached (solder reflowed) to a substrate in a package or on a PCB.

12.3.5 2.5D/2.25D MEMS and IC Integration

If the active device such as the ASIC in cases 1 through 10 is replaced by a deviceless passive TSV/RDL interposer, then it becomes the 2.5D MEMS and IC integration (packaging). If the ASIC is not there and the substrate of the MEMS is supporting the cap with lateral electrical feed-through, then cases 1 through 3 degenerate to the normal 2D MEMS packaging. If the

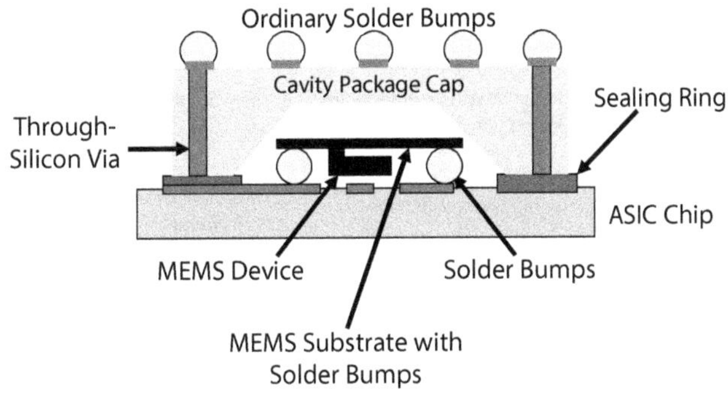

Figure 12.11 Solder-bonded MEMS device flip chip on an ASIC (electrical feed-through TSV is in the cap).

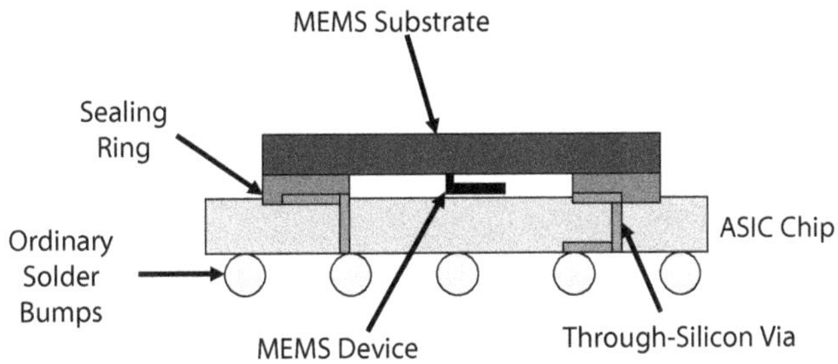

Figure 12.12 MEMS device is bonded (with a sealing ring) on an ASIC with electrical feed-through TSV.

ASIC is not there and the substrate of the MEMS is supporting the cap with vertical electrical feed-through (by TSVs), then cases 7 through 9 degenerate to the 2.25D MEMS and IC integration (packaging).

12.4 Assembly Process of 3D MEMS and IC Integration

There are many different processes to assemble these 10 3D IC and MEMS integration packages. For example, the assembly process shown in Fig. 12.13 can be applied to cases 1 through 3; Fig. 12.14, to cases 4 to 6; and Fig. 12.15, to cases 7 to 9. Case 10 is discussed later. As mentioned earlier, the MEMS wafer can be fabricated with either wirebonding pads or TSV and solder bumps or a solder-bumped flip chip, as shown in Figs. 12.13 through 12.15. These are followed by releasing (etching) the MEMS wafer and singulation. MEMS wafer to ASIC wafer bonding is not recommended because of chip yields and size differences.

12.4.1 3D MEMS and IC Integration with Lateral Electrical Feed-Through

For 3D MEMS and IC integration with lateral electrical feed-through (cases 1, 2, and 3), there is no TSV in both the ASIC and cap wafers (see Fig. 12.13). In these cases, cavities should be formed in the cap wafer by either KOH etching or use of a laser. Then one performs the bonding of the MEMS device (chip) to the ASIC wafer (C2W). This will be followed by bonding of the cavity-cap wafer to the ASIC wafer (W2W). Finally, the bonded wafers are singulated into individual units, which are ready to be wirebonded on the substrate of a package or on a PCB, as shown in Fig. 12.16. For both cases, encapsulant such as glob-top is recommended.

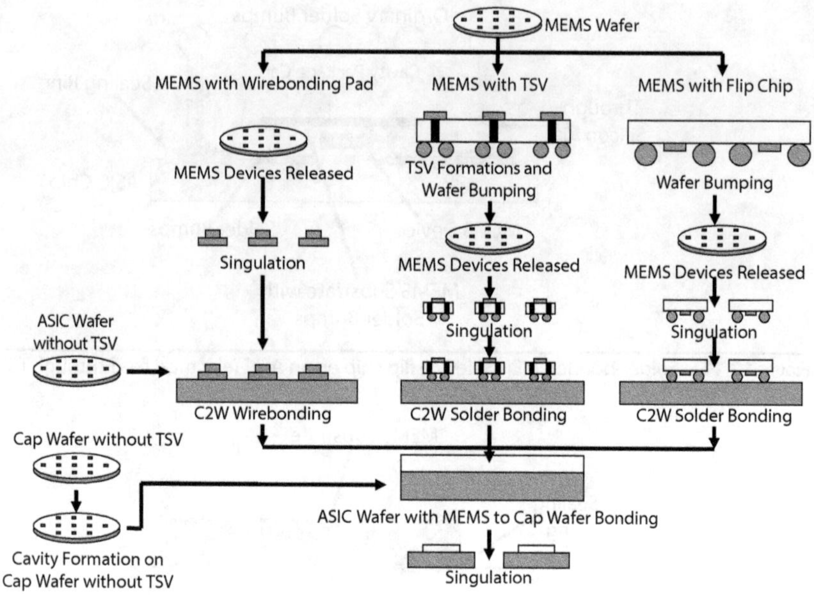

FIGURE 12.13 Assembly process for 3D MEMS packaging with lateral electrical feed-through.

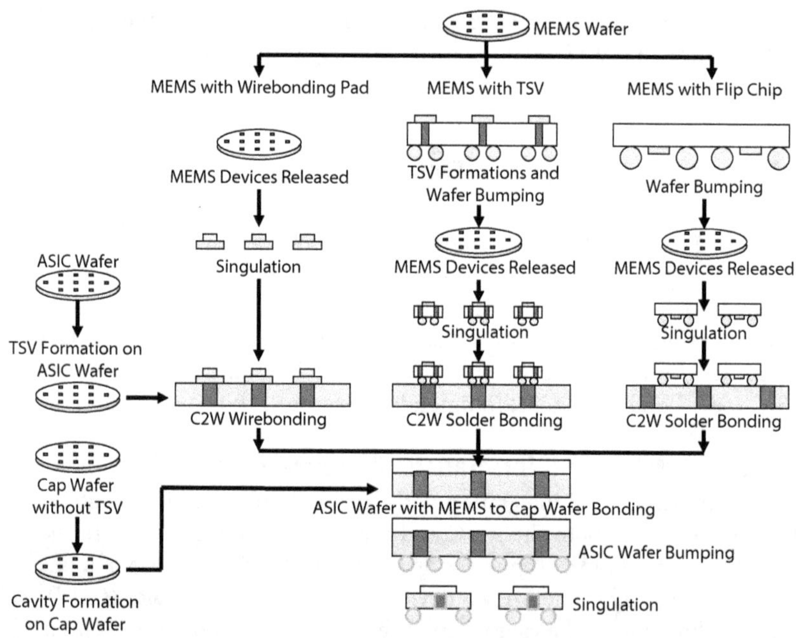

FIGURE 12.14 Assembly process for 3D MEMS packaging with vertical electrical feed-through the TSV ASIC.

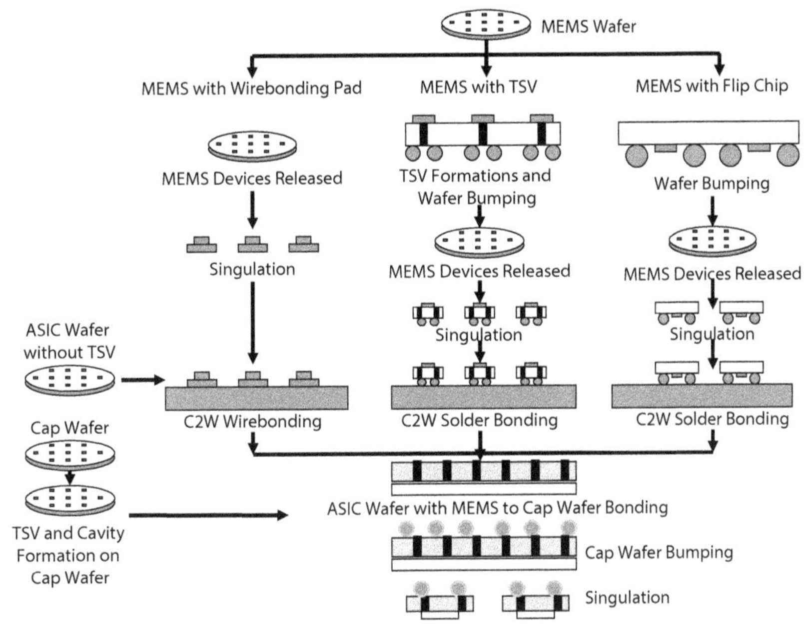

FIGURE 12.15 Assembly process for 3D MEMS packaging with vertical electrical feed-through the TSV cap.

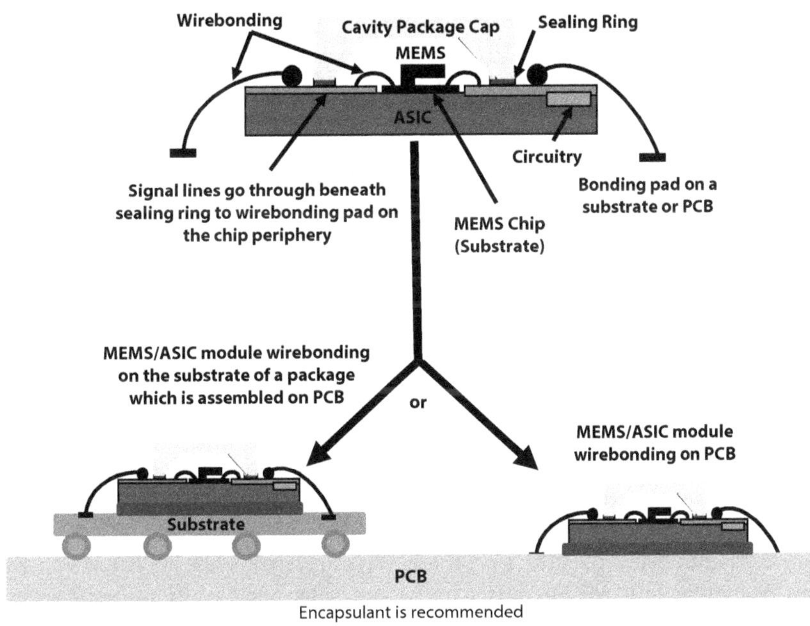

FIGURE 12.16 3D MEMS packaging with lateral electrical feed-through.

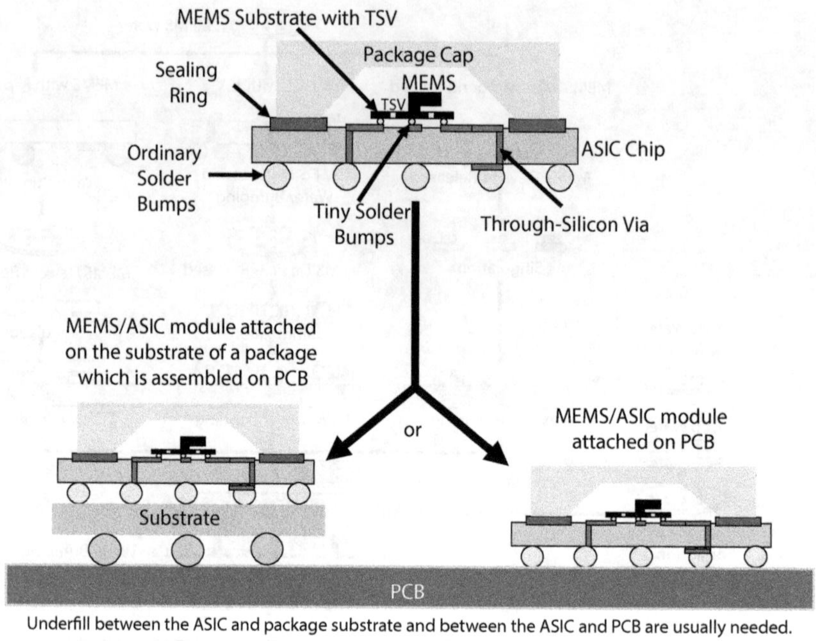

FIGURE 12.17 3D MEMS packaging with vertical electrical feed-through in TSV ASIC.

12.4.2 3D MEMS and IC Integration with Vertical Electrical Feed-Through in ASIC

For 3D MEMS and IC integration with vertical electrical feed-through in the ASIC (cases 4, 5, and 6), the TSV must be fabricated on the ASIC wafer before the MEMS chip to ASIC wafer (C2W) bonding, as shown in Fig. 12.14. After the W2W (cap-to-ASIC) bonding, wafer bumping should be performed on the bottom side of the ASIC wafer. After singulations, the individual units can be soldered on either the substrate of a package or a PCB, as shown in Fig. 12.17. For the PCB case, underfill may be necessary for the reliability of the solder joints. For the solder-bumped flip chip in-package case, use of underfill depends on the substrate material. If it is made of ceramic and the chip size is small, then underfill is optional. However, if it is an organic substrate and the chip size is not small, then underfill is a must.

12.4.3 3D MEMS and IC Integration with Vertical Electrical Feed-Through in Package Cap

For 3D MEMS and IC integration with vertical electrical feed-through in the package cap (cases 7, 8, and 9), the TSV and cavity must be fabricated on the cap wafer before W2W bonding, as shown in Fig. 12.15. After C2W bonding and W2W bonding, wafer bumping should be performed on the bottom side of the cap wafer. The rest is the same as in cases 4 through 6, and Fig. 12.18 shows an example of the complete 3D MEMS packaging with vertical electrical feed-through in the package cap. Underfill between the Si cap and the organic substrate and between the Si cap and the FR-4 PCB is needed.

As mentioned earlier, there are many ways to assemble the 3D MEMS packages. Another way is to attach the MEMS device to the cap wafer first and then perform the cap wafer to ASIC wafer bonding. In this case, however, the alignment accuracy requirements for wafer to wafer bonding are tougher.

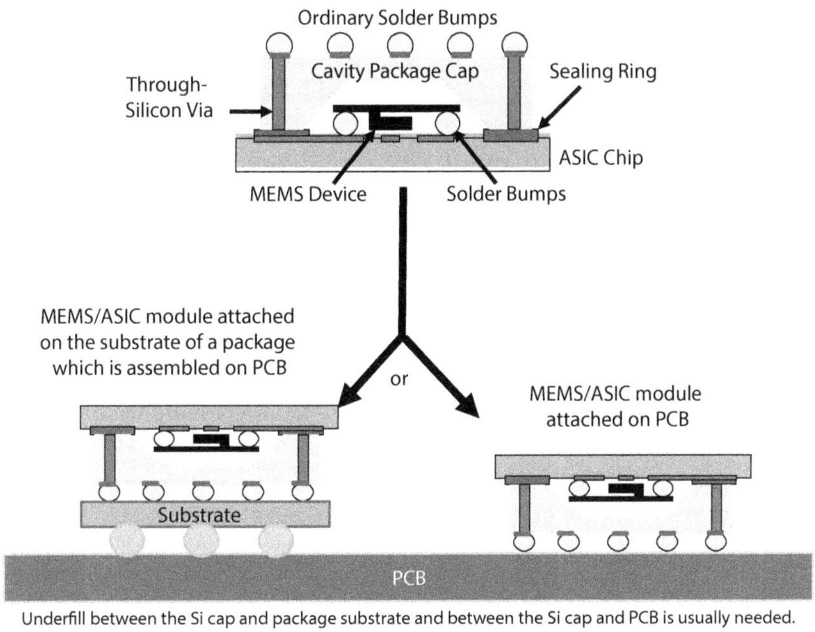

FIGURE 12.18 3D MEMS packaging with vertical electrical feed-through TSV cap.

12.4.4 A Note on Case 10—A Real 3D MEMS and IC Integration

In the food chain of electronic products, packaging is a downstream process (i.e., the packaging people cannot be or are not in a proactive position and cannot say too much), and the packaging people just package whatever is given to them by the semiconductor people (i.e., our job starts from the fabricated wafers). From packaging points of view, case 10 (see Fig. 12.12) is a very low-cost, high-performance 3D MEMS and IC integration. However, in order to make it, the MEMS device (chip) must be much larger (to make space for the sealing ring) than the packages in cases 1 through 9 with a cavity cap. Thus, from a semiconductor's point of view, this is a very "bad idea" because many fewer MEMS devices can be fabricated on the same size MEMS wafer.

12.4.5 Summary and Recommendations

By integrating the MEMS device from the MEMS wafer (with wirebonding pads, or solder-bumped TSV substrate, or solder-bumped flip chip without TSV), the ASIC chip from the ASIC wafer (with either TSV or without), and the cavity cap from the cap wafer (with either TSV or without), 10 different 3D SiPs are presented. Their assembly processes are also provided. Some important results and recommendations are summarized as follows:

- Research and development to systemically, cost-effectively, and reliably package a MEMS device are necessary in order for it to be in high-volume production. Otherwise, no matter how beautiful the MEMS device looks under the SEM, it is just hype!
- The footprints of all these 10 3D MEMS and IC integration are smaller than those of the corresponding 2D conventional (separate) packages.

- The electrical performances of all these 10 3D MEMS and IC integration are better than those of the corresponding 2D conventional (separate) packages.
- The costs of all these 10 3D MEMS and IC integration are lower than those of the corresponding 2D conventional (separate) packages.
- The footprints of all the vertical electrical feed-through 3D MEMS and IC integration (cases 4 through 10) are smaller than those of the lateral electrical feed-through 3D MEMS IC integration (cases 1 through 3).
- The electrical performances of all the vertical electrical feed-through 3D MEMS IC integration (cases 4 through 10) are better than those of the lateral electrical feed-through 3D MEMS IC integration (cases 1 through 3).
- C2W bonding is recommended for the MEMS device bonding to the ASIC wafer.
- W2W bonding is recommended for the cap wafer bonding to the ASIC wafer with the MEMS devices already attached.
- For both C2W bonding and W2W bonding, low-temperature bonding methods with solders (for example, please see Chap. 8 of [22]) are recommended. Solderless C2W and W2W room temperature bondings are better but they are not ready.

12.5 Low-Temperature Bonding of 3D MEMS Packaging with Solders

Most current C2W and W2W bonding methods use temperatures higher than 300°C. During bonding, however, the MEMS devices are already released (there are free-standing microstructures, e.g., membranes, beams, and cantilevers), and low-temperature bonding is desired to reduce damage to the microstructure owing to the thermal-expansion mismatch of the bonding structure (less bow). Fusion (e.g., Cu-to-Cu) bonding at room temperature works; however, the bonding surfaces must be very flat and clean (class 1 cleanroom), which does not permit high-volume production. On the other hand, with proper designs of bonding pads and UBM (under bump metallurgy) and selections of solder materials, 180°C or lower bonding temperatures are possible [7–20]. During/after bonding, the solders will be reacted within themselves and the UBM, and become an IMC (intermetallic compound) with a melting point much higher than that of the solders.

This feature is welcome by 3D MEMS and IC integration. For example, after the bonding of a MEMS device with an ASIC wafer with a low melting solder, all the bonding (solder-interconnect) areas become the IMC with a very high remelting temperature. When a cap wafer is bonded to the ASIC wafer (already bonded with the MEMS device), the interconnect between the ASIC and the MEMS devices will not be reflowed. Furthermore, when the whole 3D MEMS and IC integration module is attached to the PCB with SMT (surface-mount technology) lead-free soldering (260°C), the solder interconnects between the ASIC and MEMS, and cap and ASIC will not be reflowed. In this section, the feasibility of using low-temperature solder for bonding a nonfunctional 3D MEMS and IC integration package (Fig. 12.19) is demonstrated.

12.5.1 3D IC and MEMS Integration with Different Chip Sizes

Figure 12.20 shows the dimensions and materials of the nonfunctional MEMS and ASIC, and cap (cases 7, 8, and 9). It can be seen that (a) the MEMS dimensions are 3 mm × 3 mm × 200 μm and has nine pads (200 μm × 200 μm) as shown in Figs. 12.20 and 12.21; (b) the

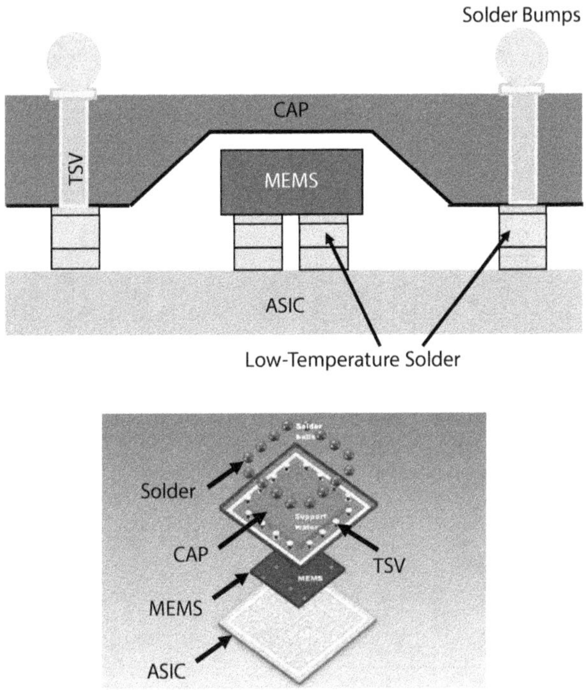

Figure 12.19 3D MEMS and ASIC integration test vehicle.

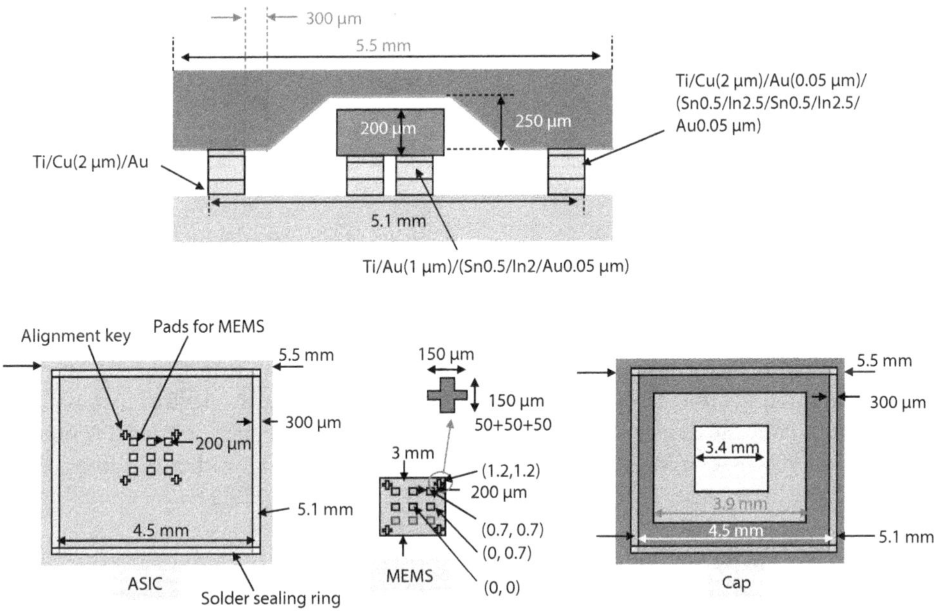

Figure 12.20 Geometry and dimensions of the test vehicle.

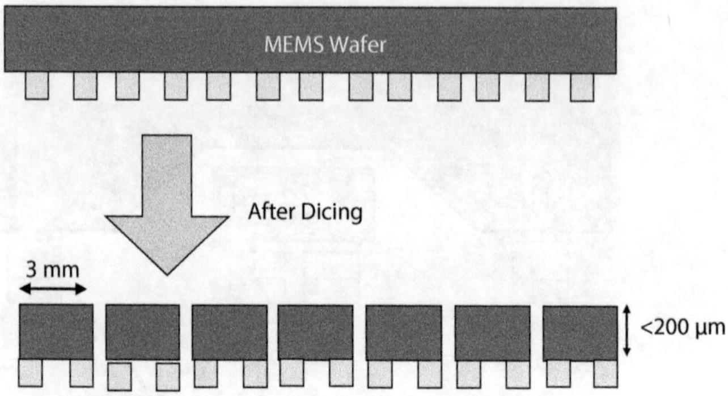

FIGURE 12.21 Individual MEMS devices after dicing.

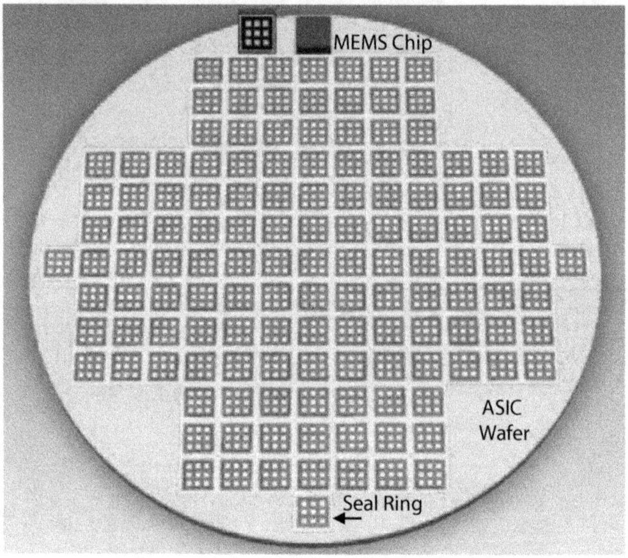

FIGURE 12.22 Schematic of ASIC wafer showing the sites for MEMS devices.

ASIC dimensions are 5.5 mm × 5.5 mm with a sealing ring which is 300 μm width and 5.1 mm on each side of the ASIC as shown in Figs. 12.20, 12.22, and 12.23; and (c) both the bonding pads (for the MEMS device) and seal ring (for the cap wafer) are coated with low-temperature solders.

As in this case, usually the chip size between MEMS and ASIC for 3D MEMS and IC integration is not the same. Thus, W2W bonding of these two devices may not be a good idea, and the interconnections generally are performed by either C2C or C2W bonding. This section discusses a MEMS chip with only the (Ti/Cu/Au) UBM bonded to an ASIC wafer with the low-temperature solders (AuInSn) as shown in Fig. 12.20.

12.5.2 Cavity and TSVs in Cap Wafer

Figures 12.24 and 12.25 show a 200-mm cap wafer with cavities fabricated by the KOH (potassium hydroxide) wet etch process and its typical cross-section. It can be seen that

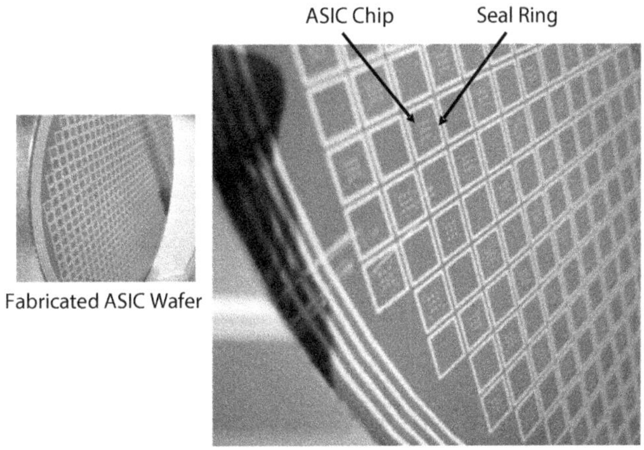

Figure 12.23 ASIC wafer with the sites for MEMS devices and sealing rings.

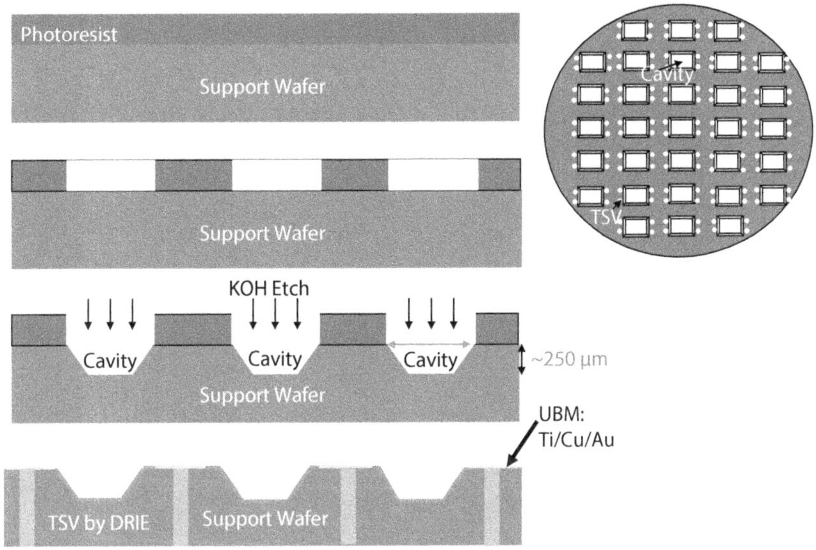

Figure 12.24 Cap wafer with cavities fabricated by KOH wet etching.

KOH etch yields a very smooth wall condition. The TSVs for the cap wafer are fabricated by DRIE (deep reactive-ion etching). The cavity and TSVs can also be made by laser machining and are shown in Fig. 12.26. Even its quality is not as good as that obtained by KOH etching; however, it is adequate for most of the applications. The surface conditions of the TSV made by laser machining are very rough and these may affect the electrical performance. Thus, in general, the TSVs for the cap wafer are made by DRIE process [33–36].

12.5.3 MEMS Chip to ASIC Wafer (C2W) Bonding

Figure 12.27 shows the assembly process of the nonfunctional 3D IC and MEMS integration. It can be seen that the ASIC wafer is fabricated with the sealing ring and pads that are

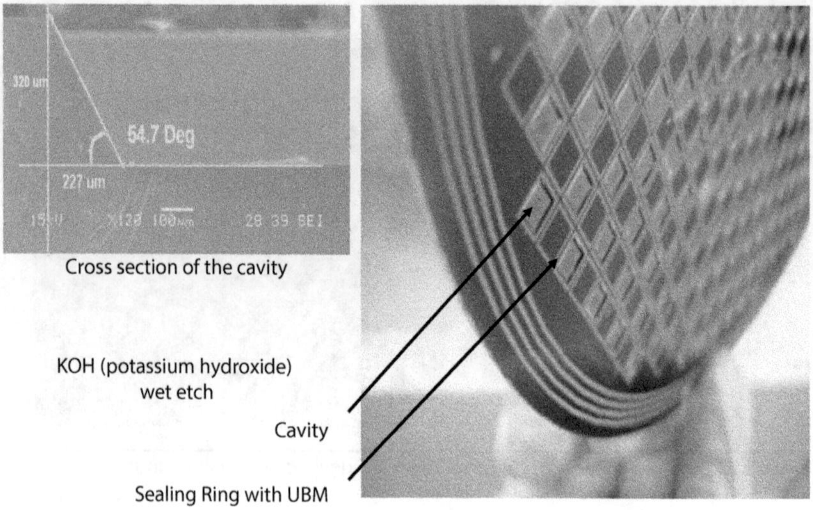

Figure 12.25 The result of cap wafer fabrication (by KOH) wet process.

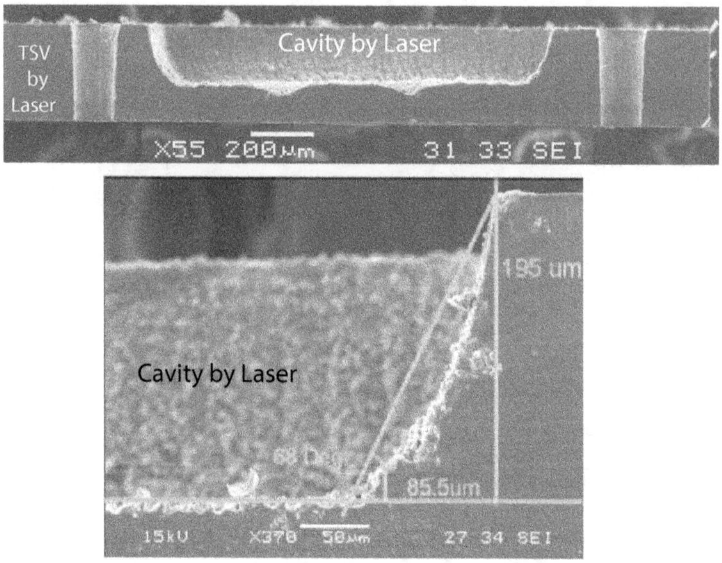

Figure 12.26 Cap wafer with cavity and TSV fabricated by laser machining.

coated with UBM and solder. Figures 12.27 and 12.28 show the MEMS chip picked up and placed on the ASIC wafer in an FC150 bonder. The bonding parameters are 6 MPa of bonding pressure at 200°C on the bonding tool (MEMS) and 90°C on the ASIC wafer, lasting for 40 seconds. The operation is performed in an enclosed nitrogen environment. Figure 12.29 shows a cross-section of the bonding. There is no misalignment.

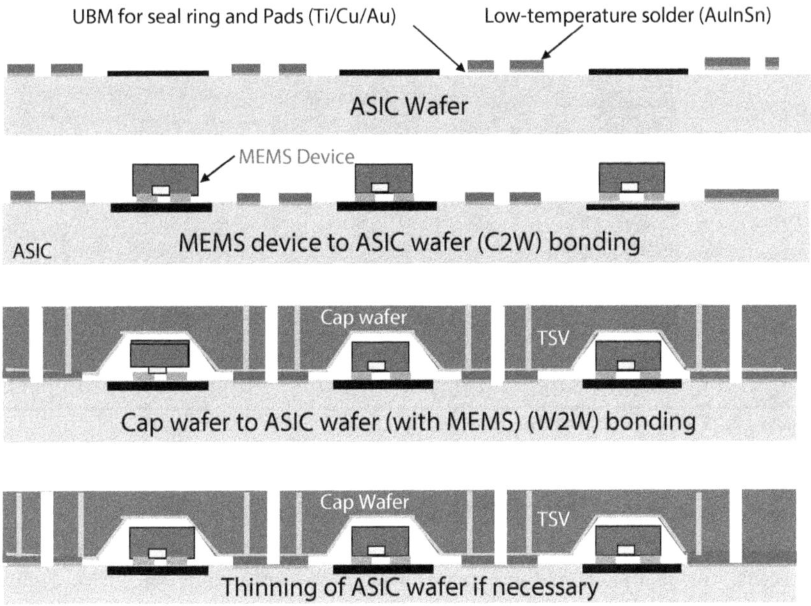

FIGURE 12.27 Summary of the process steps of the 3D MEMS and ASIC integration.

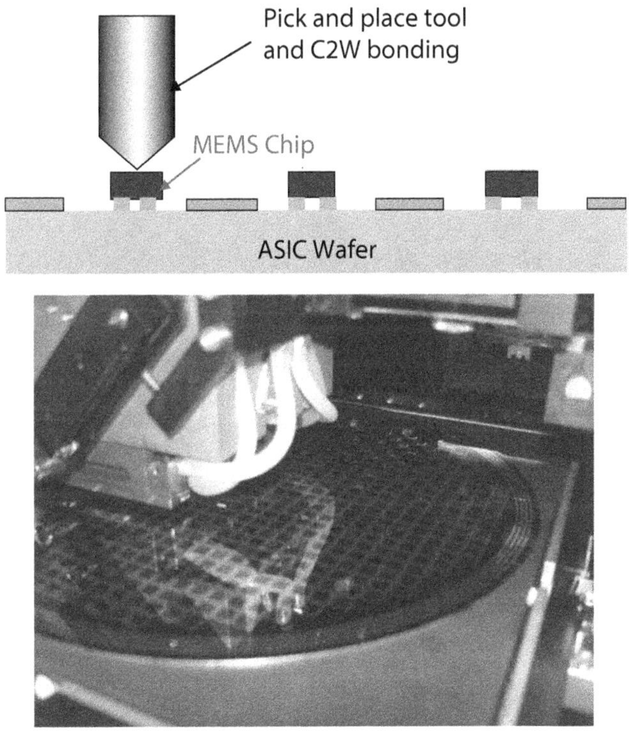

FIGURE 12.28 C2W bonding of the MEMS device on the ASIC wafer.

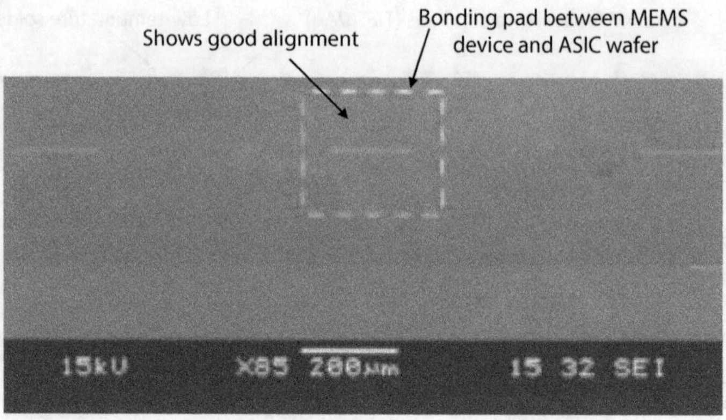

Figure 12.29 SEM image of the cross-section of the MEMS/ASIC bonding.

12.5.4 ASIC Wafer with MEMS Chips to Cap Wafer (W2W) Bonding

As mentioned earlier, the cavity on the 200-mm cap wafer is fabricated by the KOH etch process. The seal ring on the cap wafer is coated only with the UBM. Figures 12.27 and 12.30 show a schematic of the cap wafer bonding to an ASIC wafer with the MEMS device, which is accomplished in an EVG bonder with a pressure of 9 kN, temperature of 200°C, and bonding time of 20 minutes. Figure 12.31 shows an X-ray photograph and SEM images of cross-sections of the W2W bonding of the ASIC wafer with the MEMS device and the cap wafer. It can be seen that (1) bond pads between the MEMS chip and the ASIC chip are in good alignment, and (2) there is no void in the solder sealing ring. Figure 12.32 shows the shear strength test results. All the chips meet the specifications. After solder bumping and dicing, one of the individual packages is also shown in Fig. 12.32.

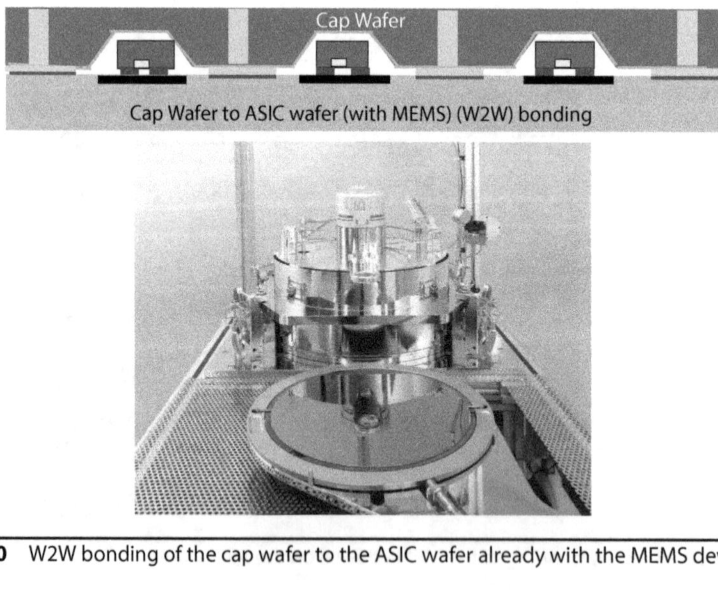

Figure 12.30 W2W bonding of the cap wafer to the ASIC wafer already with the MEMS devices on.

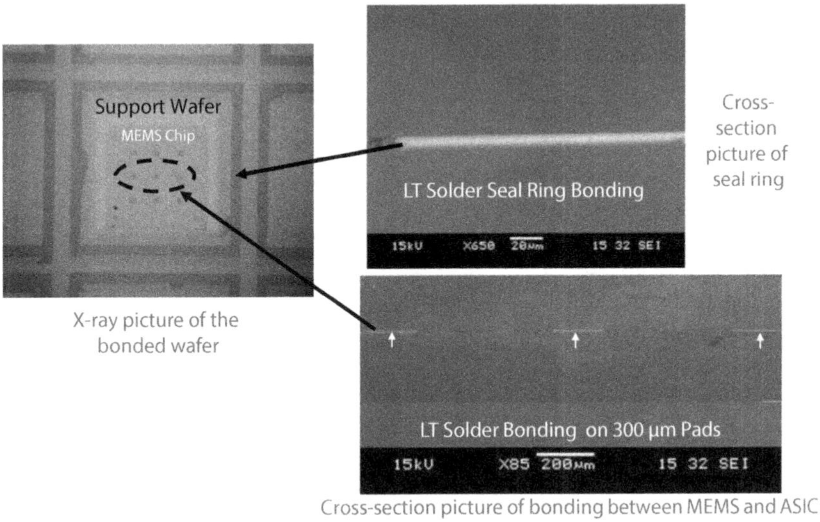

Figure 12.31 X-ray and SEM images showing the sealing ring between the cap and the ASIC.

Bond Force	Temperature	Time	Shear Strength (Cap-ASIC)	Shear Strength (MEMS–ASIC)
9 KN	200°C	20 m	71.6 MPa	9.2 MPa

Completed package with solder ball attached

Figure 12.32 The complete 3D MEMS and ASIC integration package subjected to the characterization test.

It should be ensured that the bonding chamber for W2W bonding is free from debris. Because of the high vacuum during bonding, any particles present will cause the wafer to crack or form microcracks. Also, any significant misalignment between the ASIC wafer and the cap wafer could cause the entire wafer to be unusable. EVG's Smart-View Bond Aligner for face-to-face alignment can help.

12.5.5 Summary and Recommendations

Some important results and recommendations are summarized as follows:

- The feasibility of using low-temperature solder for bonding a nonfunctional 3D MEMS and ASIC integration package is demonstrated.
- The shear strength between the ASIC and cap is 71.6 MPa and that between the MEMS and ASIC is 9.2 MPa. Both magnitudes are larger than the specification (6 MPa).
- Next step, the test vehicle should use functional MEMS deice and ASIC chip.

12.6 Recent Developments in Advanced MEMS Packaging

12.6.1 TSVs for Wafer-Level Packaging of RF MEMS Devices

Figure 12.33 schematically shows the cross-section and bird-view of the wafer-level packaging of a RF MEMS device [37, 38]. It consists of the MEMS wafer (RF MEMS device, high resistivity silicon (HR-Si) substrate with RDLs, and AuSn solder for sealing ring and bonding pads) and the cap wafer (cap with a cavity, TSVs, and RDLs, and the solder bump). It is a 2.25D MEMS and IC integration as mentioned in Sec. 12.3.5. Main objective of [37, 38] is to study the insertion loss of RF-MEMS wafer during packaging with TSV cap wafer. Two different types of CPW (coplanar waveguide) structures (1 mm CPW and 2 mm CPW), as shown in Fig. 12.33, are designed and fabricated.

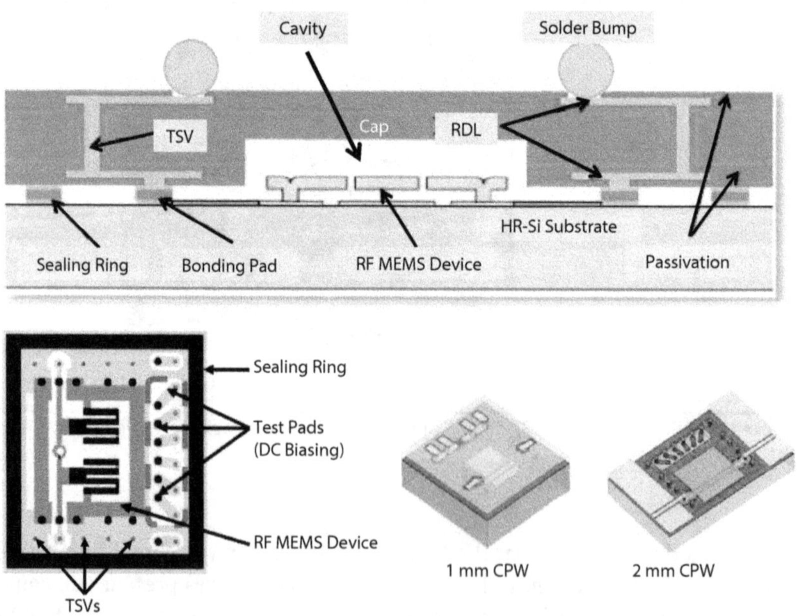

FIGURE 12.33 Schematic of the MEMS wafer-level package. Design layout of a typical RE-MEMS package with TSVs. Different length CPW structures.

The CPW lines are fabricated with Al (1.5 μm in thickness). Three passivation schemes are considered: 3 kÅ Poly Si, 2 μm of low stress oxide, and 1 kÅ SiN. The UBM on the sealing ring and bonding pads is 0.1 μm Ti/0.2 μm Pt/2μmAu/2.15 μm Sn/500 Å Au.

The fabrication process flow of the cap wafer is shown in Fig. 12.34. First, TSV (60 μm in diameter and 200 μm in depth) forming with DRIE, Fig. 12.34(a). Then, TSV filling and RDLs by Cu plating, Fig. 12.34(b). It is followed by backgrinding the cap wafer to 200 μm to expose the tip of the TSV and then passivation, Fig. 12.34(c). Cu plating of the RDLs on the back side of the cap wafer, Fig. 12.34(d). DRIE the cavity to 100 μm, Fig. 12.34(e). UBM and AuSn solder deposition, Fig. 12.34(f). Thin wafer handling is necessary. Right before the backgrinding to expose the TSV, the front side of the cap wafer is temporary bonded to a supporting wafer (carrier) with BSI HT10.10 adhesive. Right after the solder deposition, the wafers are debonded (separated) by thermal slide-off method.

Figure 12.34(g) schematically shows the MEMS wafer to cap wafer (W2W) bonding and the bonding condition is shown in Fig. 12.35. It can be seen that the bonding temperature for the AuSn solder is 280°C for 30 minutes and the bonding force is 10 kN. Figure 12.36 shows the optical image of the bonded wafer (cap wafer is on top and the MEMS wafer is at the bottom) and the X-ray images. It can be seen that there is nothing unusual. Figure 12.37 shows the X-ray and SEM images of the bonded package. It can be seen that there is no void and delamination of the sealing ring. Figure 12.38 shows the images of the bonded MEMS package. It can be seen that (a) there is no void in the TSV, and (b) there is no delamination of the dielectric, barrier, and seed layer of the TSV.

Packaging loss of the 1 mm CPW lines is measured at different frequencies before the onset of resonance at 19 GHz (Fig. 12.39). This insertion loss ranges from 0.12 to 0.2 dB over 5 to 20 GHz, and 1 dB cut-off frequency is 22.7 GHz. Insertion loss of 2 mm CPW lines ranges

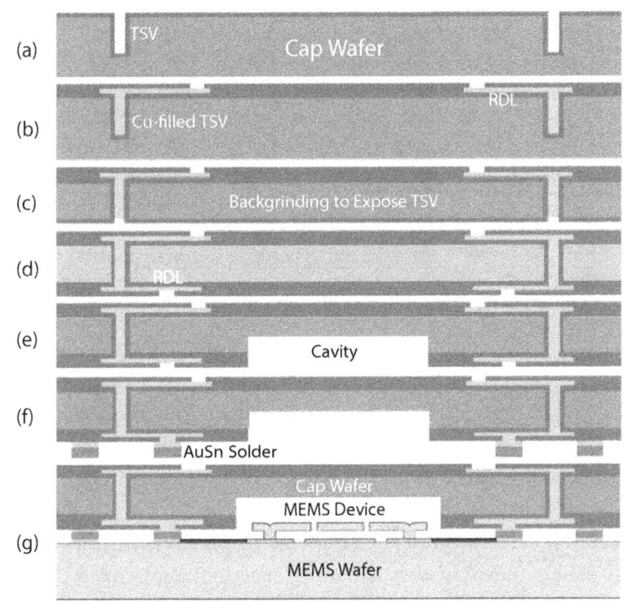

FIGURE 12.34 Schematic of TSV cap wafer fabrication process flow and W2W bonding.

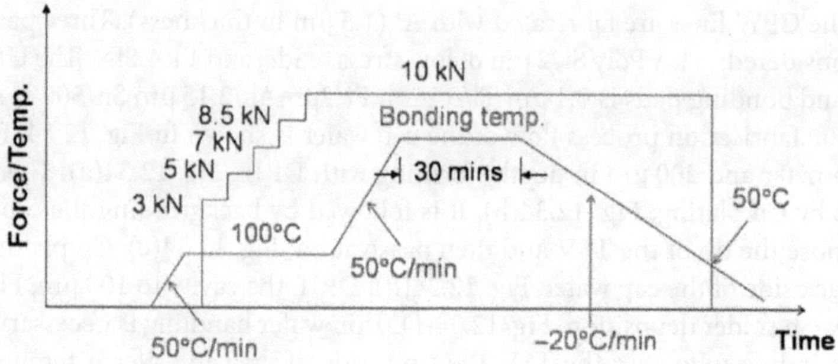

Figure 12.35 W2W (AuSn) bonding condition.

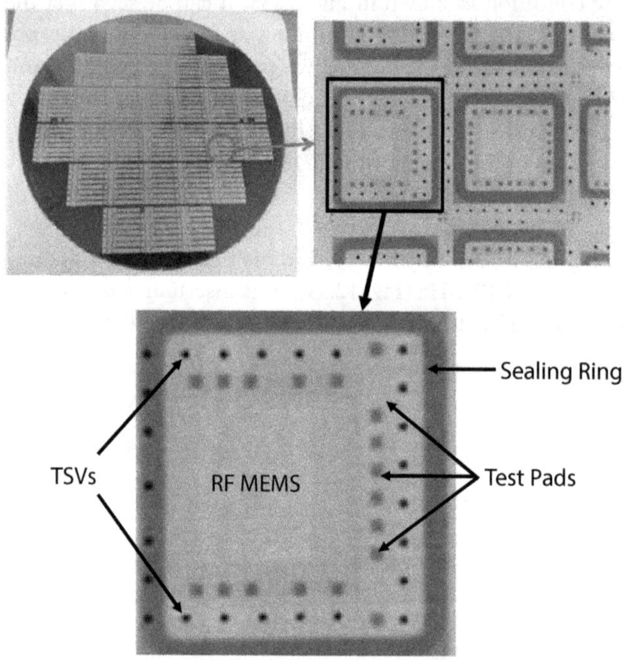

Figure 12.36 Bonded wafers. X-ray image showing the TSVs and sealing ring.

from 0.14 to 0.3 dB over 5 to 25 GHz, and 1 dB cut-off frequency is 27.4 GHz (Fig. 12.40). Since the acceptable packaging loss is <0.2 dB for <20 GHz range and 1 dB packaging cut-off frequency is >20 Hz; thus, the 1 mm CPW design is OK.

12.6.2 Zero-Level Packaging for RF-MEMS Implementing TSVs and Metal Bonding

Figure 12.41 schematically shows the zero-level packaging for RF-MEMS device [39, 40]. It consists of the RF-MEMS, HR-Si substrate with RDLs, and Cu-Sn-Cu for sealing ring and bonding pads, the HR-Si cap with TSVs and RDLs, and the solder bump. Again, it is a 2.25D

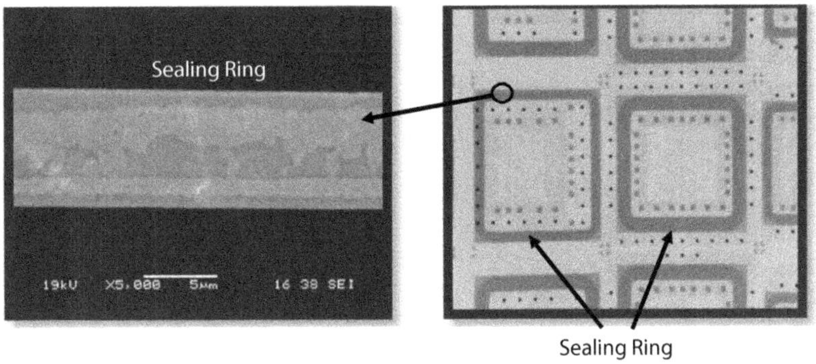

Figure 12.37 SEM and X-ray images showing the sealing ring.

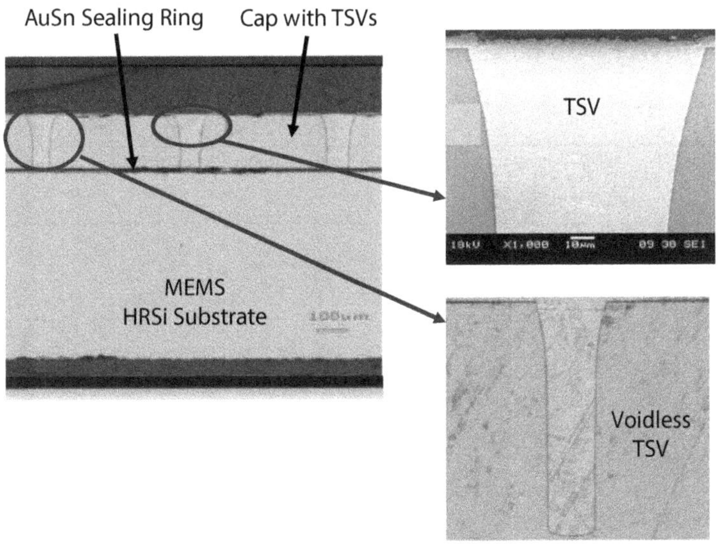

Figure 12.38 Cross-sectional view of the bonded cap and MEMS wafers showing the AuSn sealing ring and voidless TSV.

MEMS package. One of the main objectives of [39, 40] is to study the bonding characteristics (sealing ring and interconnect bumps) of the MEMS wafer and the cap wafer.

The fabrication process flow of the zero-level packaging of the MEMS device is shown in Fig. 12.42. First, photoresist, lithographing, patterning of the RDLs and UBM on the cap wafer, Fig. 12.42(a). Temporary bond the cap wafer to a supporting wafer by glue, then backgrind the cap wafer to 100 μm, Fig. 12.42(b). TSV forming by isotropic etching, DRIE, and soft landing. The TSV is not filled, as shown in Fig. 12.42(c). Then, photoresist, lithographing, patterning, plating the RDLs and it is followed by electroplating the Cu (5 μm) and Sn (3 to 5 μm), Fig. 12.42(c). Bonding the cap wafer to the MEMS wafer by thermocompression bonding, Fig. 12.42(d). Debond the supporting wafer and perform solder wafer bumping, Fig. 12.42(e).

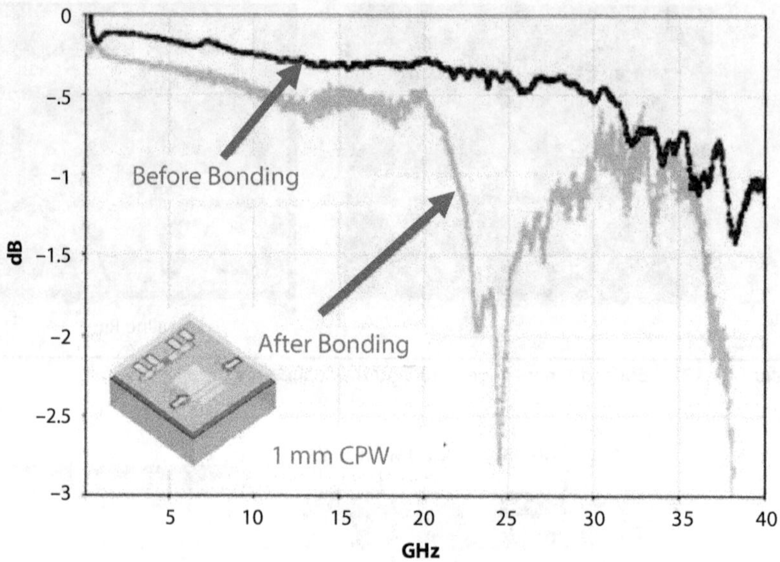

FIGURE 12.39 MEMS packaging insertion loss with 1 mm CPW design.

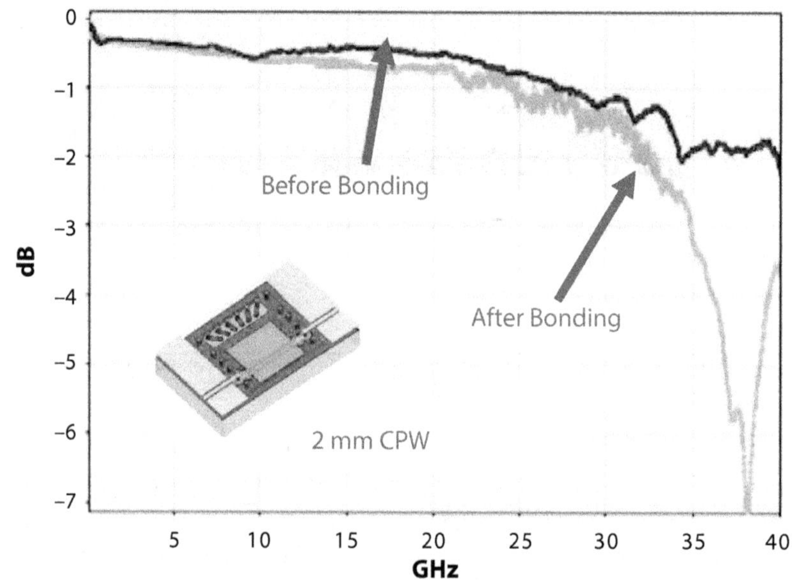

FIGURE 12.40 MEMS packaging insertion loss with 2 mm CPW design.

Figure 12.43 shows the fabricated TSV. It can be seen that a conformal spin coating polymer insulator has been applied to the TSV for the dielectric layer, as shown in Fig. 12.43(a). The curing temperature is 200°C. A seed layer of Ti (70 nm)/Cu (500 nm) is sputter deposited on the dielectric layer. Then, a conformal Cu plating process is applied as shown in Fig. 12.43(b). The TSV is not Cu filled.

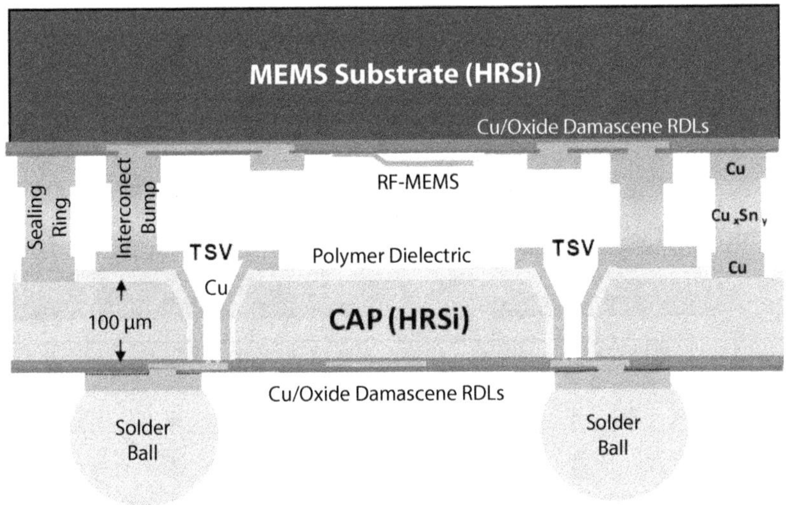

Figure 12.41 MEMS packaging schematically showing the RF-MEMS, HR-Si substrate with RDLs, and Cu-Sn-Cu for sealing ring and bonding pads, the HR-Si cap with TSVs, RDLs, and the solder bump.

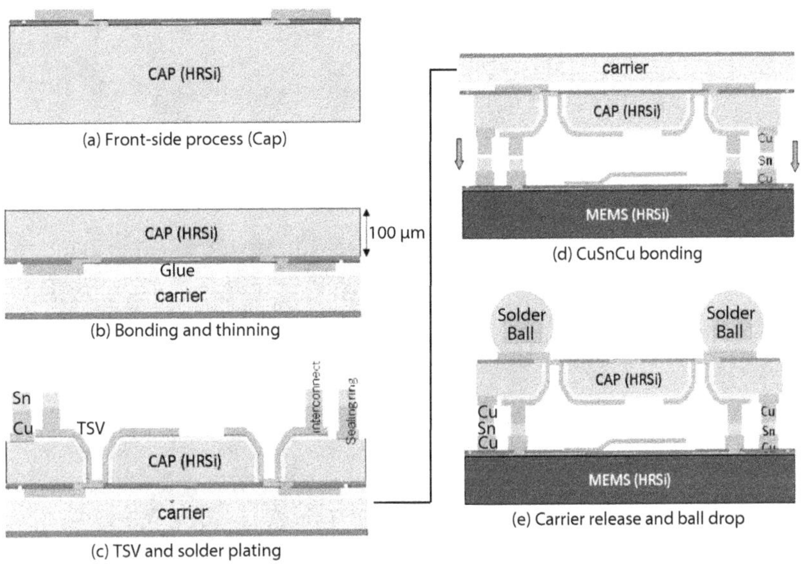

Figure 12.42 Fabrication process flow of the cap wafer. W2W (CuSnCu) bonding of the cap wafer and MEMS wafer.

Figure 12.44 shows the optical profilometer scan of a cap wafer with the As plated CuSn sealing ring and interconnect bumps before bonding. It can be seen that the average height of the interconnect bump is taller than the sealing ring by 1 μm. This uneven height can be planarized by the fly cutting (of the Sn layer) method. First, the cap wafer is coated with a resist layer as shown in Fig. 12.45. Then, a diamond cutting bit moves over the wafer

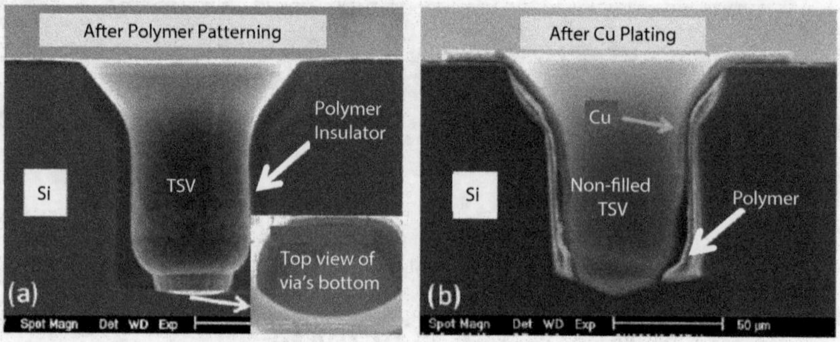

Figure 12.43 (a) TSV right after polymer patterning. (b) Right after Cu plating on the TSV side walls.

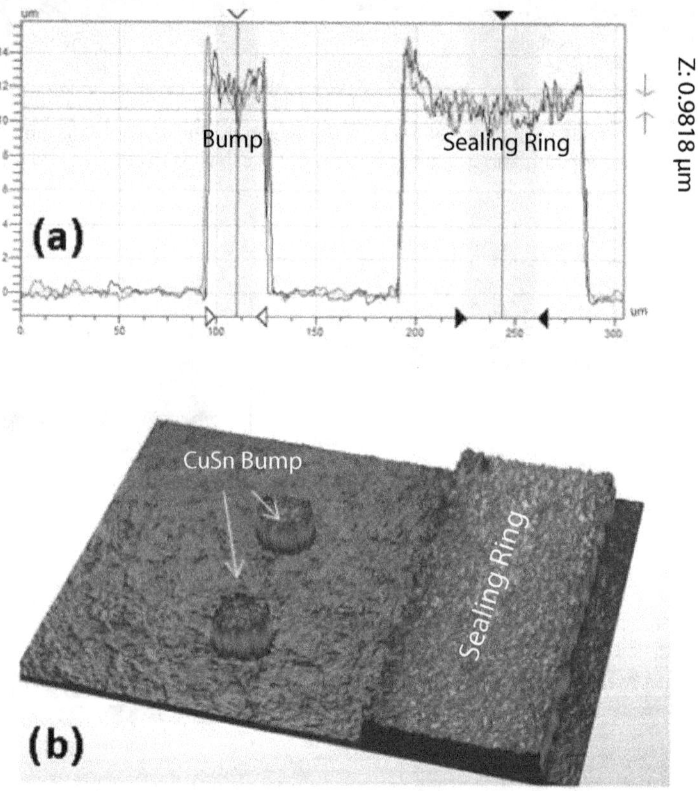

Figure 12.44 Optical profilometer scan of a cap wafer with As plated CuSn sealing ring and bump. (a) Step height profile. (b) 3D profile.

performing a milling type operation, removing all materials it comes in contact with, as shown in Fig. 12.45. Figures 12.46 and 12.47, respectively, show the optical profilometer scan and the optical images of the fly cut cap wafer with the sealing ring and interconnect CuSn bumps. It can be seen that the surface flatness has been significantly improved (the average height difference has been reduced to 0.1 µm) and becomes smooth after fly cutting.

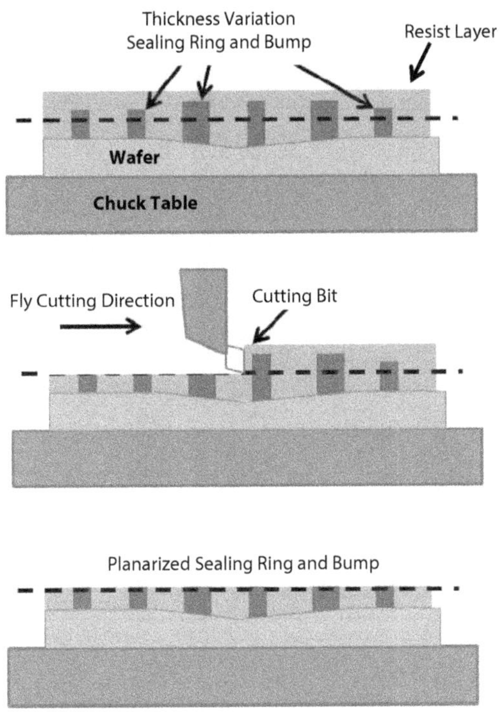

FIGURE 12.45 Fly cutting process and results.

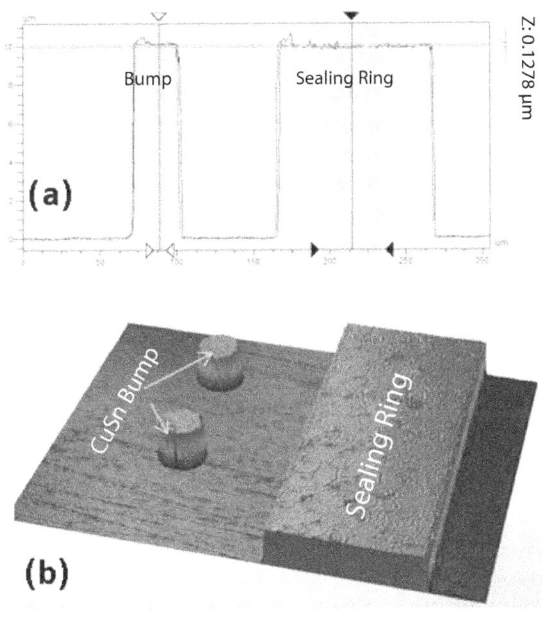

FIGURE 12.46 Optical profilometer scan of a fly cutted cap wafer with CuSn sealing ring and bump. (a) Step height profile. (b) 3D profile.

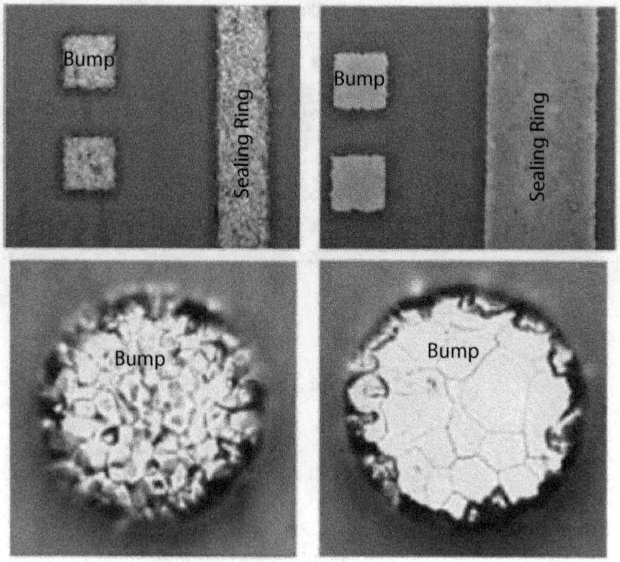

Figure 12.47 Top view of the surface of a sealing ring and bump. (a) As plated. (b) After fly cutted.

The RF performance of the MEMS package is studied by the measurements of a CPW traversing through the interconnects of the package as shown in the top of Fig. 12.48. The thickness of the CPW is 1 μm. The typical performance of the MEMS package measured in the frequency range from 100 MHz to 40 GHz is shown in the bottom of Fig. 12.48. The acceptable magnitude of the reflection coefficient ($S_{11} < 115$ dB) indicates that the actual wave impedance of the line is close to the target value (50 Ω) in a wide frequency range and that the precision of the bonding assembly is accurate enough [39, 40].

12.6.3 MEMS Package Based on Si-Interposer Wafer with Cu-Filled TSVs

Figure 12.49 schematically shows a MEMS package based on interposer wafer [41]. It can be seen that the MEMS device is attached to a Si-interposer wafer with Cu-filled TSVs and RDLs and is hermetic sealed with a cap wafer with cavity. This is a 2.5D MEMS and IC integration. The assembly process is very similar to that shown in Fig. 12.14, except that the deviced ASIC wafer is replaced with the deviceless interposer wafer. A typical cross-section image of the final assembly is shown in Fig. 12.50. It can be seen that all the key elements of the MEMS package, such as the MEMS device, Si interposer, TSV, micro-bump, and sealing ring, are in proper position. For the characterization of the MEMS package, please see [41].

12.6.4 Wafer-Scale Packaging for FBAR-Based Oscillators

Figure 12.51 show the wafer-scale packaging for FBAR (film bulk acoustic resonator)-based oscillators [42, 43]. Figure 12.51(a) shows the cap wafer (upper part) of the package. It can be seen that there is an IC device (330 μm × 285 μm), six TSVs for external connection, two internal connections (ICP), and a small cavity for the FBAR. Figure 12.51(b) shows the

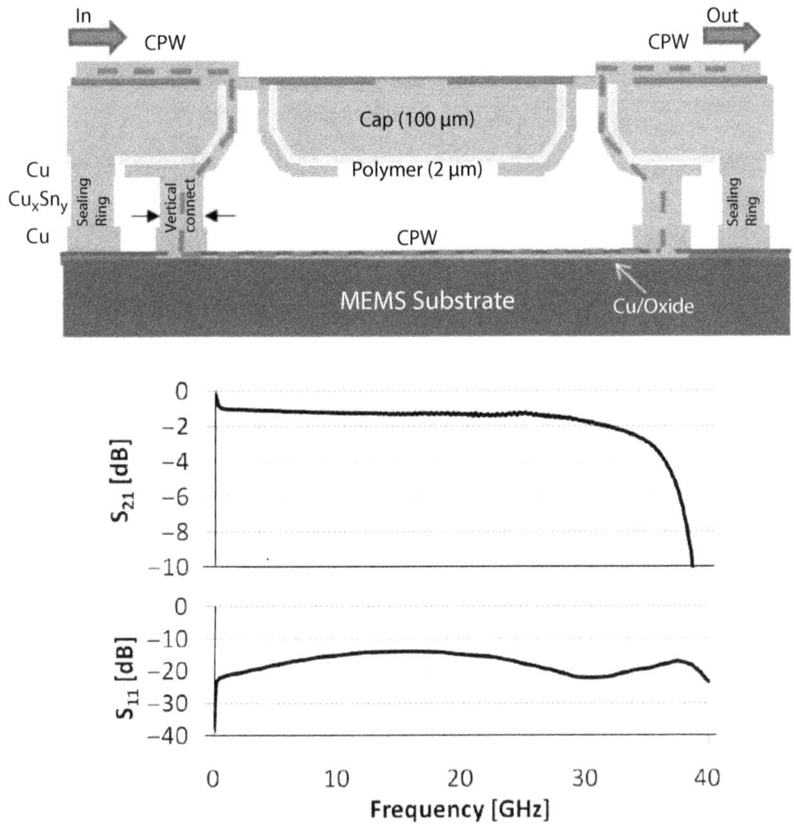

FIGURE 12.48 (Top) The dashed blue line indicates the path of the propagation of the RF signal from the input to the output. (Bottom) Typical RF performance: reflection coefficient S_{11} and transmission coefficient S_{21} of the fabricated package shown on top.

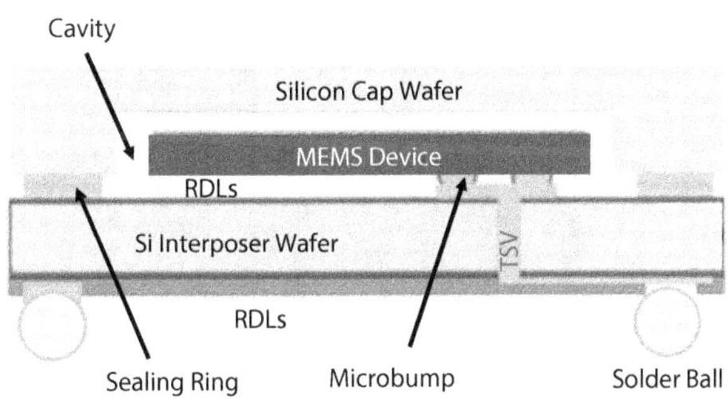

FIGURE 12.49 Schematic of MEMS wafer-level packaging with a TSV interposer.

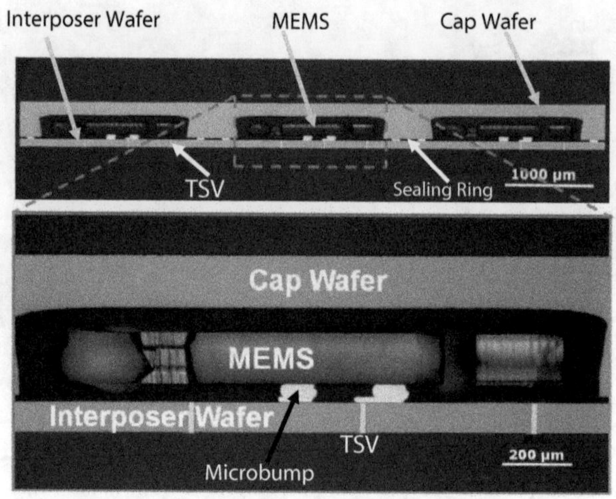

FIGURE 12.50 Cross-section image of the MEMS device on a TSV interposer and is sealed by a cap with cavity.

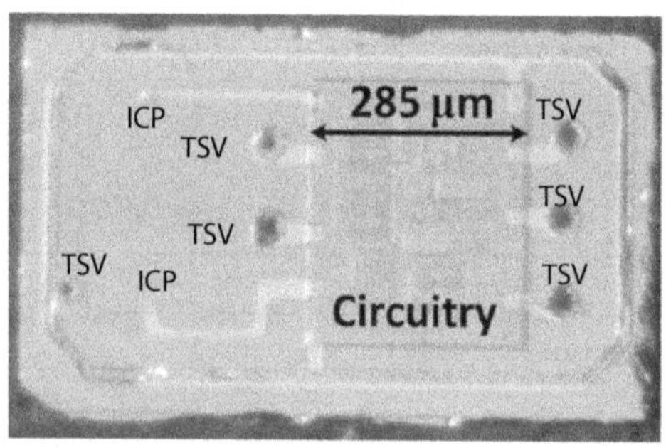

(a) Cap Wafer

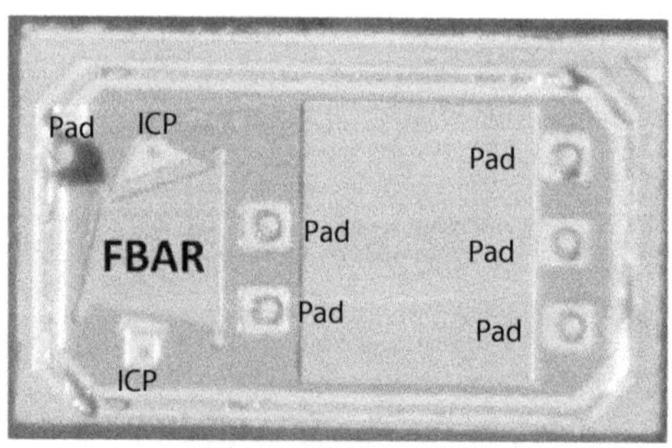

(b) FBAR Wafer

FIGURE 12.51 Photo images of the FBAR hermetic package. (a) Cap wafer with IC device, TSVs, internal connections, and cavity for the FBAR. (b) FBAR wafer with FBAR, pads, internal connections, and cavity for IC device.

FBAR wafer (lower part) of the package. It can be seen that there is an FBAR, six external pads (for those six TSVs), two internal connections (from the cap electronics to the FBAR), and a cavity (for the IC device). This is a real 3D MEMS and IC integration (very similar to case 10 shown in Fig. 12.12). Both the FBAR and cap wafers are high-impedance Si to minimize cross-talk and capacitive losses.

The fabrication process flow of the FBAR package without the IC device is shown in Fig. 12.52. For the FBAR wafer, first, define and form the cavity and then fill it with a sacrificial oxide, Fig. 12.52(a). Form the FBAR/ZDR (zero drift resonators), Fig. 12.53(b). Pattern the Au on pads and sealing ring and then remove the sacrificial oxide, Fig. 12.52(c). For the cap wafer,

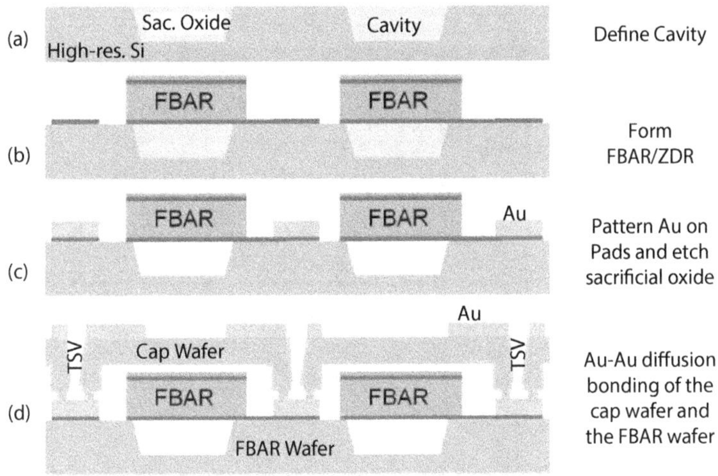

FIGURE 12.52 Fabrication process flow of the FBAR wafer and cap wafer to FBAR wafer bonding.

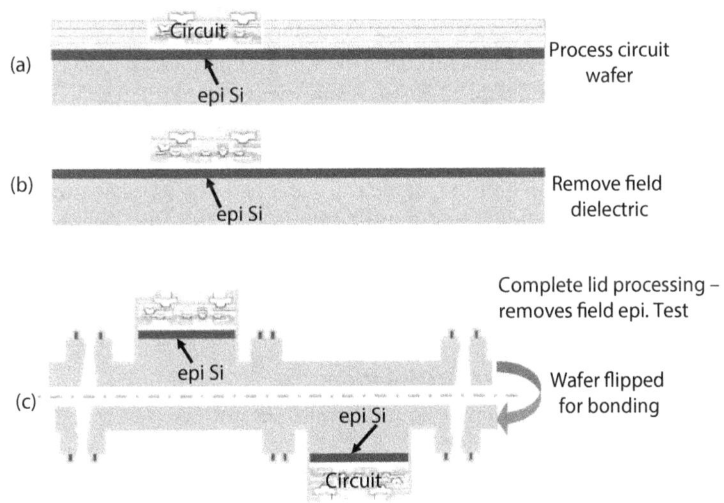

FIGURE 12.53 Fabrication process flow of the cap wafer.

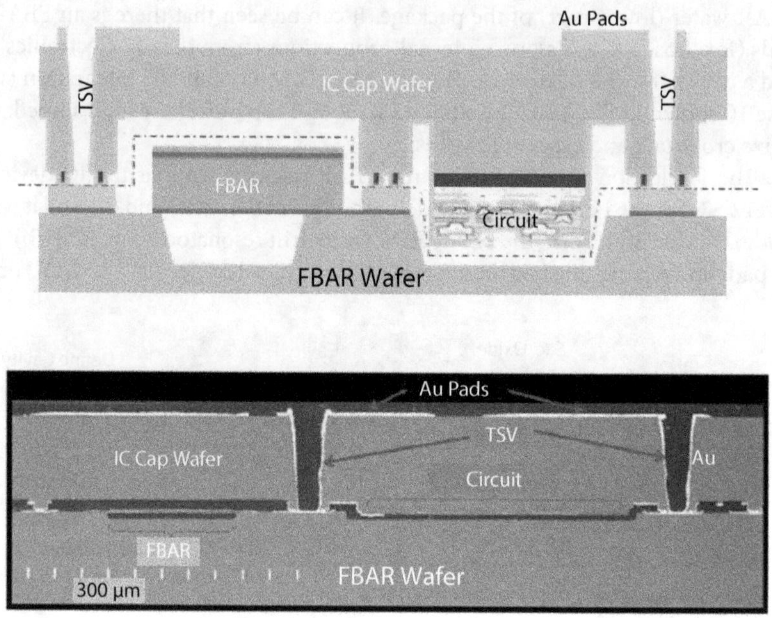

Figure 12.54 (Top) IC cap wafer to FBAR wafer Au-Au bonding. (Bottom) Cross-section SEM image of the bonded FBAR MEMS package with IC cap.

standard micromachining process is applied to make the cavity and DRIE is used to make the TSVs. Electroplate Au on external pads, internal pads, sealing ring, and the side walls of the TSVs. Finally, Au-Au diffusion bonding of the cap wafer and the FBAR wafer, Fig. 12.52(d).

The fabrication process flow of the cap wafer with IC device of the FBAR package is shown in Fig. 12.53. The devices are fabricated by a standard bipolar process, except that the interlayer dielectric is left in the field until the devices are complete, Fig. 12.53(a). Then remove the field dielectric down to Si, Fig. 12.53(b). Apply the standard micromachining process to make the cavity and DRIE to make the TSVs. It is followed by Au plating the pads and side walls of TSVs, and then flipping the wafer for bonding, Fig. 12.53(c). The TSVs are not filled.

The Au-Au diffusion bonding of the cap wafer with the IC device and TSVs and the FBAR wafer is schematically shown in the top of Fig. 12.54. The cross-sectional SEM image of the bonded package is shown in the bottom of Fig. 12.54. It can be seen that all the key elements such as the FBAR, IC device TSVs, cap, sealing ring, and pad are in proper position. For detailed characterization and mechanical and electrical performances, please read [42, 43].

12.6.5 Summary and Recommendations

Some important results and recommendations are summarized as follows [37–43]:

- TSVs (either with Cu-filled or no-filled) are applied to MEMS packaging.
- Most research papers have the TSVs in the cap without any devices (2.25 MEMS and IC integration) [37–40].

- There is only one paper with the TSV interposer supporting the MEMS device (2.5D MEMS and IC integration) [41].
- There is only one paper with the FBAR MEMS device bonded with a cap with IC device and TSVs [43] (a true 3D MEMS and IC integration).
- Hopefully, more research papers will be published in 3D MEMS and IC integration area.

12.7 References

[1] Lyshevski, S. E., *MEMS and NEMS: Systems, Devices, and Structures*, CRC Press LLC, Boca Raton, FL, 2001.

[2] Gad-el-Hak, M., *The MEMS Handbook*, CRC Press LLC, Boca Raton, FL, 2001.

[3] Gad-el-Hak, M., *MEMS: Applications*, CRC Press LLC, Boca Raton, FL, 2006.

[4] Korvink, J. G., and O. Paul, *MEMS: A Practical Guide to Design, Analysis, and Applications*, William Andrew, Inc., Norwich, NY, 2006.

[5] Hsu, T. R., *MEMS Packaging*, INSPEC, IEEE, London, UK, 2004.

[6] Gilleo, K., *MEMS/MOEMS Packaging*, McGraw-Hill, New York, NY, 2005.

[7] Lau, J. H., C. K. Lee, C. S. Premachandran, and Yu Aibin, *Advanced MEMS Packaging*, McGraw-Hill, New York, NY, 2010.

[8] Premachandran, C. S., J. H. Lau, X. Ling, A. Khairyanto, K. Chen, and M. Pa, "A Novel, Wafer-level Stacking Method for Low-chip Yield and Non-uniform, Chip-size Wafers for MEMS and 3D SiP Applications," *IEEE Proceedings of Electronic Components and Technology Conference,* Orlando, FL, May 27–30, 2008, pp. 314–318.

[9] Chen, K., C. Premachandran, K. Choi, C. Ong, X. Ling, A. Ratmin, J. H. Lau, et al., "C2W Low Temperature Bonding Method for MEMS Applications," *IEEE Proceedings of Electronics Packaging Technology Conference*, Singapore, December 2008, pp. 1–7.

[10] Made, R., C. L. Gan, L. Yan, A. Yu, S. U. Yoon, J. H. Lau, and C. Lee, "Study of Low Temperature Thermocompression Bonding in Ag-In Solder for Packaging Applications," *Journal of Electronic Materials*, Vol. 38, 2009, pp. 365–371.

[11] Yan, L.-L., C.-K. Lee, D.-Q. Yu, A.-B. Yu, W.-K. Choi, J. H. Lau, and S.-U. Yoon, "A Hermetic Seal using Composite Thin Solder In/Sn as Intermediate Layer and its Interdiffusion Reaction with Cu," *Journal of Electronic Materials*, Vol. 38, 2009, pp. 200–207.

[12] Yan, L.-L., V. Lee, D. Yu, W. K. Choi, A. Yu, S.-U. Yoon, and J. H. Lau, "A Hermetic Chip to Chip Bonding at Low Temperature with Cu/In/Sn/Cu Joint," *IEEE Proceedings of Electronics and Components Technology Conference,* Orlando, FL, May 2008, pp. 1844–1848.

[13] Yu, A., C. Lee, L. Yan, R. Made, C. Gan, Q. Zhang, S. Yoon, and J. H. Lau, "Development of Wafer Level Packaged Scanning Micromirrors," *Proceedings of Photonic West,* Vol. 6887, 2008, pp. 1–9.

[14] Lee, C., A. Yu, L. Yan, H. Wang, J. Han, Q. Zhang, and J. H. Lau, "Characterization of Intermediate In/Ag Layers of Low Temperature Fluxless Solder based Wafer Bonding for MEMS Packaging," *Journal of Sensors Actuators A: Physical,* Vol. 154, 2009, pp. 85–91.

[15] Yu, D. Q., C. Lee, L. L. Yan, W. K. Choi, A. Yu, A., and J. H. Lau, "The Role of Ni Buffer Layer on High Yield Low Temperature Hermetic Wafer Bonding Using In/Sn/Cu Metallization," *Applied Physics Letters*, Vol. 94, No. 3, 2009, pp 1–3.

[16] Yu, D. Q., L. L. Yan, C. Lee, W. K. Choi, S. U. Yoon, and J. H. Lau, "Study on High Yield Wafer to Wafer Bonding using In/Sn and Cu Metallization," *Proceedings of the Eurosensors Conference*, Dresden, Germany, 2008, pp. 1242–1245.

[17] Yu, D., C. Lee, and J. H. Lau, "The Role of Ni Buffer Layer between InSn Solder and Cu Metallization for Hermetic Wafer Bonding," *Proceedings of the International Conference on Electronics Materials and Packaging,* Taipei, Taiwan, October 22–24, 2008, pp. 335–338.

[18] Yu, D., L. Yan, C. Lee, W. Choi, M. Thew, C. Foo, and J. H. Lau,"Wafer level hermetic bonding using Sn/In and Cu/Ti/Au metallization," *IEEE Proceeding of Electronics Packaging and Technology Conference,* December 2008, pp. 1–6.

[19] Choi, W., D. Yu, C. Lee, L. Yan, A. Yu, S. Yoon, J. H. Lau, et al., "Development of Low Temperature Bonding using In-based Solders," *Proceedings* of *Electronics Component and Technology Conference,* Orlando, FL, May 2008, pp. 1294–1299.

[20] Choi, W., C. Premachandran, C., Ong, X. Ling, E. Liao, A. Khairyanto, J. H. Lau, et al., "Development of Novel Intermetallic Joints using Thin Film Indium Based Solder by Low Temperature Bonding

Technology for 3D IC Stacking," *IEEE Proceedings of Electronics Component and Technology Conference*, San Diego, CA, May 2009, pp. 333–338.
[21] Lau, J. H., *Reliability of RoHS Compliant 2D & 3D IC Interconnects*, McGraw-Hill, New York, NY, 2011.
[22] Lau, J. H., *Through-Silicon Vias for 3D Integration*, McGraw-Hill, New York, NY, 2013.
[23] Lau, J. H., *Low Cost Flip Chip Technologies for DCA, WLCSP, and PBGA Assemblies*, McGraw-Hill, New York, NY, 2000.
[24] Lau, J. H., and R. Lee, *Chip Scale Package: Design, Materials, Process, Reliability, and Applications*, McGraw-Hill, New York, NY, 1999.
[25] Lau, J. H., C. P. Wong, J. Prince, and W. Nakayama, *Electronic Packaging: Design, Materials, Process, and Reliability*, McGraw-Hill, New York, NY, 1998.
[26] Lau, J. H., *Flip Chip Technologies*, McGraw-Hill, New York, NY, 1996.
[27] Lau, J. H., *Ball Grid Array Technology*, McGraw Hill, New York, NY, 1995.
[28] Lau, J. H., *Chip On Board Technologies for Multichip Modules*, Van Nostrand Reinhold, New York, NY, 1994.
[29] Lau, J. H., *Handbook of Fine Pitch Surface Mount Technology*, VNR, New York, NY, 1994.
[30] Lau, J. H., *Handbook of Tape Automated Bonding*, VNR, New York, NY, 1992.
[31] Lau, J. H., "Design and Process of 3D MEMS Packaging," *IMAPS Proceedings of International Symposium on Microelectronics*, San Jose, CA, November 2009, pp. 1–9.
[32] Lau, J. H., "Design and Process of 3D MEMS Packaging," *IMAPS Transactions, Journal of Microelectronics and Electronic Packaging*, First Quarter Issue, Vol. 7, 2010, pp. 10–15.
[33] Khan, N., V. Rao, S. Lim, S. Ho, V. Lee, X. Zhang, J. H. Lau, et al., "Development of 3D Silicon Module with TSV for System in Packaging," *Proceedings of Electronic Components and Technology Conference*, Orlando, FL, May 27–30, 2008, pp. 550–555.
[34] Zhang, X., T. Chai, J. H. Lau, C. Selvanayagam, K. Biswas, S. Liu, D. Pinjala, et al., "Development of Through Silicon Via (TSV) Interposer Technology for Large Die Fine-pitch Cu/low*k* FCBGA Package," *IEEE Electronic Components and Technology Conference*, San Diego, CA, May 2009, pp. 305–312.
[35] Chai, T. C., X. Zhang, J. H. Lau, C. S. Selvanayagam, D. Pinjala, Y. Hoe, Y. Ong, et al., "Development of Large Die Fine-Pitch Cu/low-*k* FCBGA Package with Through Silicon via (TSV) Interposer," *IEEE Transactions on CPMT*, Vol. 1, No. 5, May 2011, pp. 660–672.
[36] Yu, A., N. Khan, G. Archit, D. Pinjalal, K. Toh, V. Kripesh, J. H. Lau, et al., "Development of Silicon Carriers with Embedded Thermal Solutions for High Power 3-D Package," *IEEE Transactions on Components and Packaging Technology*, Vol. 32, No. 3, September 2009, pp. 566–571.
[37] Sekhar, V., J. Toh, J. Cheng, J. Sharma, S. Fernando, and B. Chen, "Wafer Level Packaging of RF MEMS Devices Using TSV Interposer Technology," *Proceedings of IEEE/EPTC*, Singapore, December 2012, pp. 239–243.
[38] Chen, B., V. Sekhar, C. Jin, Y. Lim, J. Toh, S. Fernando, and J. Sharma, "Low-Loss Broadband Package Platform With Surface Passivation and TSV for Wafer-Level Packaging of RF-MEMS Devices," *IEEE Transactions on CPMT*, Vol. 3, No. 9, September 2013, pp. 1443–1452.
[39] Pham, N., V. Cherman, B. Vandevelde, P. Limaye, N. Tutunjyan, R. Jansen, N. Hoovels, et al., "Zero-level Packaging for (RF-)MEMS Implementing TSVs and Metal Bonding," *Proceedings of IEEE/ECTC*, May 2011, 1588–1595.
[40] Pham, N., V. Cherman, N. Tutunjyan, L. Teugels, D. Teacan, and H. Tilmans, "Process Challenges in 0-level Packaging Using 100μm-thin Chip Cappin with TSV," *Proceedings of IMAPS International Symposium on Microelectronics*, September 2012, San Diego, CA, pp. 276–282.
[41] Zoschke, K., C.-A. Manier, M. Wilke, N. Jürgensen, H. Oppermann1, D. Ruffieux, J. Dekker, et al., "Hermetic Wafer Level Packaging of MEMS Components Using Through Silicon Via and Wafer to Wafer Bonding Technologies," *Proceedings of IEEE/ECTC*, May 2013, Las Vegas, NV, pp. 1500–1507.
[42] Pang, W., R. Ruby, R. Parker, P. W. Fisher, M. A. Unkrich, and J. D. Larson, III, "A Temperature-Stable Film Bulk Acoustic Wave Oscillator," *IEEE Electron Device Letters*, Vol. 29, No. 4, April 2008, pp. 315–318.
[43] Small, M., R. Ruby, S. Ortiz, R. Parker, F. Zhang, J. Shi, and B. Otis, "Wafer-Scale Packaging For FBAR-Based Oscillators," *Proceedings of IEEE International Joint Conference of FCS*, 2011, pp. 1–4.

CHAPTER 13

3D CMOS Image Sensor and IC Integration

13.1 Introduction

The basic function of CMOS (complementary metal-oxide semiconductor) image sensor is to convert light (photons) into electronic signals (electrons). CIS (contact image sensor) has a huge market with portable, mobile, wearable, and automotive products and is a key element of the IoTs (Internet of Things). For example, smartphones and tablets use CISs for their cameras, and automobiles use CISs for their machine visions. In general, a CIS consists of a matrix of micro lens, transistors and metal wiring, and PD (photodiode) [1–6]. In this chapter, 3D CIS and IC integration is presented. Emphasis is placed on a 3D CIS and IC stacking [3] and a 3D integration of CIS and processor IC [5].

13.2 FI-CIS and BI-CIS

There are two different kinds of CIS, namely FI (front-illuminated) CIS and BI (back-illuminated) CIS. For FI-CIS, the matrix of micro lens is at the front, transistors and metal wiring at the middle, and PD at the deep bottom (back) of the wafer surface, as shown in Fig. 13.1. The transistors and metal wirings at the surface of the Si substrate reflect some of the light and thus the PD can only receive the remainder of the incoming light.

A BI-CIS contains the same elements, but orients the transistors and metal wiring behind the PD layer by flipping the silicon wafer during manufacturing and then thinning its reverse side so that light can strike the PD layer without passing through the transistors and metal wiring layer as shown schematically in the top of Fig. 13.2. Compared to FI-CIS, BI-CIS can improve the chance of an input photon captured from about 60% to over 90% [6]. However, BI-CIS leads to problems such as cross-talk, which causes noise, dark current, and color mixture between adjacent pixels that yield to image degradation. To overcome this issue, Sony has developed a unique PD structure [2]. Above the illuminated side, a metal light shield (as shown in the bottom of Fig. 13.2) that is formed among the photodiodes to reduce crosstalk in the pixel array achieves good color separation [2].

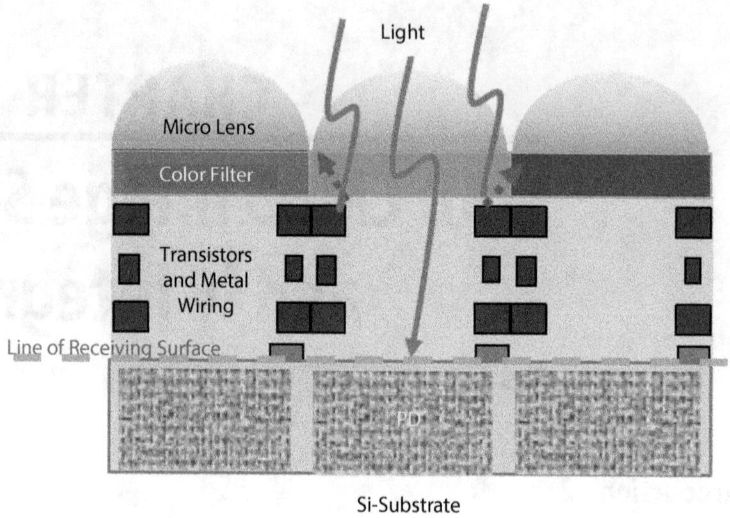

Figure 13.1 Front-illuminated CIS. Some of the lights are blocked (reflected) by the transistors and metal wirings.

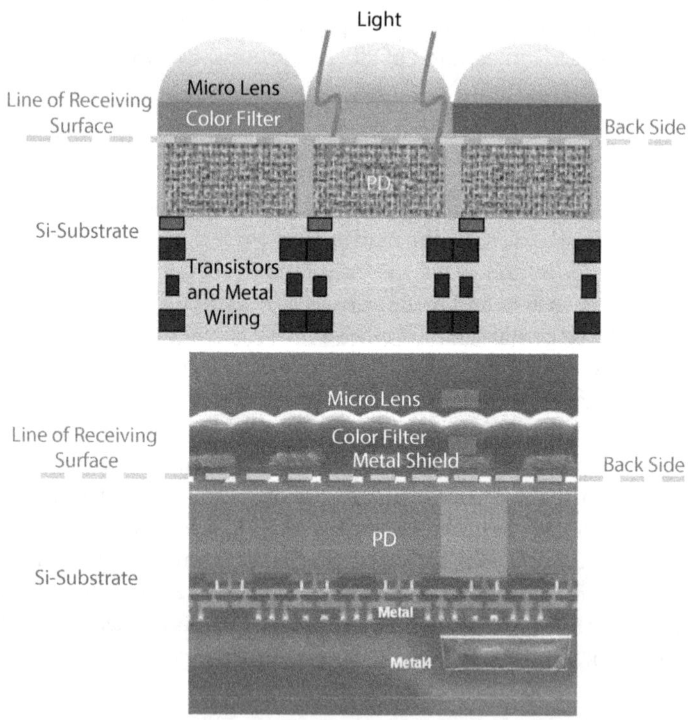

Figure 13.2 (Top) Schematic of back-illuminated CIS. (Bottom) Cross-section SEM image of a BI-CIS.

13.3 3D CIS and IC Stacking

The BI-CISs [1, 2] shown in Fig. 13.2 combine the pixel part and logic circuit part into a single chip, which is supported by a Si-substrate. Sony separates the CIS pixel chip and the logic circuit chip (you may say that the Si-substrate is replaced by the logic circuit chip) and stacks them in a 3D manner [3].

13.3.1 The Structure

Figure 13.3 shows the conventional BI-CIS [1, 2] and the new 3D stacked BI-CIS [3]. It can be seen that the new BI-CIS consists of two chips, the CIS pixel chip and the logic circuit chip and they are vertically connected through TSVs (through-silicon vias) around their edges as shown in Fig. 13.4. The advantages of this new design are that (a) more pixels can be

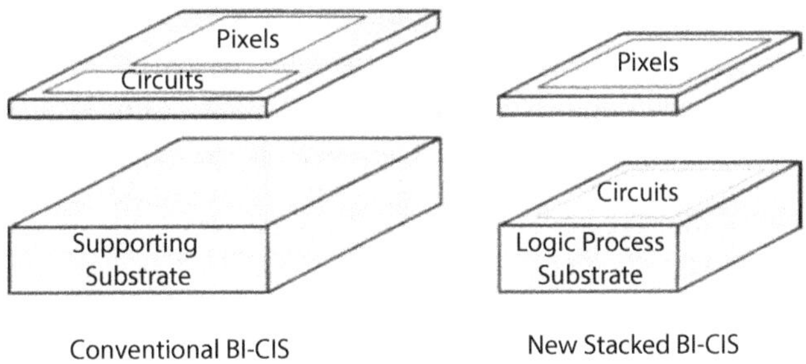

FIGURE 13.3 BI-CIS: conventional vs. new 3D stacking.

FIGURE 13.4 3D CIS pixel chip and logic IC integration.

placed on the same CIS pixel chip size (or smaller chip size can be used for the same amount of pixels) and (b) the CIS pixel chip and logic chip can be fabricated separately with different process technologies. As a result, the CIS chip size is reduced by 30% and the scaling of the logic circuit chip is increased from 500k gates to 2400k gates [3].

The number of TSVs is in the order of thousands, including signals, power supplies, and grounds. There is no TSV in the pixel array area. The column TSVs are placed in between the comparators on the pixel CIS chip and the counters of the logic circuits chip. The row TSVs are placed in between the row drivers of the CIS chip and the row decoders of the logic chip (Fig. 13.4). These arrangements of the TSVs can reduce the influence of noise and make it easy to manufacture the CIS chip. For example, to reduce the influence of noise, comparators are arranged on the pixel CIS chip, which can be manufactured by using Sony's matured process technology, rather than on the logic circuit chip.

13.3.2 Fabrication of the CIS Pixel Wafer and Logic IC Wafer

The CIS pixel chip is fabricated by the Sony conventional 1P4M BI-CIS (90 nm) process technology. The logic chip is fabricated by the matured 65 nm 1P7M logic process technology. The size of the pixel chip and the logic chip is about the same. Sony didn't disclose the TSV and assembly technologies. It is assumed herein that the CIS Si-insulator of the CIS wafer is bonded to the logic Si-insulator of the logic wafer (very similar to the SiO_2-to-SiO_2 W2W bonding) mentioned in Chap. 6 of this book. The TSVs are then formed and Cu filled after the bonding of the wafers. Figures 13.5 and 13.6 show the cross-section SEM images of the 3D CIS pixel chip and logic IC chip integration. It can be seen that (a) the top part is the BI-CIS chip, (b) the bottom part is the logic chip, (c) the BI-CIS wafer and the logic wafer are insulator-to-insulator (W2W) bonding (Fig. 13.5), and (d) the CIS chip is connected to the logic chip through TSVs (Fig. 13.6).

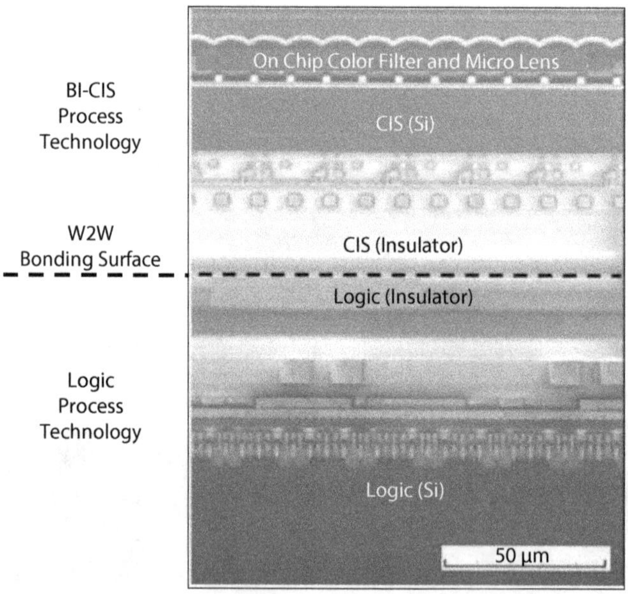

FIGURE 13.5 CIS (insulator) wafer to logic (insulator) wafer bonding.

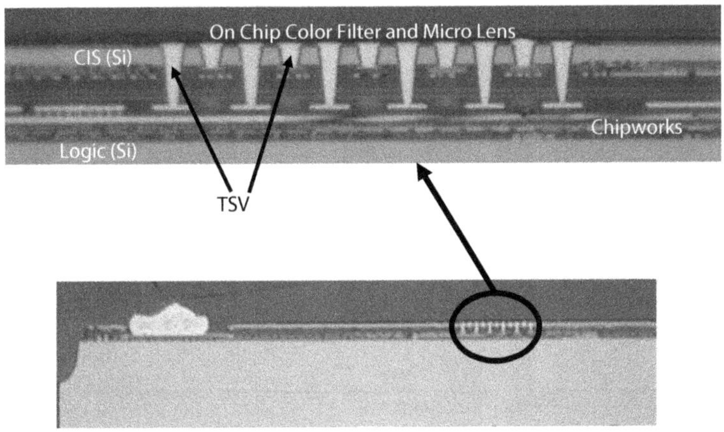

Figure 13.6 TSVs connecting the CIS pixel chip and the logic circuit chip.

13.4 3D CIS and IC Integration

13.4.1 The Structure

Figure 13.7 shows the 3D CIS and IC integration presented in [5]. It consists of the CIS, coprocessor IC, and glass carrier. The I/O count of the CIS is 80 and that of the IC is 164. The size of the CIS and the coprocessor is not the same. The dimensions of the CIS are 5 × 4.4 mm and of the IC are 3.4 × 3.5 mm. The IC and the CIS are shown, respectively, in Figs. 13.8(a) and 13.8(b) and bonded face-to-back, as shown in Fig. 13.7. The interconnects of the CIS and IC are Cu-pillar with SnAg solder cap. The TSVs are in the CIS, which are connected to the substrate with solder bump RDLs (redistribution layers) are in both CIS and coprocessor IC.

13.4.2 Fabrication Process Flow of the Coprocessor Wafer

The fabrication process flow of the coprocessor wafer is shown in Fig. 13.9. After the BEOL (back-end-of-line), sputter the barrier/seed layer (Ti/Cu) by PVD (physical vapor deposition). It is followed by making the RDLs by using a BOT (bond-on-trace) lithography and then ECD (electrochemical deposition) SiN (500 nm) and a low stress SiO_2 (600 nm)

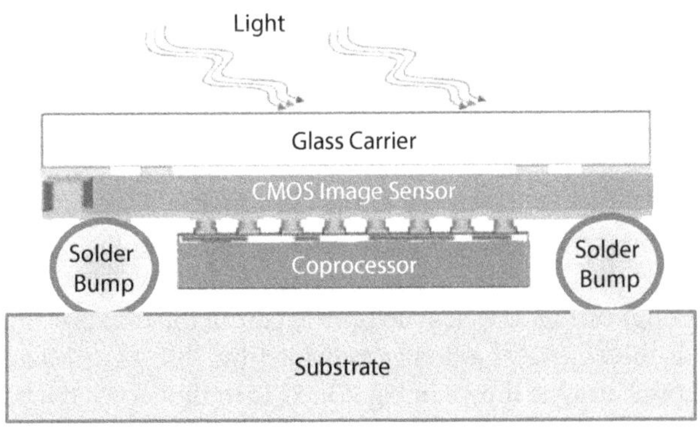

Figure 13.7 3D CIS and processor IC integration.

Figure 13.8 (a) CIS pixel chip. (b) Processor chip.

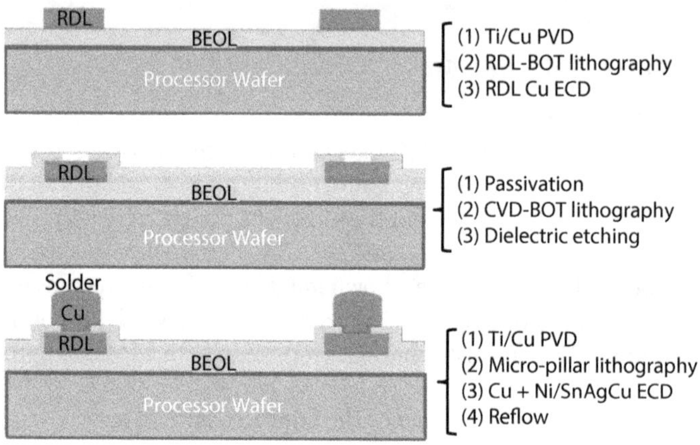

Figure 13.9 Fabrication process flow of the processor wafer.

passivation is deposited by chemical vapor deposition (CVD). The diameter of the Cu-pillar is 20 μm and the Cu-pillar (12 μm) are fabricated by electroplating on a Ti (100 nm) barrier layer and a Cu (400 nm) seed layer. Then, it is followed by electroplating the Ni (2 μm) and SnAg (2 μm). Finally, it is reflowed at 260°C. Figure 13.10 shows the SEM images of the RDLs and Cu μ-pillars on RDLs.

13.4.3 Fabrication Process Flow of the CIS Wafer

The fabrication process flow of the CIS wafer is shown in Fig. 13.11. After the BEOL, bond the front side of the CIS wafer to a glass carrier wafer. This glass wafer will act as a mechanical (supporting) carrier as well as an optical part of the package. The bonding is performed with a 7-μm-thick SiNR™ adhesive provided by Shin-Etsu MicroSi, patterned in a lace around the pixel array as shown in Fig. 13.12. Then thin down the back side of the CIS wafer to 90 μm. Form the TSVs (60 μm in diameter) from the back side of the CIS wafer and RDLs. It is followed by electroplating the 25-μm-diameter Cu μ-pillars (12 μm) and solder caps

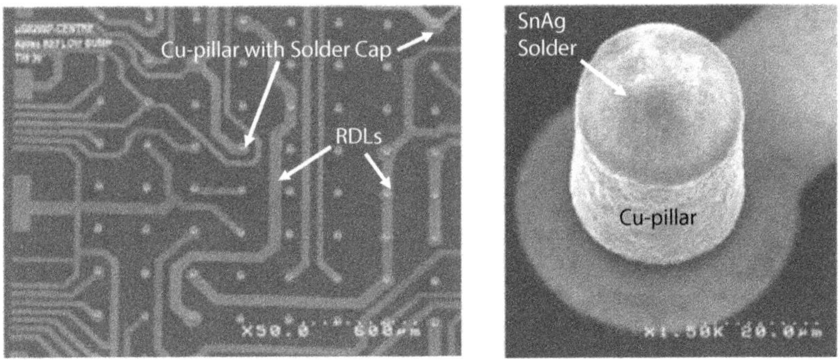

Figure 13.10 SEM images of the RDLs and Cu μ-pillar with solder cap on the processor wafer.

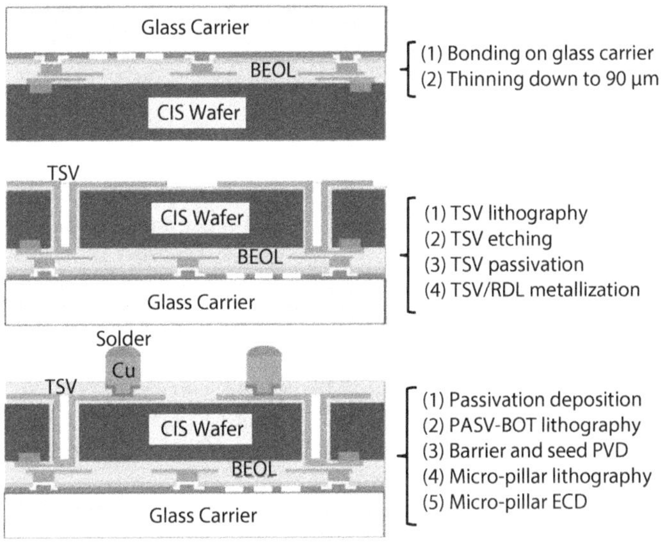

Figure 13.11 Fabrication process flow of the CIS wafer.

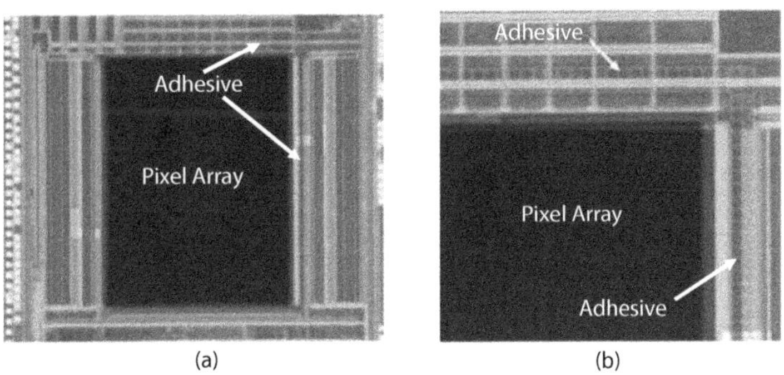

Figure 13.12 (a) CIS wafer with adhesive. (b) Zoom-in view.

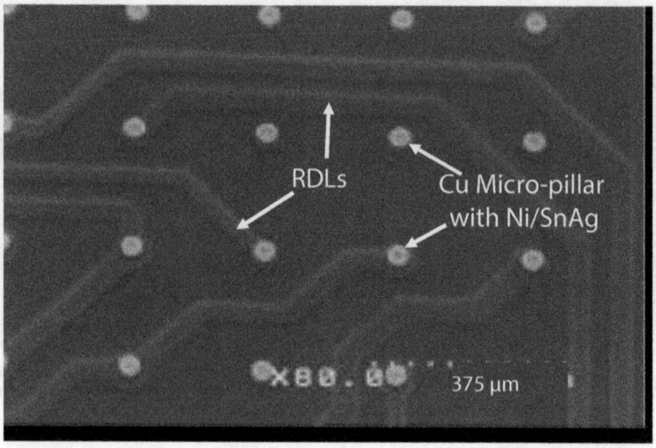

FIGURE 13.13 SEM image of RDLs and Cu µ-pillar with solder cap on the CIS wafer.

(2 µm of Ni and 2 µm of SnAg), and reflowing at 260°C. Figure 13.13 shows the SEM image of the Cu µ-pillars with solder (Ni/SnAg) caps and RDLs. (These Cu-pillars with solder caps are for the future connection with their counterparts on the top dies.)

13.4.4 Final Assembly

There are at least two ways to perform the final assembly. One is to stack up the coprocessor IC chip to the CIS wafer (chip-to-wafer or C2W) first and then do the ball mounting of the CIS wafer. The other is to perform the ball mounting on the CIS first and then do the IC chip to CIS wafer (C2W) bonding (Table 13.1). Figure 13.14 shows the 3D prototypes during the assembly process (balling first): mounted IC coprocessors and untreated sites.

TABLE 13.1 Assembly Processes of the CIS and Processor IC: Stacking First vs. Balling First

Stacking first	Balling first
Coprocessor wafer thinning	CIS flux dispense
Coprocessor wafer dicing	CIS balling
Coprocessor die flux dipping	CIS wafer reflow
Coprocessor die to CIS wafer bonding	Coprocessor wafer thinning
Reflow and cleaning	Coprocessor wafer dicing
CIS flux dispense	Coprocessor die flux dipping
CIS ball jetting	Coprocessor die to CIS wafer bonding
CIS wafer reflow	CIS wafer reflow
Capillary underfill	Capillary underfill
CIS singulation	CIS wafer singulation

Figure 13.14 3D prototypes (balling first) during the assembly process: mounted process ICs and untreated sites.

13.5 Summary and Recommendations

Some important results and recommendations are summarized as follows:

- A couple of real 3D CIS and IC integration examples are presented.
- Hopefully, more examples in this area will be published in the near future.
- Hybrid W2W (Cu-to-Cu and SiO_2-to-SiO_2 happen at the same time) bonding (Fig. 13.15) is most suitable for 3D CIS and IC integration. However, more research and development work should be performed on fabricating the TSVs (just once), since most TSVs are made before the Cu-Cu bonding and after the SiO_2-SiO_2 bonding.

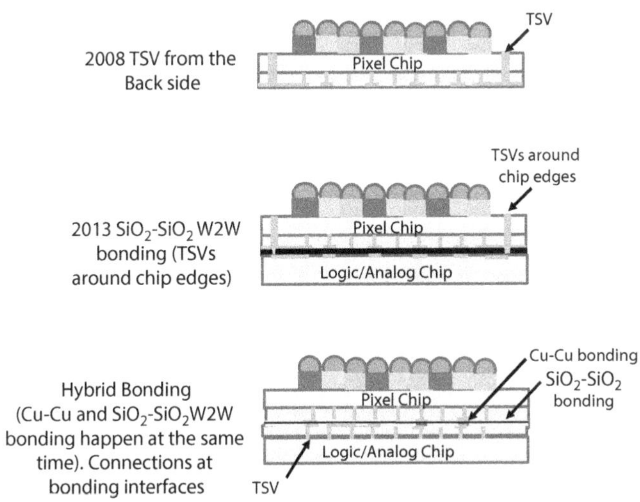

Figure 13.15 Trends in 3D CIS and IC integration.

13.6 References

[1] Iwabuchi, S., Y. Maruyama, Y. Ohgishi, M. Muramatsu, N. Karasawa, and T. Hirayama, "A Back-Illuminated High-Sensitivity Small-Pixel Color CMOS Image Sensor with Flexible Layout of Metal Wiring," *Proceedings of IEEE/ISSCC*, San Francisco, CA, February 2006, pp. 16.8.1–16.8.8.

[2] Wakabayashi, H. K. Yamaguchi, M. Okano, S. Kuramochi, O. Kumagai, S. Sakane, M. Ito, et al., "A 1/2.3-inch 10.3Mpixel 50frame/s Back-Illuminated CMOS Image Sensor," *Proceedings of IEEE/ISSCC*, San Francisco, CA, February 2010, pp. 411–412.

[3] Sukegawa, S., T. Umebayashi, T. Nakajima, H. Kawanobe, K. Koseki, I. Hirota, T. Haruta, et al., "A 1/4-inch 8Mpixel Back-Illuminated Stacked CMOS Image Sensor," *Proceedings of IEEE/ISSCC*, San Francisco, CA, February 2013, pp. 484.

[4] Rhodes, H., D. Tai, Y. Qian, D. Mao, V. Venezia, W. Zheng, Z. Xiong, et al., "The Mass Production of BSI CMOS Image Sensors," *Proceedings of International Image Sensor Workshop*, 2009, pp. 27–32.

[5] Coudrain, P., D. Henry, A. Berthelot, J. Charbonnier, S. Verrun, R. Franiatte, N. Bouzaida, et al., "3D Integration of CMOS Image Sensor with Coprocessor Using TSV last and Micro-Bumps Technologies," *Proceedings of IEEE/ECTC*, Las Vegas, NV, May 2013, pp. 674–682.

[6] Tetsuo Nomoto1 Swain, P. K., and D. Cheskis, "Back-Illuminated Image Sensors Come to the Forefront," *Photonics Spectra*, http://www.photonics.com/Article.aspx?AID=34685.

CHAPTER 14
3D IC Packaging

14.1 Introduction

As mentioned in Chap. 1, 3D integration consists of 3D IC packaging, 3D IC integration, and 3D Si integration. 3D IC/Si integrations use TSVs (through-silicon vias), but 3D IC packaging does not. Chapters 1 through 13 of this book have dealt with TSVs, which are the most important key enabling technologies of 3D IC/Si integration. In this chapter, the focus is on 3D IC packaging, which is now mature for manufacturing and some of its latest advances have been keeping 3D IC/Si integration from volume production. Emphasis is placed on (a) chip stacking by wirebonding, (b) PoP (package-on-package), (c) fan-in WLP (wafer-level package), (d) fan-out eWLP (embedded wafer-level package), and (e) ePLP (embedded panel-level package).

14.2 Chip Stacking by Wirebonding

One of the common 3D IC packaging is 3D chip stacking by wirebonding using Au and Cu wires.

14.2.1 Au Wire

Figure 14.1 schematically shows the stacking of memory chips in 3D by die attach material and Au wirebonding published by nCHIP [1] 20 years ago. Since then, memory chip (especially the NAND Flash) 3D stacking by Au wirebonding has been in high volume production. Also, the technology has been advanced to such as the one shown in Fig. 14.2.

FIGURE 14.1 Memory chips stacking by die attach and wirebonding [1].

Figure 14.2 Hynix's memory + logic + memory + logic + ··· stacking with wirebonding.

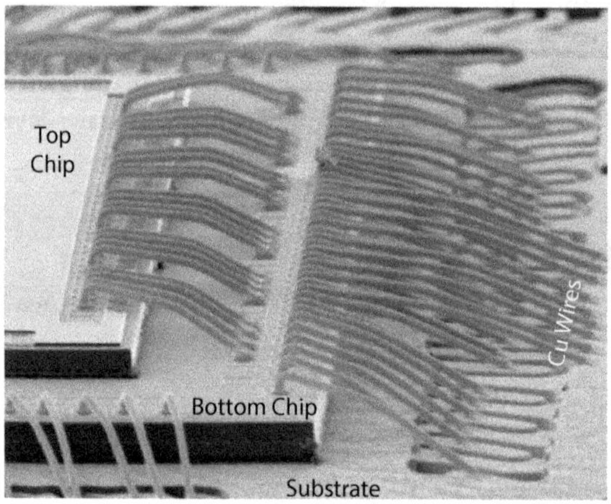

Figure 14.3 Amkor's 3D IC packaging with Cu wires.

14.2.2 Cu Wire and Ag Wire

Because of the surge in Au prices and research and development progress in Cu (or even Al) wirebonding technology, many companies have been shifting from Au to Cu wirebonding, as shown in Fig. 14.3. In 2014, the usages of Au wires and Cu (including PdCu) wires are almost even. In the past few years, Ag wires are attracting some attention.

14.3 Package-on-Package (PoP)

The other most 3D IC packaging is PoP. It comes from many different forms [2].

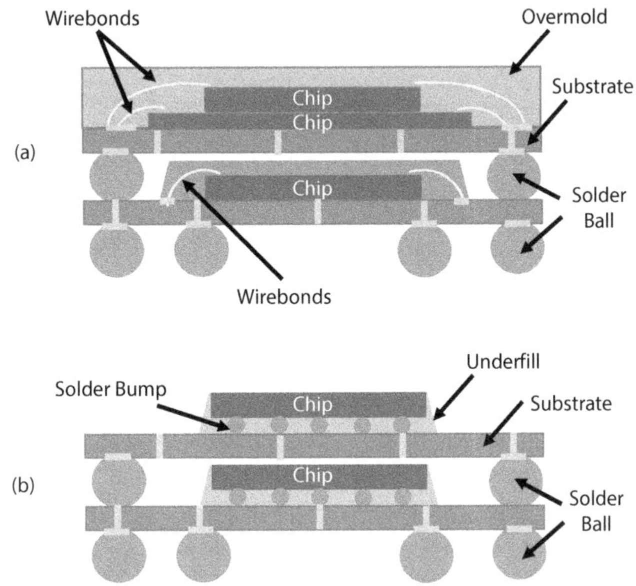

FIGURE 14.4 (a) Wirebonding PoP. (b) Flip chip PoP.

14.3.1 Wirebonding PoP

Figure 14.4(a) shows the schematic of an example of wirebonding PoP. It consists of the stacking of the top package that houses two chips cross-stacked and wirebonded on a package substrate, and the bottom package that houses one chip wirebonded on another package substrate. All these chips are over molded and all these substrates are with solder balls.

14.3.2 Flip Chip PoP

Figure 14.4(b) shows the schematic of an example of flip chip PoP. It consists of the stacking of two identical solder bumped flip chip on substrate packages. All these flip chips are with underfill and all the substrates are with solder balls.

14.3.3 Wirebonding Package on Flip Chip Package

Figure 14.5 schematically shows the cross-section of a wirebonding package on flip chip package. It can be seen that the top package houses the wirebonding chips and the bottom package houses the flip chip. A real example is shown in Fig. 14.6, which is the cross-section of iPhone 5s.

14.3.4 PoP in iPhone 5s

iPhone 5s is one of the high-end smartphones. Figure 14.7 shows the AP (application processor) chip set in a PoP inside iPhone 5s. It can be seen from the schematic cross-section that the top package houses the Elpida (now Micron)'s 1GB LPDDR3 (low-power double data rate type-3) mobile RAM (random access memory) chips ($\sim$11 $\times$ 7.8 mm), which are cross-stacked and wirebonded on a coreless FBGA (fine-pitch ball grid array) package

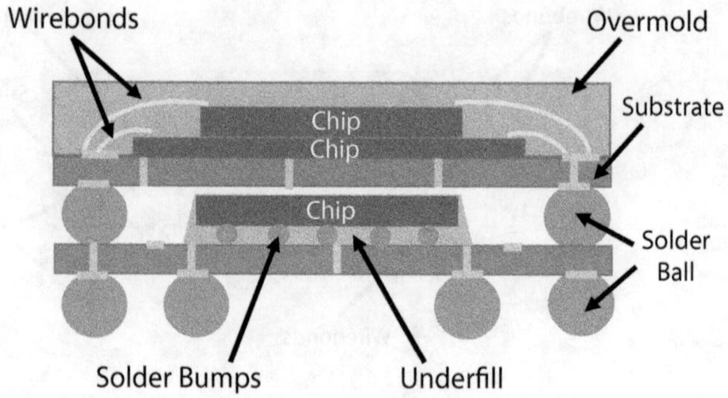

FIGURE 14.5 Wirebonding package on flip chip package.

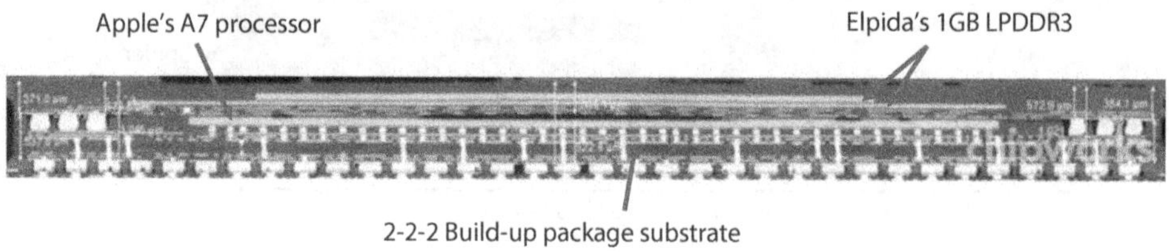

FIGURE 14.6 Cross-section image showing iPhone 5s application processor chip set.

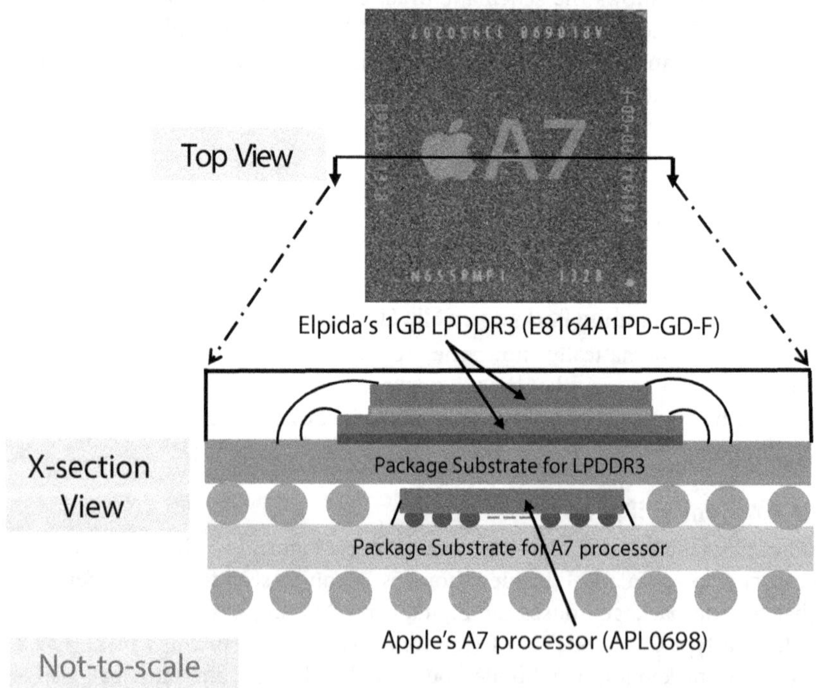

FIGURE 14.7 Top-view and cross-sectional view of the PoP (for mobile DRAM and A7 processor) inside iPhone 5s.

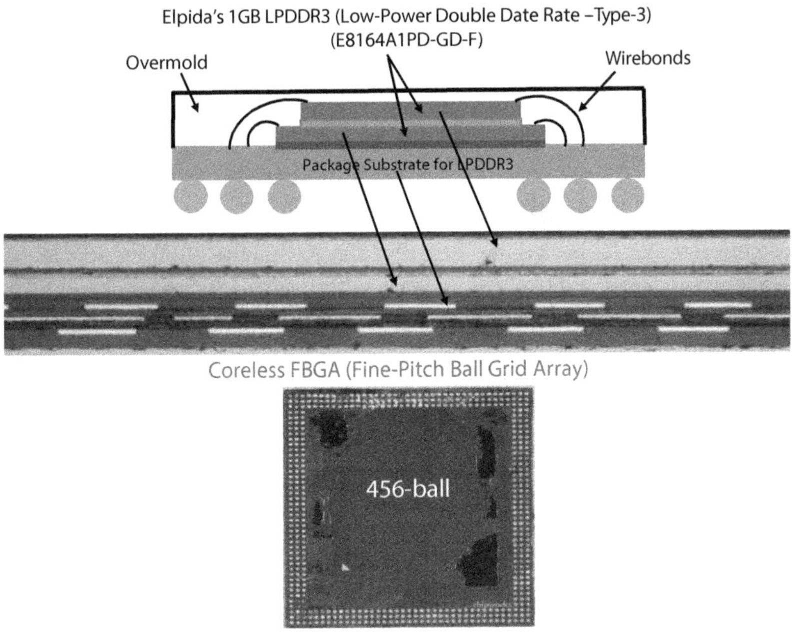

FIGURE 14.8 The mobile DRAMs in the top package of the PoP inside iPhone 5s.

substrate and then over-molded. There are three rows (456) solder balls on the FBGA (Fig. 14.8). The bottom package houses the 64-bit A7 processor chip (~10 × 10 mm), which is solder-bumped flip chip on a build-up package substrate with 38 × 34 = 1292 solder balls.

Figure 14.9 shows the cross-section of the package substrate that supports the A7 processor. It can be seen that it is a 2-2-2 (two core layers and two build-up layers on its

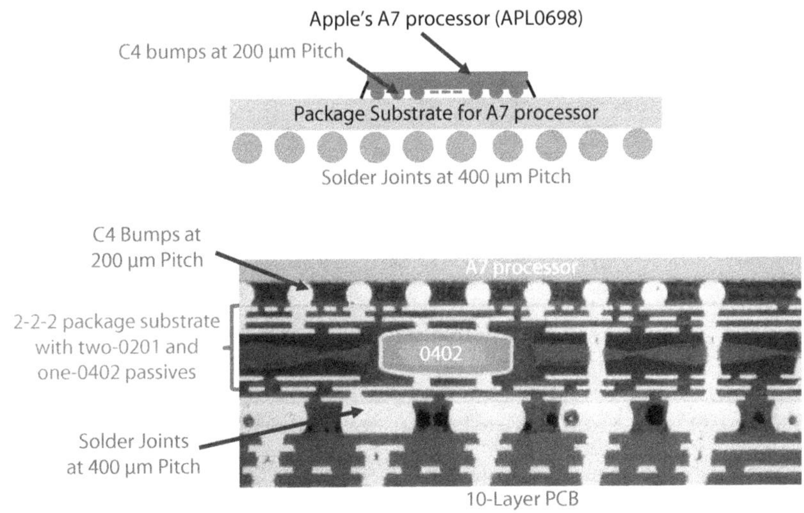

FIGURE 14.9 The A7 processor flip chip bottom package of the PoP inside iPhone 5s.

top side and bottom side) simple substrate with two 0201 and one 0402 embedded passives. The pad pitch for the C4 (controlled-collapse chip connection) bumped A7 processor chip is ~200 μm and for the ball grid array solder ball is ~400 μm. There is plenty of room for the build-up package substrate to grow.

14.4 Wafer-Level Packaging

For conventional IC packaging, first dice the IC wafer into individual IC chips and then package them. For WLP (wafer-level packaging), during the processes of IC packaging such as solder bumping, the IC wafer is intact. When all the processes are completed, dice the wafer into individual package. Thus, WLP technology leads to WLCSP (wafer-level chip scale package) [3, 4].

14.4.1 Fan-In WLP

Most of the chips on a wafer are with peripheral pads, as shown in Fig. 14.10. The ordinary flip chip technology is to put bumps on the peripheral pads by wafer bumping. The unique feature of fan-in WLP (WLCSP) is the use of a metal layer to redistribute the very fin-pitch peripheral-arrayed pads on the chip to much larger-pitch area-array pads with much larger solder bumps.

Figure 14.11 shows a square chip (9.64 × 9.64 mm) with 144 peripheral-arrayed pads on a wafer. The pad size is 0.1 × 0.1 mm and the pitch is 0.25 mm. Adding a metal layer on top of the wafer means that the fine-pitch peripheral-arrayed pads on the chip can be redistributed to a much larger-pitch and area-array pad in the interior of the chip. In this case, the pitch is 0.75 mm and the pad size is 0.3 mm in diameter. The right-hand side of Fig. 14.11 shows the details of redistribution. It can be seen that the solder bump is supported by a Cu stud (core), which is connected to the redistributed Cu-Ni pad through the TiCu UBM (under bump metallurgy). The redistributed metal layer is made of Cu-Ni, as shown in Fig. 14.12 [5, 6].

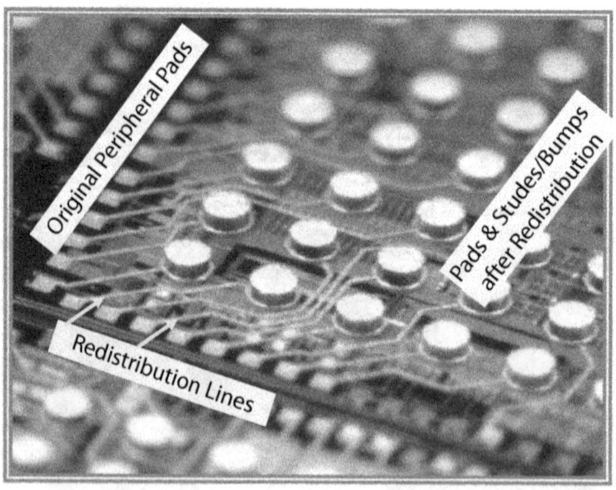

FIGURE 14.10 Wafer-level (redistribution) chip scale package.

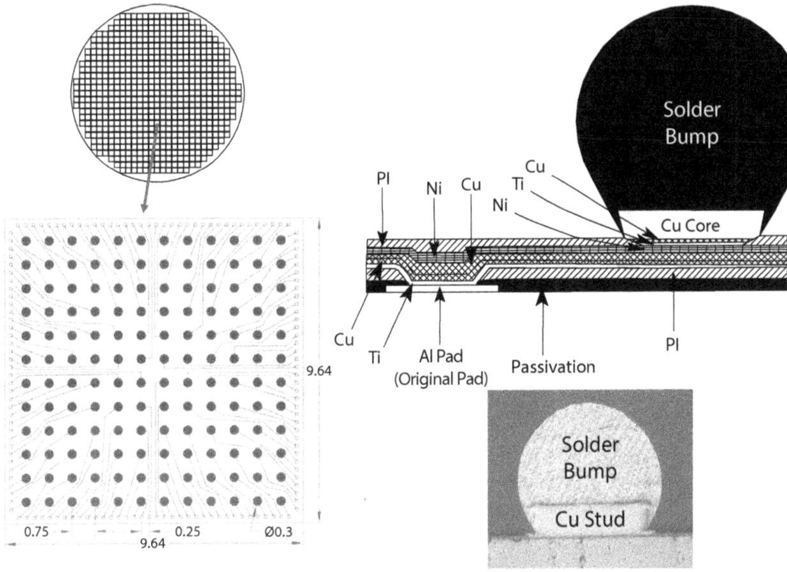

FIGURE 14.11 Fan-in wafer level packaging.

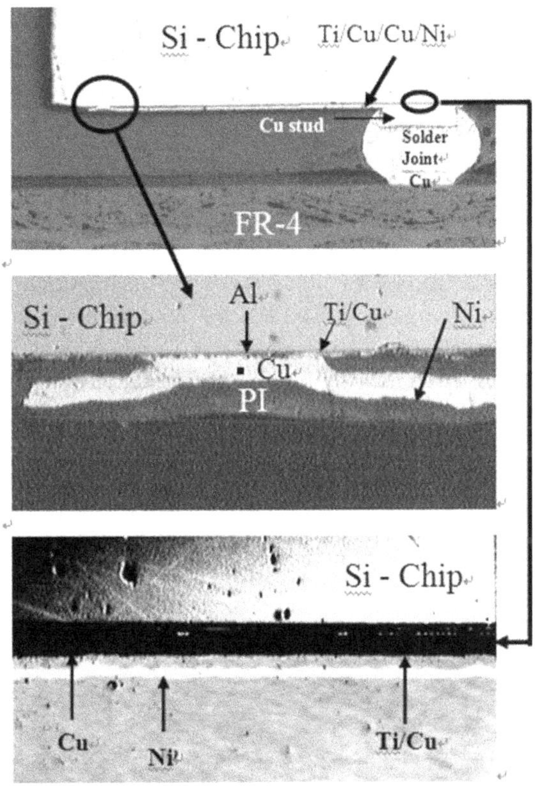

FIGURE 14.12 Cross-section SEM image of a fan-in wafer level package.

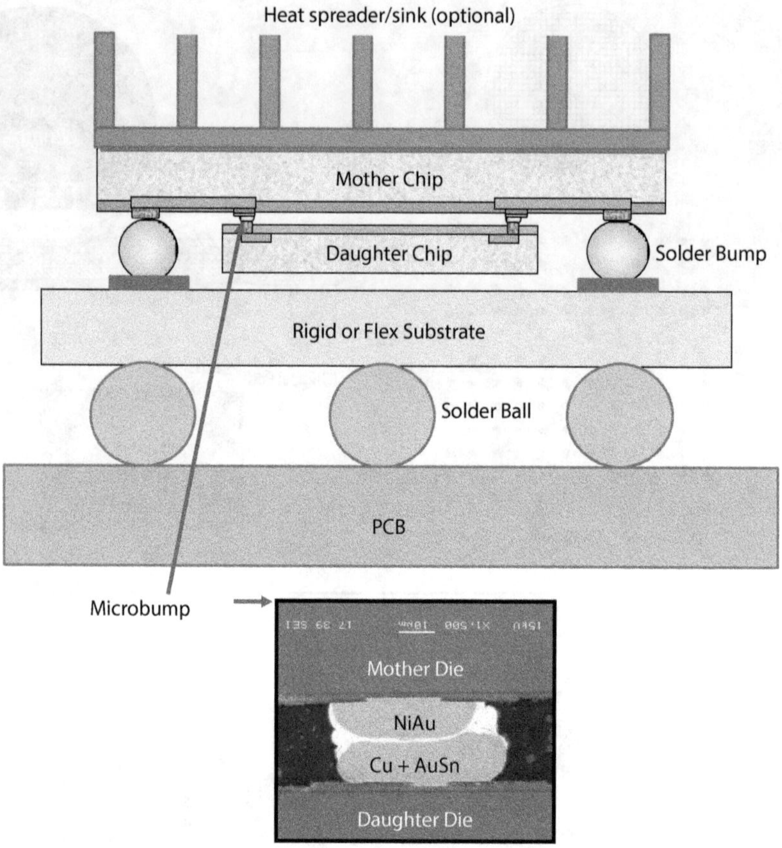

Figure 14.13 Chip-to-chip and face-to-face interconnection.

14.4.2 3D Chip-to-Chip WLP

Figure 14.13 shows the schematic of a 3D chip-to-chip interconnect [7, 8]. It consists of the mother chip that is face-to-face connected to a daughter chip. The back side of the mother chip can be attached to a heat spreader (and a heat sink if necessary). The whole module is attached (through the flip chip mother die) to a rigid or flexible substrate. It is a 3D IC packaging without using the TSVs [7, 8].

Figure 14.14 shows Amkor's POSSUM™ assembly where the daughter die is mounted face-to-face with the larger mother die [9]. The mother die is then flip chip mounted onto a substrate. Figure 14.15 shows Amkor's double POSSUM™ multi-stacked die configurations without the use of TSVs. It can be seen that the grandma die is supporting the mother die, which is supporting the three daughter dies. The grandma die is solder bumped flip chip on a package substrate, which is attached on to a PCB (printed circuit board).

14.5 Fan-Out eWLP

There are a few different names of this technology. Infineon called it eWLB (embedded wafer-level ball grid array) [10–17]. Freescale called it RCP (redistributed chip package) [18, 19]. The industry called it fan-out eWLP (embedded wafer-level package) or simply

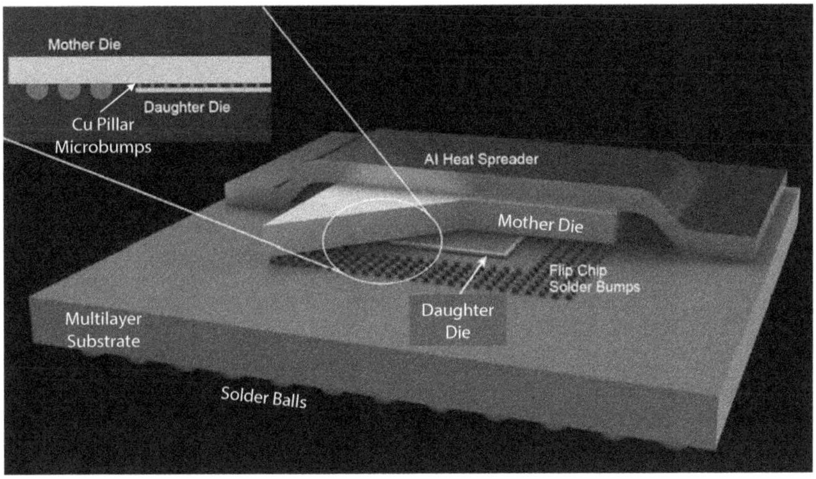

Figure 14.14 Amkor's POSSUM assembly.

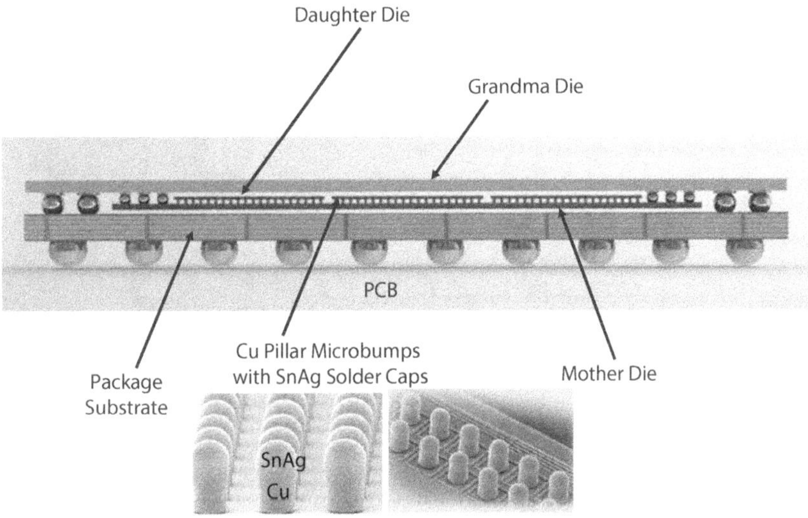

Figure 14.15 Amkor's double POSSUM assembly.

embedded wafer-level package (eWLP). ASE, STATSChipPAC, STMicroelectronics have been collaborating with Infineon on the eWLB technology, and NEPES has been collaborating with Freescale on the RCP technology.

14.5.1 Fan-Out eWLP

Figures 14.16 and 14.17 show, respectively, the key process steps from Infineon and Freescale and it can be seen that they are very similar. First, the wafer is tested for KGDs (known-good dies) and then singulated into individual dies. It is followed by picking up the KGDs and placing them face-down on a reconfigured (temporary) wafer (substrate) which can be

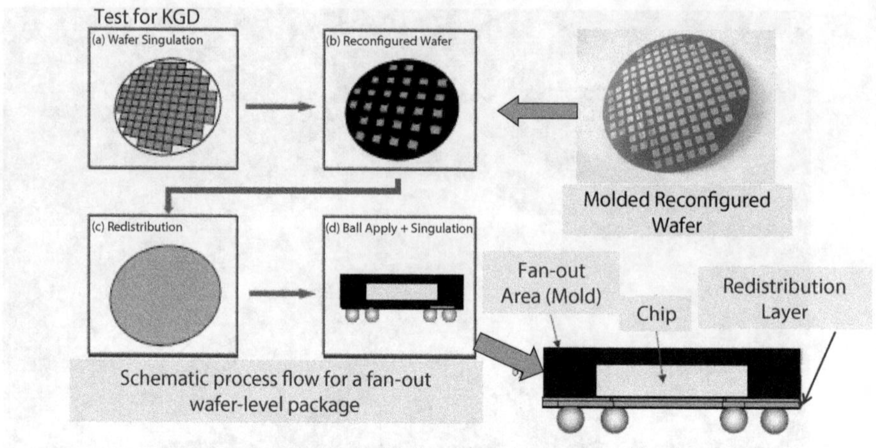

Figure 14.16 Infineon's embedded wafer-level ball grid array.

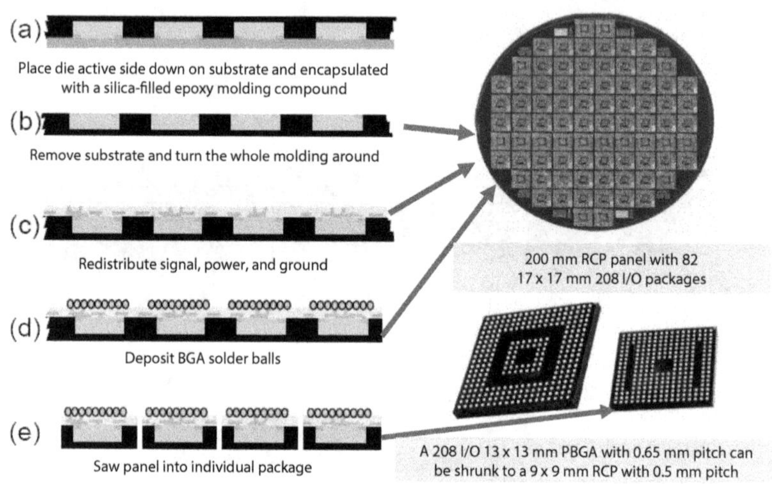

Figure 14.17 Freescale's redistributed chip package.

round or rectangular. Then, over-mold the reconfigured wafer. It is followed by removing the substrate and turning the whole molding (with KGDs) around, and making the RDLs (redistribution layers) for signals, powers, and grounds. Finally, solder balls are mounted and diced. Figure 14.18 shows a dissection of a typical eWLP.

As their names imply, both WLP and eWLP are working on wafers. The biggest difference between them is for WLP, good dies and bad dies on a wafer will be subjected to the same treatments during processing. On the other hand, for eWLP, only the KGDs will be processed on a reconfigured wafer.

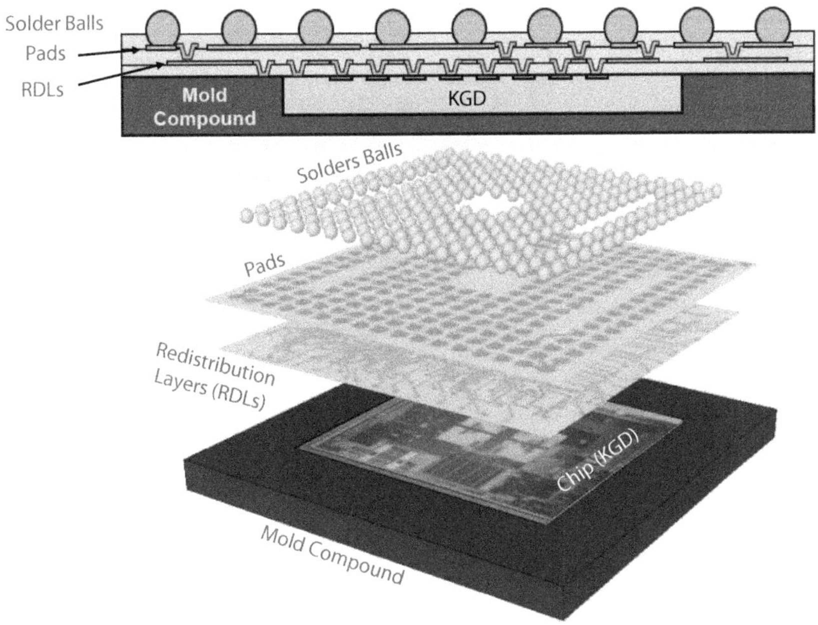

FIGURE 14.18 A dissection of a fan-out wafer level package.

The eWLP is compared with the popular solder bumped flip chip in PBGA (plastic ball grid array), as shown in Fig. 14.19. It can be seen that (a) the solder bump is eliminated, (b) the underfill is eliminated, (c) the organic substrate is eliminated and replaced by the RDLs, (d) the stiffness of the eWLP is provided by the molding and the chip, and (e) the size of the eWLP is smaller (lower profile) than that of the PBGA.

Warpage control is a big challenge of the eWLP due to thermocompression molding. Higher silica filler particles content (>85%) of epoxy molding compound (EMC) should be used to increase the stiffness to resist warpage.

14.5.2 3D eWLP—Two-Chip Stacking

Figure 14.20 shows two chips that are stacked and packaged in a eWLP format [20–26]. It can be seen that (a) chip 1 (7 × 7 mm) is larger than chip 2 (5 × 5 mm), (b) chip 1 and chip 2 are attached with a dicing die attach film (0.025 mm), (c) the interconnects of chip 1 and chip 2 are Cu pillars and they are connected through the RDLs, and (d) chip 1, chip 2, and Cu pillars are over-molded.

14.5.3 3D eWLP—Chip on eWLP (Face-to-Face)

Figure 14.21 shows the top and cross-sectional views of a 3D RCP [19] with the core 3D RCP elements as a foundation. To add to this foundation, contained within the RCP system base package, is an embedded MEMS device and its paired controller die. The portion of the RCP containing the functional MEMS and ASIC devices is approximately 600 μm in thickness. Also, included is a near face-to-face chip-to-chip assembly with faces separated by only the

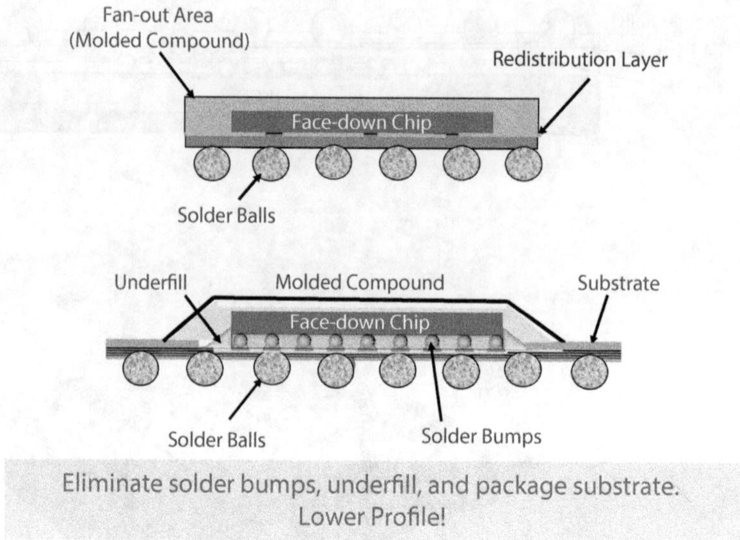

Figure 14.19 Comparison between the PBGA and eWLP.

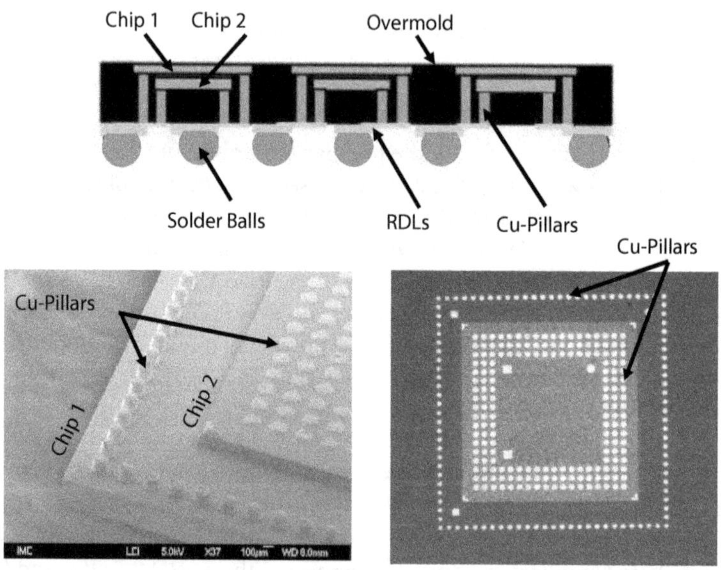

Figure 14.20 Two chips stacked in a eWLP.

solder bumps and a single layer for RCP routing. The interconnection is from the PCB, solder balls, through package via (TPV), RDLs, and to the solder bumps and chips.

14.5.4 3D eWLP—Chip on eWLP (Face-to-Back)

Figure 14.22 shows a 3D eWLP. It consists of an eWLP [15, 16], which is housing chip 1 and chip 2 with over-molded and RDLs, and a solder bumped flip chip that is face-down

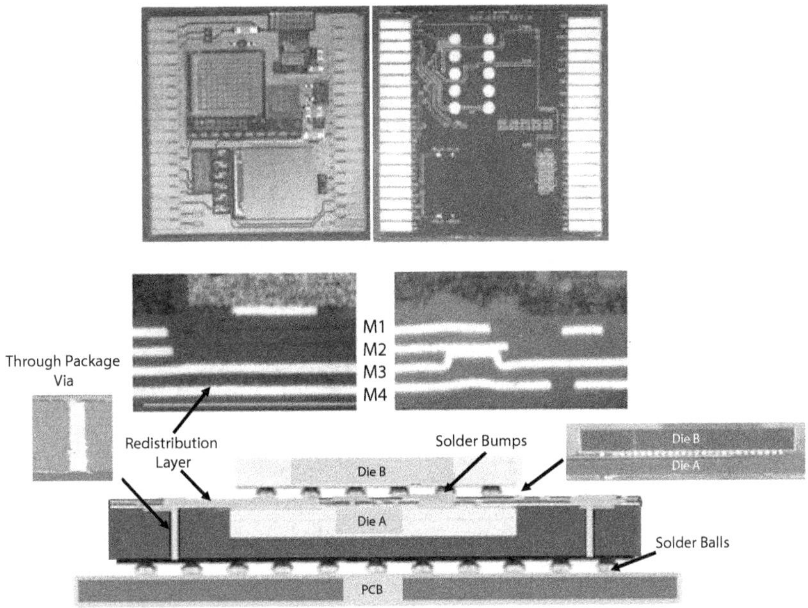

FIGURE 14.21 3D eWLP with a chip face-to-face.

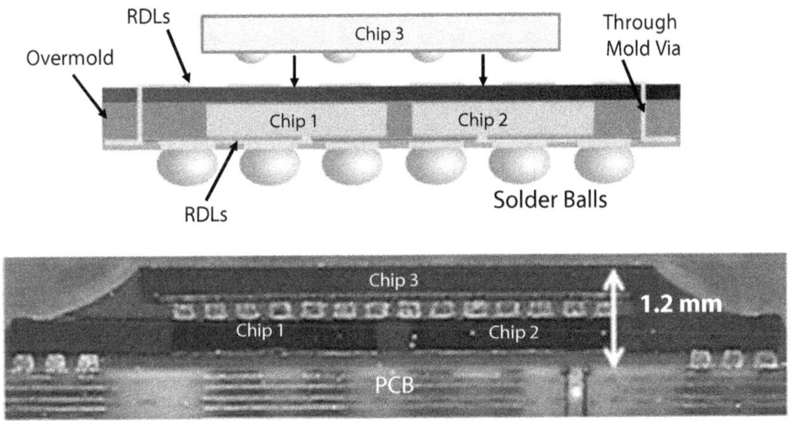

FIGURE 14.22 3D eWLP with a chip face-to-back.

attached to the eWLP. The interconnection is from the PCB, solder balls, to the lower RDLs, chip 1 and chip 2, and upper RDLs and chip 3 through the molded vias.

14.5.5 3D eWLP—Package on eWLP

In order to get a lower profile package than the PoP shown in Figs. 14.6 and 14.7, Fig. 14.23 shows the cross-section SEM images of a 3D eWLP [17]. It consists of a bottom package,

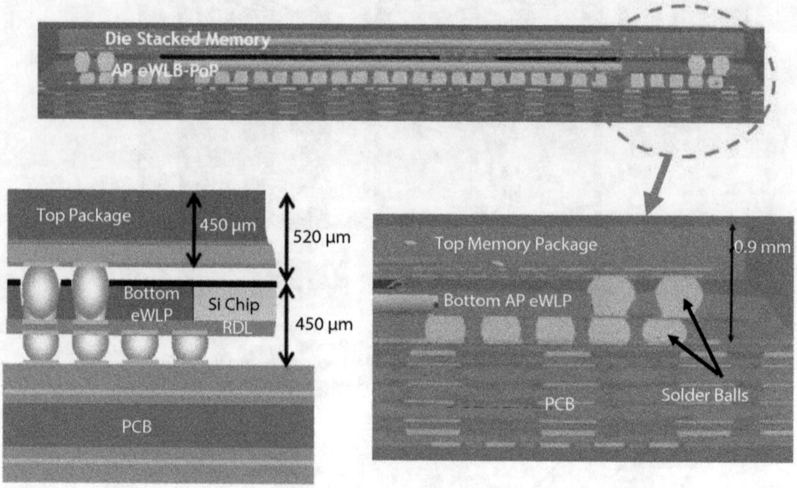

FIGURE 14.23 3D eWLP in a PoP format (the eWLP is the bottom package and the memory chips with wirebondings is the top package).

which is an eWLP (i.e., to replace the solder bumped flip chip package of Figs. 14.6 and 14.7), and a top package, which is a memory package. It can be seen that (a) the eWLP is only 450 μm thick, (b) the eWLP houses an application processor, (c) the top package is 520 μm thick and it houses the memory chips with wirebonding, and (d) the interconnection is from the PCB, solder balls, RDLs, to processor, and solder balls, RDL, to the memory chips.

14.5.6 3D eWLP—eWLP on eWLP

Figure 14.24 schematically shows Infineon's eWLP on eWLP [27, 28]. The vertical interconnects between the eWLPs are through the solder joints and TEVs (through encapsulant vias). The TEV is made by laser drilling through the molding compound. The adhesion layer is Ti, which is made by sputtering. It is followed by a thin sputtered copper layer as plating base and an electrode for the electrochemical deposition of Cu. The cross-sections and SEM images are shown in Fig. 14.24. It can be seen that the TEV is not filled. The TEV diameter is 150 μm and the plated Cu thickness ranges from 15 to 30 μm.

14.6 Embedded Panel-Level Packaging

14.6.1 Advantages and Disadvantages

Figure 14.25 schematically shows the embedded passive component and chip. It can be seen from Fig. 14.25(a) that the passive component is connected to an inner layer by either solder or adhesive first, and then embedded. On the other hand, the chip is mounted on an inner layer/foil first, embedded, and then connected by drilling and Cu plating, as shown in Fig. 14.25(b). The advantages of embedding passive components and chips are (a) low profile, (b) low cost with large panel, (c) better electrical performance with low inductances, and (d) easy to extend to 3D chip stacking. The disadvantages are (a) cannot rework, (b) infrastructure, and (c) supply chain.

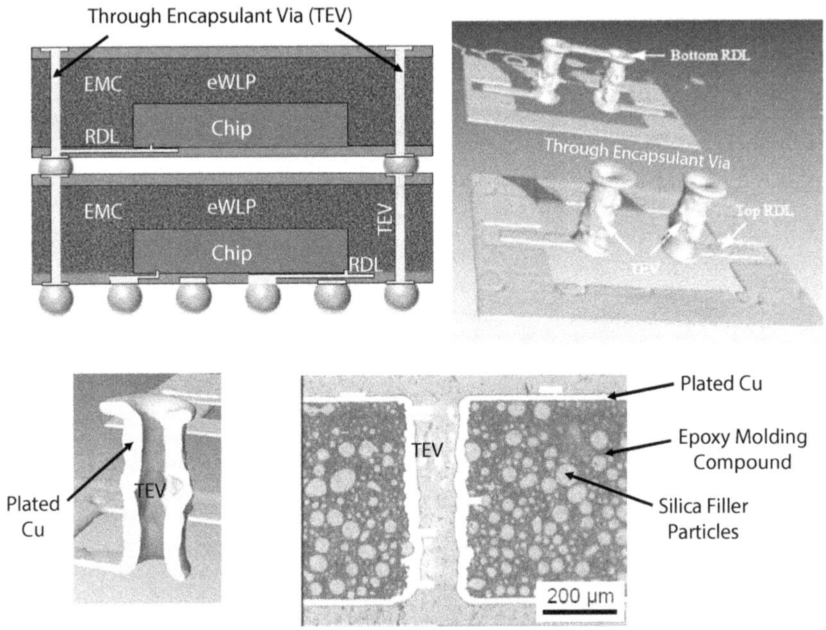

Figure 14.24 3D eWLP on eWLP.

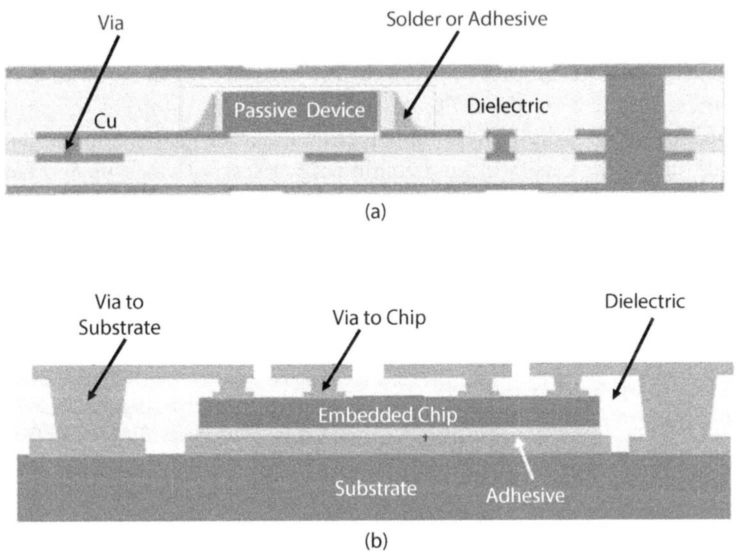

Figure 14.25 (a) Embedded passive component. (b) Embedded chip.

14.6.2 Various Chip-Embedding Processes

There are at least two ways to embed the chip in a panel, namely (a) face-up and (b) face-down. Figure 14.26 show the schematic of the face-up and face-down processes. It can be seen that (a) during the chip attachment process, for the face-up fashion the back

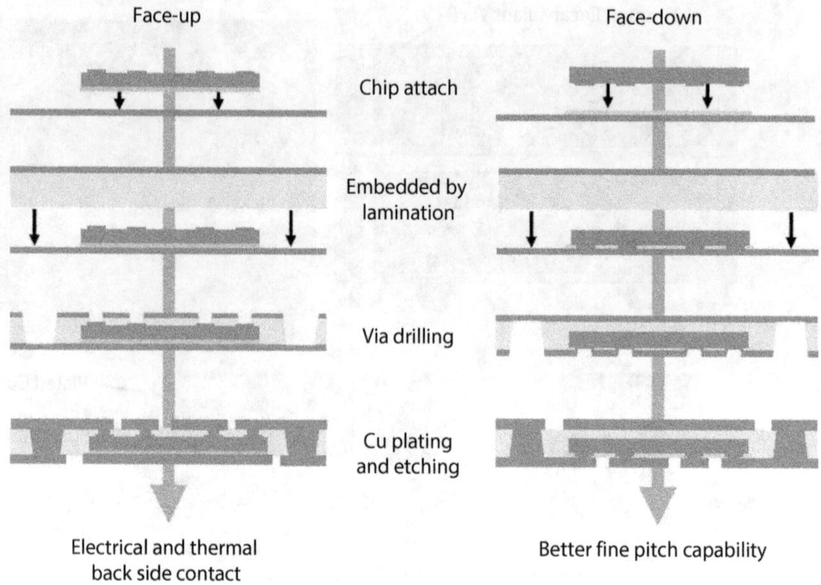

Figure 14.26 Fabrication processes for embedded chip in face-up and face-down fashions.

side of the chip is attached to the foil and for the face-down fashion the front side of the chip is attached to the foil; (b) during the embed lamination process, the front side of the face-up fashion is embedded into the dielectric layer and the back side of the face-down fashion is embedded into the dielectric layer; (c) during the via drilling process, for the face-up fashion the drilling is from the top and stops at the pads of the chip and for the face-down fashion the drilling is from the bottom and stops at the pads of the chip; and (d) during the Cu plating and etching processes, both face-up and face-down fashions electroplate the Cu to fill the vias and make/etch the circuit traces. Figure 14.27 shows a

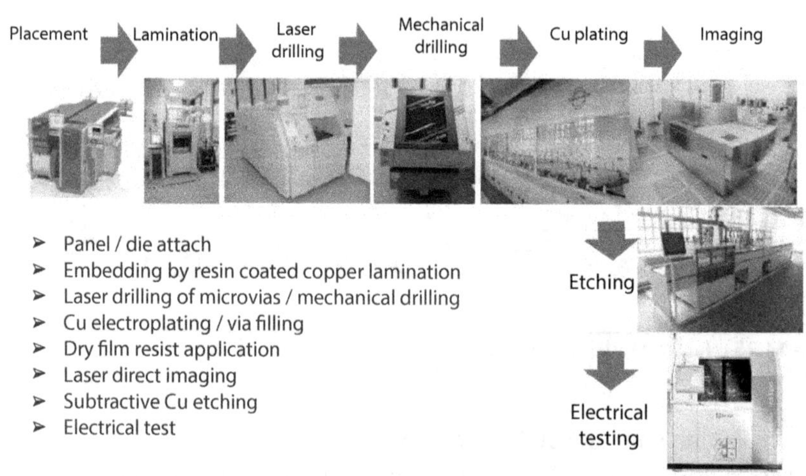

Figure 14.27 Embedded chips in substrate/board assembly line/equipment.

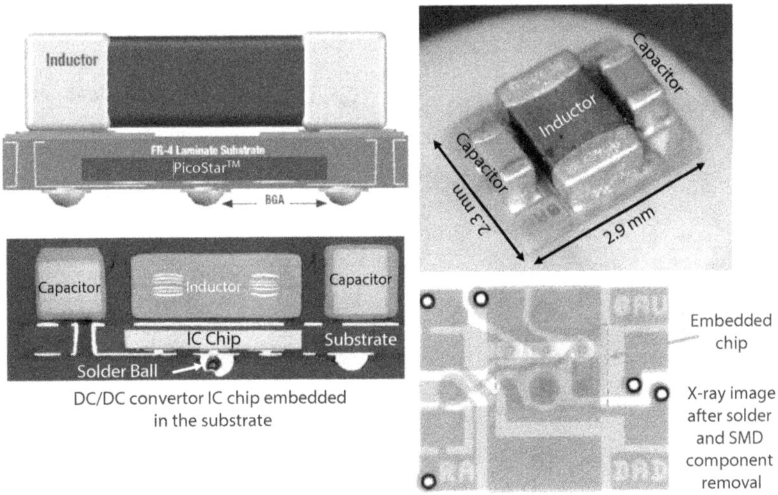

FIGURE 14.28 TI's embedded chip in a rigid substrate of a DC/DC convertor.

typical chip embedding line [29]. The key equipment is for placement, lamination, laser drilling, mechanical drilling, Cu plating, imaging, etching, and electrical testing.

14.6.3 Embedded Chip in SiP Rigid Substrate

Figure 14.28 shows TI's MicroSiP™ [30] manufactured by AT&S. It is a DC/DC convertor with an IC chip (PicoStar™) embedded in the substrate (2.9 × 2.3 mm) with solder ball to be attached on PCB. It can be seen from Fig. 14.28 that the IC chip is in a face-down fashion. The inductor and two capacitors are on the top of the substrate.

14.6.4 3D Embedded Chip in SiP Flexible Substrate

Figure 14.29 shows schematically the multichip SiP with embedded chip. It can be seen from the left-hand side of Fig. 14.29 that for the conventional SiP there are five chips (U1 through U5) side-by-side on the substrate (3.5 × 3.5 mm). On the left-hand side of Fig. 14.29, it shows that the U1 chip is embedded into the flexible substrate (2.5 × 2.5 mm) by Fujikura's WABE (wafer- and board-level embedded package) technology [31]. The other four chips are on the top of the substrate surface. The package size is reduced by 50%. The cross-section in Fig. 14.29 shows the embedded chip (ASIC) and one sensor chip (the other three chips are not in the photo image). This is a 3D IC packaging with embedded chip. All the chips are molded together.

14.6.5 3D Embedded Stacking Chips in SiP Flexible Substrate

Figure 14.30 shows the cross-section photo image of a multilayer board embedding two chips in stacked configuration [32] fabricated by the WABE technology. The minimum thickness of these chips is 85 μm and they are interconnected through the Cu traces and vias. There are nine layers of the board and its total thickness is 0.55 mm.

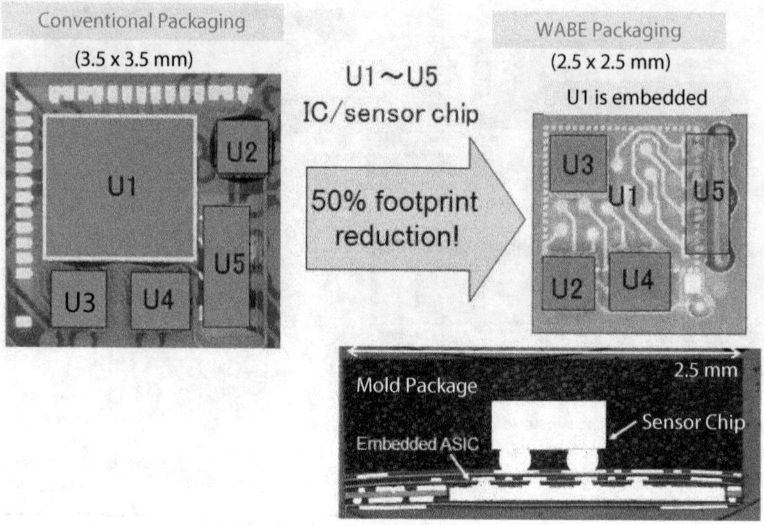

FIGURE 14.29 Fujikura's embedded chip in a flexible substrate of a 3D SiP.

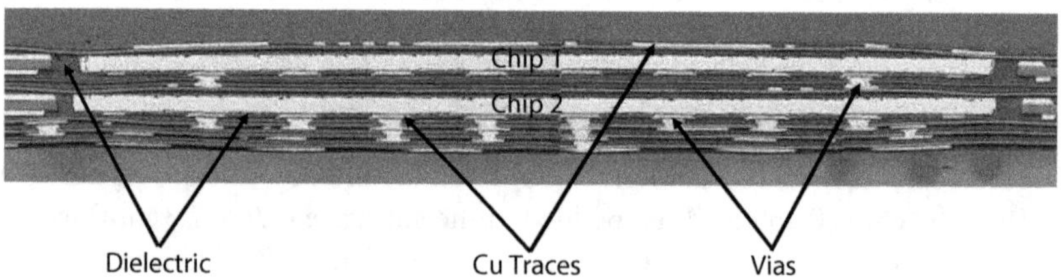

FIGURE 14.30 Fujikura's embedded two chips in stacked configuration.

14.7 Summary and Recommendations

Some important results and recommendations are summarized as follows:

- Memory chip (especially the NAND Flash) stacking by wirebonding and PoP (especially the mobile DRAMs package on the top of the AP flip chip package) are used in high-volume manufacturing.
- Chip-to-chip and face-to-face 3D IC packaging are not yet popular (except SONY's CXD53135GG); however, their time will come soon.
- Fan-out eWLP is used in low-volume manufacturing now, but it will be used in high-volume manufacturing for mobile products such as smartphones and tables.
- Embedded chip(s) in rigid/flexible substrate/PCB is suitable for wearable products.

14.8 References

[1] Tuckerman, D. B., L. O. Bauer, N. E. Brathwaite, J. Demmin, K. Flatow, R. Hsu, P. Kim, et al., "Laminated Memory: A New 3-Dimensional Packaging Technology for MCMs," *Proceedings of IEEE/MCM Conference*, 1994, pp. 58–63.

[2] Eslampour, H., M. Joshi, S. Park, H. Shin, and J. Chung, "Advancements in Package-on-Package (PoP) Technology, Delivering Performance, Form Factor & Cost Benefits in Next Generation Smartphone Processors," *Proceedings of IEEE/ECTC*, Las Vegas, NV, May 2013, pp. 1823–1828.

[3] Lau, J. H., *Low Cost Flip Chip Technology for WLCSP*, McGraw-Hill Book Company, New York, NY, 2000.

[4] Lau, J. H., and S. W. R. Lee, *Chip Scale Package*, McGraw-Hill Book Company, New York, NY, 1999.

[5] Lau, J. H., T. Chung, R. Lee, C. Chang, and C. Chen, "A Novel and Reliable Wafer-Level Chip Scale Package (WLCSP)," *Proceedings of the Chip Scale International Conference*, SEMI, September 1999, pp. H1–H8.

[6] Lau, J. H., R. Lee, C. Chang, and C. Chen, "Solder Joint Reliability of Wafer Level Chip Scale Package (WLCSP): A Time-Temperature-Dependent Creep Analysis," *ASME Paper NO. 1999-IMECE/EEP-5*.

[7] Lim, S., V. Rao, W. Hnin, W. Ching, V. Kripesh, C. Lee, J. H. Lau, et al., "Process Development and Reliability of Microbumps," *IEEE Transactions on CPMT*, Vol. 33, No. 4, December 2010, pp. 747–753.

[8] Lim, S., V. Rao, H. Yin, W. Ching, V Kripesh, C. Lee, J. H. Lau, et al., "Process Development and Reliability of Microbumps," *IEEE/EPTC Proceedings*, Singapore, December 2008, pp. 367–372.

[9] Sutanto, J., "POSSUM™, "Die Design as a Low Cost 3D Packaging Alternative," *3D Packaging*, Issue No. 25, November 2012, pp. 16–18.

[10] Brunnbauer, M., E. Fürgut, G. Beer, T. Meyer, H. Hedler, J. Belonio, E. Nomura, et al., "An Embedded Device Technology Based on a Molded Reconfigured Wafer," *Proceedings of IEEE/ECTC*, May 2006, pp. 547–551.

[11] Brunnbauer, M., E. Furgut, G. Beer, and T. Meyer, "Embedded Wafer Level Ball Grid Array (eWLB)," *Proceedings of IEEE/EPTC*, Singapore, December 2006, pp. 1–5.

[12] Meyer, T., G. Ofner, S. Bradl, M. Brunnbauer, and R. Hagen, "Embedded Wafer Level Ball Grid Array (eWLB)," *Proceedings of IEEE/EPTC*, Singapore, December 2008, pp. 994–998.

[13] Brunnbauer, M., T. Meyer, G. Ofner, K. Mueller, and R. Hagen, "Embedded Wafer Level Ball Grid Array (eWLB)," *Proceedings of IEEE/IEMTS*, 2008, pp. 1–6.

[14] Pressel, K., G. Beer, T. Meyer, M. Wojnowski, M. Fink, G. Ofner, and B. Römer, "Embedded Wafer Level Ball Grid Array (eWLB) Technology for System Integration," *Proceedings of Japan IEEE/CPMT Symposium*, 2010, pp. 1–4.

[15] Yoon, S., J. Caparas, Y. Lin, and P. Marimuthu, "Advanced Low Profile PoP Solution with Embedded Wafer Level PoP (eWLB-PoP) Technology," *Proceedings of IEEE/ECTC*, May 2012, pp. 1250–1254.

[16] Jin, Y., J. Teysseyrex, X. Baraton, S. Yoon, Y. Lin, P. Marimuthu, "Development of Next General eWLB (Embedded Wafer Level BGA) Technology," *Proceedings of IWLPC*, San Jose, CA, November 2011, pp. 7.1–7.7.

[17] Yoon, S., P. Tang, R. Emigh, Y. Lin, P. C. Marimuthu, and R. Pendse, "Fanout Flipchip eWLB (embedded Wafer Level Ball Grid Array) Technology as 2.5D Packaging Solutions," *Proceedings of IEEE/ECTC*, Las Vegas, NV, May 2013, pp. 1855–1860.

[18] Keser, B., C. Amrine, T. Duong, O. Fay, S. Hayes, G. Leal, W. Lytle, et al., "The Redistributed Chip Package: A Breakthrough for Advanced Packaging," *Proceedings of IEEE/ECTC*, May 2007, pp. 286–291.

[19] Hayes, S., N. Chhabra, T. Duong, Z. Gong, D. Mitchell, and J. Wright, "System-in-Package Opportunities with the Redistributed Chip Package (RCP)," *Proceedings of IWLPC*, November 2011, pp. 10.1–10.7.

[20] Sharma, G., V. Rao, A. Kumar, Y. Lim, K. Houe, S. Lim, J. H. Lau, et al., "Design and Development of Multi-Die Laterally Placed and Vertically Stacked Embedded Micro-Wafer-Level Packages," *IEEE Transactions on CPMT*, Vol. 1, No. 5, May 2011, pp. 52–59.

[21] Sharma, G., S. Vempati, A. Kumar, N. Su, Y. Lim, K. Houe, J. H. Lau, et al., "Embedded Wafer Level Packages with Laterally Placed and Vertically Stacked Thin Dies," *IEEE/ECTC Proceedings*, San Diego, CA, May 2009, pp. 1537–1543.

[22] Lim, Y., X. Xiao, R. Vempati, S. Nandar, K. Aditya, J. H. Lau, J. H. Lau, et al., "High Quality and Low Loss Millimeter Wave Passives Demonstrated to 77-GHz for SiP Technologies Using Embedded Wafer-Level Packaging Platform (EMWLP)," *IEEE Transactions on Advanced Packaging*, Vol. 33, 2010, pp. 1061–1071.

[23] Lim, Y., S. Vempati, N. Su, X. Xiao, J. Zhou, A. Kumar, J, H, Lau, et al., "Demonstration of High Quality and Low Loss Millimeter Wave Passives on Embedded Wafer Level Packaging Platform (EMWLP)," *IEEE/ECTC Proceedings*, San Diego, CA, May 2009, pp. 508–515.

[24] Kripesh, V., V. Rao, A. Kumar, G. Sharma, K. Houe, X. Zhang, J, H, Lau, et al., "Design and Development of a Multi-Die Embedded Micro Wafer Level Package," *IEEE Proceedings of IEEE/ECTC*, Orlando, FL, May 2008, pp. 1544–1549.

[25] Khong, C., A. Kumar, X. Zhang, S. Gaurav, S. Vempati, V. Kripesh, J. H. Lau, et al., "A Novel Method to Predict Die Shift During Compression Molding in Embedded Wafer Level Package," *IEEE/ECTC Proceedings*, San Diego, CA, May 2009, pp. 535–541.

[26] Kumar, A., D. Xia, V. Sekhar, S. Lim, C. Keng, S. Gaurav, J, H, Lau, et al., "Wafer Level Embedding Technology for 3D Wafer Level Embedded Package," *IEEE/ECTC Proceedings*, San Diego, CA, May 2009, pp. 1289–1296.

[27] Wojnowski, M., G. Sommer, K. Pressel, and G. Beer, "3D eWLB – Horizontal and Vertical Interconnects for Integration of Passive Components," *Proceeding of IEEE/ECTC*, May 2013, pp. 2121–2125.

[28] Wunderle, B., J. Heilmann, S. G. Kumar, O. Hoelck, H. Walter, O. Wittler, G. Engelmann, et al., "Accelerated Reliability Testing and Modeling of Cu-Plated Through Encapsulant Vias (TEVs) for 3D-Integration," *Proceeding of IEEE/ECTC*, May 2013, pp. 372–382.

[29] Lang, K.-D., "Advanced Embedding Technologies – Core Process for Panel Level Packaging," *Embedded Technology Forum, SEMICON Taiwan*, September 2012.

[30] Texas Instruments, *Design Summary for MicroSiP™-enabled TPS8267xSiP*, Dallas, TX, First Quarter, 2011.

[31] Itoi, K., M. Okamoto, Y. Sano, N. Ueta, S. Okude, O. Nakao, T. Tessier, et al., "Laminate Based Fan-Out Embedded Die Packaging Using Polyimide Multilayer Wiring Boards," *Proceedings of IWLPC*, San Jose, CA, November 2011, pp. 7.8–7.14.

[32] Munakata, K., N. Ueta, M. Okamoto, K. Onodera, K. Itoi, S. Okude, O. Nakao, et al., "Polyimede PCB Embedded with Two Dies in Stacked Configuration," *Proceedings of IWLPC*, San Jose, CA, November 2013, pp. 5.1–5.6.

Index

Note: Page numbers followed by f denote figures; page numbers followed by t denote tables.

A

B

C

F

G

M

N

O

P

Q

R

S

U

V

W

X

Y

Z

www.ingramcontent.com/pod-product-compliance
Lightning Source LLC
LaVergne TN
LVHW080846300926
849162LV00003B/343
* 9 7 8 0 0 7 1 8 4 8 0 6 0 *